Author Index

physica status solidi (a)

ISSN 0031-8965 ∗ VOL. 79 · NO. 1 · SEPTEMBER 1983

AKADEMIE-VERLAG · BERLIN

THEORIA CUM PRAXI

Classification Scheme

1. Structure of Crystalline Solids
   1.1 Perfectly Periodic Structure
   1.2 Solid-State Phase Transformations
   1.3 Alloys. Metallurgy
   1.4 Microstructure (Magnetic Domains See 18; Ferroelectric Domains See 14.4.1)
   1.5 Films
   1.6 Surfaces
2. Non-Crystalline State
3. Crystal Growth
4. Bonding Properties
5. Mössbauer Spectroscopy
6. Lattice Dynamics. Phonons
7. Acoustic Properties
8. Thermal Properties
9. Diffusion
10. Defect Properties (Irradiation Defects See 11)
    10.1 Metals
    10.2 Non-Metals
11. Irradiation Effects (X-Ray Diffraction Investigations See 1 and 10)
12. Mechanical Properties (Plastic Deformations See 10)
    12.1 Metals
    12.2 Non-Metals
13. Electron States
    13.1 Band Structure
    13.2 Fermi Surfaces
    13.3 Surface and Interface States
    13.4 Impurity and Defect States
    13.5 Elementary Excitations (Phonons See 6)
        13.5.1 Excitons
        13.5.2 Plasmons
        13.5.3 Polarons
        13.5.4 Magnons
14. Electrical Properties. Transport Phenomena
    14.1 Metals. Semi-Metals
    14.2 Superconductivity. Superconducting Materials and Devices
    14.3 Semiconductors
        14.3.1 Films
        14.3.2 Surfaces and Interfaces
        14.3.3 Devices. Junctions (Contact Problems See 14.3.4)
        14.3.4 High-Field Phenomena, Space-Charge Effects, Inhomogeneities, Injected Carriers (Electroluminescence See 20.3; Junctions See 14.3.3)
    14.4 Dielectrics
        14.4.1 Ferroelectrics
15. Thermoelectric and Thermomagnetic Properties
16. Photoconductivity. Photovoltaic Effects
17. Emission of Electrons and Ions
    17.1 Field Emission Microscope Investigations
18. Magnetic Properties
    18.1 Paramagnetic Properties
    18.2 Ferromagnetic Properties
        18.2.1 Ferromagnetic Films
    18.3 Ferrimagnetic Properties
    18.4 Antiferromagnetic Properties

(Continued on cover three)

physica status solidi (a)

applied research

Volume 79 · Number 1 · Pages 1 to 336, K1 to K112, and A1 to A8

September 16, 1983

PSSA 79(1) 1—336, K1—K112, A1—A8 (1983)

ISSN 0031-8965

AKADEMIE-VERLAG · BERLIN

Subscriptions and orders for single copies should be directed

in the GDR:

to the Postzeitungsvertrieb or
to the Akademie Verlag, DDR-1086 Berlin, Leipziger Str. 3–4:

in the other socialist countries:

to a book-shop for foreign language literature or
to the competent news-distributing agency;

in the FRG and BERLIN (WEST):

to a book-shop or
to the wholesale distributing agency Kunst und Wissen, Erich Bieber, OHG, D-7000 Stuttgart 1, Wilhelmstr. 4–6;

in the other Western European countries:

to Kunst und Wissen, Erich Bieber GmbH, CH-8008 Zürich, Dufourstr. 51;

in USA and CANADA:

to Verlag Chemie International, Inc., Plaza Centre, Suite E, 1020 N. W. 6th Street, Deerfield Beach, FL 33441, USA;

in other countries:

to the international book- and journal-selling trade,
to Buchexport, Volkseigener Außenhandelsbetrieb der Deutschen Demokratischen Republik, DDR-7010 Leipzig, Postfach 160, or
to the Akademie-Verlag, DDR-1086 Berlin, Leipziger Str. 3–4.

Editorial Note:

"physica status solidi (a)" undertakes that an original paper *accepted for publication* before the 23rd of any month will be published within 50 days of this date unless the author requests a postponement. In special cases there may be some delay between *receipt* and *acceptance* of a paper due to the review and, if necessary, revision of the paper.

Schriftleiter und verantwortlich für den Inhalt:
Professor Dr. Dr. h. c. P. Görlich, DDR-1020 Berlin, Neue Schönhauser Str. 20 bzw. DDR-6900 Jena, Schillbachstr. 24.

Verlag:
Akademie-Verlag, DDR-1086 Berlin, Leipziger Str. 3–4; Fernruf: 2236221 und 2236229; Telex-Nr.: 114420; Bank: Staatsbank der DDR, Berlin, Kto.-Nr.: 6836-26-20712.

Chefredakteur:
Dr. H.-J. Hänsch.

Redaktionskollegium:
Prof. Dr. E. Gutsche, Dr. H.-J. Hänsch, Dr. H. Lange, Dr. S. Oberländer.

Anschrift der Redaktion:
DDR-1020 Berlin, Neue Schönhauser Str. 20; Fernruf: 2823380.

Veröffentlicht unter der Lizenznummer 1620 des Presseamtes beim Vorsitzenden des Minsterrates der Deutschen Demokratischen Republik.

Gesamtherstellung:
VEB Druckerei „Thomas Müntzer", DDR-5820 Bad Langensalza.

Erscheinungsweise:
Die Zeitschrift „physica status solidi (a)" erscheint jeweils am 16. eines jeden Monats. Jährlich erscheinen 6 Bände zu je 2 Heften.

Bezugspreis:
Je Band 200,— M zuzüglich Versandspesen (Preis für die DDR: 130,— M).

Bestellnummer dieses Bandes: 1085/79.

AN (EDV) 20735

Contents

Review Article

Original Papers

Short Notes

Pre-Printed Titles

Systematic List

Subject classification: Corresponding papers begin on the following pages (pages given in italics refer to the principle subject classification):

Contents of Volume 79 Continued on Page 339

Review Article

phys. stat. sol. (a) 79, 11 (1983)

Subject classification: 9; 11; 14.3.3; 14.3.4; 16; 22

Institute of Physics, Academy of Sciences of the Azerbaidzhan SSR, Baku[1])

Photostimulated Diffusion in Semiconductors

By

T. D. Dzhafarov

Contents

1. Introduction

The effect of laser radiation on semiconductors is accompanied by heat, thermoelastic, thermoplastic, electrical, and photoelectrical phenomena. Absorption of high-power laser radiation from the region of intrinsic absorption of a semiconductor leads to the creation of a high concentration of free charge carriers and internal electrical fields in the near-surface region, rapid and local heating of this region, appearance

[1]) Narimanov pr. 33, 370143, Baku, USSR.

of thermoelastic and plastic deformations in it. At high densities of radiation it can even bring about the local melting of the surface. In turn, the high density of electrons and holes generated by the electromagnetic radiation, can exert an influence on many peculiarities of the semiconductor (e.g. diffusion and recombination of charge carriers, optical absorption, etc.). Electromagnetic radiation from the intrinsic and impurity absorption regions can condition recharging of the impurity and other defective centres. Concurrent with these phenomena, light absorption (both for the impurity and the intrinsic lattice atoms) can amplify the atomic oscillations, thus raising the probability of atomic transitions to an activated state. In the final analysis, these illumination-stimulated processes encourage the formation (or annihilation) of defects, creation of electrical, elastic, temperature fields in such a manner as to change the rates of atomic diffusion.

Illumination of p–n homojunction and heterojunction, accompanied by a change in the contact electric fields, can accelerate the redistribution of the charged impurities. Nonradiative recombination of the charge carriers generated by light, electrical or other effects, contributes to the formation of defects and atomic diffusion.

In this paper we review the results of theoretical and experimental works, dedicated to the research on photostimulated processes (mainly diffusion, formation of defects, annealing of defects) in crystalline and amorphous semiconductors and their structures.

2. Some Peculiarities of Laser Radiation Effect on Diffusion Processes and Properties of Semiconductors

The mechanism of the electromagnetic radiation effect on the processes of defect formation and atomic displacement mainly amounts to four processes: 1. ion drift in the electric field; 2. change in the atomic charge state; 3. excitation of the atomic oscillation modes at nonradiative recombination of the electron [1], and 4. photon–phonon interaction through infrared absorption [2].

Internal electrical fields in semiconductors may appear when they are illuminated. The appearance of such fields is determined by light generation of free charge carriers and nonuniformity in the distribution of the latter due to diffusion [3]. This problem will be dealt with in Section 2.2.

The internal electrical fields (along with the external ones, for instance, the electrical component of the laser radiation) can change the crystal lattice potential barrier, decreasing the potential barrier in the direction of the applied electrical field, thus increasing the probability of an ion jump in that direction (Fig. 1a). For the actual attainment of the external electrical field ($\mathscr{E} \approx 10^6$ V/cm) the change in the height of the potential barrier may be $\Delta E = e\mathscr{E}\delta \approx 10^{-1}$ eV (where δ is the lattice parameter, e the electron change).

Under the effects of optical excitation, injection of charge carriers, doping temperature variation, etc., changes in the charge state of atoms (positions of Fermi level or quasi-Fermi level) in semiconductors may be accompanied by a change in their equilibrium geometric position in the lattice (vacancy, interstitial site, etc.) or by the

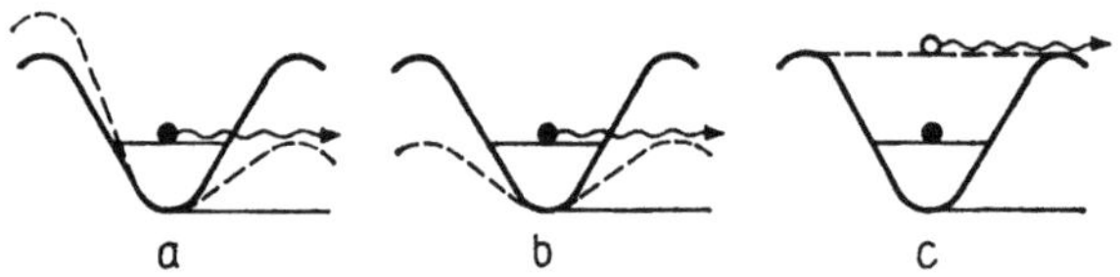

Fig. 1. Variation of the potential barrier before illumination (solid line) and at illumination (dashed line) for various mechanisms of diffusion: a) drift, b) variation in the charge state (equilibrium position of the impurity), c) excitation of the atom oscillation energy at electron capture

formation of complexes, finally causing a variation in the rate of atomic migration. From Fig. 1 b, representing the potential pattern, one can see that the illumination-induced changes in the atomic charge state (e.g., acceptor (in the vacancy)–donor (in the interstice) transition) reduce the activation energy of migration, i.e. stimulate the diffusion process. In this case, the diffusion activation energy should decrease by the magnitude of the impurity atom ionization energy (see Section 2.5).

The third mechanism of illumination-induced diffusion is connected with the transfer of the energy released at electron non-radiative recombination to the atom, and excitation of its additional oscillations (see Fig. 1 c). In such a situation the decrease in the potential barrier height for migration should be equal to the energy difference between free state and defect-captured electron states. The recombination-induced diffusion will be discussed in Section 2.6.

The fourth mechanism of the photoinduced migration consists in the excitation of atomic vibrations by means of the electromagnetic radiation (infrared absorption), i.e. in photon–phonon energy transfer (through compound states) (see Section 2.3).

2.1 Heat and thermoelastic effects

The effect of high-power laser radiation on solids is attended by heating of the surface. It can also lead to the appearance of mechanical stresses.

First, we shall discuss heat effects, arising from heavy absorption of laser radiation by the isotropic solid [4]. Let us assume that an axial-symmetrical light beam falls on the surface of a solid in the direction of the z-axis. The energy distribution in the transverse direction of the light beam obeys the Gaussian relation

$$I(r) = I_0 \exp(-K_0 r^2), \tag{1}$$

where $K_0 = 2/R$, R is the radius of the surface region under illumination. Assuming that the laser radiation is completely absorbed on the surface under illumination and $R \ll L$ (L is the characteristic size of the solid), energy exposure time $t_0 \ll L^2/a$ (a is the coefficient of temperature conductivity of the material) (see Fig. 2), one can determine the temperature field from the solution of the heat conductivity equation

$$\frac{\partial T}{\partial t} = a \nabla^2 T \tag{2}$$

under the following initial and boundary conditions:

$$\left.\frac{\partial T}{\partial z}\right|_{z=0} = -\frac{I_0(1-\tau)}{K} \exp(-K_0 r^2), \tag{3}$$

$$T(r, z, 0) = 0, \qquad \lim_{z\to\infty} T(r, z, t) < N, \qquad \lim_{r\to\infty} T(r, z, t) < M. \tag{4}$$

Here K is the heat conductivity coefficient, τ the reflection coefficient of the solid surface.

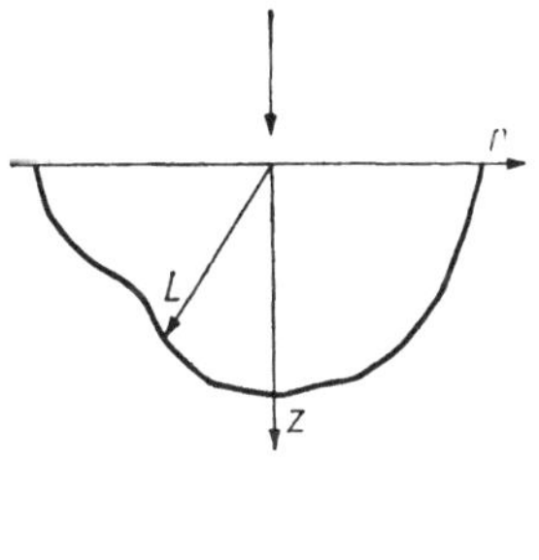

Fig. 2. Schematic picture of the axial-symmetric light beam falling on the solid surface

The solutions of (2) to (4) take the form

$$T(r, z, t) = \frac{I_0(1-\tau)}{4KK_0} \int_0^\infty J_0(\xi r)\, e^{-\xi^2/4K} \left[e^{-\xi z} \Phi^*\left(\frac{z}{2\sqrt{at}} - \sqrt{at}\,\xi\right) - \right.$$

$$\left. - e^{\xi z} \Phi^*\left(\frac{z}{2\sqrt{at}} + \sqrt{at}\right)\right] d\xi\,, \tag{5}$$

where $J_0(\xi r)$ is the zero-order Bessel function, $\Phi^*(x) = 1 - \Phi(x)$; $\Phi(x)$ is the error function.

At the exposure time satisfying the inequality $R^2/a \ll t_0 \ll L^2/a$, when the stationary temperature field is established, expression (5) assumes the form

$$T(r, z, \infty) = \frac{I_0(1-\tau)}{2KK_0} \int_0^\infty J_0(\xi r) \exp(-\xi^2/4K_0) \exp(-\xi z)\, d\xi\,. \tag{6}$$

At the point $r = z = 0$, where the maximum temperature is reached, (6) is appreciably simplified,

$$T(0, 0, \infty) = \frac{\sqrt{\pi}}{2\sqrt{2}} \frac{I_0(1-\tau) R}{K}\,. \tag{7}$$

Fig. 3 shows the stationary distribution of temperatures in aluminium and gold-plated quartz samples, calculated by formula (6). For aluminium the maximum permissible value of laser radiation intensity, involving melting of the surface, is equal to 9×10^7 W cm^{-2} at a pulse duration $t_0 = 10^{-6}$ s. Increase in t_0 brings about a decrease in the value of I.

It might be well pointed out that under the influence of high-power laser radiation, which generates an electron–hole plasma with concentration up to 10^{20} to 10^{22} cm^{-3} in the near-surface region, before giving the excess energy to the lattice, the charge carriers diffuse to an appreciable depth (of the order of their diffusion length). That is why the energy of the incident radiation diffuses in a crystal volume far greater than the region where the absorption occurs. As a result of this fact the surface region gets heated up to temperatures much lower than those predicted by (5) and (6).

The presence of an inhomogeneous temperature field due to laser radiation absorption along the thickness and the surface of the sample should lead to the appearance of compressive stresses, plastic flow of the material with the formation of residual stresses. When a particular limiting intensity of the laser radiation, resulting in the limiting flow of the material in a radiation time t_0 is attained, plastic deformations appear in the surface layer. The maximum permissible value of the intensity, causing

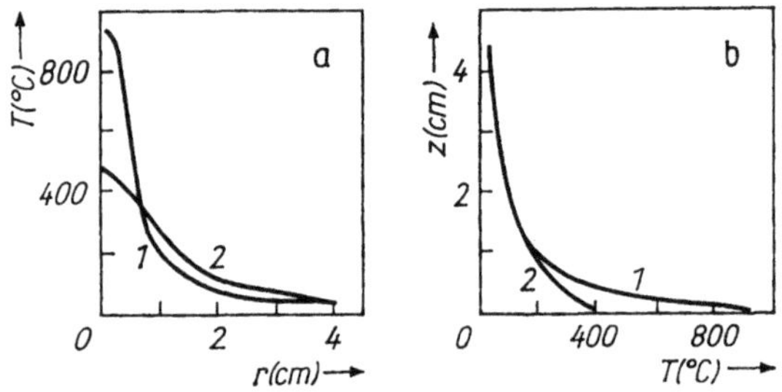

Fig. 3. Temperature distribution a) along radius and b) along the thickness in a gold-plated quartz sample ($\tau = 98.6$ %, $P = 1$ kW, $L = 0.5$ (1) and 1 cm (2))

plastic flow of the material, is by some orders smaller than that of laser-stimulated melting of the material. At this the values of maximum permissible intensities decrease linearly with increasing duration of the radiation.

The problem of atomic diffusion in solids with elastic deformation gradient under isothermal conditions is discussed in [3]. Under the effect of laser radiation on the surface of a solid the appearance of elastic deformation fields is also accompanied by the formation of temperature field gradients in the sample. In this case the mathematical problem of atomic diffusion with gradients of temperature and elastic fields is strongly complicated.

The power of laser radiation P_0 required to heat the surface of a semi-infinite semiconductor sample starting from the initial temperature T_0 to some final temperature T_f is determined by the expression [5]

$$P_0 = \frac{\varrho c(T_0)\, \Delta T L_H}{[1 - \tau(T_0)]\, t_0}, \tag{8}$$

where ϱ is the density of the sample, c the specific heat, $\Delta T = T_f - T_0$, τ the reflection coefficient, t_0 the duration of the laser pulse, the value of L_H may be roughly assumed to be the thickness of the heated region of the sample.

2.2 *Appearance of internal electical fields*

The effect of laser radiation on semiconductors leads to the appearance of an internal electrical field ($\mathscr{E} \approx 10^4$ to 10^7 V/cm) in the near-surface region. The internal electrical field can initiate desorption (adsorption) of ions and electrons [6] and stimulate migration processes of these particles in the crystal volume. It modifies the electronic structure of the surface and creates a remarkable folding of the energy bands in the near-surface region of the sample.

In the case when the energy of laser radiation quanta exceeds the band gap of the intrinsic semiconductor ($h\nu \geqq E_g$), the appearance of the electrical field is caused by nonuniform light absorption through the thickness of the sample, which is attained by an increase in the charge carrier density in the region under illumination, subsequent diffusion and separation of these charges due to the difference in the mobility of electrons and holes. Formation of the charged particle diffusion flux leads to the disturbance of the electrical neutrality and origination of a space charge and an electrical field hindering the diffusion, and thus to further increase in the deviation from neutrality. As far as the charge carrier lifetime generated by laser radiation (hence, that of the electric field) is well above ω^{-1} (ω is the electromagnetic wave frequency) the field in question may be treated as a static one.

When creating electron–hole pairs near the surface of the semiconductor with the help of illumination, the electron and hole components of current have the form

$$j_n = -\mu_n kT \frac{\partial n}{\partial x} + e\mu_n n\mathscr{E}, \tag{9}$$

$$j_p = -\mu_p kT \frac{\partial p}{\partial x} - e\mu_p p\mathscr{E}. \tag{10}$$

Here μ_n and μ_p are the mobilities of electrons and holes, respectively, n and p are the electron and hole concentrations, respectively, e is the electron charge, $\mathscr{E}$ is the mean value of the electrical field density originated by the gradients of the electron and hole concentrations. Under the condition of absence of electric current in the semiconductor ($j_n = j_p$), one can determine the intensity of the internal electric field

$$\mathscr{E} = \frac{kT\left(\mu_n \frac{\partial n}{\partial x} - \mu_p \frac{\partial p}{\partial x}\right)}{\sigma_n + \sigma_p}. \tag{11}$$

Therefore, the value of the electric field density increases with the increase of the difference between the electron and hole mobilities. It is also inversely proportional to the full conductivity of the sample.

In the above discussion we assumed that the value of the electric field intensity is independent of the coordinate. In fact, the density of the electrical field changes along the thickness of the sample. For a correct discussion of this problem it is essential that the continuity equations (9), (10) and the Poisson equation are solved simultaneously

$$\frac{\partial \mathscr{E}}{\partial x} = \frac{4\pi e}{\varepsilon} \varrho_c, \tag{12}$$

where ϱ_c is the volume charge density, ε the dielectric constant of the semiconductor.

2.3 Photostimulated volume diffusion

The temperature dependence of the atomic diffusion coefficient in solids is known to be determined by an Arrhenius equation

$$D = D_0 \exp\left(-\frac{W}{kT}\right). \tag{13}$$

Here W is the diffusion activation energy, D_0 the pre-exponential factor. One can see from (13) that for increasing diffusion rate it is necessary either to increase the temperature or to decrease the activation energy.

The electromagnetic effect is one of the possible non-thermal ways of exciting the phonons which stimulate atomic diffusion in solids. Photon-to-phonon energy transfer (through an intermediate electronic state) can be realized through infrared absorption, stimulated Raman scattering, direct and indirect absorption at defective centres, etc.

The mechanism of photon–phonon interaction through infrared absorption, stimulating the migration of atoms and ions, can be represented as consisting of the following stages (Fig. 4) [2]:

1. conversion of the photon to an optical phonon with wave vector $\boldsymbol{k}_0$;
2. anharmonic interaction, giving longitudinal (LA) and transverse (TA) acoustic phonons with wave vectors $\boldsymbol{k}_{LA}$ and $\boldsymbol{k}_{TA}$;
3. phonon transmission of pseudomomentum to atoms and ions in longitudinal and transverse directions.

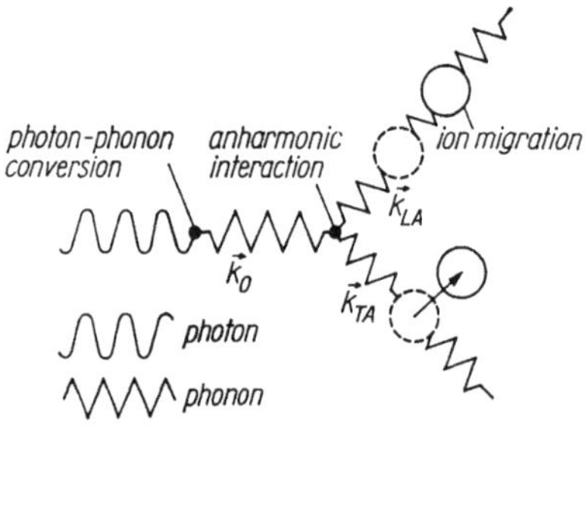

Fig. 4. Mechanism of photon–phonon interaction at IR absorption

In the majority of ideal crystals the optical phonon lifetime is limited by the anharmonic interaction. A transverse optical phonon (TO) of frequency ω and momentum $\hbar k \approx 0$ disintegrates into two acoustic phonons of frequencies ω_2(LA) and ω_3(TA) and propagation vectors $\boldsymbol{k}_2$ and $\boldsymbol{k}_3 = -\boldsymbol{k}_2$. As a consequence of photon–phonon interaction the electromagnetic field energy transfer to phonons stimulates the increase in the amplitude of atomic oscillations thus increasing the probability of their migration.

The contribution to the acceleration of the atomic migration by photoexcited phonons is due to effective temperature changes (13) rather than to the decrease of activation energy stimulated by phonons.

For the case of single-photon absorption with subsequent decay of the optical phonon into acoustic phonons, the relation W/kT_{eff} in the harmonic approximation is determined by the expression

$$\frac{W}{kT_{\text{eff}}} \approx \frac{1}{\sum\limits_{s} \hbar[2(n_s + n_e) + 1]}, \tag{14}$$

where T_{eff} is the effective temperature embracing the photo-excited phonons and thermally-excited phonons, n_s and n_e are the numbers of thermally and photo-excited phonons, respectively. One can see from (14) that excitation of n_e phonons by the electromagnetic radiation increases T_{eff}, thus increasing the atomic diffusion coefficient. In this case the ratio of diffusion coefficients, with and without photoexcited phonons is determined by the formula

$$\frac{D_e}{D} = \exp \frac{W}{k}\left(\frac{1}{T} - \frac{1}{T_{\text{eff}}}\right), \tag{15}$$

where T involves the thermally excited phonons.

To observed the infrared absorption effect on diffusion in solids one needs fluxes N_e, described by the relation

$$N_e = n_e^c\, C/V\ 10^{26} \text{ to } 10^{27} \text{ photons cm}^{-2}\text{s}^{-1}, \tag{16}$$

where n_e^c is the critical number of laser radiation-excited phonons, C is the sound velocity, V the atomic volume. It should be noted that a nonfocused ruby laser (10^8 W cm^{-2}) produces a flux of 3×10^{26} photons cm^{-2} s^{-1} (a focused one 10^{31} photons cm^{-2} s^{-1}) which is sufficient for detecting laser radiation-stimulated diffusion in solid states. The theory predicts that for accelerating the reorientation of Na: F_A centres in KCl at 110 K by one order it is enough to have $n_e = 0.2$ of photostimulated phonons

During inhomogeneous absorption of light along the crystal thickness there should exist a nonuniform distribution of the photostimulated phonons, too.

At high temperatures and with only the phonon concentration gradient n_e in the cubic crystal, the diffusion flux in the x-direction is determined by

$$J = -D\frac{QN}{RT^2}\left(\frac{\hbar\omega_e}{k}\right)\frac{dn_e}{dx}. \tag{17}$$

Here Q is the heat conductivity, N the concentration of migrating particles, ω_e the frequency of acoustic phonons.

For lower temperatures the standard equation of thermal diffusion must be used

$$J = -\boldsymbol{K}\nabla T, \tag{18}$$

where

$$K_{ij} = D_{ij}\frac{QN}{RT^2}. \tag{19}$$

The mechanisms of atomic migration discussed above can provide the basis for the laser radiation-stimulated formation of point defects, clusters, microcracks, etc.

2.4 Photostimulated surface diffusion

The experimental investigations [7] disclose that the laser illumination of solid surfaces influences the atomic migration even at low-power illumination (10 W cm^{-2}). The observed effect does not stem from a general heating of the surface. In [8] the authors develop the theory of multiphoton excitation of adsorbed atom vibrations through the laser radiation, leading to the atomic migration at the surface. Under the condition of equilibrium laser excitation the migration exhibits diffusion character.

Let us assume that the electromagnetic wave falls on the surface of the solid having the optic phonon frequency ω_A

$$v(t) = \frac{\omega_R}{2} \exp(-i\omega_L t) , \tag{20}$$

where ω_L and ω_R are the frequency of the incident radiation and the Rabi frequency, respectively. Here the surface temperature is assumed to be sufficiently low, that is why heat excitation of the vibrations can be disregarded in comparison with the coherent multiphoton excitation. In the case of resonance ($\omega_L = \omega_A$) the diffusion coefficient of the adsorbed atoms is determined by the expression

$$D_L = 4\gamma_T d^2(\omega_R/\gamma)^2 [\exp(\hbar\omega_A)/kT) - 1]^{-1} . \tag{21}$$

Here d is the distance between the nearest equilibrium positions, γ the constant of the adsorbed atomic vibration damping, $\gamma = \gamma_v + \gamma_T$, where γ_v is the constant of the atomic vibration damping due to the energy exchange with the lattice in the given position, γ_T is the constant of damping due to the energy exchange between the positions.

In the case of laser-stimulated migration the diffusion coefficient of the adsorbed atoms is proportional to the intensity of field radiation, and similarly it depends on the surface temperature, frequency of the optical vibrations of the adsorbed atoms, relaxation constants, and laser radiation frequency.

2.5 The effect of photostimulated impurity recharging on diffusion

Irradiation of semiconductors with low-power light, inadequate to remarkably increase of the temperature, in some cases can bear an essential influence on the diffusion displacement of impurity atoms. This effect should be clearly defined for the impurities diffusing through the dissociation mechanism, when the impurity travels both by vacant sites and by interstitial sites being in different charge states, which have a rather quick exchange between each other. In this case the effective coefficient of impurity diffusion D_{eff} is determined by the relation [3]

$$D_{eff} = D_i \frac{\partial N_i}{\partial N} + D_s \frac{\partial N_s}{\partial N} . \tag{22}$$

Here D_i and N_i are the diffusion coefficient and concentration of interstitial site particles, D_s and N_s are the diffusion coefficient and concentration of particles in vacant sites, $N = N_i + N_s$ is the total impurity concentration. In the case that the flux of the particles by vacant sites is negligible with respect to the flux by interstitital sites (i.e. $D_s\,(\partial N_s/\partial N) \ll D_i(\partial N_i/\partial N)$) from (22) one has

$$D_{eff} \approx D_i \frac{\partial N_i}{\partial N} . \tag{23}$$

Now it may be put forward that the impurity in vacant sites arranges as single-ionized acceptor ions (A_s^-) and in the interstitial sites as single-ionized donor (A_i^+). The reaction of the impurity transition from the vacant site to the interstitial site under an external effect (illumination) can be written as

$$A_s^- + e^+ \rightleftarrows A_i^+ + e^- + V, \tag{24}$$

where e^-, e^+, and V are electron, hole, and neutral vacant lattice site, respectively. Applying the law of mass action to (24) and the relation $np = n_i^2$ for the concentration of the impurity donor component in the interstitial sites, one has

$$N_i = K_r \frac{N_s p^2}{N_v n_i^2}, \tag{25}$$

where K_r is the reaction equilibrium constant, N_v the concentration of vacant sites. Considering that the hole concentration in the valence band is connected with the acceptor concentration N_s and with the ionization energy E_a, then p is described by the expression [9]

$$p = \sqrt{2N_s N_v} \exp\left(-\frac{\Delta E_a}{2kT}\right), \tag{26}$$

where N_v is the state density in the valence band, and the concentration of the intrinsic charge carriers is determined by

$$n_i^2 = N_c N_v \exp\left(-\frac{E_g}{kT}\right), \tag{27}$$

where E_g is the band gap of the semiconductor, N_c is the state density in the conduction band, and (25) is transformed to

$$N_i = \frac{2K_r N_s^2}{N_v N_c} \exp\left(-\frac{\Delta E_a - E_g}{kT}\right). \tag{28}$$

Assuming that $N \approx N_s$ ($N_s \gg N_i$), determining $\partial N_i/\partial N$ from (28) and substituting it into (23) we get the relation for the effective diffusion coefficient,

$$D_{\text{eff}} \approx D_i \frac{4K_r N_s}{N_v N_c} \exp\left(-\frac{\Delta E_a - E_g}{kT}\right). \tag{29}$$

The temperature dependence of the effective impurity diffusion coefficient (with consideration of the temperature dependence of the diffusion coefficient of the interstitial impurity component $D_i = D_{i0} \exp(-W_i/kT)$) can be represented in the form

$$D_{\text{eff}} \approx \frac{4K_r D_{i0}}{N_v N_c} N_s \exp\left[-\frac{W_i - (E_g - \Delta E_a)}{kT}\right], \tag{30}$$

where W_i is the activation energy of the impurity diffusion by interstitial sites, D_{i0} is the pre-exponential factor.

Therefore, at the dissociative mechanism of migration the effective diffusion coefficient depends upon the charge state of the migrating impurity ions. An illumination-stimulated change in the charge state of the ions may bear an influence on the effective rate of diffusion. As it follows from the case of the externally stimulated recharging of the diffusing impurity (acceptor–donor transition) the diffusion activation energy decreases by the magnitude of $E_g - \Delta E_a$.

For impurities with diffusion via vacant lattice sites, the illumination-stimulated recharging of the vacant lattice sites is also expected to exert an effect on diffusion.

At this, the electromagnetic radiation may change the character and strength of the interaction, such as the interaction of impurity–vacancy type connected with complex formation or with the change of the electrical neutrality condition, thus influencing the processes of diffusion and solubility.

2.6 Recombination-accelerated diffusion

Injection of minority charge carriers (through light, electrical, and radiation effects) into inhomogeneous semiconductors (p–n homojunctions, heterojunctions, Schottky barriers) may stimulate radiationless recombination processes along with the radiative recombination of charge carriers. Charge carrier capture by point defects (vacancies, interstitial atoms, impurities, complexes, etc.) is one of the possible channels of radiationless recombination. The energy released in the recombination of charge carriers at the defect (or near it) is attended by the excitation of additional vibrational energy of these centres. Such a process of radiationless recombination is the basis of the mechanism of recombination-accelerated defect formation and diffusion of impurities and defects in semiconductors [1, 10].

Let us consider a defect creating a deep-lying energy level in the forbidden band of the semiconductor. An electron strongly connected with such a defect, should be localized within a small region around the defect. Because of the difference in the masses and bond constants of the defect centres and atoms of the main lattice, some localized vibrational modes connected with the defect should come into being. Electron and vibrational properties of the defect and its environment resemble those of a "defect molecule" in the matrix of the main lattice. On the basis of this model the authors of [10] develop the theory of recombination-accelerated diffusion of defects.

Radiationless capture of an electron from the conduction band to the deep level ΔE_c below the conduction band bottom can increase the defect vibration energy. This event, in turn, should bring about the conversion of the defect to an activated state and subsequently to a diffusion jump into a neighbouring free equilibrium position in the crystal lattice. The rate of the recombination-accelerated diffusion is approximately determined by the expression

$$K_{ra} \approx A R_c \exp\left[-(E^* - \Delta E_c)/kT\right], \tag{31}$$

where E^* is the energy required for activating "the defect molecule", R_c the capture rate of charge carriers by the defect, A a parameter depending upon the number of the effective oscillators in the "defect molecule" as well as upon the scattering rate of the local defect vibration energy to the crystal lattice. This dependence is valid for the case, when

$$E^* - \Delta E_c \gg (S - 1)\, kT, \tag{32}$$

where S is the number of effective oscillators in the complex forming the defect (in the "defect molecule").

One can see from (31) that the recombination-stimulated diffusion rate is proportional to the charge carrier capture (recombination) rate, while the activation energy of such a diffusion is decreased by the energy ΔE_c transmitted to the defect by capturing the electron as compared with a common thermal diffusion.

2.7 Photostimulated displacement of p–n junction

Light from the region of the p–n homojunction intrinsic absorption can cause a diffusive redistribution of the impurity forming the p–n junction. This event is connected with the appearance of the illumination-stimulated directed photoelectromotive force reducing the p–n junction field. For the impurity, creating the p–n junction, such a decrease in the electric field intensity in the region of the space

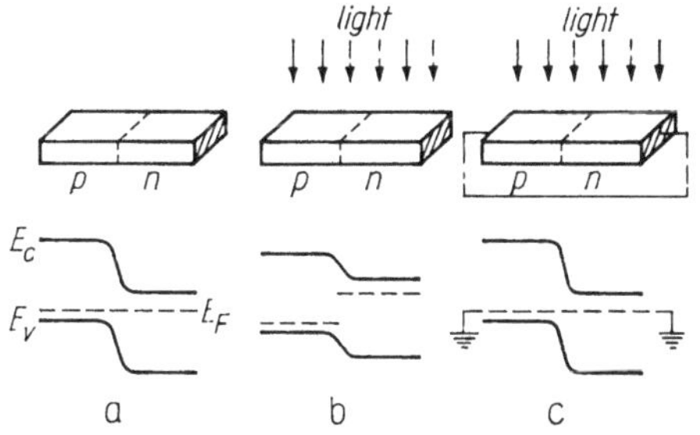

Fig. 5. Energy diagram of a p–n junction a) in the dark and b, c) under illumination. a, b) are open-circuited junctions, c) is a short-circuited junction

charge should cause a redistribution of this impurity in the region of p–n junction localization [11].

Let us discuss a p–n junction in an n-type semiconductor, created by the acceptor impurity diffusion. As an example of such a diffusion we take the impurity diffusing by the dissociative mechanism mainly locating in vacant sites (as acceptor) and in interstitial lattice sites (as donor). In the absence of illumination the diffusion flux of a nonuniformly distributed charged impurity J_D in the p–n junction region is compensated by the field flux J_E (Fig. 5a)

$$-D\frac{\partial N}{\partial x} = \mu E N\,, \tag{33}$$

where D, μ, and N are the diffusion coefficient, mobility and concentration of the non-uniformly distributed impurity forming the p–n junction, E is the p–n junction electric field intensity.

Light illumination of the p–n junction in the region of the semiconductor intrinsic absorption leads to the appearance of the directed photoelectromotive force at open-circuit terminals U_{oc} (see Fig. 5b) decreasing the p–n junction field [12],

$$U_{oc} = \frac{kT}{e}\ln\left(\frac{j_{ph}}{j_s} + 1\right), \tag{34}$$

where j_{ph} is the photocurrent density, j_s the saturation current density determined by

$$j_s = j_0 \exp\left(-\frac{E_g}{\beta kT}\right). \tag{35}$$

The magnitude of the directed photoelectromotive force increases with increasing illumination power. It reaches saturation at $U_{oc} \approx E_g/e$, when the energy barrier at the p–n junction for the charge carriers disappears, i.e. the p–n junction field density becomes equal to zero. Therefore, illumination promotes the decrease in the p–n junction electrical field density, thus resulting in the disturbance of the flux equilibrium condition (33) and the diffusion flux of the nonuniformly distributed impurity.

Experimentally the light-stimulated displacement of the p–n junction was first observed in gallium arsenide, the p–n junction being produced by copper diffusion [11].

3. Photostimulated Diffusion in Crystalline Semiconductors

In this section we shall discuss the results of the experimental study of the impurity photostimulated diffusion in crystalline semiconductors at low-level optical excitation, when thermal effects of the crystals may be either neglected or considered to be very small.

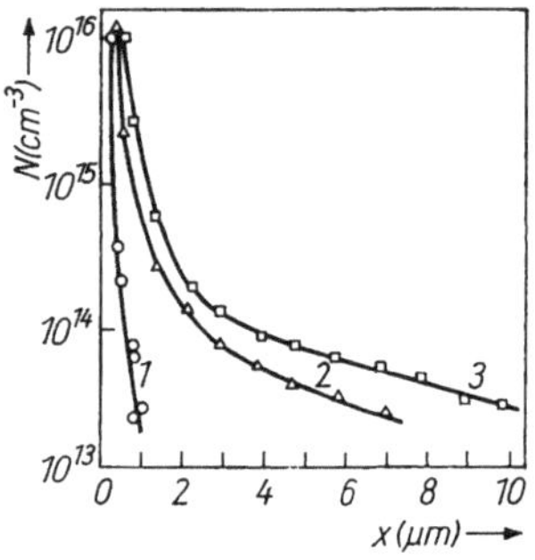

Fig. 6. Concentration profiles for gold in silicon (1) check sample (non illuminated), (2) and (3) samples illuminated for 30 and 50 h, respectively, $T = 120$ K

The attempts to detect the photostimulated diffusion in silicon were first done in 1970 [13]. In these experiments a gold film, labelled by the radioactive isotope ^{198}Au and vacuum deposited on the p-type silicon surface at 120 K, was used as a diffusate. Illumination of the sample from the side opposite to the gold film, was performed by means of a 340 W electric lamp. In this case the temperature of the surface under illumination did not exceed 120 K.

Fig. 6 represents gold concentration profiles in the nonilluminated check sample (curve 1) and in the samples illuminated for 30 and 50 h (curves 2 and 3, respectively). The gold concentration distribution profiles were determined by the method of successive etching of thin layers and measurement of the residual activity of the sample. It is to be mentioned that in the process of etching, layers of different thicknesses (about 0.15 and 1 μm, respectively) were removed from the nonilluminated and illuminated samples. As may be seen from Fig. 6 in the illuminated samples the gold concentration profile propagates to a depth of approximately 10 μm, while in the nonilluminated sample to a depth of approximately 1 μm. The authors of [13] consider the photostimulated diffusion to be responsible for the deeper penetration of gold in the illuminated samples and the mechanism of the mentioned process to be connected with the expansion or even with the break of the chemical bonds in the hexagonal ring under the effect of illumination, reducing the barrier for the migrating gold atoms or ions [14].

The authors of [15] believe the observed difference in the gold concentration distribution in the illuminated and check silicon samples to be due to the methodical disadvantages connected with the absorption of gold by silicon in the process of etching. Considering that layers of various thicknesses were removed from the check and illuminated samples, the quantity of the radioactive gold atoms absorbed from the etchant at the surfaces of these samples should be different. In turn, this fact can bring about a greater residual radioactivity of the illuminated sample (where the thicknesses of the removed layers are greater) and hence a deeper "penetration" of gold in the illuminated samples as compared to the nonilluminated ones.

A photostimulated variation of mobility μ and hole concentration p was observed in gallium-doped and in lithium-compensated germanium samples [16]. The variations of the electrical parameters were detected for the samples fastly cooled from 100 °C to liquid nitrogen temperature and subsequently illuminated. As it takes place, an increase in the illumination time (from 0 to 1 h) leads to an increase in the concentration of holes and to a drop in their mobility saturating (at $t > 1$ h). In slowly cooled samples the variations of the electrical parameters due to the subsequent illumination are negligible. The observed light-stimulated variations in the concentration and mobility of holes in sharply cooled samples are conditioned by the photostimulated migration of lithium to complexes of gallium–lithium type, accompanied by gallium decomposition, increasing the concentration of the holes and decreasing their mobility [16]. The authors of [16] consider that in slowly cooled samples pair formation of lithium ions with gallium ions increases the activation energy of lithium diffusion. That is why illumination effects play a minor role in this case. It is felt that in such samples subsequent illumination should have little influence on the concentration and mobility of the holes due to the lithium–germanium and gallium–germanium solid solution precipitation, reducing the concentration of the electrically active impurity atoms and their complexes. Lower magnitudes of p for the samples in question at the instant of illumination ($t = 0$) compared with sharply cooled samples, where the decay processes in solid solutions should be negligible, is likely to be indicative of solid solution precipitation.

It should be particularly emphasized that in the discussed experiments the light with energy higher than the germanium gap energy bears an essential influence on the electrical characteristics of rapidly cooled germanium samples. Proceeding from these facts and considering that the samples with thickness approximately equal to the charge carrier diffusion length (≈ 0.2 mm) were used for the experiments, it is believed that the photostimulated acceleration of lithium diffusion acts upon the internal electric field, arising due to inhomogeneous light absorption from the region of germanium intrinsic absorption.

The mechanism of the effect of illumination from the region of quantum energies below the bandgap on the lithium drift in germanium, leading to the expansion of the p–n junction, is studied in [17]. The p–n junction is created by the diffusion of lithium into p-type germanium at 400 °C. Lithium ion drift (at constant voltage) in these samples was subsequently studied under illumination and dark conditions in the temperature range from 13 to 18 °C. The sample was illuminated from a silite rod through a germanium filter. As a result of lithium ion drift the i-region width in the illuminated samples was 1.2 times as high as that in the nonilluminated samples. The temperature of these samples exceeds that of the dark samples by 5 K even though special care to avoid heating of the samples was taken in the process of these experiments. In this case light absorption and increase of photocurrent through the illuminated sample being under constant voltage (photodiode condition), make a contribution to the increase of the temperature. The estimates show that an increase in the illuminated sample temperature by 5 K can bring about an increase in their i-region width by 1.2 times as compared with the nonilluminated sample. The results of these experiments speak in favour of the fact that the acceleration of lithium ion drift in the germanium p–n junction field at illumination from the impurity absorption region is motivated by heating effects of the illumination, which increase the lithium ion mobility.

In [18] the effect of illumination on copper diffusion in n-type cadmium sulphide, involving an electrical compensation of the near-surface region of the sample, was studied by an electrical method. Cadmium sulphide samples with a sputtered copper layer (10 μm) was illuminated by means of a He–Ne laser ($\lambda = 0.6328$ μm, $P = 200$ mWcm^{-2}) and annealed simultaneously at a temperature ranging from 200 to 600 °C. In doing so the contribution of illumination to the temperature rise of the sample was negligibly small ($T \leq 1$ K). The diffusion process was studied in the direction of the C-axis. Basing on measurements of the resistance distribution along the cadmium sulphide sample thickness the illumination of the samples in the process of diffusive annealing was shown to give rise to a wider compensated region in them than in the check samples annealed at the same temperature in the dark. For instance, at an annealing temperature 200 °C during 2 h the depth of the compensated region in the illuminated and non-illuminated samples amounted to 31 and 18 μm, respec-

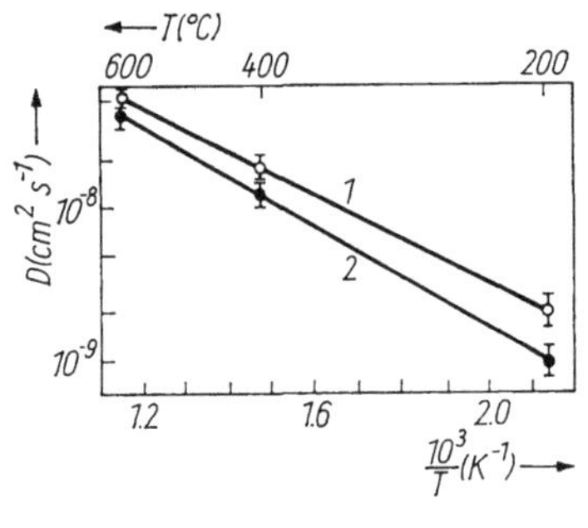

Fig. 7. Temperature dependence of diffusion coefficient for copper in cadmium sulphide in the illuminated (1) and test (2) samples

tively. Temperature dependences of copper diffusion coefficients in light and dark samples, determined from the measured profiles of the resistance distribution, are represented in Fig. 7. The activation energy of copper diffusion in cadmium sulphide, estimated from the data in Fig. 7 for the illuminated and nonilluminated samples, are equal to $W_i = 0.46$ eV and $W_{ni} = 0.55$ eV, respectively. The decrease in copper diffusion activation energy when illuminating cadmium sulphide (by 0.1 eV) is attributed to the variation of the copper charge state, leading to the acceleration of the migration process of these ions.

The experimental study of the external electrical field effect on the photostimulated diffusion of copper in [19] allowed to show that in illuminating cadmium sulphide (from the region of copper impurity absorption $h\nu \approx 1.9$ to 2.0 eV) the negative charge of copper acceptor ions decreases by 15 to 18% ($q = -(1.2 \pm 0.6)\, e$, where e is the electron charge). This result could be treated as an increase of the fraction of negatively charged copper ions (in the process of illumination), which are characterized by fast migration via interstitial sites of the crystal lattice.

It is to be mentioned that the value of diffusion activation energy for the cadmium sulphide check samples (0.55 eV) obtained in [18], is well below the values (0.8 to 1.2 eV) given in previously published papers [20]. This difference may be due to the fact that determining the diffusion coefficient only by the electrical measurement (without knowing the law of distribution of the impurity atom concentration along the crystal thickness) is connected with errors. Moreover, the procedure of copper diffusion from a relatively thick layer, when the thickness of the sputtered copper layer (≈ 10 µm) is comparable with the diffusion region thickness (≈ 30 µm) adopted in [18], may give rise to additional errors connected with the ambiguity of the boundary between the copper layer and cadmium sulphide after thermal (or photothermal) annealing. At such a copper thickness it is conceivable that new phases (of $Cu_{2-x}S$ type) are formed at the boundary between copper layer and cadmium sulphide substrate, which may change the electrical resistance of the sample. It is of some interest to estimate the variation in the activation energy of copper diffusion in cadmium sulphide in the process of illumination. In accordance with relation (30) at photostimulated diffusion of the impurity by a migration mechanism having dissociative character, the activation energy should decrease by the value $E_g - \Delta E_a$ over the ordinary (heat) diffusion. For cadmium sulphide at room temperature $E_g \approx 2.4$ eV, and the positions of two energy levels of the acceptor copper in the forbidden band are $\Delta E_{a1} = 1.88$ eV and $\Delta E_{a2} = 2.06$ eV. From this it is inferred that at photostimulated diffusion of copper into cadmium sulphide, when the charge state of copper ions changes ($Cu^- \rightarrow Cu^+$ transition) under the action of illumination, the diffusion activation energy should decrease by ≈ 0.4 eV. This value differs from that (0.1 eV) reported in [18]. At the same time the value 0.4 eV corresponds to the diffusion activation energy differences determined for the nonilluminated check samples and the illuminated samples in [20].

Therefore, the (qualitative) detection of the illumination effect on copper diffusion in cadmium sulphide may be looked upon as being substantiated. This is also testified by the subsequent experimental study of the photostimulated diffusion of silver into cadmium sulphide by direct methods (in particular, by the radioactive isotope method). The quantitative data on determining the parameters of the photostimulated (as well as nonilluminated) diffusion represented in [18] are likely not to be conclusive.

The author and his coworkers [21] have investigated the effect of illumination on silver diffusion in cadmium sulphides by the tracer techniques. The experiments were performed at room temperature, at which the role of thermal diffusion processes should be considered negligibly small.

N-type cadmium sulphide plates ($10 \times 10 \times 0.5$ mm^3) with resistivity $\varrho = 10^9$ Ωcm aligned along the crystallographic direction [0001] served as the initial samples. ^{110m}Ag was diffused from a silver layer (approximately 0.5 μm thick) deposited in vacuum (6×10^{-3} Pa) on the surface of a cadmium sulphide plate. The obtained Ag–CdS structures were illuminated (from the side of the silver plate) by blue laser radiation ($\lambda = 0.4410$ μm) of power 11 mWcm^{-2} or white light (from an electric filament lamp) of power 90 mWcm^{-2} at room temperature. The samples were fixed on massive copper holders to avoid heating of the samples in the process of illumination. The heating of the samples under the action of laser radiation or electric lamp did not exceed 1 to 2 K and 4 to 5 K, respectively. Ag–CdS structures served as the check samples and were not illuminated. After exposure the ^{110m}Ag layer which did not diffuse in the sample, was removed from cadmium sulphide by etching in 1 HNO_3 + + 2 H_2O solution. The silver concentration profiles in cadmium sulphide were analysed by the method of removing (etching) thin layers and measuring the residual radioactivity of the sample [22]. To remove cadmium sulphide layers concentrated or diluted HCl was utilized. It should be noted that the thicknesses of the layers removed from the illuminated and check samples were approximately the same.

Fig. 8 shows the silver concentration distribution in cadmium sulphide samples, illuminated by light from a blue laser of power 11 mWcm^{-2} during 4 h (1) and 2 h (2) and in the check (nonilluminated) sample (3) as well. White light illumination of the samples results in the same shape of silver concentration profiles. One can see from

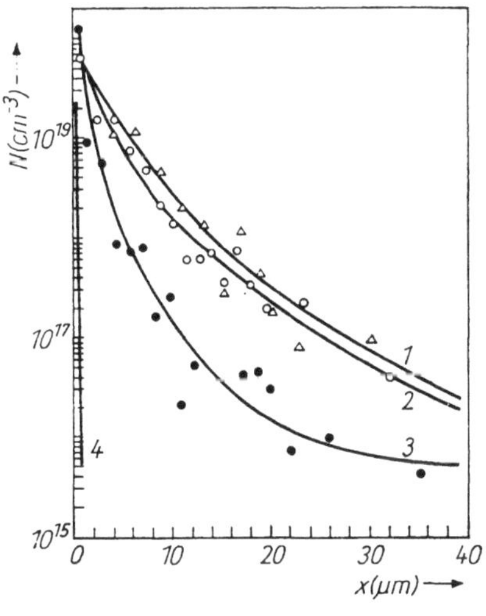

Fig. 8. Concentration profiles for silver in cadmium sulphide (1) and (2) samples illuminated by blue laser light for 4 (1) and 2 h (2); (3) check sample; (4) estimated silver concentration distribution for $T = 30$ °C ($D = 1.3 \times 10^{-14}$ cm^{2}s^{-1}) and $t = 4$ h

Fig. 8 that in the illuminated samples the concentration and penetration depth for silver are well above those in the check sample. In this case the silver concentration in cadmium sulphide slightly increases with increasing exposure time. It must be emphasized that the silver penetration observed in the check samples may be due to uncontrollable illumination at silver deposition and subsequent analysis of the impurity concentration profile.

The silver concentration profiles in cadmium sulphide, summarized in Fig. 8, are described by the empiric dependence

$$N = Bt^{1/3}x^{-n}\,, \tag{36}$$

where t is the illumination time, $n = 2.3$ to 2.9 and $n = 2.6$ in the case of illumination with blue laser and white light, respectively. B is a constant depending upon the illumination intensity. The deeper penetration of silver into cadmium sulphide observed in the process of these experiments should not be attributed to heat effects of illumination. Indeed, if we take advantage of the fast component of the silver diffusion coefficient in cadmium sulphide [20] extrapolated to room temperatures ($D = 7 \times 10^{-15}$ and 1.3×10^{-14} cm^{2}s^{-1} at the temperatures 25 and 30 °C, respectively), the estimated value of the depth of silver diffusive penetration as a result of thermal treatment at 30 °C, is of the order of 1 μm (see Fig. 8, curve 4), and the temperature-rise-induced (from 21 to 30 °C) displacement of the silver concentration profiles should be in the neighbourhood of 0.1 μm. During illumination the experimentally measured values of silver penetration depth amount to significantly greater values ($>$ 40 μm) as compared with those expected from the laws of ordinary thermal diffusion. It might be well pointed out that an estimation of the value of effective silver diffusion coefficient by the concentration profiles 2 and 3 represented in Fig. 8 demonstrates that $D \approx 1 \times 10^{-11}$ cm^{2}s^{-1}, which corresponds to a common thermal diffusion at $\approx$ 110 °C. From this it is inferred that the photostimulated diffusion at room temperature is equivalent to thermal diffusion at 110 °C.

Impurity recharging may be one of the reasons of such an acceleration in silver migration in cadmium sulphide at room temperature and copper migration in CdS at higher temperatures (200 to 600 °C) [18]. There exist some other mechanisms of photostimulated diffusion connected with the photoreaction diffusion [22], the influence of metal–semiconductor barrier contact electric field on diffusion [23], etc.

Table 1

Displacement of p–n junction (in μm) at the illuminated (i), non-illuminated (ni) surfaces, and in the check samples (ch)

time (h)	open-circuited p–n junction			short-circuited p–n junction		experimental conditions
	i	ni	ch	i	ni	
5	440	75	8	—	—	in water
15.5	—	—	285	−375*)	330	sample with graduated junction in water
10.5	260	130	—	60	80	sample with "sharp" junction in air

*) Minus is an indication of p–n junction displacement in the direction opposite to the concentration gradient, which is connected with the electrical transfer of copper.

Combined investigations on silver photostimulated diffusion and the effect of this process on the properties of cadmium sulphide and structures on its basis are hoped to explain unambiguously the mechanism of this phenomenon in CdS.

The effect of illumination on the p–n junction displacement created by copper diffusion in n-type gallium arsenide was observed at room temperature [11]. The depth of the p–n junction amounted to 1 to 5 mm. The samples were illuminated by means of an iodine–tungstem lamp. The power of the light flux incident on the sample surface was equal to approximately 1 Wcm^{-2}. The estimations show that such a power is not sufficient for creating an essential (above some degrees) temperature difference between the illuminated and dark surfaces of the sample during experiments. For better heat dissipation the surfaces under illumination are usually blown by an air flux. In the other set of experiments the samples were placed in distilled water, mechanically mixed during the illumination process. The check sample (also with p–n junction) shielded from light was located adjacent to the illuminated one. Some experiments were conducted on samples with short-circuited p–n junction. After illumination p–n junctions appeared at both surfaces of the samples (illuminated and nonilluminated). The data on p–n junction displacement under various experimental conditions are summarized in Table 1. As one can see from the table the open-circuit p–n junction at the illuminated surface always migrates over a larger distance as compared with that at nonilluminated or check sample surfaces. The displacement of the p–n junction at the nonilluminated surface and at the surfaces of the check sample is attributed to high rates of copper diffusion in gallium arsenide at room temperature.

Copper in gallium arsenide plays the role of an acceptor impurity and is located in the crystal lattice vacant sites. The copper concentration in the interstitial sites is negligible with respect to that in the vacant sites. Copper migrates mainly in the form of positive ions which move over interstitial sites and occasionally settle down in substitutional sites. The p–n junction field created in gallium arsenide by copper is directed so that it prevents the penetration of positive copper ions from p-inton-region (see Fig. 5a). In the process of illumination the p–n junction field intensity decreases. This fact should bring about an increase in the interstitial copper flux to the n-region and to a displacement of the p–n junction in the same direction (Fig. 5b). This mechanism of p–n junction light-stimulated displacement in gallium arsenide (see Section 2.7) is confirmed by the experimental results obtained for the displacement of the short-circuited p–n junction. In the process of these experiments two samples were illuminated: one sample with open-circuited and the other with short-circuited p–n junctions. It is felt that in the short-circuited sample the p–n junction displacement will be smaller than that in the open-circuited one as far as the short-circuited junction field remains the same as that of the non-illuminated one (see Fig. 5c).

Indeed, as one can see from Table 1, the displacement of the short-circuited junction is smaller than that of the open-circuited junction. During one experiment the short-circuited p–n junction displaces even in the direction of p-type semiconductor. This may be associated with the electrical transfer of the interstitial copper in the field created by the short-circuit current (1.5 mA) to the compensated region of the sample with smooth p–n junction and is directed opposite to the copper concentration gradient.

It should be mentioned that the illumination effect on p–n junction displacement in gallium arsenide is naturally observed only in the layer, where the main absorption of light occurs. As the layers are removed from the illuminated surface, the p–n junction displacement value decreases and at a thickness equal to 10 to 20 μm assumes the same value as that for the nonilluminated surface of the sample.

Consequently, summarizing the above-discussed review one can conclude that the problem of studying the photostimulated diffusion in real crystal semiconductors is still at an early stage. Principally, an investigation of the effect of illumination on the migration of fast diffusing group I impurities (gold, silver, copper, and lithium) in some semiconductors (silicon, cadmium sulphide, gallium arsenide, and germanium) is made. On the basis of these investigations it can be established that the illumination exerts an influence (nonthermal) on the processes of atomic diffusion in semiconductors. But the problems of regularities and peculiarities of the photostimulated impurity diffusion mechanism still remain unsolved. To solve them one needs to accumulate new experimental data, using integrated methods of measurements, which include direct measurements of phase formation, concentration profiles of the impurity atoms, etc. As might be expected, the photostimulated diffusion effects are easily detected in studying the migration of fast diffusing impurities. Besides the group I elements the transition metal group (chromium, manganese, iron, cobalt, nickel, etc.) is similarly related to such impurities. Up to the present time, the literature contains very little information on impurity diffusion under the action of high-power laser radiation in semiconductor crystals, which were not previously exposed to radiation or ion bombardment [24, 25].

Usually to study the photostimulated annealing of the defects created in the process of ion implantation and photostimulated diffusion in the implanted semiconductor layers use is made of pulsed laser radiation with the typical duration of $\tau_0 \approx 10^{-3}$ s (free generation condition) and $\tau_0 \approx 10^{-8}$ s (Q-spoilt mode). Under the action of a ms-laser pulse on the semiconductors, thermal effects propagate into a greater part of the sample ($\approx$ 10 to 50 μm) and the processes of redistribution of impurities and defects are determined by heating. Under the influence of ns-pulses the thickness of the sample layer subjected to heating is negligible ($\approx$ 1 μm), and in this case the effectiveness of the laser annealing of the implanted layers is also determined by other effects of radiation (internal electrical field, variation of the charge state of impurities and defects, coherent interaction of light with the lattice, etc.) [26]. Below we discuss the results of studying the photostimulated diffusion of impurities with the use of laser radiation having pulse duration in the range of tens of ns.

In studying the photostimulated diffusion of boron from a silicon layer with inhomogeneous distribution of boron previously created by means of diffusion in the furnace ($T = 1000$ °C, $t = 10$ min) a single pulse of a ruby laser ($\lambda = 0.694$ μm, $P = 1.5$ Jcm^{-2}, $\tau_0 = 60$ ns) is shown to give rise to an appreciable redistribution of the initial boron diffusive distribution (Fig. 9a) [24]. As a consequence of laser irradiation the electrically nonactive boron initially concentrated in a thin near-surface silicon layer ($\approx$ 0.05 μm) transfers into the volume of the sample, behaving as an

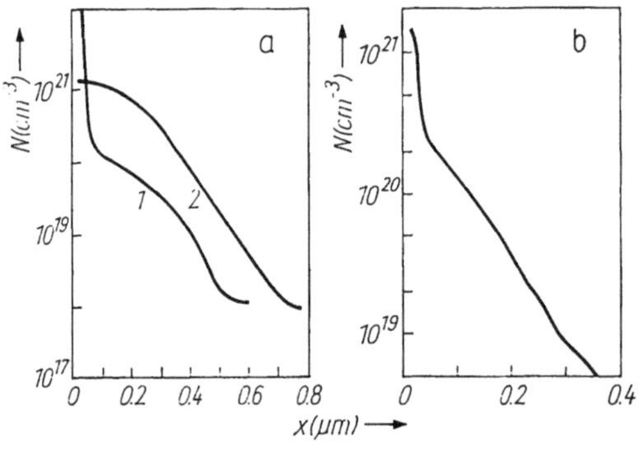

Fig. 9. Concentration profiles for boron in silicon at laser illumination ($\lambda = 0.694$ μm, $P = 1.5$ Jcm^{-2}, $\tau_0 = 60$ ns) from a) the diffusively doped layer and b) the boron layer deposited on the sample surface. (1) and (2) before and after laser illumination, respectively

ionized acceptor. After a single laser pulse the concentration and penetration depth of boron increase markedly (from 0.6 to 0.8 μm).

The same results are obtained in the case of laser radiation-stimulated diffusion of boron and aluminium from the diffusate film (≈ 10 nm) deposited on a silicon surface (under identical experimental conditions). The concentration profile of boron in silicon after the laser radiation is represented in Fig. 9b [24]. One can see that during a shorter period of irradiation ($\approx 6 \times 10^{-8}$ s) the boron penetration depth amounts to approximately 0.4 μm. Such a fast penetration of boron in silicon is tolerable, if the diffusion coefficient is accepted to be of the order of 10^{-3} cm^{2}s^{-1}, which is characteristic of the diffusion in molten silicon. As far as the laser radiation power used in [24] is sufficient to melt the near-surface region of the sample (≈ 1 μm), one can assume the fast migration of boron (and aluminium, too) to be due to diffusion in the molten region of silicon with subsequent recrystallization of this region.

Laser irradiation-stimulated diffusion of impurities in silicon allowed to produce various rectifying junctions and ohmic contacts [25]. One manages to obtain rectifying n–p and p–n junctions having sharp profiles of the diffusing impurity concentration distribution through the illumination of p- and n-type silicon with arsenic and gallium layers (≈ 0.5 μm), respectively, by a neodymium laser operating in the Q-switched mode ($\lambda = 1.06$ μm, $P = 150$ to 300 MW cm^{-2}). In this case the best electrical characteristics of the resultant structures are attained at the illumination power 150 MW cm^{-2}. Laser annealing of n- and p-type silicon with phosphorus and boron layers, respectively, allows to create ohmic contacts with fairly low transient resistance. In these experiments high-power radiation is likely to encourage melting of the silicon surface as for boron and aluminium laser diffusion [24].

The results of a number of experiments on photostimulated diffusion are represented in [27].

4. Photostimulated Annealing of Radiation Defects in Crystalline Semiconductors

4.1 Photostimulated annealing of radiation defects

The interaction of high-power nuclear radiation with solids is known to result in the formation of structural defects in the crystalline lattice. At room temperature most radiation defects are thermally stable. But at subsequent annealing of the samples they recover or transform to other types of defects due to the increase of mobility. Annealing of radiation defects is mainly determined by the magnitudes of their mobility substantially depending upon temperature and other types of external effects (illumination, electrical field, elastic and plastic deformation, etc.). At present the results of research on the effect of high power pulses of laser radiation on the annealing of defects implanted at ion doping of semiconductors are intensively discussed in the literature.

The problems of the photostimulated annealing and transformation of the defects produced by neutron, electron, and γ-radiation in semiconductors are treated only in a few works. First we shall discuss the results of studying the influence of low-power illumination (insufficient for pronounced heating of the sample) on the processes of formation, transformation, and migration of defects in semiconductors. The processes responsible for the formation of point defects and more complicated defects under the action of low-power light radiation (below the threshold energy) are poorly investigated.

In [28] the authors reported that pre-illumination of indium antimonide at 78 K from the region of intrinsic absorption led to a variation in the sample electro-conductivity. They believe point defects in the form of Frenkel pairs to be produced in indium antimonide under the effect of illumination. It is probable that illumination of indium antimonide with quantum energy well below the threshold energy required to produce defects by a percussive mechanism, cannot give rise to defects. It is felt that under illumination the processes of recharging of defect centres can give a contribution to the variations in the electrical properties of indium antimonide.

A light sensitivity effect is observed in p-type germanium exposed to electron radiation at liquid nitrogen temperature and subsequently annealed at temperatures from 130 to 220 K [29, 30]. The phenomenon consists in the fact that white light illumination of these samples, after cooling them to liquid nitrogen temperature, leads to an increase in the hole concentration, which remains unchanged even after switching off the illumination. Reheating of the sample to temperatures from 130 to 220 K and cooling to 77 K reduce the hole concentration to the previous level.

The photosensitivity effect observed in electron-irradiated germanium is attributed to the presence of traps with smallest capture cross-section of holes in the upper half of the forbidden band [31].

Light from the impurity absorption region ($h\nu \gtrapprox 0.32$ eV) also bears an influence on radiation defect annealing in germanium exposed to fast electrons (4 MeV) at 77 K [30]. In this case the curves obtained for the variation of the specific resistance at isochronic annealing under illumination essentially differ from those in the dark over the temperature range from 77 to 300 K.

Within the range from 77 to 210 K the specific resistance of the illuminated samples passes through a minimum, while that of the nonilluminated samples through a maximum. The illumination-stimulated variations in the specific resistance, being due to reversible and irreversible variations in radiation defect structures (tentatively vacant site–impurity complexes), are attributed to variations in the charge state of the radiation defects.

Illumination produces a similar effect upon the development of the radiation defects implanted in germanium by means of electron radiation (1 MeV) also at rather lower temperatures (10 K) [32].

Such defects cannot be detected by electrical methods immediately after radiation exposure. Subsequent to IR light illumination ($h\nu < E_g$) the radiation defects are subjected to ionization and then they are detected. At this two types of ionized donor defects, differing by the ionization rates and annealing temperatures (40 to 55 K and 50 to 70 K) appear in p-type germanium. The annealing rate of these defects depends on the charge state.

In p-type silicon, doped by the donors of group V elements (phosphorus, arsenic, and antimony) due to electron radiation, a donor–vacancy complex (E-centre) arises which creates an acceptor level $E_c - 0.4$ eV in the forbidden band. The E-centre is annealed at temperatures from 150 to 200 °C. In this case the rate of E-centre thermal annealing depends on the charge state of the complex. With the Fermi level below the level with $E_c - 0.4$ eV (E-centre is neutral) the activation energy for annealing the complex is smaller than that with Fermi level above the complex level (E-centre is negatively charged) [33]. Qualitatively this is explained by the fact that the geometric radius of the negatively charged complex is larger than that of the neutral one and, hence, it is less mobile. In these experiments the Fermi level position was controlled by means of doped silicon. The position of the Fermi level (quasi-level) can also be changed with the aid of illumination and in this case one should expect variations in the annealing rate of the E-centre in silicon.

The effect of illumination on E-centre annealing in electron-irradiated silicon has been studied in [34] measuring charge carrier concentrations by means of the Hall

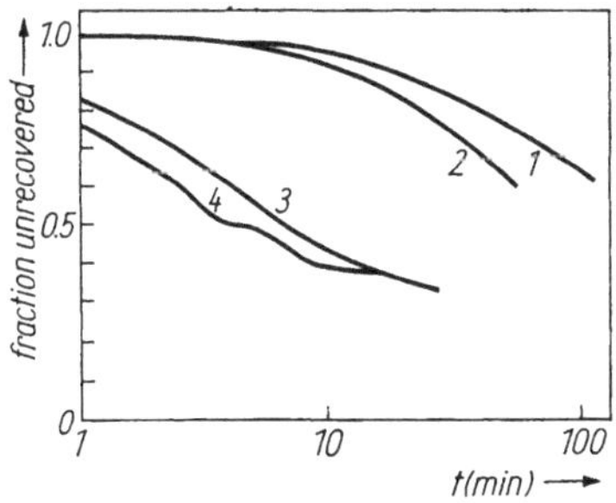

Fig. 10. Curves for isothermal annealing of E-centres in silicon at 134 °C (1, 2), 156 °C (3), and 155 °C (4); (1, 3) in the dark; (2, 4) under illumination

effect. In the process of thermal annealing (in the temperature range 130 to 160 °C) the samples were simultaneously illuminated with white light from a tungsten-iodide lamp. To avoid an appreciable heating of the samples ($\Delta T \leqq 3$ K) annealing was conducted during 1 min. The illumination power ensured the generation of charge carriers (1×10^{14} cm^{-3}) in silicon with specific resistance 10 Ωcm. Fig. 10 shows the curves of isothermal annealing at two temperatures (134 and 155 °C) of the E-centre in the irradiation silicon ($\varrho = 10$ Ωcm) under illumination (curves 1, 3) and in the dark (curves 2, 4). As one can see from Fig. 10, illumination of the samples stimulated a faster annealing of the E-centres as compared with the samples annealed in the dark. At the temperatures in question the E-centre level in silicon is partially filled by electrons, i.e. the Fermi level is located in the region of the complex acceptor level. In illuminating such samples with white light ($h\nu > E_g$) prevalent hole capture in the E-centre effectively lowers the Fermi level below the E-centre level, thus increasing the fraction of the neutral complexes and their migration rate.

The photostimulated annealing of neutron irradiation defects in silicon and gallium arsenide by means of optical measurements is discussed in [35]. Neutron or electron radiation of silicon is known to change the vibrational spectra of the impurities and defects as well as the energy spectrum connected with the interband transitions and with the transitions between the localized levels of defects, impurities, and the respect-

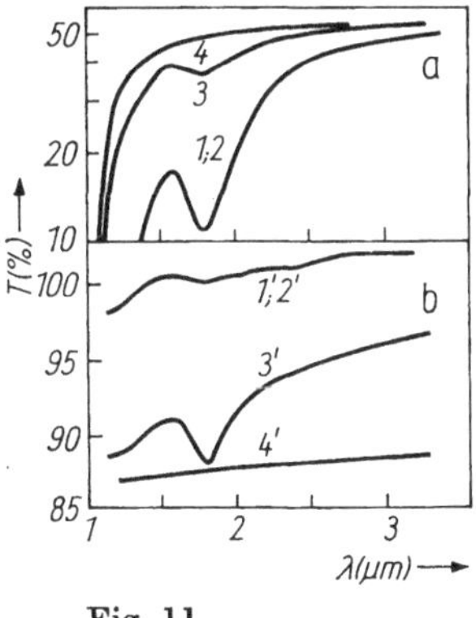

Fig. 11

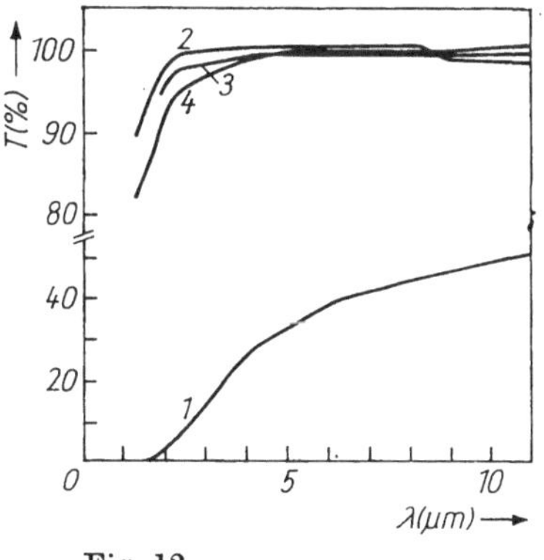

Fig. 12

Fig. 11. a) Transmission and b) differential transmission spectra for neutron-irradiated silicon samples after various stages of photostimulated isochronal annealing ($t = 30$ min) (1, 1′) before annealing; (2, 2′), (3, 3′), and (4, 4′) $T = 100$, 220, and 260 °C, respectively

Fig. 12. Transmission (1) and differential transmission spectra (2 to 4) for neutron-irradiated gallium arsenide samples after various stages of photostimulated isochronal annealing ($t = 5$ h), (1) before annealing; (2), (3), and (4) $T = 200$, 250, and 300 °C, respectively ive band. In the absorption spectrum of the exposed silicon the peaks at 1.8, 3.3, and 3.9 μm due to the ionization of singly, doubly negatively charged and singly positively charged divacancies, respectively, were observed. During thermal annealing under illumination and in the dark of silicon the observations were made for the absorption peak at 1.8 μm, of gallium arsenide for the wavelength region between 0.9 and 22 μm, disclosing additional nonstructured absorption at neutron irradiation.

In these experiments the samples of phosphorus-doped n-type silicon and tellurium-doped n-type gallium arsenide ($n = 6 \times 10^{17}$ cm^{-3}, $\mu = 3340$ cm^{2}V^{-1}s^{-1}) were exposed to a fast neutron flux (2×10^{18} neutron cm^{-2}) at temperature below 50 °C. As a consequence of the neutron irradiation the samples became high-ohmic with hole conductivity. The transmission spectra of irradiated silicon and gallium arsenide are given in Fig. 11 and 12, respectively. It is seen from Fig. 11 (curve 1) that neutron-irradiated silicon shows a selective absorption at 1.8 μm due to divacancies, while gallium arsenide (Fig. 12, curve 1) exhibits a shift of the main absorption edge in the direction towards the long-wavelength region of the spectrum. Subsequently the irradiated sample was cut into halves, one half was subjected to isochronic annnealing under the effect of light (for 30 min and 5 h for silicon and gallium arsenide, respectively), the second half (the check sample) was annealed in a furnace in dark at the same temperature. A filament lamp (1 kW) was used as the light source. For each individual anneal the temperature of the illuminated and check samples was kept at the same level. To observe the illumination-stimulated variations in the optical spectrum of the radiated samples after each individual isochronic anneal the differential transmission spectrum was taken for the sample annealed in the dark with respect to the illumination-annealed sample. Concurrent with the above the measurements were also done for the transmission spectra of individual samples (at $T = 300$ K).

One can see from Fig. 11 that heating of silicon samples in the process of illumination promotes a rapid decrease of the divacancy peak at 1.8 μm as compared with the nonilluminated samples. Isochronic annealing at 260 °C leads to the disappearance of the difference in the transmission of the samples in question. As it follows from the single spectra of the sample (see Fig. 11a, curve 4); this situation is connected with a complete annealing of divacancies.

Heating of the radiated gallium arsenide samples with light, as in the case of silicon, enhances their transmission in comparison with the check samples annealed in the dark (Fig. 12). In this case the annealing temperature rise (illumination intensity) brings about an increase in the difference between the annealing kinetics of the illuminated sample and that of the nonilluminated one. Considering that complete reduction of the additional absorption in gallium arsenide occurs at higher temperature 600 °C than in silicon, still no complete smoothing of transmission of the illuminated and nonilluminated gallium arsenide samples is observed in these experiments after annealing at 300 °C.

Summarizing the described results one can conclude that the photostimulated annealing of radiation defects is observed in silicon and gallium arsenide samples. In the case of annealing of divacancies in silicon, light-generated charge curriers are thought to recharge the vacancies from the neutral state to the negative one. By virtue of the fact that the negatively charged vacancies in silicon are more mobile [36], such recharging of vacancies promotes their faster annealing under illumination. The accelerated photostimulated annealing of the disordered regions responsible for the additional nonstructural absorption of the irradiated GaAs from 0.86 to 10 μm is likely to be due to the recharging of defects (vacancies).

Laser annealing (high-power single pulses of a ruby laser of 2 to 20 Jcm^{-2} in the ms and ns ranges) of electron radiation produced (3.5 MeV) point defects in silicon and gallium arsenide is reported in [26]. The annealing temperature of the E-centres in silicon turned out to increase as the ruby laser pulse duration is decreased. Annealing of the E-centres in silicon at temperatures below the melting point is possible only by maintaining the heating duration above 10^{-6} s. To anneal E-centres in silicon by means of ns pulses the temperature of the sample should be brought above the melting point.

4.2 Laser annealing of implanted semiconducting layers

High-power laser radiation is also used to anneal the defects produced by ion implantation. In modern practice ion implantation finds extensive application in local doping of semiconductors. But as a consequence of ion doping, when the impurity ions accelerated in a high electrostatic field impinge into the crystalline targets, a high concentration of defects (point defects, dislocations, dislocation loops, etc.) is observed in the implanted layer. The latter may even become amorphous. Owing to the interaction with such defects the implanted impurity atoms are not electrically active. The defects of ion implantation are annealed in the temperature range 600 to 900 °C and in this case the implanted impurities become electrically active. In the last few years owing to the outstanding works by Soviet physicists [37, 38] high-power laser radiation with quantum energy exceeding the forbidden band width of the semiconducting target is used for local heating of the implanted layers of the semiconducting crystal and annealing of the defects, too. One great advantage of laser radiation compared to the common heating of the whole crystal is the local heating of the implanted layers, whereas the major part of the semiconducting matrix is not subjected to a thermal effect which is usually accompanied by undesired consequences such as changes in the properties of the semiconducting substrate. As a consequence of the effect of laser pulses having sufficient power, the implanted layers recrystallize and the implanted impurities activated electrically during sufficiently short times (up to 10^{-8} s).

Comparative studies of doping the ion implantation defects by the pulsed laser radiation ($\lambda = 0.694$ µm, $P = 1.5$ Jcm^{-2}, $\tau_0 = 60$ ns) and a common heating ($T = 900$ to 1100 °C, $t = 30$ min), made by the methods of transmission microscopy and ion backscattering, demonstrated that in the first case the point defects, dislocations, dislocation loops are completely annealed, whereas after thermal annealing a major portion of the defects are conserved in the form of dislocation loops [24].

Here we do not aim at reviewing the laser annealing of the implanted layers in semiconductors, because there are some fine reviews on this problem (see [39, 40]). Below we shall briefly deal only with one peculiarity in the variation of the concen-

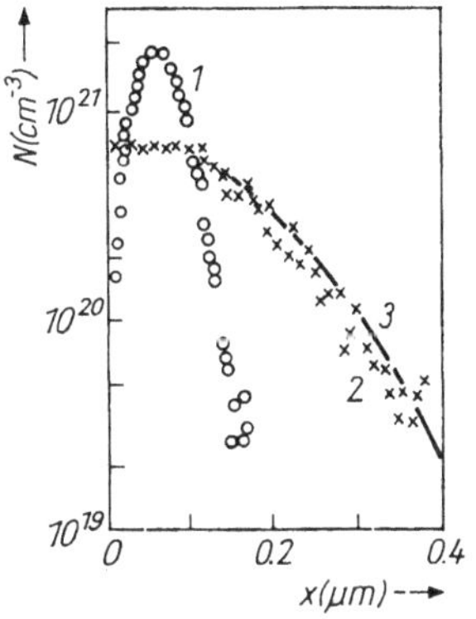

Fig. 13. Concentration profiles for arsenic in implanted silicon, (1) after ion implantation (○), (2) after laser annealing (×), (3) estimated curve (solid line)

tration profiles arising at laser annealing, and with the mechanisms of laser annealing of the defects.

As a consequence of ion doping of semiconductors the concentration profile of the doping impurity is known to have the form of a curve with a maximum located approximately 0.1 μm from the surface. At subsequent laser annealing of the implanted layers a redistribution of the implanted impurity leading to a broadening of the concentration distribution maximum is observed. Fig. 13 represents arsenic (^{75}As) concentration profiles in silicon after ion implantation (ion energy 100 keV, implanted at a dose 1.4×10^{16} cm^{-2}, curve 1) and after laser radiation ($\lambda = 0.694$ μm, $P = 25$ MW cm^{-2}, $\tau_0 = 60$ ns, curve 2) [24]. One can see that due to the ns laser effect the initial As concentration profile transforms to the gradual diffusive distribution without concentration maximum. The same graph shows (curve 3) the estimated profile of arsenic concentration for the implanted ion diffusion in the melted silicon, which is well described by the experimental data. The same concentration profile is obtained at laser annealing of silicon layers implanted by other ions (^{11}B, 35 keV, 1×10^{16} cm^{-2}; ^{31}P, 80 keV, 1.16×10^{16} cm^{-2}). In accordance with [24] the mechanism of the laser-stimulated redistribution of implanted impurities in semiconductors is the following. Under high-power laser radiation of ns duration the near-surface region of the sample (slightly thicker than the implanted one) melts and rapid diffusion of the implanted particles occurs in the liquid phase of the matrix. Then, owing to liquid-phase epitaxial growth, the molten near-surface layer recrystallizes at the substrate. In this case a defect-free recrystallized layer, where the impurity atoms occupy equilibrium positions in the crystal lattice, is observed to grow.

In [41] the authors observed slightly differing arsenic concentration profiles in silicon implanted layers (energy 40 keV, implanted at a dose 10^{16} cm^{-2}). Fig. 14 gives arsenic concentration profiles in the implanted silicon (curve 1) and after laser radiation ($\lambda = 0.694$ μm, $\tau_0 = 20$ ns) of various powers 23, 44, and 70 MW cm^{-2} (curves 2, 3, and 4, respectively). Here, in contrast to Fig. 13, the character of the arsenic concentration profile (curve with maximum) remains the same. The increase in the power of the ruby laser from 23 to 70 MWcm^{-2} gives rise to a decrease in the concentration profile maximum and its position shifts into the bulk of the sample. It is of interest that analogous changes in the diffusing impurity concentration profile in semiconductors are observed at forced diffusion (in an electric field) from an infinitely thin impurity layer into a semiinfinite solid [42]. The variations in the concentration profiles at laser radiation seen in Fig. 14 may be explained by the effect of internal fields. The internal electric field, appearing in silicon due to the absorption of light quanta with energy higher than the forbidden bandwidth, subsequent generation of electron–hole pairs and their separation owing to the difference in their mobilities, is directed from the surface to the volume of the sample. Such fields should accelerate

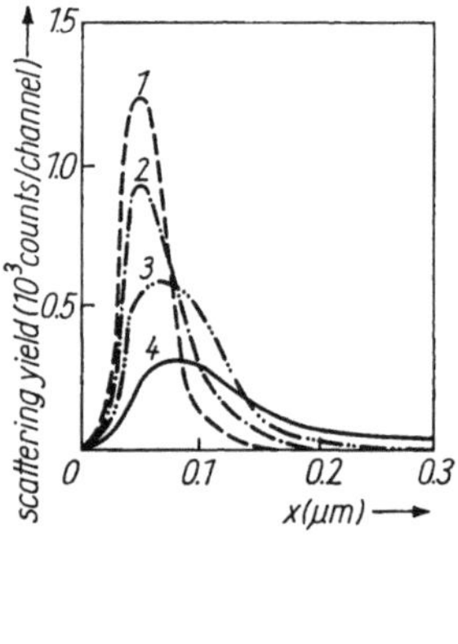

Fig. 14. Concentration profiles for arsenic in implanted silicon, (1) after ion implantation, (2), (3), and (4) after laser radiation of 23, 44, and 70 MWcm^{-2}, respectively

the positively charged arsenic ions and by the drift the concentration distribution of arsenic ions may change in the manner shown in Fig. 14.

At the present time no uniform view on the mechanisms of laser annealing of the ion implantation defects and redistribution of the implanted impurities in semiconductors is available in the literature.

The authors of [26] consider the thermal effect of the laser to be the decisive factor in annealing silicon-implanted layers by ms pulses. But the results of the experimental study of the annealing of implanted layer defects at laser excitation of ns duration cannot be explained only by thermal effects.

The analysis of the experimental data published on pulsed laser annealing of ion implantation defects in silicon given in [43] allows to single out five factors, which cannot be explained in the framework of the traditional model involving the processes of melting and recrystallization of the monocrystalline substrate layer. They are as follows: 1. no liquid phase is observed to appear in the reflection spectra measured in the process of laser annealing; 2. at fast heating of the silicon sample surface (up to 1400 °C during 10 ns) mechanical stresses arising from thermal expansion should bring about pulverization or deformation of the layer, but experimentally no such phenomenon was observed; 3. no peculiarities due to melting of the silicon surface were observed in the beam energy dependences of thermoelastic waves created by the incident laser beam at the free surface of silicon; 4. the redistribution of the impurities near surface does not correspond to the equilibrium coefficient of the distribution between liquid and solid phases; 5. it is not clear how at such a high rate of recrystallization ($\approx 2\ \mathrm{ms}^{-1}$) there appears such a perfect structure of silicon surface layer.

It should be noted from [44] recently published that the results of the measurements of the time dependence of the reflection coefficient (at $\lambda = 0.6328\ \mu\mathrm{m}$) for the silicon surface in the process of high-power pulsed neodymium laser radiation in the Q-switched mode ($P = 1$ to $6\ \mathrm{Jcm}^{-2}$, $\tau_0 = 80$ ns, $\lambda = 1.06\ \mu\mathrm{m}$) speak in favour of a melting of the silicon surface under the influence of laser radiation. The results of electron beam diffraction indicate that a high-power laser pulse ($P = 3\ \mathrm{Jcm}^{-2}$, $\tau_0 = 30$ ns) incident on polycrystalline silicon (0.3 μm), precipitated on sapphire, also leads to the epitaxial monocrystalline growth of a silicon layer on sapphire. In the case of polycrystalline silicon, precipitated on the amorphous substrate SiO_2, a laser pulse does not promote the formation of a monocrystalline layer in silicon. These results presuppose that laser stimulates complete melting of the silicon layer and the sapphire substrate acts as a nucleus for the recrystallization of the layer. The data given in [44] do not confirm the first conclusion made in [43].

Hence, the above-discussed facts allow to conclude that so far the literature does not contain a unique opinion on the annealing mechanism of the semiconducting layer defects implanted by high-power ns laser pulses. One can suppose that some other factors, determining the laser radiation effect discussed in Section 2, may have an essential influence upon the annealing of the defects alongside with the thermal gradients created by laser radiation in the near-surface region of the semiconductor.

High-power laser radiation allows to change directly the equilibrium state of the impurities in the semiconductors doped in the process of growth, thus generating near-surface p–n junctions without depositing additional impurities at the surface of the semiconductors. The mechanism of the laser radiation effect on the distribution of doping elements is as follows. If a liquid phase is formed at the semiconductor surface in the process of interaction of radiation with a substance, the melt is observed to crystallize at subsequent cooling. As a result of displacing the crystallization front from the monocrystalline substrate to the surface of the sample, the uniformly distributed impurity atoms are displaced to the surface of the sample. It is interesting that this effect is more appreciable when the impurity segregation coefficient in the semiconductor is smaller. This procedure was adopted to produce p–n junctions in cadmium antimonide doped with Al, In, Ga, and Te donor impurities under the effect of laser radiation [45]. In this case a thin layer of highly doped n-type semiconductor is formed at the CdSb surface and the conductivity of the exposed layer changes from electron type to hole type due to impurity segregation to the surface. The thickness of the recrystallized layer can be changed depending on the condition of laser radiation (from approximately 1 μm at ns pulse radiation to approximately 1 mm at ms pulses). This method can also be used to obtain n–p–n junctions. Heating the substrate to 250 to 300 °C simultaneously with the laser radiation decreases the temperature gradient along the thickness of the molten layer, thus decreasing the thermal stresses and the accompanying defects in the recrystallized layer. It should be mentioned that the photovoltaic characteristics of n–p junctions in CdSb, produced by laser annealing, differ from those of photoresistors based on cadmium antimonide.

The process of laser radiation-stimulated displacement of the impurities to the surface is observed not only in semiconductors doped during the growth procedure, but they are also detected in the ion-implanted layers of semiconductors. In copper-ion-implanted gallium arsenide the initial concentration distribution of copper (with maximum at 0.1 μm) changes to the distribution where the impurity concentration increases from the depth of 0.05 μm to the surface as a result of laser radiation-stimulated displacement of the impurity to the surface [24]. Redistribution of copper to the surface is likely to be due to the low segregation coefficient of copper in gallium arsenide.

5. Thermal and Photostimulated Diffusion in Glassy Semiconductors

Chalcogenide glass semiconductors discovered by the Soviet scientists Kolomiets and Goryunova [46] hold the greatest promise for the electronic engineering because of a variety of remarkable peculiarities (switching, memory, radiation stability, etc.). Chalcogenide glasses comprise a wide range of semiconducting materials, created on the basis of chalcogens (sulphur, selen, tellurium) and the elements of the groups IVA and VA of the Periodic Table. Prior to discussing the results of the photostimulated diffusion in chalcogenide glass semconductors, a brief review of the data for thermal diffusion of the materials in question is given.

5.1 Thermal diffusion of group I impurities

Until the present time investigations have been conducted for the diffusion and electron transfer of group I impurities only (copper, silver, and gold) in arsenic selenide and arsenic sulphide by the direct radioactive method.

During thermal diffusion (110 to 200 °C) in As_2Se_3 and As_2S_3, the impurities of group I migrate in the form of positively charged ions with different charges, reflecting the different characters of the interaction of the diffusing impurity with the matrix atoms.

Within the temperature range from 110 to 175 °C silver in arsenic selenide migrates mainly via the interlayer spaces as positive ions with effective charge $q = +1.0e$ [47], gold via a two-flux mechanism (both via interlayer spaces and via the glass set) with effective charge $q = +$ (0.5 to 0.6) e [48, 49].

The temperature dependences of the diffusion coefficients of silver (at 110 to 160 °C) [47] and gold (at 130 to 172 °C) [48, 49] in As_2Se_3 are, respectively, described by

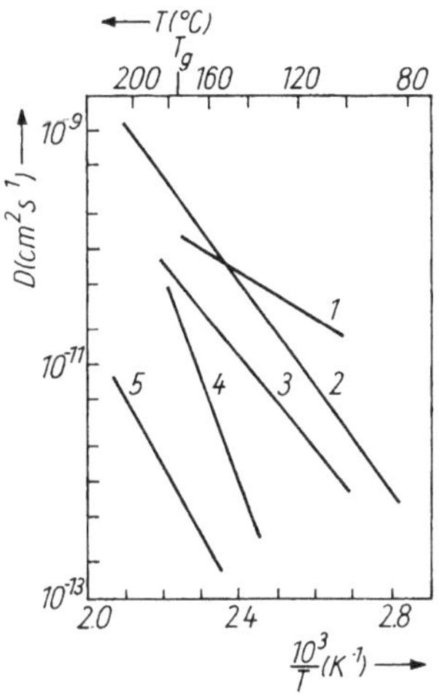

Fig. 15. Temperature dependence of diffusion coefficients for silver (1, 2), copper (3), and gold (4, 5) in glassy As_2Se_3 (1) after [47], (2, 5) [50], (3) [51], (4) [48, 49]

(Fig. 15)

$$D_{Ag} = 5.9 \times 10^{-6} \exp\left(-\frac{0.41}{kT}\right), \tag{37}$$

$$D_{Au} = 8.0 \times 10^{8} \exp\left(-\frac{1.74}{kT}\right). \tag{38}$$

The investigations on thermal diffusion of silver ($T = 80$ to 210 °C) and gold ($T = 150$ to 210 °C) in As_2Se_3 made in [50] under boundary conditions corresponding to the diffusion from an infinitely thin layer, give the following temperature dependences for the diffusion coefficients (see Fig. 15):

$$D_{Ag} = 3 \exp\left(-\frac{0.89}{kT}\right), \tag{39}$$

$$D_{Au} = 6 \exp\left(-\frac{1.15}{kT}\right). \tag{40}$$

The differences in the diffusion activation energies for silver and gold reported in [47 to 50] are likely to be due to various prehistories of these glasses. It should be emphasized that the softening temperature of As_2Se_3 is known to amount to $T_g \approx \approx 174$ °C. In this respect glass softening may have an effect upon the diffusion measurements at $T > T_g$, undertaken in [50].

Within the temperature range from 120 to 180 °C the copper diffusion coefficient varies according to the law [51] (see Fig. 15)

$$D_{Cu} = 5 \times 10^{-2} \exp\left(-\frac{0.79}{kT}\right). \tag{41}$$

In arsenic selenide copper migrates in the form of positive ions ($q = +1.0e$) via a mechanism similar to that of interstitial migration of impurities in crystalline semiconductors.

The uncommon relations of the diffusion coefficients of silver, copper, and gold ($D_{Ag} > D_{Cu} > D_{Au}$) (see Fig. 15), consisting in the breaking of the inequalities ($D_{Cu} > D_{Ag} > D_{Au}$) expected from the relations of ion radii ($r_{Cu^+} < r_{Ag^+} < r_{Au^+}$), may be attributed to the chemical interaction of silver with the glassy matrix due to the remarkable affinity between silver and selenium.

The diffusion behaviour of silver and gold impurities in As_2S_3 differs from that in As_2Se_3. As the measurements of silver diffusion and electron transport with the use of labelled atoms [52] demonstrated the concentration profiles of silver in arsenic sulphide, resulting from the diffusion at temperatures from 130 to 194 °C, consist of two parts: a) the near-surface one (up to 30 µm deep) characterized by a rapid decrease in the concentration of silver and by the independence of silver migration upon the external electric field; b) the volume one ($x > 30$ µm) described by the solution of Fick's equation with temperature-dependent of diffusion coefficient [52],

$$D_{Ag} = 6 \times 10^{-2} \exp\left(-\frac{0.8}{kT}\right), \tag{42}$$

where silver migrates via the interlayer spaces in the form of positively charged ions. The appearance of the near-surface part is attributed to the formation of a new phase Ag_2S [52]. The formation of the new phase in the near-surface region during thermal diffusion of silver in arsenic sulphide is confirmed by the studies of silver electron spectra.

Gold diffusion in arsenic sulphide is rather sophisticated. The values for diffusion coefficient and ion mobilities of gold in the temperature range from 154 to 192 °C depend weakly on temperature and amount to $D_{Au} \approx (1 \text{ to } 2) \times 10^{-13}\ \mathrm{cm^2 s^{-1}}$ and $\mu_{Au} \approx 4 \times 10^{-12}\ \mathrm{cm^2 V^{-1} s^{-1}}$ [52]. The observed small values of D and μ for gold in arsenic sulphide are explained by the "dense" structure of this compound as compared to that in As_2Se_3. In spite of the fact that arsenic selenide and arsenic sulphide are isostructural, in the latter the distance between the layers and the angles between the bonds are smaller and this hinders the migration of the impurities.

The authors of [53] came to the same conclusion when studying the migration of silver and gold in arsenic sulphide. Thus, for the activation energy of silver diffusion in the bulk of As_2S_3 samples the value of $W \approx 1$ eV is obtained, which is close to the activation energy of silver in As_2S_3 ($W = 0.8$ eV) given in [52].

5.2 Photostimulated diffusion

Of all photostimulated phenomena observed in glassy semiconductors — photostructural variations, photocrystallization, and photodoping (see [54, 55]) — only the latter will be discussed in the present section.

The phenomenon of impurity atom penetration into chalcogenide glass semiconductors at room temperature under the action of illumination from the intrinsic absorption region, subsequently called photodoping, photodiffusion, or photostimulated diffusion, was first observed by Kostishin et al. [56]. It is found that as a consequence of illumination layers of a variety of chalcogenides (As_2S_3, As_2Se_3, Sb_2S_3, PbS, CdS, ZnTe) and halogens (CuI, SbI_3, PbI_2), deposited on metallic substrates (Ag, Cu, Zn, Bi, Pb, Cd, Al, Fe, Ni, etc.), produce a visible image. Photodoping involves irreversible variations in the properties of the semiconductor and significantly influences the optical, photoelectrical, and electrical properties as well as the microstructure of the glassy semiconductor.

Therefore, this phenomenon is being investigated very actively in the laboratories throughout the world. In spite of the fact that the mechanism of chalcogenide glass semiconductor photodoping is still to be determined finally, one fact is considered to be reliably determined. The results of numerous experiments studying the influence of the illumination on the properties of metal (Ag, Cu)–chalcogenide glass semiconductor systems produced by depositing the metal film on the semiconducting substrate, indicate that owing to the illumination the metal penetrates into the substrate.

We now turn our attention to the discussion of the mechanisms of photodoping or photostimulated diffusion and then to a brief description of the experimental results.

The authors of [56] believe the mechanism of the photosensitivity of thin semiconducting layers on metal substrates to consist in the photostimulated dissociation of the semiconducting compound, mutual diffusion of layer and substrate components, and formation of new compounds. Considering that the activation energy from the temperature dependence of light sensitivity of the Ag–As_2Se_3 system is close to the activation energy of silver ion diffusion in α-Ag_2Se, the authors of [57] infer that α-Ag_2Se is the product of a chemical reaction. In this process the illumination is essential both for silver ionization and for dissociation of the semiconducting layer [56, 57].

To provide support for the model developed in [56] the authors of [58] observed the formation of Ag_2S and As_2S_2 phases by X-ray and microprobe measurements in the illuminated Ag–As_2S_3 system.

The authors of [59] hold a different viewpoint based on the results of electronographic investigations: no Ag_2S crystalline phase is formed during photodoping, silver is simply dissolved (to 30%) in the amorphous matrix, which may be schematically represented as

$$(Ag)_c + (As_2S_3)_a \xrightarrow{h\nu} (Ag, As, S)_a , \qquad (43)$$

where the indices c and a relate to the crystalline and amorphous phases, respectively.

Dissolution of silver in an amorphous matrix is advantageous thermodynamically, but in the dark it is restrained by a certain kinetic barrier. Illumination stimulates photoexcitation of arsenic sulphide and generation of nonequilibrium electrons and holes captured by traps. Such states at the surface of a semiconductor play the role of chemical catalysts clearing the barrier.

The different conclusions drawn by the authors of [58] and [59] on the formation of the Ag_2S crystal phase in the Ag–As_2S_3 system during the photodoping process are likely to be due to the difference in the structure of the glasses, which is known to be appreciably dependent on the thermal evaporation rate. Slow, low-temperature evaporation leads to the formation of an As_2S_3 stoichiometric film, whereas high-temperature spraying gives arsenic-enriched films.

Another model is based on the influence of the illumination on the potential barrier decrease at the metal–chalcogenide glass semiconductor boundary, accelerating the penetration of metal into the chalcogenide glass semiconductor [23].

Even in the dark, stimulated by chemical interaction at the metal–chalcogenide glass semiconductor boundary, a barrier appears for the positively charged particles. The electron–hole pair generated in arsenic sulphide due to the illumination with quantum energy $h\nu \geqq E_g$, is divided by the electrical field available in the near-boundary region of the semiconductor (Fig. 16). In this process the electric field ejects

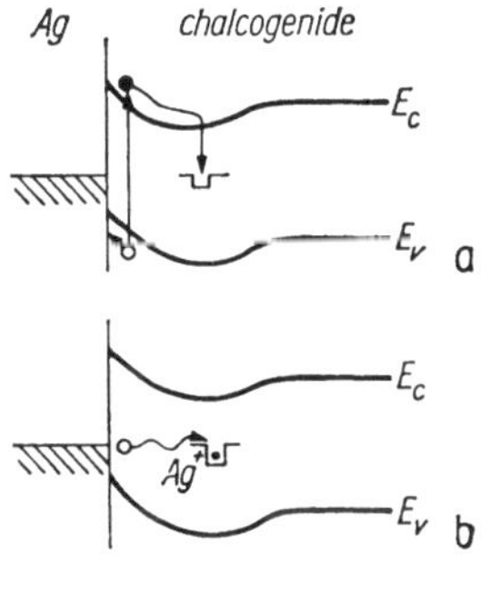

Fig. 16. Energy diagram for an Ag–chalcogenide glass semiconductor structure, a) separation of light-generated charge carriers, b) attraction of silver ions to the negatively charged chalcogens

the holes from the semiconductor into the metal layer thus positively charging the silver atoms and displaces the electrons into the volume of the semiconductor.

Further the traps (chalcogen atoms may play the role of traps) capture the electrons and charge them negatively. The Coulomb interaction between the positively charged silver ions at the surface and the negatively charged chalcogen ions in the volume should decrease the boundary barrier and enhance the migration of the more mobile silver ions.

The experimental investigations of the photoelectrical properties of the Ag–As_2S_3 boundary conducted by the authors of [60] provide the possibility for further development of the photodoping mechanism suggested in [23]. In the dark part silver atoms are supposed to diffuse into the arsenic sulphide near-boundary region creating a deep level $E_c - 1.9$ eV (see Fig. 16c) in the forbidden band. Silver atoms located in arsenic sulphide ($Ag \rightarrow Ag^+ + e^-$) become ionized at light energies below the arsenic sulphide forbidden bandwidth but above the ionization energy of the silver donor level ($1.9\ \text{eV} \leqq h\nu < 2.5$ eV). The released electron goes into the volume of the glass negatively charging the sulphur atoms ($Se + e^- \rightarrow Se^-$). Then, as in [23] the positively charged silver ions should be attracted into the volume of the semiconductor by the negatively charged sulphur ions.

The next model of photodoping is also based on the absence of chemical interaction between silver and semiconductor [61]. In this case the kinetics of silver photodoping (or photodissolution) in arsenic sulphide is supposed to consist of two independent processes. During the first process the critical concentration of defects (charged centres) forms and accumulates in arsenic sulphide either at preliminary illumination of the semiconductor (before precipitation of silver) or after precipitation. The second process consists in direct photodissolution of silver in arsenic sulphide at constant rate, conditioned by photostimulated jumps of silver atoms over the barrier at the boundary.

The investigations of photodoping of silver in glassy arsenic sulphide by helium ion backscattering demonstrated that the silver concentration is almost constant up to a definite depth. Beyond this depth it drops rapidly [62]. Consider also that in the photodoping region the sulphur concentration is equivalent to that in arsenic sulphide. Preliminary illumination of arsenic samples accelerates the diffusion of silver giving rise to "tails" in the concentration. In [62] the steep silver concentration profile observed at photodoping is attributed to a smooth change in silver diffusion coefficient versus coordinate. The concentration of the charged centres set up by the illumination must be distributed along the thickness of the sample in accordance with the exponential law, because the light absorption coefficient varies with thickness in the same manner. If at photodiffusion these centres interact with silver which is also ionized, one should expect a silver concentration distribution corresponding to a decrease in the diffusion coefficient versus arsenic sulphide depth.

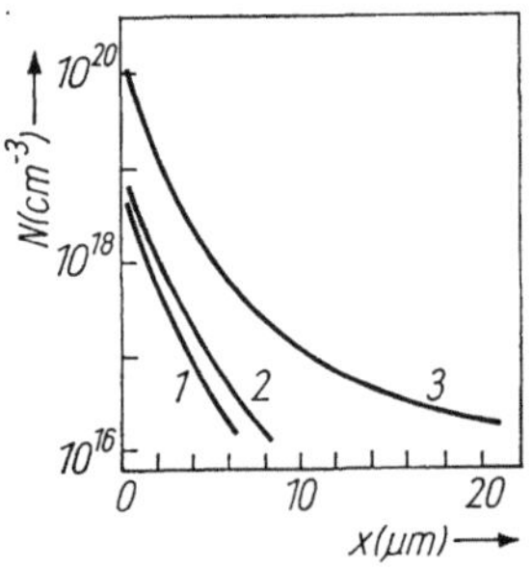

Fig. 17. Concentration profiles for silver in As_2S_3 (1) test check sample, (2) preliminary illumination of the sample before depositing ^{110m}Ag, (3) illumination after depositing ^{110m}Ag

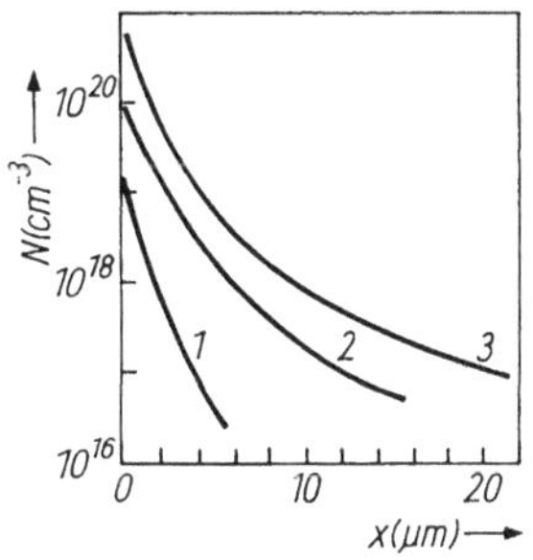

Fig. 18. Concentration profiles for silver in As_2S_3 (1) check (nonilluminated) sample, (2) and (3) red and blue laser illuminated samples, respectively, $I = 5 \times 10^{15}$ photons $cm^{-2}s^{-1}$

The majority of the investigations of photodoping in $Ag–As_2Se_3$ and $Ag–As_2S_3$ systems are conducted by indirect methods: optical, electrical, and photoelectrical measurements. Incidentally, the most valuable information on the photodoping mechanism may be gained using direct diffusive investigations of silver penetration in chalcogenide glass semiconductors under illumination by radioactive isotopes, by means of X-ray and electronographic analyses. Below we shall discuss the results of these investigations.

First we deal with the data on the effect of preliminary illumination of arsenic sulphide upon silver penetration at room temperature [52]. Fig. 17 shows the concentration profiles of silver in arsenic sulphide determined by a tracer technique. In these experiments the virgin matrix of arsenic sulphide was preliminarily illuminated by monochromatic light with the wavelength $\lambda = 0.44\ \mu m$ and intensity 3.3×10^{15} photons $cm^{-2}s^{-1}$ for 2 h. After the illumination radioactive silver ^{110m}Ag was deposited on the virgin matrix (sample 2) and the check (nonilluminated) sample (1). Both samples (1 and 2) were washed, sectioned, and analysed. Fig. 17 also gives the concentration profiles for silver in an $Ag–As_2S_3$ structure subjected to illumination equal to the exposition after depositing silver (3) for comparison. One can see from Fig. 17 that preliminary illumination of arsenic sulphide increases the boundary concentration by 1.5 times and the penetration depth compared to the check (nonilluminated) sample. Illumination of the $Ag–As_2S_3$ structure (curve 3) further increases the boundary concentration and penetration depth for silver. Therefore, the results of these experiments may be explained in the framework of the model which suggests that preliminary illumination of glassy arsenic sulphide samples leads to the formation of defects stimulating the subsequent migration of silver.

In the case of the $Ag–As_2Se_3$ system preliminary illumination of arsenic sulphide sample has no influence upon subsequent thermal diffusion of silver at temperatures from 132 to 144 °C [47].

Illumination of arsenic sulphide with the deposited silver layer significantly influences the silver concentration profile [52, 63, 64]. An investigation of the effect of light wavelength (at equal intensity) on the silver concentration in As_2S_3 points to

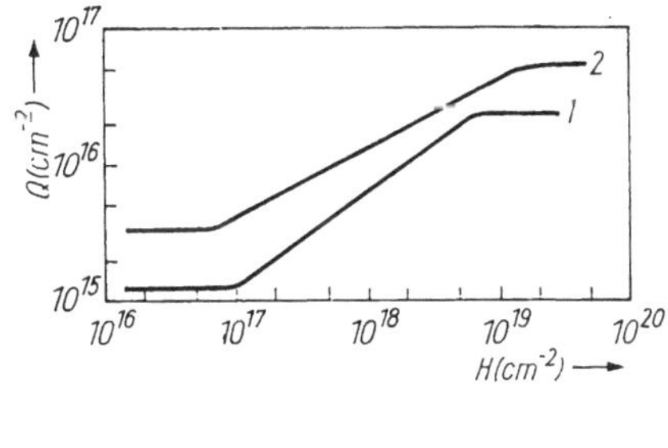

Fig. 19. Dependence of the quantity of photodoped silver Q upon the radiation exposition H (at constant intensity $I = 2.4 \times 10^{15}$ photons $cm^{-2}s^{-1}$) in bulk samples of arsenic sulphide (1) and arsenic selenide (2)

the fact that blue laser light ($\lambda = 0.44$ µm) increases the silver concentration in the near-surface region of the sample with respect to that for the sample exposed to red light ($\lambda = 0.63$ µm) and the check (nonilluminated) sample (Fig. 18). Increase in intensity I and exposition time involves an increase in the boundary concentration and silver penetration depth in the sample. It is interesting that this effect is observed up to some critical values of the intensity (at I = const), above which no variations in the concentration profiles are observed.

Research of the kinetics of photodoping (i.e. the dependence of the photodoped silver quantity on the value of blue light exposition $H = It$) for As_2S_3 and As_2Se_3 disclosed three regions (Fig. 19) [52]: the first region ($H \leqq 10^{17}$ photons cm^{-2}), where the quantity of photodoped silver weakly depends on the exposition, the second region where the quantity of silver is proportional to the exposition (10^{17} photons $cm^{-2} <$ $< H < 10^{19}$ photons cm^{-2}), the third region ($H > 10^{19}$ photons cm^{-2}), where the quantity of photodoped silver weakly varies with exposition and tends to saturation. In accordance with these regions the photodoping process may be supposed to consist of three stages. The first stage is likely to be connected with the process of accumulating the glass defects (broken bonds) at illumination. The second stage may be ascribed to the photochemical reaction between silver ions and the virgin matrix. The third stage, characterized by the trend to saturation of the amount of photodoped silver may be caused by the decrease in the intensity of the penetrating radiation due to light absorption in the intermediate layer.

The distinguishing feature of the observed silver concentration profile in arsenic sulphide is their distribution along the depth of the sample and the dependence of the boundary concentration N_0 on time and intensity,

$$N = AN_0(I, t)\, x^{-n} . \tag{44}$$

The observed illumination-stimulated concentration profiles for silver in arsenic sulphide cannot be explained by free diffusion of silver, because they do not obey the solution of the diffusion equation for the respective initial and boundary conditions. In [65] the authors suppose the photostimulated penetration of silver to be promoted by the reaction diffusion giving rise to a new phase in the near-surface region of the sample. X-ray and electronographic analyses of light-exposed As_2S_3–Ag samples allowed to reveal the presence of $AgAsS_2$, Ag, As_2O_3, As_2S_3 phases and in some samples of the Ag_2S monoclinic phase. The illumination of a chalcogenide glass semiconductor in the intrinsic absorption region ($h\nu \geqq E_g$) is probably encouraging the photodisintegration of As_2S_3 and the formation of a new compound $AgAsS_2$ due to the remarkable affinity between sulphur and copper subgroup metals. The formation of a new phase may be accompanied by heat liberation (exothermic reaction) which, in turn, must stimulate diffusive jumps of atoms.

On the basis of the analysis of the time dependence of the silver quantity diffused in arsenic sulphide at illumination (characterized by a curve increasing to saturation) the authors of [66] also consider that silver diffusion is accompanied by the formation of a new compound $AgAsS_2$ and metallic arsenic. The presence of metallic arsenic capable of oxidizing dictates the nonstability of a chalcogenide film in the Ag–As_2S_3 system formed as a result of illumination.

Investigations of silver electrodiffusion ($\mathcal{E} = 4.3$ kV/cm) from a thin layer deposited at the As_2S_3 surface at temperatures above 150 °C with simultaneous illumination demonstrate that an uncommon concentration profile with broad plateau (≈ 1.5 mm) is observed instead of the expected bell-shaped concentration profile. The authors of [67] think that silver in arsenic sulphide may exist in two states with different diffusion velocities. In the near-surface region silver atoms are neutral, but they may turn to positively charged ones and quickly diffuse into the volume particularly in the presence of a constant electric field. To put it differently photoionization of silver atoms creates favourable conditions for fast drift of silver ions in the electric field, resulting in the observed silver concentration profile in As_2S_3.

In Al–As_2S_3 structures photodiffusion of aluminium is independent of the electrical field at room temperature [68, 69]. At the same time in Al–chalcogenide glass semiconductor–Al (where As_2Se_3 layers and the dopants of As–S–Se, As–Se–Te, As–S–Ge systems are used as the chalcogenide glass semiconductor) the electrophotodoping effect was observed at room temperature [69]. On simultaneous exposure to light and constant electric field the optical properties of chalcogenide glass semiconductors turned out to change (transmission increases and reflection decreases). The effect grows with increasing electric field intensity. One possible way of explaining the observed electrophotodoping effect may be the photostimulated ionization of aluminium atoms ($Al \xrightarrow{h\nu} Al^+ + e^-$) and subsequent electro transport of aluminium ions to the cathode in the electric field.

6. Conclusion

The review presents the physical background of the photostimulated processes of diffusion of impurities and annealing of defects in crystalline and amorphous semiconductors as well as the systematization of the experimental results obtained in studying this problem. The analysis of the available works on the photostimulated diffusion in semiconductors, when the heat effect of radiation may be ignored, showed that for the time being the mechanism of the light-stimulated diffusion of impurities into the sample is poorly understood. It requires new experimental investigations with the use of precise methods of measuring the concentration profiles and electrical state of the impurities, processes of phase formation in a thin near-surface region of the semiconductor, etc. As regards results of the research in the diffusion of the impurities (as well as annealing of the implanted layers) under the effect of high-power pulse laser radiation, when the heat effects are essential, the problem of the mechanism of the diffusion of impurities and defects with sharp temperature gradient, electrical field, high concentration of the electron–hole pairs also waits for further investigations. To tackle this problem one needs to perform a mathematical analysis of the diffusion equation with due regard of different factors of the effect of high-power laser radiation on diffusion.

Acknowledgement

The author wishes to express his sincere thanks to Prof. B. I. Boltaks for helpful discussions of the problem.

References

[1] L. C. Kimerling, Solid State Electronics **21**, 1391 (1978).
[2] W. M. Franklin and P. Sengupta, IEEE J. Quantum Electronics 8, 393 (1972).
[3] G. B. Abdullaev and T. D. Dzhafarov, Atomic Diffusion in Semiconducting Structures, Atomizdat, Moscow 1980 (in Russian).
[4] V. V. Apollonov, A. I. Barchukov, N. V. Karlov, A. M. Prokhorov, and E. M. Shefter, Kvantovaya Elektronika **2**, 380 (1975).
[5] J. R. Meyer, M. R. Kruer, and F. Y. Bartoli, J. appl. Phys. **51**, 5513 (1980).
[6] M. Wautelet, Science **95**, 299 (1980).
[7] M. S. Djidjoev, R. V. Khokhlev, A. V. Kiselev, V. I. Lygin, V. A. Namoit, A. I. Osipov, V. I. Panchenko, and B. I. Provotorov, in: Tunable Lasers and Applications, Ed. Mooradian, T. Jaeger, and P. Stokseth, Springer-Verlag, Berlin 1976 (p. 100).

[8] M. S. SLUTSKY and T. F. GEORGE, J. chem. Phys. 70, 1231 (1979).
[9] V. L. BONCH-BRUEVICH and S. G. KALASHNIKOV, Physics of Semiconductors, Izd. Nauka, Moscow 1977 (p. 194) (in Russian).
[10] J. D. WEEKS, J. C. TULLY, and L. C. KIMERLING, Phys. Rev. B 12, 3286 (1975).
[11] B. I. BOLTAKS, S. M. GORODETSKI, T. D. DZHAFAROV, G. S. KULIKOV, and SH. MALKOVICH, Fiz. tverd. Tela 13, 3420 (1971).
[12] S. M. RIVKIN, Photoelectric Phenomena in Semiconductors, Fiz. mat. giz. Moscow 1963 (p. 409) (in Russian).
[13] L. N. ZUZ, A. E. KIV, O. R. NIYAZOVA, and F. T. UMAROVA, Zh. eksper. teor. Fiz., Pisma 12, 213 (1970).
[14] A. E. KIV, F. T. UMAROVA, and Z. I. ISKANDEROVA, Fiz. Tekh. Poluprov. 4, 3, 1798 (1970).
[15] R. SH. MALKOVICH and I. V. NISTIRYUK, Zh. eksper. teor. Fiz. Pisma 17, 3 (1973).
[16] A. A. KASTALSKI, I. S. SHLIMAK, and A. R. GADGIEV, Fiz. Tekh. Poluprov. 7, 1020 (1973).
[17] B. I. BOLTAKS, L. F. KONOROVA, and R. SH. MALKOVICH, Fiz. Tekh. Poluprov. 9, 1203 (1975).
[18] YU. A. ZAGORUYKO, B. L. TIMAN, and M. SH. FAYNER, Fiz. Tekh. Poluprov. 12, 669 (1978).
[19] YU. A. ZAGORUYKO and B. L. TUMAN, Fiz. Tekh. Poluprov. 14, 2283 (1980).
[20] B. I. BOLTAKS, Diffusion and Point Defects in Semiconductors, Izd. Nauka, Leningrad 1972 (in Russian).
[21] B. I. BOLTAKS, T. D. DZHAFAROV, V. KH. KUDOYAROVA, A. SH. MEKHTIEV, and T. V. TSIGANOVA, Fiz. Tekh. Poluprov. 16, 383 (1981).
[22] B. I. BOLTAKS, T. D. DZHAFAROV, V. KH. KUDOYAROVA, N. F. KARTENKO, R. M. IMAMOV, and A. A. OBRAZTSOV, Fiz. Tekh. Poluprov. 13, 41 (1979).
[23] H. KOKADO, I. SHIMIZU, and E. INONE, J. non-crystall. Solids 20, 131 (1976).
[24] C. W. WHITE, J. NARAYAN, and R. T. YOUNG, Science 204, 461 (1976).
[25] M. R. SIREGAR, W. LÜTHY, and K. AFFOLTER, Z. angew. Math. Phys. 30, 839 (1979).
[26] G. A. KACHURIN and E. V. NIDAEV, Fiz. Tekh. Poluprov. 11, 2012 (1977); 14, 424 (1980).
[27] V. S. VAVILOV, A. E. KIV, and O. R. NIYAZOVA, Mechanisms of Migration of Defects in Semiconductors, Izd. Nauka, Moscow 1981 (p. 236) (in Russian).
[28] N. A. VITOVSKII, G. A. VAKHLII, V. V. GALAVANOV, T. V. MASHOVETS, and K. KHANSEVAROV, Fiz. Tek. Poluprov. 3, 132 (1969).
[29] A. R. BASMAN, A. B. GERASIMOV, N. G. KAKHIDZE, B. M. KONOVALENKO, and A. A. TSERTSVADZE, Fiz. Tekh. Poluprov. 7, 1347 (1973).
[30] A. B. GERASIMOV, N. D. DOLIDZE, N. G. KAKHIDZE, B. M. KONOVALENKO, and N. B. CHELIDZE, Fiz. Tekh. Poluprov. 1, 982 (1967).
[31] H. SAITO, N. FUKUOKA, and Y. TATSUMI, Japan. J. appl. Phys. 10, 389 (1971).
[32] T. M. FLANAGAN and E. E. KLONTZ, Phys. Rev. 167, 789 (1968).
[33] L. C. KIMERLING, H. M. DE ANGELIS, and C. P. CARNES, Phys. Rev. B 3, 427 (1971).
[34] L. C. KIMERLING and C. P. CARNES, J. appl. Phys. 42, 3548 (1971).
[35] T. D. DZHAFAROV, Proc. Internat. Conf. Radiation Effects in Semiconductors, Dubrovnic 1976, Ed. N. B. URLI and J. W. CORBETT, Bristol/London, The Institute of Physics, 1977.
[36] B. L. GREGORY and H. H. SANDER, Proc. IEEE 58, 1328 (1970).
[37] G. A. KACHURIN, N. B. PRIDACHIN, and L. S. SMIRNOV, Fiz. Tekh. Poluprov. 9, 1428 (1975).
[38] E. I. SHTYRKOV, I. B. KHAIBULIN, M. M. ZARIPOV, M. F. GALYAUTDINOV, and P. M. BAYAZITOV, Fiz. Tekh. Poluprov. 9, 2000 (1975).
[39] I. B. KHAIBULIN, E. I. SHTYRKOV, M. M. ZARIPOV, R. M. BAYAZITOV, and M. F. GALYAUTDINOV, Radiat. Eff. 36, 225 (1978).
[40] G. A. KACHURIN, V. A. BOGATYRIOV, S. I. ROMANOV, and L. S. SMIRNOV, in: Ion Implantation in Semiconductors 1976, Ed. F. CHERNOW, J. A. BORDERS, and D. K. BRICE, Plenum Press, New York, 1977 (p. 455).
[41] H.-D. GEILER, G. GÖTZ, K.-D. KLINGE, and N. TRIEM, phys. stat. sol. (a) 41, K171 (1977).
[42] B. I. BOLTAKS, Diffusion in Semiconductors, Fiz. mat. giz. Moscow 1961 (p. 148) (in Russian).
[43] J. A. VAN VECHTEN, R. TSU, F. W. SARIS, and D. HOONHOUT, Phys. Letters A 74, 417 (1979).
[44] W. LÜTHY, K. AFFOLTER, H. P. WEBER, M. E. ROULET, M. FALLAVIER, J. P. THOMAS, and J. MACKOWSKI, Appl. Phys. Letters 35, 873 (1979).
[45] G. V. PLYATSKO, S. G. KIYAK, A. F. SEMIZOROV, and M. I. MOISA, Fiz. Tekh. Poluprov. 14, 404 (1980).

[46] B. T. KOLOMIETS and N. A. GORYUNOVA, Zh. Tekh. Fiz. **25**, 984, 2069 (1955); Izv. Akad. Nauk SSSR, Ser. fiz. **20**, 1496 (1956).
[47] B. I. BOLTAKS, Z. U. BORISOVA, V. KH. BIKTIMIROVA, T. D. DZHAFAROV, A. A. OBRAZTSOV, and E. A. SHENYAVSKAYA, phys. stat. sol. (a) **30**, 731 (1975).
[48] B. I. BOLTAKS, V. KH. BIKTIMIROVA, T. D. DZHAFAROV, E. A. SHENYAVSKAYA, Z. U. BORISOVA, and A. A. OBRAZTSOV, Proc. VI Internat. Conf. Amorphous and Liquid Semiconductors, Leningrad, Izd. Nauka, Leningrad 1976 (p. 269).
[49] V. KH. BIKTIMIROVA, B. I. BOLTAKS, Z. U. BORISOVA, T. D. DZHAFAROV, and A. A. OBRAZTSOV, Fiz. Tekh. Poluprov. 8, 2174 (1974).
[50] P. SÜPTITZ and I. WILLERT, phys. stat. sol. (a) **28**, 223 (1975).
[51] E. NEBAUER, P. SÜPTITZ, and I. WILLERT, phys. stat. sol. (a) **43**, 451 (1977).
[52] V. KH. KUDOYAROVA, Avtoref. Kand. Diss. Baku 1981.
[53] A. BUROFF, E. NEBAUER, P. SÜPTITZ, and I. WILLERT, phys. stat. sol. (a) **40**, K195 (1977).
[54] J. P. DE NEUFVILLE, Amorphous and Liquid Semiconductors, Ed. J. STUKE and W. BRENIG, Taylor and Francis, London 1974 (p. 1351).
[55] V. S. MINAEV, M. A. PLISHEVSKI, and L. I. TOPALOVA, Elektron. Tekh. (Mater.) **6**, No. 6, 3 (1977).
[56] M. T. KOSTISHIN, E. V. MIKHAILOVSKAYA, and P. F. ROMANENKO, Fiz. tverd. Tela **8**, 571 (1966).
[57] M. T. KOSTISHIN, P. F. ROMANENKO, S. A. DEMBOVSKI, and G. Z. VINOGRADOVA, Izv. Akad. Nauk SSSR, ser. neorg. Mater. 7, 210 (1971).
[58] A. MATSUDA and M. KIKUCHI, J. Japan. Soc. appl. Phys. Suppl. **42**, 239 (1973).
[59] J. P. DE NEUFVILLE, S. C. MOSS, and OVSHINSKY, J. non-crystall. Solids **13**, 191 (1973/74).
[60] H. KOKADO, I. SHIMIZU, T. TATSUNO, and E. INOUE, J. non-crystall. Solids **21**, 225 (1976).
[61] D. GOLDSCHMIDT and P. S. RUDMAN, J. non-crystall. Solids **22**, 229 (1976).
[62] Y. YAMAMOTO, T. ITOH, Y. HUROSE, and H. HIROSE, J. appl. Phys. **47**, 3603 (1976).
[63] B. I. BOLTAKS, T. D. DZHAFAROV, V. N. KUDOYAROVA, Z. U. BORISOVA, and A. A. OBRAZTSOV, Proc. Conf. Amorphous Semiconductors 76, Akademiai Kiado, Budapest 1976 (p. 123).
[64] B. I. BOLTAKS, T. D. DZHAFAROV, V. KH. KUDOYAROVA, F. S. NASREDINOV, and P. P. SEREGIN, Proc. Conf. Amorphous Semiconductors 78, Inst. Solid State Phys. Pardubice 1978 (p. 166).
[65] B. I. BOLTAKS, T. D. DZHAFAROV, V. KH. KUDOYAROVA, N. F. KARTENKO, R. M. IMAMOV, and A. A. OBRAZTSOV, Fiz. Tekh. Poluprov. **13**, 41 (1979).
[66] V. V. PETROV and A. A. KRUCHIN, Dokl. Akad. Nauk SSSR **230**, 326 (1976).
[67] P. SÜPTITZ, E. NEBAUER, and A. BUROFF, Proc. Conf. Amorphous Semiconductors 78, Inst. Solid State Phys. Pardubice 1978 (p. 138).
[68] A. M. ANDRIESH, V. M. GANIN, B. T. KOLOMIETS, V. M. LYUBIN, and D. I. TSIULYANU, Zh. Tekh. Fiz. Pisma **1**, 578 (1975).
[69] V. M. GANIN, Avtoref. Kand. Diss., Leningrad 1979.

(Received January 18, 1982)

Original Papers

phys. stat. sol. (a) 79, 47 (1983)

Subject classification: 10.2 and 14.3; 22.1.1

Institute of Solid State Physics, Academy of Sciences of the USSR, Chernogolovka[1])

The Effect of Annealing on the Defect Structure and Electrical Properties of Deformed Single Crystals of Ge

By

V. V. Aristov, Yu. A. Osipyan, R. Scholz, I. I. Snighireva, I. I. Khodos, and S. A. Shevchenko

The dissociation of dislocations in deformed and annealed specimens of p-type germanium is studied using the weak-beam method of electron microscopy. Both, the Hall e.m.f. and dc conductivity are measured within the temperature range 4.2 to 300 K. It is shown that the splitting values of dislocations with a predominant screw component depend on the cooling rate and subsequent annealing of the deformed specimens. The effect of annealing on the concentration and the mobility of current carriers is also found. It is concluded that the splitting parameters of dislocations in the specimens under study are connected with the behavior and state of point defects.

Die Dissoziation von Versetzungen in verformten und getemperten Proben von p-leitendem Germanium wird mit der Schwachstrahltechnik der Elektronenmikroskopie untersucht. Sowohl die Hallspannung als auch die Gleichstromleitfähigkeit werden im Temperaturbereich von 4,2 bis 300 K gemessen. Es wird gezeigt, daß die Aufspaltungswerte von Versetzungen mit überwiegender Schraubenkomponente von der Abkühlgeschwindigkeit und der darauffolgenden Ausheilung der deformierten Proben abhängen. Ein Einfluß der Temperung auf die Konzentration und Beweglichkeit der Stromträger wird ebenfalls gefunden. Es wird angenommen, daß die Aufspaltungsparameter der Versetzungen der untersuchten Proben mit dem Verhalten und dem Zustand der Punktdefekte verknüpft sind.

1. Introduction

The defect structure of germanium crystals continues to be the subject of much investigation despite a great number of publications on this point. In order to clearly understand the properties of the main elements of a defect structure — dislocations and point defects — and for a better insight into the mechanism of their interaction we need detailed information on the dislocation structure as well as on the distribution and charge state of point defects and their complexes. Recently considerable progress in the study of the dislocation structure in germanium and silicon has been achieved due to the weak-beam electron microscopy [1 to 12]. As a result, the resolution of structure defects was enhanced, and important information on the dissociation and the local features of the structure of dislocations was obtained. For example, the dislocations at and near screw orientations have been found to be dissociated anomalously widely [6, 11]. It is of indubitable interest to study the nature of the diversified dissociation of dislocations.

On the other hand, to study the point defects in semiconductor crystals, electric measurements have since long been used rather effectively [13]. Therefore, the present study was intended to obtain and compare the data on the dislocation dissociation and on the state and concentration of point defects in deformed and then annealed germanium single crystals using weak-beam electron microscopy and measurement of the Hall e.m.f. and dc conductivity.

---

[1]) 142432 Chernogolovka, Moscow district, USSR.

2. Experimental Technique

The electron microscopy studies and most of the electric measurements were carried out on single crystals of p-type germanium (the main acceptor impurity is gallium) with specific resistivity at room temperature $\varrho_0 = 10\ \Omega$cm. To define the nature of the acceptors produced by deformation we also used single crystals of n-type germanium (the main donor impurity is antimony) with $\varrho_0 = 45\ \Omega$cm. The difference concentration of electrically active impurities in p-type germanium is $N_a = 4 \times 10^{14}\ \text{cm}^{-3}$ (the total concentration is below $10^{15}\ \text{cm}^{-3}$), while in n-type germanium $N_d = 5 \times \times 10^{12}\ \text{cm}^{-3}$ (the total concentration is below $10^{13}\ \text{cm}^{-3}$).

The specimens were deformed in compression under a constant load at $T_g = 420$ °C in a vacuum of 5×10^{-5} Torr for some hours. One of the faces was parallel to the $(1\bar{1}0)$ plane, the other two deviated by 10° from the (110) and (001) planes. The axis of compression was normal to a face deviating by 10° from the (110) plane.

After deformation, the specimens were cooled under various conditions (Table 1).

Table 1
Cooling and annealing conditions of germanium specimens

series	cooling rate	annealing (h)	dissociation
A	rapid	—	normal
B	rapid	0.5	normal
C	slow	—	anomalous
D	slow	0.5	normal

Some specimens were furnace cooled with a rate of about 10 K/min at 350 °C (series C and D), others were cooled more rapidly in an argon flow with a rate of about 50 K/min at the same temperature (series A and B). Then part of the specimens were coated with gold and annealed at 520 °C (see Table 1).

Both the Hall e.m.f. and the dc conductivity were measured within the temperature range 4.2 to 300 K as described in [14].

For the electron microscope studies the specimens were sectioned into slices parallel to the primary glide plane (111). After polishing they were thinned electrolytically in a KOH solution. The foils were studied in a JEM-100 CX electron microscope fitted with a side-entry ±45° double tilt goniometer and in a JEM-100 C one with a top-entry ±10° double tilt goniometer.

The dislocation dissociations were determined from weak-beam images taken in 220 reflections. The separation of single dislocations, Δ, was measured in the corresponding weak-beam reflection $\boldsymbol{g} = \langle 220 \rangle$ for which the condition $\boldsymbol{gb} = 2$ was fulfilled, the value of the weak-beam deviation parameter being $s_g = -2 \times 10^{-2}\ \text{Å}^{-1}$ [15] ($\boldsymbol{b}$ is the Burgers vector of the full dislocation, $\boldsymbol{g}$ the diffraction vector). For a more precise evaluation of the dissociation width we used the known correction formulae [16], putting the value of the Poisson ratio equal to $\nu = 0.2$ [17]. To distinguish between the images of dipoles and those of dissociated dislocations and to exclude thereby the errors in determining the separations of the latter we used the difference in the behaviour of the contrast on these defects. The image of the faulted dipole vanishes in one of the 220 reflections whose diffraction vector is parallel to the dipole, since in this case the condition $\boldsymbol{gb}_i^p = 0$ and $\boldsymbol{gb} \times \boldsymbol{u} = 0$ are met for each of the four partial dislocations of the dipole [18] ($\boldsymbol{b}_i^p$ is the Burgers vector of the i-th partial dislocation, $\boldsymbol{u}$ the vector of the dislocation line direction). The image of the ordinary dipole made up of two dislocations of opposite sign, each of which is dissociated into two Shockley partials, does not vanish in any of the 220 reflections.

3. Experimental Results

Typical elements of dislocation structure in germanium are shown in Fig. 1 to 4. The dislocation structure in all the specimens consists of single dissociated dislocations, dipoles of dissociated dislocations at orientations from 60° to 90°, and faulted dipoles. Besides, the unannealed specimens contain a great quantity of small defects which appear on the weak-beam images as light dots of some tens of angstroms in size. Basically these defects are seen as rows, often rectilinear, and seem to be the fragments of dipoles.

Fig. 4 gives a weak-beam image of the dislocation dipole. The vertical part F of the defect represents a faulted dipole, and the upper part O is an ordinary dipole made up of two dislocations dissociated into Shockley partials. Perhaps, dislocation configurations like this [19, 20] illustrate either an unfinished process of the formation of a faulted dipole from an ordinary one or a partial return to the unfaulted variety. All four partial dislocations of the faulted dipole are out of contrast at $\boldsymbol{g} = \pm 2\bar{2}0$

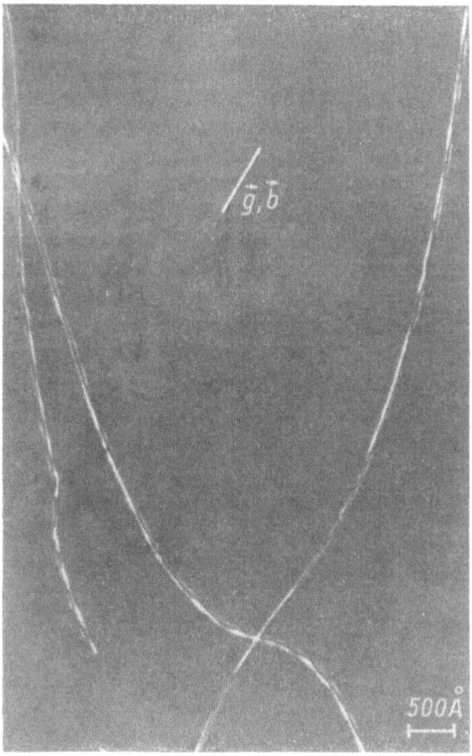

Fig. 1

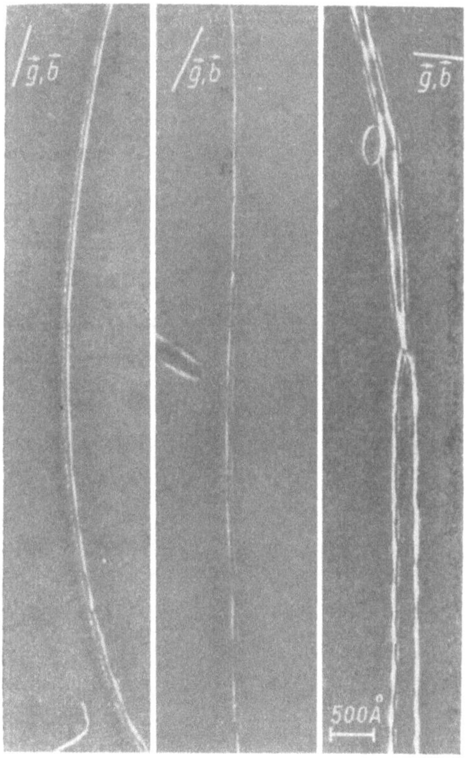

a b c

Fig. 2

Fig. 1. Weak-beam $2\bar{2}0$ images showing segments of dislocations of various orientations relative to their total Burgers vector $\boldsymbol{b} = a/2\,[1\bar{1}0]$ (irrespective of sign) in a germanium specimen which has been deformed at 420 °C and rapidly cooled (series A). $|\boldsymbol{gb}| = 2$; $s_g \approx -2 \times 10^{-2}$ Å^{-1}

Fig. 2. Weak-beam images of dislocation segments in germanium annealed for 0.5 h after rapid cooling (series B). $|\boldsymbol{gb}| = 2$; $s_g \approx -2 \times 10^{-2}$ Å^{-1}

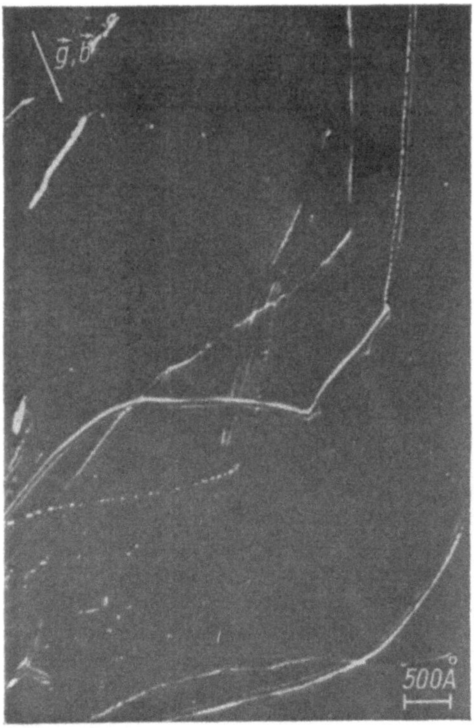

Fig. 3. Weak-beam image of dislocation segments in germanium which has been deformed at 420 °C and slowly cooled (series C). $|\boldsymbol{g}\boldsymbol{b}| = 2$; $s_g \approx -2 \times 10^{-2}$ Å^{-1}

(Fig. 4d). In other micrographs (Fig. 4a to c) the image contrast depends upon the sign of $\boldsymbol{g}$. A strong contrast in these images is associated with two Shockley partials [18]. In some images two sharp lines of low intensity are visible between the Shockley partials for a particular sign of $\boldsymbol{g}$. These lines may be due to the two stair-rod partial dislocations with $\boldsymbol{b}_2 = -\boldsymbol{b}_3 = a/6\,[110]$ and to the stacking fault between them in the plane $(11\bar{1})$. Thus, the different behaviour of the contrast on dissociated dislocations and on dipoles permits to surely distinguish between them.

The measured partial dislocation separations, Δ, for the isolated dislocations of different orientation (the latter is determined by the value of the angle α between the dislocation line and its total Burgers vector $\boldsymbol{b}$) is shown in Fig. 5 having regard to the correction for the displacement of the weak-beam images of the partial dislocations [16]. It follows from this figure that the results obtained are not described by the theoretical dependence $\Delta(\alpha)$ (curves 1 and 2). Moreover, dislocations of different

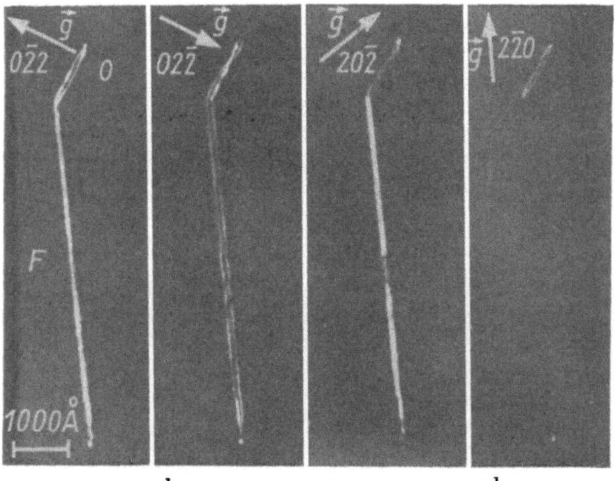

a b c d

Fig. 4. Weak-beam images of a faulted dipole showing the behaviour of the diffraction contrast typical for such a structure defect. The portion O is a segment of an ordinary dipole

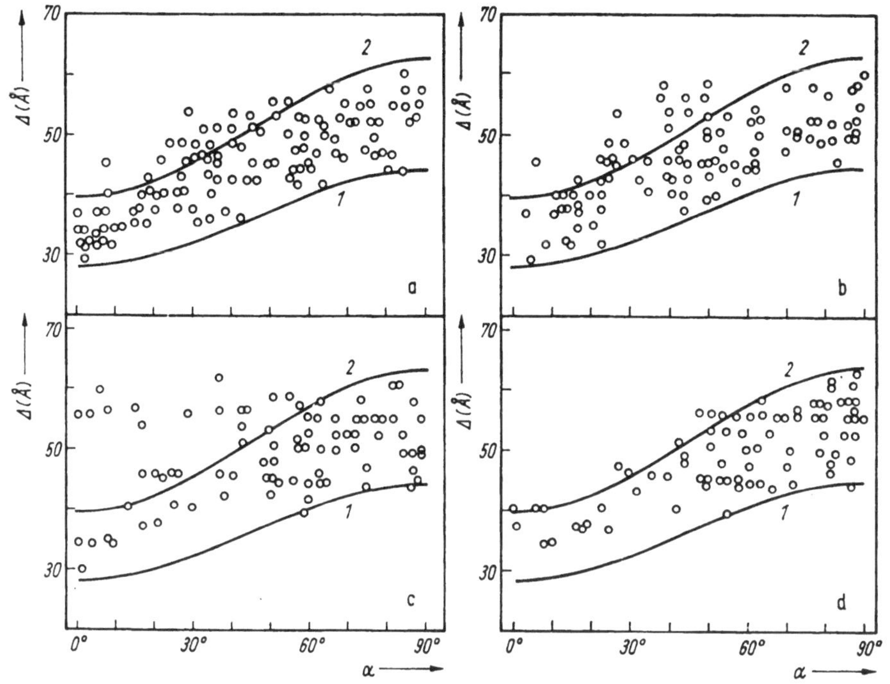

Fig. 5. The measured dissociations of dislocations, Δ, in germanium, and the theoretical curves as a function of line orientation for $\gamma = 75$ mJm^{-2} (curves 1) and 53 mJm^{-2} (curves 2). Rapid cooling: a) without and b) with annealing; show cooling: c) without and d) with annealing

orientations are dissociated differently as the cooling rate is changed and the subsequent annealing is performed.

The dislocations with $55° \lessapprox \alpha \leqq 90°$ are dissociated equally in all specimens. Their dissociation values lie within the range limited by curves 1 and 2 calculated using the isotropic elasticity theory [17] for two values of stacking fault energy: $\gamma = 75$ and 51 mJm^{-2}, respectively.

However, the dissociation of dislocations with a predominant screw component ($\alpha \leqq 25°$) was found to be dependent on the cooling rate after deformation and on the subsequent annealing. In rapidly cooled specimens (series A) these dislocations are dissociated in confirmity with the range limited by curves 1 and 2 (the "normal" dissociation), while in slowly cooled specimens (series C) a considerable portion of such dislocations is dissociated "anomalously" widely (Fig. 5a, c). Subsequent annealing of slowly cooled specimens materially affects the dissociation of dislocations with $\alpha \leqq 25°$; in the specimens of series D these dislocations proved to be "normally" dissociated (Fig. 5d). The dissociation in the specimens annealed after a rapid cooling (series B) practically remains "normal" (Fig. 5b). Thus, in contrast to the specimens of series A, B, and C, a portion of dislocations with $\alpha \leqq 25°$ in specimens D is dissociated "anomalously" widely.

As is seen from Fig. 5, the separations of the dislocations of mid-orientations ($25° \leqq \alpha \leqq 50°$) exceed the "normal" values in the specimens of all series, and most considerably in the specimens of series A, B, and C.

For the isotropic case, the dissociation values within the above range (Fig. 5, curves 1 and 2) correspond to a stacking fault energy of $\gamma = (64 \pm 11)$ mJm^{-2}. The same value of γ was found for the anisotropic case. In computing it we used the dependence $\gamma(\Delta)$ obtained in anisotropic elastic theory for 60° dislocations [17], and the experimental value of $\Delta = (49 \pm 9)$ Å.

It is apparent from Fig. 6 that plastic deformation of 4.5% (curves A, C) in germanium gives rise to acceptor centres whose concentration drops off considerably on annealing for 30 min at $T = 520$ °C (curves B and D). It is impossible to determine the position of levels of these centres in the forbidden zone of germanium either after deformation or annealing due to the absence of clearly defined linear parts in the dependence $\lg p = f(1/T)$ and to the large concentration of small acceptors in the starting crystal. These levels can be determined in slightly deformed specimens (with electronic conductivity before deformation) at a rather ordered dislocation structure, as for example in specimen C″, where after deformation we can observe two levels with activation energies $E_1 = (0.010 \pm 0.002)$ eV (acceptors A_1) and $E_2 = (0.022 \pm 0.002)$ eV (acceptors A_2).

Plastic deformation also affects the mobility of free holes — the higher the degree of deformation the lower is the dependence $\lg \mu = f(\lg T)$ (Fig. 7, curves A, C, C′). At $T > 70$ K the hole mobility in the annealled specimens exceeds that of unannealed ones, while at $T < 70$ K it becomes lower and its temperature dependence decreases more abruptly on lowering the temperautre below 90 K.

The main qualitative changes in hole concentration and mobility occur under the above conditions of annealing. Further increase in the temperature or in the annealing time no longer causes considerable changes of these parameters. The conditions of cooling (rapid or slow) practically have weak influence on the magnitude and behaviour of the temperature dependence of the hole concentration and mobility either after deformation or annealing (Fig. 6, 7, curves A to D).

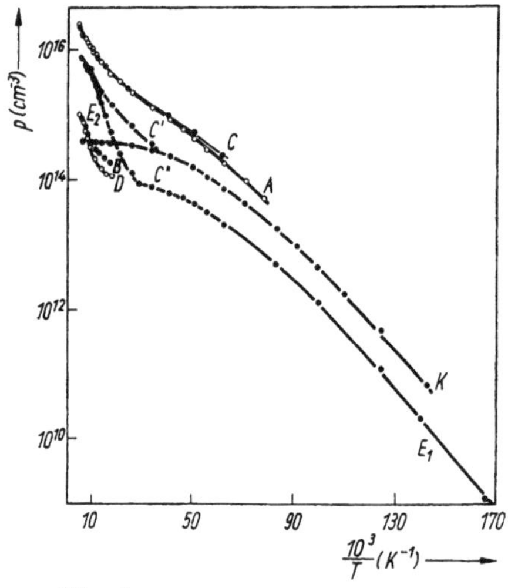

Fig. 6

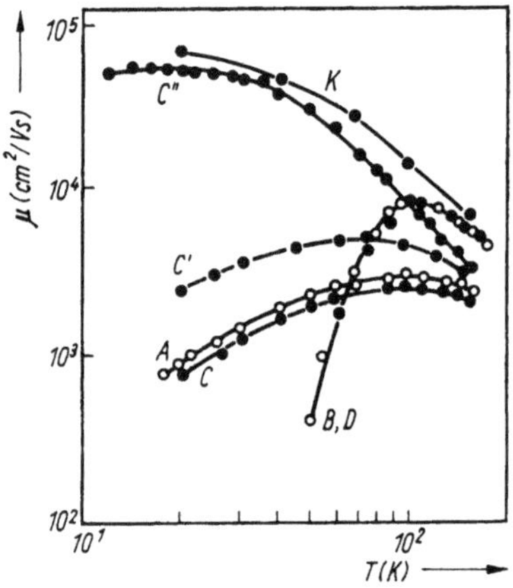

Fig. 7

Fig. 6. The concentration of free holes vs. temperature in the reference specimen (R) ($N_a = 4 \times 10^{14}$ cm^{-3}) and in the deformed specimens (A, B, D, C, C′, C″). The degree of deformation is 3.6 and 4.5% for specimens C′ and A, B, C, D, respectively. The anisotropy of the dislocation distribution along [110] and [1$\bar{1}$0] measured by the etch pits is 1:20 in specimen C″. The notations of curves correspond to the specimen series in Table 1

Fig. 7. The hole mobility as a function of temperature in the specimens studied. The notations are the same as in Fig. 6

4. Discussion

1. The electron microscope studies revealed the dissociations of dislocations in germanium to be dependent on cooling conditions after deformation, namely, after rapid cooling, the dislocations at screw orientation were dissociated "normally", but after slow cooling, a considerable part of such dislocations was dissociated "anomalously widely. At the same time, dislocations with $55° \leqq \alpha \leqq 90°$ had equal ("normal") separations, irrespective of the cooling rate and following annealing.

As to the dislocations at and near screw orientation, their "anomalous" dissociation may be assumed to be due to a thermal effect directly on the dislocation structure of the crystal during its slow cooling, with the result that the dislocation, overcoming a potential barrier, can lower its energy by changing from "normal" dissociation into "anomalous" one. But at $T > T_D$ the influence of the following annealing contradicts this assumption: the "anomalous" dissociation for part of dislocations in slowly cooled specimens C turned to the "normal" dissociation in specimens D. Note, that in silicon the "anomalous" dissociation observed in furnace-cooled specimens [6] was changed to the "normal" one in specimens cooled very slowly [21].

Thus, the transition from the initial normal/or anomalous dissociation to the anomalous/or normal one does not go on in one direction, as the duration of the thermal action on the crystal increases. From this it comes out that other structural defects of the crystal studied need be analyzed in order to take account of their possible influence on the dislocation structure. Among these are point defects (PD), i.e. impurity atoms and their complexes with vacancies and/or interstitials produced by plastic deformation.

2. Some information on the origin of the deformation-induced acceptors can be obtained in specimens with a low dislocation density and ordered dislocation structure (specimen C″, Fig. 6). It follows from this figure that the deformation-induced acceptors are not associated with the dangling bonds of dislocations because the total concentration of centres A_2 exceeds the possible concentration of the dangling bonds of introduced dislocations by a factor nearly 10^2 ($N_D = 1 \times 10^6$ cm^{-2}). Taking into account the high values of the low-temperature hole mobility ($T < 100$ K) in the specimen at a rather high concentration of introduced acceptors ($\approx 10^{16}$ cm^{-3}) we believe that acceptors A_2 are deformation-produced PD. By using the known formulas for the computation of the hole mobility due to scattering on lattice vibrations, ionized and neutral impurities [14], and then comparing between theoretical and experimental values (Fig. 7, curve C″) we conclude that acceptors A_2 are neutral at low temperatures. This is indicative of a low degree of the compensation of these acceptors with donors. In this case the activation energy E_2 corresponds to the level $E_2 = E_V + (0.040 \pm \pm 0.004)$ eV, which coincides with that of the copper atoms localized on lattice sites [22].

The concentration of acceptors A_1 is much below that of acceptors A_2, therefore the degree of their compensation with donors is considerably higher. As a consequence, we connect the level $E_1 = E_V + 0.01$ eV with these acceptors A_1, but their nature is not yet clear.

3. When the degree of deformation grows, the defect structure of crystals becomes more complicated. As a result, we fail to determine the activation energy of the acceptors from the dependences $\lg p = f(1/T)$ for specimens A, C, and C′ (Fig. 6). Taking into account the proximity of values E_1 and E_2 for specimen C″ to the values E_1 and E_2 obtained in [20] for the deformation-produced PD as well as the fact that the annealing of the deformation-induced acceptors in specimens A and C is performed within the same temperature range as in plastically deformed n-type germanium [23],

the same PD produced by deformation are also expected to exist in specimens A, C, and C′.

4. The point defects also influence the mobility of free holes, μ, in deformed specimens. This is shown by the decrease in μ with increasing degree of deformation in unannealed specimens, and by its increase at $T > 70$ K after annealing of PD (curves B and D in Fig. 7). Therefore, it was hoped that some information on the distribution of PD in deformed specimens could be gained from the data on the influence of cooling conditions upon hole concentration and mobility. Since for the specimens cooled at different rates these parameters differ insignificantly (Fig. 6, 7), one can assume that the distribution of Pd outside the regions of the space charge around the dislocations remains invariable under the above conditions. Estimations show that the value of radius R of these regions which screen the dislocation exceeds the value $R = 10^{-5}$ cm for specimens A to D at $T \leqq 300$ K. Consequently, the changes in the distribution of PD at distances less than 10^{-5} cm from the dislocation cannot be detected by electric measurements.

5. Thus, the electric measurements evidence that dislocations in deformed specimens are surrounded by PD and can interact with the latter either during deformation or annealing. According to [21] the point defects can affect the separation of dislocations. Since in the specimens studied the width of the stacking fault (SF) for dislocations at different orientations depends differently upon cooling and annealing conditions (Fig. 5), it is not inconceivable that each type of dislocations has its own mechanism of anomalous dissociation.

According to [21] the deformation temperature of specimens A to D corresponds to a range where the Coulomb interaction between partial dislocations can show itself, because at $T \leqq 420$ °C the screening distance of partial dislocations exceeds the SF width. With a specific amount of charged PD around the partial dislocations, this interaction is increased. After the deformation is completed the distribution pattern of PD round the dislocations may depend, for example, on the cooling rate. Therefore, the effect of cooling conditions on the SF width observed in specimens A and C for dislocations with $\alpha \leqq 25°$, as well as the absence of the anomalous dissociation of these dislocations after the annealing of PD both testify to a mechanism associated with the influence of the atmosphere of PD around the partial dislocations on the Coulomb interaction between them [21]. That the cooling conditions have no influence on the concentration and mobility of holes is not at variance with this mechanism, because the changes in the distribution of PD are essential for it at distances comparable with the SF width, i.e. at $R \leqq 10^{-6}$ cm. Such changes could not be detected by electric measurements.

The anomalous dissociation (for the part of dislocations with $\alpha \approx (35 \pm 15)°$) is noteworthy, for it exists under different cooling conditions and is observed even after the annealing of PD. Therefore, it may be connected with the effect of the impurities precipitated in the SF (the Suzuki effect) [24]. The dissociation feature of these dislocations is possibly due to the fact that only dislocations with $\alpha = 30°$ have a screw partial dislocation.

References

[1] D. J. H. COCKAYNE, I. L. F. RAY, and M. J. WHELAN, Phil. Mag. **20**, 1265 (1969).
[2] I. L. F. RAY and D. J. H. COCKAYNE, Proc. Roy. Soc. **A325**, 543 (1971).
[3] R. MEINGAST and H. ALEXANDER, phys. stat. sol. (a) **17**, 229 (1973).
[4] F. HÄSSERMANN and H. SCHAUMBURG, Phil. Mag. **27**, 745 (1973).
[5] I. L. F. RAY and D. J. H. COCKAYNE, J. Microscopy **98**, 170 (1973).
[6] A. GOMEZ, D. J. H. COCKAYNE, P. B. HIRSCH, and V. VITEK, Phil. Mag. **31**, 105 (1975).

[7] G. Packeiser and P. Haasen, Phil. Mag. **35**, 821 (1977).
[8] K. Wessel and H. Alexander, Phil. Mag. **35**, 1523 (1977).
[9] D. J. H. Cockayne and A. Hons, J. Physique **40**, C6-11 (1979).
[10] H. Alexander, H. Eppenstein, H. Gottschalk, and S. Wendler, J. Microscopy **118**, 13 (1980).
[11] S.-W. Chiang, C. B. Carter, and D. L. Kohlstedt, Phil. Mag. **A42**, 103 (1980).
[12] H. Föll and C. B. Carter, Phil. Mag. A **40**, 497 (1979).
[13] F. J. Blatt, Physics of Electronic Conduction in Solids, McGraw-Hill Publ. Co., 1968.
[14] Yu. A. Osipyan and S. A. Shevchenko, Zh. eksper. teor. Fiz. **65**, 698 (1973).
[15] D. J. H. Cockayne, Z. Naturf. **27a**, 452 (1972).
[16] D. J. H. Cockayne, M. L. Jenkins, and I. L. F. Ray, Phil. Mag., **24**, 1383 (1971).
[17] J. P. Hirth and J. Lothe, Theory of Dislocations, Atomizdat, Moscow 1972 (p. 221, 327).
[18] C. B. Carter and S. M. Holmes, Phil. Mag. **32**, 599 (1975).
[19] J. W. Steeds, Phil. Mag. **16**, 785 (1967).
[20] C. B. Carter, Phil. Mag. **36**, 147 (1977).
[21] P. B. Hirsch, J. Physique **40**, C6-27 (1979).
[22] B. I. Boltaks, Diffusion and Point Defects in Semiconductors, Izd. Nauka, Leningrad 1972.
[23] S. A. Shevchenko and A. I. Kolubakin, Fiz. Tekh. Poluprov. **13**, 1046 (1979).
[24] H. Suzuki, J. Phys. Soc. Japan **17**, 322 (1962).

(Received March 28, 1983)

phys. stat. sol. (a) 79, 57 (1983)

Subject classification: 1.3 and 18.3; 21.1; 21.4

Institute of Metal Physics, Academy of Sciences of the USSR, Ural Scientific Centre, Sverdlovsk[1])

Anisotropy and Exchange Effects in Heavy Rare-Earth–Cobalt Compounds of the RCo_5 Type

By

V. V. KELAREV, V. V. CHUEV, A. N. PIROGOV, and S. K. SIDOROV

A series of compounds of Co and heavy rare-earth elements of the RCo_5 type (where R = Tb, Dy, Ho) is investigated by the neutron diffraction method in a wide temperature range (from 4.2 to 1050 K). The atomic and magnetic structures are investigated. According to the investigations these compounds are collinear ferrimagnetics before as well as after the spin-orientation transition. The data obtained are interpreted in terms of the phenomenological theory of magnetocrystalline anisotropy of the RCo_5 compounds by Irkhin and Rosenfeld. In terms of this theory the anisotropy constants of the R-sublattice (at 0 K) and the parameters of intersublattice exchange are obtained.

Eine Reihe von Verbindungen mit Co und schweren Elementen der Seltenen Erden vom Typ RCo_5 (mit R = Tb, Dy, Ho) wird mittels Neutronenbeugung in einem großen Temperaturbereich (von 4,2 bis 1050 K) untersucht. Die atomaren und magnetischen Strukturen werden untersucht. Danach sind diese Verbindungen kollineare Ferrimagnete sowohl vor als auch nach dem Spin-Orientierungsübergang. Die erhaltenen Ergebnisse werden mit der phänomenologischen Theorie der magnetokristallinen Anisotropie der RCo_5-Verbindungen von Irkhin und Rosenfeld interpretiert. Mit dieser Theorie werden die Anisotropiekonstanten des R-Untergitters (bei 0 K) und die Parameter des Interuntergitteraustauschs erhalten.

1. Introduction

Compounds of the RCo_5 type where R = Ho, Tb, Dy are the most interesting representatives in the series of highly anisotropic materials. They can be treated as models for the solution of a number of general problems associated with the nature of exchange interactions and magnetocrystalline anisotropy. The phenomenological theory of the magnetic anisotropy of the RCo_5 compounds [1, 2] created recently provides wide possibilities for investigating the origin of magnetism of the above-mentioned magnetics. This theory is the first to estimate the numerical relations between the main magnetic characteristics of these intermetallic compounds. For example, it gives an expression for the temperature dependence of magnetization of the rare-earth (R) sublattice,

$$\langle m_R \rangle = \frac{\sum\limits_{n=-j}^{n=j} n \exp D}{\sum\limits_{n=-j}^{n=j} \exp D}\,, \tag{1}$$

$$D = [I^0_{R\text{-}Co} n \cos(\theta_R - \theta_{Co}) + \frac{1}{2} K^0_R n^2 (1 - 3\cos^2\theta_R)]\,(k_B T)^{-1}\,,$$

[1]) ul. S. Kovalevskoi 18, 620219 Sverdlovsk GSP-170, USSR.

where n is the projection of the R ion moment on the direction of the magnetization vector of the R sublattice; I^0_{R-Co} the parameter of intersublattice exchange; $m_{Co}(T)$ the relative magnetization of the Co sublattice; K^0_R the microscopic constant of the second-order anisotropy per ion of the rare-earth element; θ_R and θ_{Co} the angles between the c-axis and the sublattice magnetizations.

From the point of view of adequacy of the model representations of the real situation it is of interest to test the possibility of describing the experimental curves $m_R(T)$ for different R using the theoretical relation (1) throughout the whole temperature range of magnetic ordering. Simultaneously from the experimental relations $m_R(T)$, $m_{Co}(T)$, $\theta_R(T)$, and $\theta_{Co}(T)$ the parameters I_{R-Co} and K^0_R are obtained. The relations mentioned can be obtained only using the neutron diffraction measurements. These data either are not available in the present day literature, or their description leaves much to be desired [3 to 6]. For example, there exist quite opposite views concerning the reasons for the unusual behaviour of the mean magnetic moment of $DyCo_5$ when changing the temperature. While explaining the anomalous behaviour of $\overline{m}_{DyCo_5}(T)$ the authors of [2] assume the magnetic moments in the temperature range above the spin-orientation transition to have collinear ferrimagnetic ordering; in [7] it is suggested that the R sublattice is divided into two non-collinear sub-sublattices. This assumption, evidently, concerns other alloys of the RCo_5 type, the easy axis of which changes its direction when increasing the temperature. The choice of the adequate model is of primary concern here because the divergence mentioned reflects two incompatible physical conceptions; in the first model a sharp increase of the mean crystal magnetic moment observed experimentally in the region of spin orientation is due to the relation between the magnetization of the R sublattice and the orientation of vector $\boldsymbol{m}_R$ with respect to the c-axis, i.e. due to the effect of anisotropy; as for the second conception, it ignores anisotropy completely.

Therefore, we need detailed diffraction measurements in a wide temperature range and a comprehensive analysis of the obtained data.

2. Experimental Procedure

The neutron diffraction measurements were carried out on a powder, the size of a grain being less than 10^5 nm. The compositions of the alloys $TbCo_{5.1}$, $DyCo_{5.2}$, $HoCo_{5.5}$ correspond to those of the homogeneity region [8]. The preparation procedure of alloys of the RCo_5 type and their thermal treatment are described in a number of papers, for example, in [2]. After homogeneous annealing for 24 h at 1200 °C and subsequent fast cooling the sample had $CaCu_5$-type structure. In each of the alloys the admixture of the R_2Co_{17} phase did not exceed 3%. The measurements were made in the interval 4.2 to 1000 K. In the region of low temperatures the temperature was controlled by a Cu–(Fe + Cu) thermocouple and in the region of high temperatures by a chromel–alumel one. The wavelength of neutrons is $\lambda = 0.129$ nm. The form factors for Tb, Ho, Co were taken from [9 to 11]. As far as we know, the literature gives no information on the form factor dependences of the Dy^{3+} ion in $DyCo_5$. As it turned out the replacement of the unknown form factor dependence of Dy^{3+} by the known ones of Gd^{3+}, Tb^{3+}, Er^{3+} taken from [9, 11] within the investigated interval, $\sin\theta/\lambda = 0.1$ to 0.3, brings about the same results. The best agreement between calculated and experimental intensities (2 to 3%) is obtained for the Ho^{3+} form factor at $T = 4.2$ K. Therefore, it has been used in the analysis of the magnetic structure of $DyCo_5$. The nuclear scattering amplitudes are $b_{Tb} = 0.76$, $b_{Ho} = 0.85$, $b_{Dy} = 1.69 \times \times 10^{-5}$ nm. Normalization of reflex intensities has been performed on the basis of the powdered sample of Cu. Neutron absorption by $DyCo_{5.2}$ is known to be comparatively large. However, evaluation of the angle dependence of the absorption of thermal neutrons with the wavelength $\lambda = 0.129$ nm by the cylindrical $DyCo_{5.2}$ sample has shown the transmission factor $A(\mu)$ to change slightly over the investigated region, $\sin \theta/\lambda = 0.1$ to 0.3. This fact allows one to neglect the absorption effect at relative measurements.

Magnetic moments m_R, m_C, m_g and the orientation angles have been obtained by the neutron diffraction patterns taken at different temperatures over the interval 4.2 to 1000 K. Each of the alloys has been repeatedly measured at the corresponding temperatures in order to elucidate the character of magnetic ordering before and after the spin-orientation transition. To increase the accuracy of the first T_{SR}^1 and second T_{SR}^2 temperature limits of the orientation transition, the temperature dependences of the peak intensities of the (100), (001), (101), (102) reflections have been taken because their magnetic components depend strongly upon the orientation of the magnetic moments with respect to the crystallographic axes.

3. The Atomic Structure

It is noteworthy that none of the RCo_5 compounds has been investigated by neutron diffraction methods above 450 K. However, to analyse the magnetic state by the neutron diffraction results one needs the data concerning the magnitude of the nuclear contribution to the reflexes and also a qualified structure determination of the samples. In [12 to 14] the neutron diffraction patterns of the $TbCo_{5.1}$, $HoCo_{5.5}$, $DyCo_{5.2}$ compounds in the paramagnetic region ($T > 1000$ K) are obtained. The reflexes agree with the hexagonal crystal system, the parameters of the cells are close to those given in the literature. The best agreement with experiment is obtained on the assumption (as in [15]) that the structure of the compounds under discussion differs from the ideal $CaCu_5$ one. Some of the rare-earth atoms in these compounds are randomly replaced by pairs of Co and this fact results in the deformation of the structure (the distances between the nearest atoms located in the basic plane and the site of the pair become shorter).

Table 1 shows the intensities of the nuclear reflections measured at $T > T_C$ and calculated by the above models.

Table 1

The observed and the calculated intensities of nuclear reflections (10^{-10} nm²/formula unit). The obtained coordinates of the displaced Co atoms are similar to those for YCo_5 [15]

$TbCo_{5.1}$

hkl	J_{obs}	J_{calc} RCo_5 ideal	RCo_5 distorted
100	4 ± 2	5	5
001	9 ± 2	9	7
101	64 ± 4	73	64
110	48 ± 4	49	47
200 + 111	210 ± 5	231	219
002	45 ± 4	47	44
201	9 ± 2	7	9
102	0 ± 2	2	1
R%		9.8	3.8

$R = \sum_i |J_{obs}^i - J_{calc}^i| / \sum_i J_{obs}$

4. Magnetic Structure

For all the compounds mentioned the predessesor of the magnetic reflections is (000). This implies that the atomic and magnetic unit cells are the same over the entire investigated temperature interval (4.2 to 1000 K). The wave vector of the magnetic structure $\boldsymbol{K}$ equals zero.

The problem of the recovery of the magnetic structure of crystals, especially in polycrystalline samples, cannot be correctly solved without the symmetrical analysis. According to [16, 17] the group theoretical analysis of the magnetic structures possible in crystals consists in the calculation of the composition of the magnetic representation and the basic functions of the irreducible representations incorporated in the magnetic one. The magnetic representation for crystals with $CaCu_5$ structure (space group D_{6h}^1) and $\boldsymbol{K} = \boldsymbol{0}$ calculated by formulas from [16, 17] is as follows [18]:

$$1(a): d_M = \tau_3 + \tau_9 ,$$
$$2(c): d_M = \tau_3 + \tau_6 + \tau_9 + \tau_{12} ,$$
$$3(g): d_M = \tau_3 + \tau_5 + \tau_7 + \tau_9 + \tau_9' .$$

Representations are numbered according to Kovalyov [19]. The magnetic modes of atoms in (1a), 2(c), and 3(g), given in Table 2, allow one to enumerate all the magnetic

Table 2
Magnetic modes of D_{6h} ($K = 0$) for atoms in 1(a), 2(c) and 3(g)

	1(a)	2(c)		3(g)		
	1	1	2	1	2	3
τ_3	001	001	001	001	001	001
τ_5	—	—	—	100	010	$\bar{1}\bar{1}0$
τ_6	—	$00\bar{1}$	$00\bar{1}$	—	—	—
τ_7	—	—	—	120	$\bar{2}\bar{1}0$	$1\bar{1}0$
τ_9	$\bar{v}10$	$\bar{v}10$	$\bar{v}10$	$1w0$	200	$\overline{\sqrt{3}}\,\bar{1}\,0$
	$vu0$	$vu0$	$vu0$	$\sqrt{3}\,\bar{v}0$	$\bar{2}\bar{2}0$	$\bar{1}\sqrt{3}\,0$
τ_9'	—	—	—	$\bar{v}00$	020	$ww0$
	—	—	—	$\bar{w}00$	020	$vv0$
τ_{12}	—	$\bar{v}\bar{u}0$	$vu0$	—	—	—
	—	$\bar{v}10$	$v\bar{1}0$	—	—	—

$u = 2 - \sqrt{3}$, $v = 1 - \sqrt{3}$, $w = 1 + \sqrt{3}$.

structures with $\boldsymbol{K} = \boldsymbol{0}$ permitted by the symmetry of the crystal with $CaCu_5$ structure. In [12 to 14] the calculated diffraction patterns for $HoCo_{5.5}$, $DyCo_{5.2}$, and $TbCo_{5.1}$ are compared with the experimental ones. From the analysis it could be inferred that the magnetics under discussion are collinear ferrimagnetics over the entire temperature range 4.2 to 1000 K. This fact completely refutes the hypothesis [7] about the non-collinear orientation of the rare-earth magnetic moments in $DyCo_{5.2}$ (and apparently, in other RCo_5 compounds with spin-orientation transition) above the temperature of the spin-orientation transition. It is to be noted that the non-collinear magnetic structure suggested in [7] must have inevitably led to magnetic reflexes with fractional indexes in the neutron diffraction patterns (the size of the magnetic unit cell along the c-axis is twice the size of the atomic one in the given structure) which has not been observed experimentally.

5. The Behaviour of the Rare-Earth Lattice in Compounds of the RCo_5 Type ($R = Ho, Dy, Tb$)

Fig. 1 shows the $m_R(T)$ curves for the three RCo_5 ($R = Ho$, Dy, Tb) compounds taken from [13, 14]. According to the character of the temperature dependence each curve consists of three well distinguished parts: the weak change of m_R corresponds to the region below the first boundary of the spin-orientation transition; the sharp fall of m_R (the so-called negative "jump" of the rare-earth lattice magnetization) corresponds to the transition interval with increasing temperature. The flattest part of the curve corresponds to a comparatively narrow region above the transition.

It is of importance that the magnetization of the Rsublattice is essential up to T_C. The neutron diffraction patterns of all the three compounds show the appearance and growth of the background due to magnetic incoherent neutron scattering on the rare-earth ions [20] when increasing the temperature.

With increasing temperature the value of the background increment and the R sublattice magnetization (this is shown on the example of $TbCo_5$ in [21]) change proportional to each other within the measurement error, which indicates the constancy of the rare-earth ion magnetic moment (absolute value) over the observed temperature range. This condition has been implicitly borne in mind while writing relation (1). Using the obtained dependences $\theta_R(T)$, $\theta_{Co}(T)$, and $m_{Co}(T)$, one can choose parameters $I^0_{R\text{-}Co}$ and K^0_R (for each of the compounds) which allow good agreement between the experimental $m_R(T)$ curves and those calculated by (1). In this case we obtained K^0_R values higher than in [1]. The reason for this discrepancy is as follows: in [1] K^0_R and $I^0_{R\text{-}Co}$ are estimated using the relation connecting these parameters with the starting and final temperatures of the spin-orientation T^1_{SR} and T^2_{SR}, respectively. It is likely that the values of K^0_R obtained in [1] are more real, since they do not depend upon the value and orientation of the magnetic moment of the rare-earth ion. According to [22] during the plane-axis transition the Tb^{3+} magnetic moment increases slightly. Consideration of this fact in calculating K^0_R values by (1) favours the agreement between the calculated and the experimental data and yields K^0_R values similar to [1].

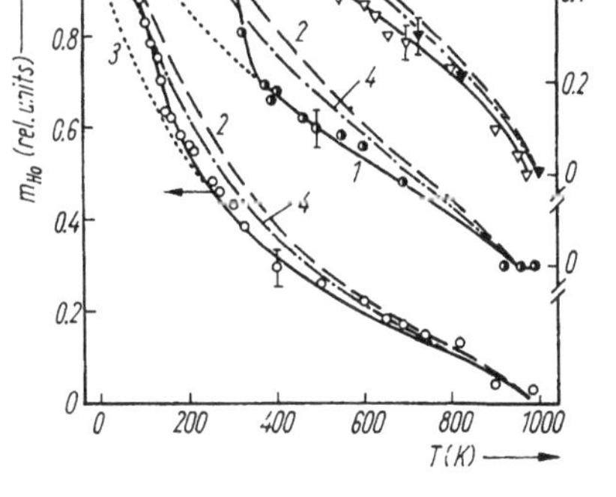

Fig. 1. Temperature dependence of the rare-earth sublattice magnetization in $TbCo_{5.1}$, $DyCo_{2.5}$, $HoCo_{5.5}$. Experimental: ▽ $TbCo_{5.1}$; ◑ $DyCo_{5.2}$; ○ $HoCo_{5.5}$; ▼ $Tb_{0.2}Y_{0.8}Co_5$; □ the treatment of data from [31]. Calculation by (1): (1) experimental relation $\theta_R(T)$ is used; (2) $\theta_R = 90°$ ($m_R \perp c$); (3) $\theta_R = 0°$ ($m_R \parallel c$); (4) $K^0_R = 0$

Table 3

Experimental values of the constants K_R^0, $I_{R\text{-}Co}^0$, E_{an}, and $E_{R\text{-}Co}$ per ion of R (R = Tb, Dy, Ho)

RCo_5	K_R (10^{-23} J)	$I_{R\text{-}Co}$ (10^{-23} J)	E_{an} (10^{-23} J)	$E_{R\text{-}Co}$ (10^{-23} J)
$TbCo_5$	6.9 ± 0.7	200 ± 14	226	1200
$DyCo_5$	8.3 ± 0.7	110 ± 11	435	828
$HoCo_5$	4.1 ± 0.7	76 ± 7	242	662

Parameters $I_{R\text{-}Co}^0$ and K_R^0 selected in such a manner are given in Table 3. The possibility of a satisfactory description of the experiments on compounds with different rare-earth elements can be apparently, considered to be a proof of the correctness of the model representations used in (1). It is of interest that for any of the RCo_5 compounds discussed the obtained complex experimental dependence could not match the calculated one, if $K_R^0 = 0$ in (1).

Fig. 1 shows the experimental $m_R(T)$ curves and those calculated by (1) for each of the three RCo_5 compounds for three cases: (i) $\theta_R = 90°$ ($\boldsymbol{m}_R \perp \boldsymbol{C}$); (ii) $\theta_R = 0°$ ($\boldsymbol{m}_R \parallel \boldsymbol{C}$) (iii) $K_R^0 = 0$. It is easily seen that the experimental points are well described by the calculated curve 2 up to the beginning of the transition. They coincide with curve 3 above the second transition boundary. Thus, the spin-orientation transition region is characterized by the change in the character of the temperature dependence of the rare-earth lattice magnetization shown by the magnetization jump on the real curve $m_R(T)$. It should be noticed that in the case of $Tb_{0.2}Y_{0.8}Co_5$ which, unlike $TbCo_5$, has no transition ($\boldsymbol{m}_{Tb} \parallel \boldsymbol{C}$), the experimental values of m_R (per Tb^{3+}) within the interval 4.2 to 990 K are also described by curve 3 (see Fig. 1). The sum of all the above factors evidences an appreciable contribution of the magnetic crystalline anisotropy to the R-sublattice magnetization. It also shows that the value of the rare-earth sublattice anisotropy energy ($E_{an.}^R$) can be compared with the value of the intersublattice exchange energy ($E_{R\text{-}Co}$),

$$E_{an} = K_R^0 P\left(P - \tfrac{1}{2}\right), \tag{2}$$

$$E_{R\text{-}Co} = I_{R\text{-}Co}^0 P, \tag{3}$$

where P is the total angle moment ($P = 6$, 15/2, and 8 for Tb^{3+}, Dy^{3+}, and Ho^{3+}, respectively). Table 3 shows the results for $HoCo_{5.5}$. $DyCo_{5.2}$, $TbCo_{5.1}$ calculated by (2) and (3).

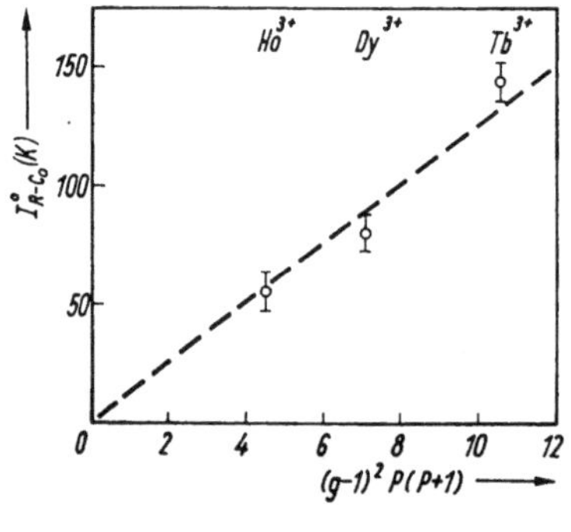

Fig. 2. Parameter of intersublattice exchange I_{R-Co}^0 as a function of $(g - 1)^2 P(P + 1)$ in the compounds RCo_5 (R = Ho, Dy, Tb)

As seen from Fig. 2 $I^0_{R\text{-}Co}$ depends upon $(g-1)^2 P(P+1)$ and their function is close to a linear one. Thus, in the investigated series of "heavy" RCo_5 rare-earth compounds the parameters of intersublattice exchange can be considered to be proportional to the square of the projection of the rare-earth ion spin on the direction of its total magnetic moment [20]. The reason for the divergence is known (taking into account that the values of the magnetic moment of Co in the compounds are practically the same we should expect $I^0_{R\text{-}Co} \sim (g-1)P$). It might be that in the observed function the divergence from linearity is due to the differences in the R–Co distance in different compounds.

6. The Behaviour of the d-Sublattice in the RCo_5 Compound

Fig. 3 shows the values of the relative magnetization of the Co-sublattice m_{Co} in YCo_5 (calculated by the results of [23] assuming that $m_Y = 0$) and the corresponding m_{Co} values for $TbCo_{5.1}$ obtained in [12]. As seen from Fig. 3 $m_{Co}(T)$ for $TbCo_5$ unlike $m_{Co}(T)$ for YCo_5 is a complex function of temperature: within the range 100 to 390 K the velocity of change of the former is two times higher; within the range 420 to 600 K the character of the $m_{Co}(T)$ dependence in $TbCo_{5.1}$ is different: magnetization falls appreciably slower (about 2 times) than between 100 and 390 K and about 2.5 lower than within the same interval in YCo_5. This difference in the behaviour of the Co sublattice in $TbCo_5$ before and after the spin-reorientation transition reveals the relation between the value of its magnetization and the orientation of the vector m_{Co} relative to the crystal axes [24]. The relation between the Co sublattice magnetic moment and the magnetization direction (along or normal to the c-axis) is also found in YCo_2Ni_3, YCo_3Ni_2 [25] at 4.2 and 100 K, respectively, and in YCo_5 [26] at 4.2 and 300 K. The analysis of the concentration dependence of the spontaneous magnetic moment in $Tb_xY_{1-x}Co_5$ at 4.2 K [22] also leads to the conclusion supporting the relation between the modulus of the Co magnetic moment and its orientation. The above experiments show that the magnetic crystalline anisotropy of the d-sublattice plays an important role in the formation of its magnetization in compounds of the RCo_5 type.

In fact, in the papers mentioned all the atoms of the Co sublattice are believed to be equally responsible for the observed effects. However, the conditions including also the local symmetry, of the Co atoms in 2(c) and 3(g) are different. This fact allows one to suggest that the behaviour and properties of these atoms as well as their roles in the formation of the entire crystal properties may be appreciably differerent. In literature there are data supporting this suggestion. The analysis of the data obtained on $Y(Co_{1-x}Fe_x)_5$, $Tb(Co_{1-x}Fe_x)_5$ [27, 28], and YCo_5 [15] lead to the following conclusion: the atoms responsible for the anisotropy of the entire d-sublattice are the Co atoms

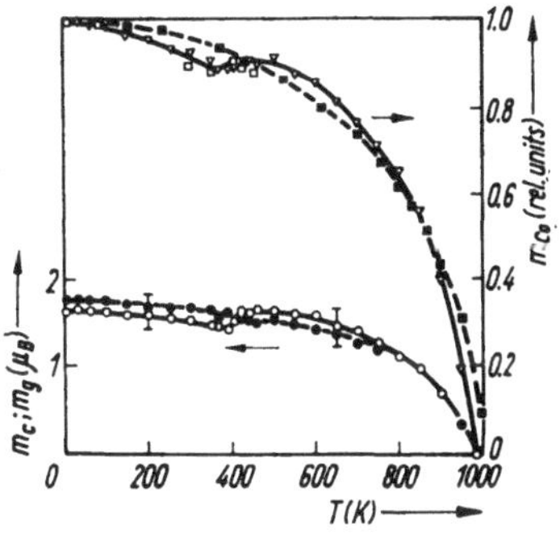

Fig. 3. Temperature dependence of the Co sublattice magnetization in $TbCo_{5.1}$ (▽ mean magnetization, our data; □ data of [5]; ● magnetization of the Co atoms in 3(g); ○ magnetization of the Co atoms in 2(c)–our data) and YCo_5 (■ calculated according to the data of [23] in [12])

located in 2(c) sites. The data obtained in the present paper lead to a similar conclusion. Really, the temperature dependences of 2(c) and 3(g) sublattice magnetizations $m_c(T)$ and $m_g(T)$ for $TbCo_{5.1}$ and $HoCo_{5.5}$ differ appreciably. Fig. 3 shows experimental curves $m_c(T)$ and $m_g(T)$ for $TbCo_{5.1}$. It is seen that in this alloy the magnetization of the 2(c) sublattice changes with temperature more rapidly before the spin-orientation transition than after it. The behaviour of $m_c(T)$ can be understood on the whole only assuming that at any temperature $|m_c|$ is larger if orientations of the vector $\boldsymbol{m}_c$ and the c-axis coincide ($\boldsymbol{m}_c \parallel \boldsymbol{C}$), and smaller, if $\boldsymbol{m}_c$ is in the basic plane, so that the $m_c^{\parallel}(T)$ curve is higher than the $m_c^{\perp}(T)$ one within the whole temperature range. The reorientation region is characterized by the transition from curve $m_c^{\perp}(T)$ to curve $m_c^{\parallel}(T)$ and this leads to an appreciable positive jump of magnetization. The value of the jump Δm_c estimated by the sum of measurements made on the right and on the left of the transition does not exceed $0.15\mu_B$ per atom of Co in the 2(c) sublattice, that is 15% of the magnitude of the 2(c) sublattice magnetization ($\approx 1.8\mu_B$) at 500 to 550 K.

The curve $m_g(T)$ shown in Fig. 3 is practically monotonous. It is of interest that before the transition $m_c < m_g$ ($m_g - m_c \approx 0.2$ to $0.15\mu_B$). Alternatively, above the transition the reverse inequality takes place.

As shown in [29] on the example of Ni, the magnetization anisotropy in d metals is due to the spin–orbit interaction. The dissimilar behaviour of $m_c(T)$ and $m_g(T)$ indicates, seemingly, the fact that magnetocrystalline anisotropy of the Co atoms located in various sites is appreciably different. In 2(c) it has apparently a great value and in 3(g) it is significantly lower. Thus, practically all the magnetization anisotropy of the d sublattice can be deemed to be due to atoms of Co in 2(c), which also dominate in the contribution to the magnetic crystalline anisotropy of this sublattice.

High neutron absorption and an appreciable nuclear contribution to the reflections on the neutron diffraction patterns of $DyCo_5$ make a detailed investigation of the temperature dependence of the Co sublattice magnetization in $DyCo_5$ (similar to those made on $TbCo_{5.1}$ and $HoCo_{5.5}$) difficult. In this case, the rather high scattering of the experimental values of m_{Co} makes it possible to describe the experimental dependences $m_{Co}(T)$ using a wide enough family of curves, which incorporates also the one corresponding to YCo_5.

7. Definition of the Spin-Orientation Transition Intervals

An essential advantage of the neutron diffraction method is the possibility to investigate the magnetic state without applied magnetic field, the effects of which are difficult to consider. θ_R as a function of T plotted according to the data of [12 to 14] for RCo_5 (R = Ho, Dy, Tb) is shown in Fig. 4 (θ_R and θ_{Co} are experimentally indistinguishable at any temperature). As seen from the discussed series of compounds the temperature limits within which the spin reorientation occurs, are shifted to high temperatures with the number of the rare-earth element, while the transition interval is narrowed. This can easily be explained in terms of a two-sublattice model of the magnetic crystalline anisotropy of the compounds RCo_5 [1], which suggests the orientation of the crystal easy magnetization axis to be defined by the macroscopic sublattice constant (K_R or K_{Co}) prevailing at the given temperature. In the ground state for each of the three compounds under investigation K_R calculated by formula [30] $K_R = K_R^0 NP(P - 1/2)$ (see Table 3) is higher than K_{Co} (K_{Co} according to [24] is about 2.1×10^{-21} J/5Co, N is the number of the rare-earth atoms).

As already mentioned in [30] the reason for the spin-orientation transition lies in the faster (compared to K_{Co}) decrease of K_R with temperature and at some temper-

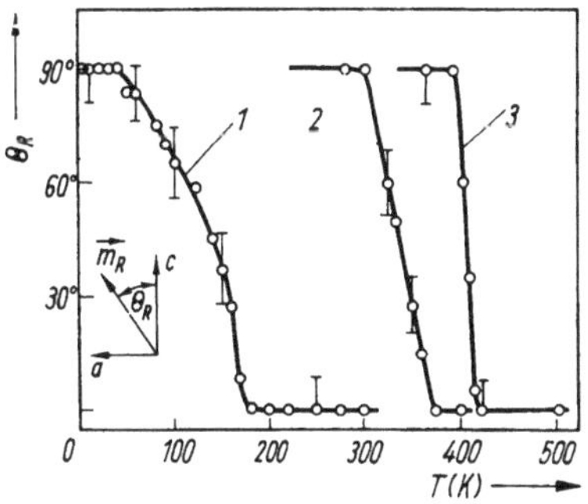

Fig. 4. Temperature dependence of the orientation angle of the magnetization vector of the R sublattice $\theta_R(T)$ in $HoCo_{5.5}$ (1), $DyCo_{5.2}$ (2) and $TbCo_{5.1}$ (3)

ature the state is achieved where K_R becomes equal to and then smaller than K_{Co} (since for all compounds of the RCo_5 type investigated within the temperature range 0 to 600 K m_{Co} changes slightly, as a rough approximation, it can be assumed constant); this implies that within this interval K_{Co} also can be assumed constant. Since $K_{Ho} < K_{Dy} < K_{Tb}$, it is evident that at equal variation of K_R with temperature the inequality $K_R < K_{Co}$ is realized first for Ho^{3+}, then for Dy^{3+}, and at last for Tb^{3+}, which is observed experimentally. The localization of the spin-orientation interval is strongly related to the variation of K_R with temperature, which depends upon the value of intersublattice exchange $I^0_{R\text{-}Co}$. Apparently, the larger parameter $I^0_{R\text{-}Co}$ is, the higher is the transition temperature. Thus, both factors determining the position of the transition act in the same direction.

The comparison of the data obtained in [12 to 14] with those obtained by magnetic measurements (available in the literature) shows that a magnetic field shifts the boundaries of the orientation transition towards lower temperatures, the value of the shift (at constant magnetic field) being dependent upon the transition temperature. $HoCo_{5.5}$ is marked by the most appreciable manifestation of this effect, because the spin reorientation in this compound is realized in the low-temperature region. According to [13] in the case of $H = 0$ and $T = 4.2$ the angle θ_R is close to 90° in $HoCo_{5.5}$ and the orientation change of the magnetic moment begins at about 50 K. According to [15, 16] in the field of 1.27×10^6 A/m ($T = 4.2$ K) the state of this compound corresponds practically to the centre of the transition region ($\theta_R \approx 50°$).

8. Conclusion

The theory of magnetic crystalline anisotropy in compounds of the RCo_5 type developed in [1, 2] is supported experimentally in this paper. The results obtained give an impulse to investigations of magnetism in the rare-earth compounds with high magnetic anisotropy in terms of this theory [1, 2]. In the near future similar investigations are planned to be carried out on the isomorphous compounds of the $R(Co_{1-x}M_x)_5$ type (where M is a d metal) as well as on compounds with a derivative RCo_5 structure in order to study how the properties of the R and d ions are connected with the local symmetry and the surrounding composition.

Acknowledgements

The authors would like to thank Yu. P. Irkhin, R. Z. Levitin, V. V. Druhzinin, and E. V. Rosenfeld for helpful discussions.

References

[1] Yu. P. Irkhin and E. V. Rosenfeld, Fiz. tverd. Tela **16**, 485 (1974).
[2] A. C. Ermolenko, Yu. P. Irkhin, E. V. Rosenfeld, V. V. Kelarev, A. F. Rohzda, S. K. Sidorov, A. N. Pirogov, and A. P. Vokhmyanin, Zh. eksper. teor. Fiz. **69**, 1743 (1975).

[3] W. JAMES, R. LEMAIRE, and F. BERTAUT, C. R. Acad. Sci. (France) **255**, 896 (1962).
[4] H. BARTHOLIN, B. VAN LAAR, R. LEMAIRE, and J. SCHWEIZER, J. Phys. Chem. Solids **27**, 1287 (1966).
[5] R. LEMAIRE and J. SCHWEIZER, J. Physique **28**, 216 (1967).
[6] H. KATSURAKI and S. YOSHII, J. Phys. Soc. Japan **24**, 1171 (1968).
[7] M. OHKOSHI, H. KOBAYASHI, T. KATAYAMA, M. HIRANO, and T. TSUSHIMA, Physica (Utrecht) **86–88B**, 195 (1977).
[8] K. H. J. BUSCHOW and W. A. J. J. VELGE, Z. angew. Phys. **26**, 157 (1969).
[9] J. W. CABLE, W. KOEHLER, and E. O. WOLLEN, Phys. Rev. **136A**, 240 (1964).
[10] W. C. KOEHLER, E. O. WOLLEN, and M. R. WILKINSON, Phys. Rev. **110**, 37 (1958).
[11] R. M. MOON, Phys. Rev. **136A**, 195 (1964).
[12] V. V. KELAREV, A. N. PIROGOV, V. V. CHUEV, A. P. VOKHMYANIN, and S. K. SIDOROV, Fiz. Metallov i Metallovedenie **50**, 59 (1980).
[13] V. V. CHUEV, V. V. KELAREV, and S. K. SIDOROV, Fiz. tverd. Tela **23**, 1760 (1981).
[14] V. V. CHUEV, V. V. KELAREV, and A. N. PIROGOV, Fiz. Metallov i Metallovedenie **52**, 80 (1981).
[15] J. SCHWEIZER and F. TASSET, J. Phys. F **10**, 2799 (1980).
[16] YU. A. IZYUMOV, V. E. NAISH, S. V. PETROV, and V. N. SYROMYATNIKOV, Fiz. Metallov i Metallovedenie **47**, 231 (1979).
[17] YU. A. IZYUMOV, V. E. NAISH, S. V. PETROV, and V. N. SYROMYATNIKOV, Fiz. Metallov i Metallovedenie **47**, 455 (1979).
[18] V. N. SYROMYATNIKOV, A. N. PIROGOV, V. V. CHUEV, and V. V. KELAREV, Fiz. Metallov i Metallovedenie **52**, 323 (1981).
[19] O. V. KOVALYOV, Irreducible Representations of the Space Groups, Izd. Akad. Nauk SSSR, Kiev (1961).
[20] V. V. KELAREV, A. N. PIROGOV, V. V. CHUEV, and S. K. SIDOROV, Thesis, Kharkov (1979).
[21] V. V. KELAREV, A. N. PIROGOV, S. K. SIDOROV, and A. P. VOKHMYANIN, Zh. eksper. teor. Fiz., Pisma **26**, 330 (1970).
[22] A. S. ERMOLENKO and E. V. ROSENFELD, Fiz. Metallov i Metallovedenie **48**, 505 (1979).
[23] T. OKOMOTO, H. FUJII, C. INOUS, and E. TATSUMOTO, J. Phys. Soc. Japan **34**, 835 (1973).
[24] V. V. KELAREV, A. N. PIROGOV, and S. K. SIDOROV, Thesis Internat. Conf. Magnetism 1976, Amsterdam 1976 (p. 142).
[25] J. M. ALAMEDA, B. GIVORD, and R. LEMAIRE, J. appl. Phys. **52**, 2079 (1981).
[26] D. GIVORD, J. LAFOREST, and R. LEMAIRE, J. appl. Phys. **50**, 7489 (1979).
[27] F. ROTHWARF, H. A. LEUPOLD, J. GREEDAN, W. E. WELLACE, and K. DILIP, Internat. J. Magnetism **V4**, 267 (1973).
[28] J. LAFOREST and J. S. SHAH, IEEE Trans. Magnetics **9**, 217 (1973).
[29] E. I. KONDORSKII and E. STRAUBE, Zh. eksper. teor. Fiz., Pisma **17**, 41 (1973).
[30] E. V. ROSENFELD, Thesis, Institute of Metal Physics, Akad. Nauk SSSR Sverdlovsk 1978.
[31] I. NOWICK and J. H. WARNICK, Phys. Rev. **140**, 131 (1965).

(Received March 1, 1983)

phys. stat. sol. (a) 79, 67 (1983)

Subject classification: 10.1; 21; 21.7

Institute of Metal Physics,
Academy of Sciences of the USSR, Ural Scientific Centre, Sverdlovsk[1])

Atomic Configurations of Interstitials in Intermetallic Compound Nb_3Sn

By

N. V. MOSEEV, V. G. CHUDINOV,
B. N. GOSHCHITSKII, and V. I. PROTASOV

Interstitials in Nb_3Sn are investigated by molecular dynamics technique. It is shown that around the interstitials the atomic displacements are appreciably larger than around the vacancies and antisite replacements. It is concluded that this phenomenon must lead to amorphization of Nb_3Sn under irradiation with light charged particles at low temperatures.

Zwischengitterstörstellen in Nb_3Sn werden mit der Methode der Molekulardynamik untersucht. Es wird gezeigt, daß um die Zwischengitterstörstellen die atomaren Verlagerungen wesentlich größer als um Leerstellen und Antiplatzverlagerungen sind. Es wird angenommen, daß diese Erscheinung zu einer Amorphisierung des Nb_3Sn unter Einstrahlung mit leichten geladenen Partikeln bei niedrigen Temperaturen führt.

In the previous paper [1] atomic configurations of vacancies and antisite replacements in Nb_3Sn were investigated. The results obtained allowed to give a qualitative description of the experimentally observed differences in the structural properties of this compound after irradiation with fast neutrons and light charged particles at room temperatures. It is to be noted that Nb_3Sn is one of the most promising materials to be used in superconducting solenoids acting under radiation at ≈ 4 K. Therefore, it is of interest to investigate the properties of the interstitials which do not anneal at a given temperature and hence, may affect the properties of the compound.

Calculations have been carried out with a molecular dynamics technique using the program Moldyn [2]. The cubic crystallite consisted of 1332 atoms located at the sites of the A-15 structure characteristic of Nb_3Sn. The calculated cell consisted of 144 atoms and was surrounded by a layer of fixed atoms, the depth of which being equal to the cut-off radius of interatomic potentials. The lattice constant a_0 corresponded to the minimum of the potential energy of the crystallite, i.e., it was the equilibrium one. Interatomic interactions of Nb–Nb, Sn–Sn, Nb–Sn were given by centrosymmetrical pair potentials, calculated in terms of the pseudopotential theory [1] using the Heine-Abarenkov-Animalu model pseudopotential [3] and the dielectric screening function by Vashishta and Singvi [4]. The cut-off radius of these three potentials was taken equal to a_0. At the point of cutting the potentials smoothly tended to zero. To find an equilibrium configuration an artificial energy damping was used. The calculations were carried out with a computer ES-1022.

Three types of stable configurations of Nb interstitials have been investigated. They are as follows:

(i) an interstitial atom is placed in the geometrical centre of the triangle formed by two Nb atoms and one Sn atom;

---

[1]) ul. S. Kovalevskoi 18, 620219 Sverdlovsk GSP-170, USSR.

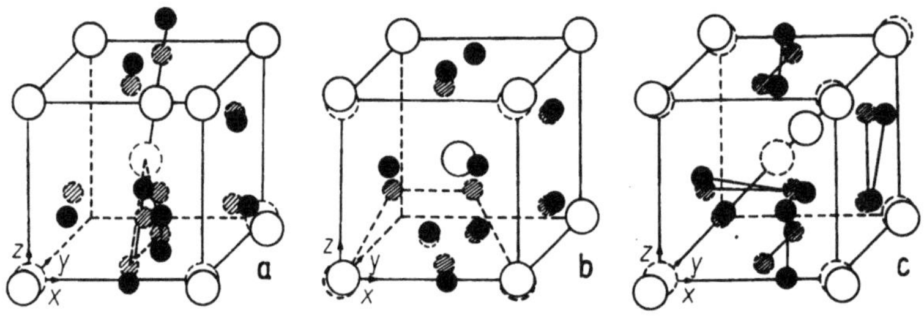

Fig. 1. Atomic configuration of interstitial Nb a) of type I, b) of type II, and c) of type III. ●, ○ positions of the Nb and Sn atoms after relaxation; hatched and dashed circles show the positions of the Nb and Sn atoms in the ideal lattice

(ii) an interstitial is placed in the geometrical centre of the trapeze formed by two Sn atoms and two Nb atoms;

(iii) an interstitial is placed on the main diagonal of the cube formed by the atoms of the unit cell.

Fig. 1a to c show the equilibrium configurations of the above types of interstitials, respectively. Here are shown also the displacements of atoms of the cube that forms the unit cell, and for simplicity, the displacements of the Nb atoms on the three mutually normal faces only. Tables 1, 2, 3 contain the values of atom displacements in per cents of a_0. Configurations of the interstitial Sn have been expected to be similar to those of Nb, since the potentials of interatomic interactions of Nb–Nb and Sn–Sn

Table 1

The values of atoms displacements in the unit cell around interstitial Nb of type I

atom coordinates	$\frac{\Delta x}{a_0}$ (%)	$\frac{\Delta y}{a_0}$ (%)	$\frac{\Delta z}{a_0}$ (%)
0, 0, 0	−2.0	0.5	− 4.4
1, 0, 0	2.0	0.5	− 4.4
0, 1, 0	−2.0	−0.9	− 4.5
1, 1, 0	1.9	−0.9	− 4.5
0, 0, 1	0.4	−0.2	0.1
1, 0, 1	−0.4	−0.2	0.1
0, 1, 1	0.5	−0.2	− 1.0
1, 1, 1	−0.5	−0.2	− 1.0
1/2, 1/2, 1/2	0.0	19.1	25.8
1/2, 1/4, 0	0.0	−2.0	−11.1
1/2, 3/4, 0	0.0	1.5	−10.4
1/2, 1/4, 1	0.0	2.1	11.1
1/2, 3/4, 1	0.0	8.3	17.8
0, 1/2, 1/4	−8.6	−0.2	− 0.8
0, 1/2, 3/4	−2.3	−1.1	− 3.6
1, 1/2, 1/4	8.6	−0.2	− 0.8
1, 1/2, 3/4	2.3	−1.1	− 3.6
1/4, 0, 1/2	−2.5	0.2	−12.7
3/4, 0, 1/2	2.5	0.2	−12.7
1/4, 1, 1/2	−2.4	−0.3	−12.5
3/4, 1, 1/2	2.4	−0.3	−12.5
interstitial	0.0	−1.0	17.4

Table 2

The values of atoms displacements around interstitial Nb of type II in the unit cell

atom coordinates	$\frac{\Delta x}{a_0}$ (%)	$\frac{\Delta y}{a_0}$ (%)	$\frac{\Delta z}{a_0}$ (%)
0, 0, 0	0.7	0.4	0.0
1, 0, 0	−0.7	0.4	0.0
0, 1, 0	−0.1	0.8	0.0
1, 1, 0	0.1	0.8	0.0
0, 0, 1	−1.2	0.2	4.0
1, 0, 1	1.2	0.2	4.0
0, 1, 1	0.0	0.5	0.4
1, 1, 1	0.0	0.5	0.4
1/2, 1/2, 1/2	0.0	−0.4	0.6
1/2, 1/4, 0	0.0	2.8	−11.4
1/2, 3/4, 0	0.0	2.0	1.0
1/2, 1/4, 1	0.0	1.9	7.2
1/2, 3/4, 1	0.0	0.0	0.2
0, 1/2, 1/4	−1.1	2.2	0.0
0, 1/2, 3/4	−0.5	1.3	1.1
1, 1/2, 1/4	1.1	2.2	0.0
1, 1/2, 3/4	0.5	1.3	1.1
1/4, 0, 1/2	−1.8	0.3	13.7
3/4, 0, 1/2	1.8	0.3	13.7
1/4, 1, 1/2	0.0	1.1	0.6
3/4, 1, 1/2	0.0	1.1	0.6
interstitial	0.0	−0.4	0.8

Table 3

The values of atoms displacements around interstitial Nb of type III in the unit cell

atom coordinates	$\frac{\Delta x}{a_0}$ (%)	$\frac{\Delta y}{a_0}$ (%)	$\frac{\Delta z}{a_0}$ (%)
0, 0, 0	− 4.1	− 4.1	− 4.1
1, 0, 0	3.1	− 1.2	− 1.8
0, 1, 0	− 1.8	3.1	− 1.2
1, 1, 0	− 0.2	2.5	− 3.3
0, 0, 1	− 1.3	− 1.8	3.0
1, 0, 1	2.5	− 3.3	− 0.2
0, 1, 1	− 3.3	− 0.2	2.5
1, 1, 1	− 2.0	− 2.0	− 2.0
1/2, 1/2, 1/2	10.8	10.8	10.8
1/2, 3/4, 0	− 5.7	2.9	10.7
1/2, 1/4, 1	10.3	− 2.4	0.2
1/2, 3/4, 1	− 4.0	0.0	11.2
0, 1/2, 1/4	−10.3	11.5	3.0
0, 1/2, 3/4	10.8	− 5.7	2.8
1, 1/2, 1/4	0.2	10.2	− 2.4
1, 1/2, 3/4	11.3	− 4.0	0.0
1/4, 0, 1/2	3.1	−10.3	11.5
3/4, 0, 1/2	2.9	10.8	− 5.8
1/4, 1, 1/2	− 2.4	0.2	10.2
3/4, 1, 1/2	0.0	11.3	− 4.0
1/2, 1/4, 0	11.5	3.1	−10.3
interstitial	1.8	1.8	1.8

Table 4

The values of atoms displacements around interstitial Sn of type III

atom coordinates	$\frac{\Delta x}{a_0}$ (%)	$\frac{\Delta y}{a_0}$ (%)	$\frac{\Delta z}{a_0}$ (%)
0, 0, 0	− 3.1	− 3.0	− 3.1
1, 0, 0	3.1	− 1.0	− 1.8
0, 1, 0	− 1.9	3.1	− 1.0
1, 1, 0	− 0.2	2.4	− 3.3
0, 0, 1	− 1.1	− 1.8	3.1
1, 0, 1	2.4	− 3.3	− 0.2
0, 1, 1	− 3.3	− 0.2	2.4
1, 1, 1	− 2.1	− 2.1	− 2.1
1/2, 1/2, 1/2	10.8	10.8	10.8
1/2, 1/4, 0	11.0	2.9	−10.2
1/2, 3/4, 0	− 0.6	2.4	10.9
1/2, 1/4, 1	10.2	− 2.5	0.0
1/2, 3/4, 1	− 4.1	− 0.1	11.2
0, 1/2, 1/4	−10.2	11.0	3.0
0, 1/2, 3/4	10.9	− 6.0	2.4
1, 1/2, 1/4	0.0	10.1	− 2.5
1, 1/2, 3/4	11.2	− 4.1	− 0.1
1/4, 0, 1/2	2.9	−10.2	11.0
3/4, 0, 1/2	2.5	10.9	− 6.0
1/4, 1, 1/2	− 2.5	0.0	10.1
3/4, 1, 1/2	0.0	11.1	− 4.1
interstitial	2.0	2.0	2.0

are qualitatively the same [1]. Actually, atom displacements for the interstitial Sn of type III (Table 4) turned out to be close to those of Nb of the same type.

Let us note some typical features of the investigated interstitials. The atomic configuration of the interstitials of type III (Fig. 1c) presents a statical crowdion in the ⟨111⟩ direction. The values of atom displacements turned out to be appreciably larger than those around the antisite replacements [1], with their maximum value being ≈ 25% around the interstitial of type I. It is noteworthy that these large statical displacements of atoms must greatly distort the crystal structure. This results in the loss of the translational symmetry and, finally, in the amorphization, assuming that the observed defects appear under irradiation of Nb_3Sn at low temperatures. In this case the results obtained can be used for the calculation of both the density of electron states on the Fermi surface in a cluster approximation [5] and the superconducting transition temperature of Nb_3Sn involving defects of the type mentioned.

References

[1] V. G. Chudinov, N. V. Moseev, B. N. Goshchitskii, and V. I. Protasov, phys. stat. sol. (a) **73**, 325 (1982).

[2] V. I. Protasov and V. G. Chudinov, Modelirovanie na EVM defektov v kristallakh, Prepint LIJP Leningrad 1980 (p. 105 to 106).

[3] V. Heine and I. V. Abarenkov, Phil. Mag. **9**, 541 (1964).
A. O. E. Animalu, Phys. Rev. B **8**, 3542 (1973).

[4] P. Vashishta and K. S. Singvi, Phys. Rev. B **6**, 875 (1972).

[5] E. V. Kuznetzov, V. I. Anisimov, V. A. Crubanov, B. N. Goshchitskii, and E. X. Kurmaev, Solid State Commun. **38**, 507 (1981).

(Received March 22, 1983)

phys. stat. sol. (a) 79, 71 (1983)

Subject classification: 13.4 and 16; 22.5

Laboratoire Cristaux et Couches Minces C.E.N., Alger[1]) *(a)*
and S.A.I.-D.E.I.N. C.E.N. Saclay, Gif-sur-Yvette[2]) *(b)*

Characterization of Deep Energy Levels in Mercuric Iodide

By

T. Mohammed-Brahim (a), A. Friant (b), and J. Mellet (b)

The past few years have seen an increasing interest in HgI_2 detectors for room temperature γ- and X-ray spectrometry. Performance and effective thickness of these detectors are presently limited by carrier trapping which results in incomplete charge collection. Characterization of the trapping levels is performed by several photoelectronic methods (photoconductivity, thermal and optical quenching of the photoconductivity, TSC, lifetime measurement). A model is proposed taking into account the results obtained by these techniques in both, vapor phase and solution grown crystals.

Depuis quelques années, les détecteurs HgI_2 présentent un intérêt croissant pour la spectrométrie des photons γ et X à la température ordinaire. Les performances et l'épaisseur utile de ces détecteurs sont actuellement limitées par des phénomènes de piégeage qui affectent la collecte des charges créées par les photons. La caractérisation des niveaux localisés dans la bande interdite, responsable de ce piégeage, a été menée par différentes méthodes photoelectroniques (photoconductivité, quenching thermique et optique de la photoconductivité, TSC, mesure de durée de vie). Un modèle est proposé prenant en compte les résultats obtenus par ces méthodes sur des cristaux réalisés soit à partir de la phase vapeur, soit en solution.

1. Introduction

The high sensibility of the layered material mercuric iodide to light is known since the early twenties of this century [1, 2]. In the fifties and sixties this large band gap semiconductor ($E_g = 2.1$ eV at 300 K) was studied with a view to understand its excitonic properties [3 to 5]. During the last decade, it gained further importance in research because of its possible potential use as detectors in the room temperature X-ray and γ-ray spectroscopy [6 to 31]. The performance of these detectors is actually limited by the carrier trapping processes preventing the complete collection of charge carriers created by the photons in the material. The characterization of the localized trapping levels within the band gap has been done mostly by studying thermally stimulated currents (TSC) in the system. In spite of the large number of publications in this direction from various laboratories, there seems to be only a very rare agreement between any two of them. The aim of the present work is to try to establish those results on all crystals available at present, which are in general agreement. We are glad to report reasonable success which has in fact lead us to propose an energy level scheme apparently valid for solution-grown crystals [32] as well as those grown from vapour phase [33].

2. Experimental

The high resistivity of this material (10^{13} Ωcm) precludes the use of the methods based on capacitive measurements to characterize these levels. However, one can take advantage of the excellent photoconducting properties of this material to use

[1]) BP 1017, Alger-Gare, Algeria.
[2]) 91191 Gif-sur-Yvette, France.

electro-optical methods involving the measurements of photoconductivity, thermally stimulated currents (TSC), and studying the effects of thermal and optical quenching on the former.

From the experiments on TSC, one can deduce reasonable values for the parameter E (trap depth) of the energy levels only after taking certain precautions like varying the experimental conditions and using different methods of analysis of the TSC spectrum [34]. Another difficulty encountered in the analysis of the TSC spectra is the existence of many peaks close to each other on the temperature scale. We have used numerical methods described in detail earlier [35] which enabled us to show the existence of eight peaks in the TSC spectrum of HgI_2. We have also used the "initial rise" method. In order to compare our results with those of Bube [36] we have calculated the activation energies corresponding to the peaks by the method used by Bube to determine the position of the Fermi level at the maximum of the peak.

Thermal and/or optical quenching of photoconductivity assumes that the current in the material is due to only one type of carriers and the contribution of the other type is negligible. Following this assumption Rose [37] has discribed a conceptual model, made quantitative later by Bube [38], based on the existence of two types of centres: centres E_r (class I), having equal capture cross-sections for electrons and holes and centres E (class II) having capture cross-sections for holes very large compared to those for electrons (in the case where the current is due to only electrons). Liberation of holes from the traps of depth E thermally by heating the sample or optically by exciting with photons of energy E, increases the capture of holes and consequently that of electrons by the levels E_r. The current mostly carried by the electrons will therefore be reduced. Mercuric iodide, in which the mobility and lifetime of electrons are much higher than those of holes, can be considered as a semiconductor with a single type of carriers. The existence of hole trapping levels in HgI_2 can therefore be revealed by such experiments.

The lifetime of electrons has been measured at room temperature by determining the initial slope of the decrease in current consecutively by exciting with a pulsed laser creating electron–hole pairs uniformly in the sample by "two-photon" processes.

3. Samples and Preparation

The crystals studied are grown by Nicolau (CEN Grenoble) from solution as described by Nicolau and Joly [32] and supplied. In order to compare with the results on crystals grown from the vapour phase, crystals supplied from Strasbourg and from EG & G Laboratory at Santa Barbara (USA), are also used in the present studies.

The crystals are cut with the aid of a thread wetted by a solvent of HgI_2 and then etched with a solution of 5% KI. The mercuric iodide poses difficult problems with regard to the deposition of electrodes (because of high reactivity of mercury with most of the metals and also its high vapour pressure), we could prepare the contacts either by depositing graphite points (aquadag or ecccoccoat) or by depositing palladium without heating the sample. The latter is achieved via an intermediate target of palladium bombarded by an ion beam. These injecting electrodes are either of the form of two parallel gates on the same face of the crystal or of a sandwich geometry.

4. Results and Discussion

The experimental results presented here are all reproducible whether they are measurements made on (i) the same crystal, (ii) on different crystals grown by the same method, or (iii) on crystals grown by different methods. Some crystals, particularly a majority of those grown from solution give varying results explaining why the TSC spectra reported in literature till now differ largely from one another. However, we found that these crystals could be made to give reproducible results either by a prolonged illumination with light of energy corresponding to the band gap of the material under an electric field or by having a MIS structure at the positive electrode. The details of these processes will be published later separately.

4.1 Photoconductivity spectra

The photocurrent as a function of excitation wavelength shows (Fig. 1) the normal exciton peak at 5900 Å (2.1 eV). It also reveals a weak peak around 7100 Å (1.74 eV) which is likely to be due to a localized level in the band gap.

4.2 Thermal quenching of photoconductivity

When the sample temperature is increased from 80 K the photocurrent falls initially rapidly and is followed by two more reductions around 120 and 200 K. These reductions could be interpreted in the case of HgI_2, where the photocurrent is mostly carried by electrons, as due to the hole trap levels (class II in the model of Rose). The photocurrent variation with temperature has been obtained for seven different intensities of illumination (Fig. 2). We use the formula of Bube [38],

$$\ln n = \ln\left(N_v \frac{s}{s'}\right) - \frac{E}{kT} \tag{1}$$

giving the concentration n of free electrons responsible for the photocurrent at the temperature where the current begins to fall and

$$\ln n = \ln\left(N_v \frac{N_r}{N}\right) - \frac{E}{kT} \tag{2}$$

giving the concentration n at the temperature T at which the fall of current terminates. In these formulae N_v, the density of states in the valence band, is given by $N_v = 6.3 \times 10^{15} T^{3/2}$ cm^{-3} when the effective mass of the holes in HgI_2 is taken as $1.2m_0$

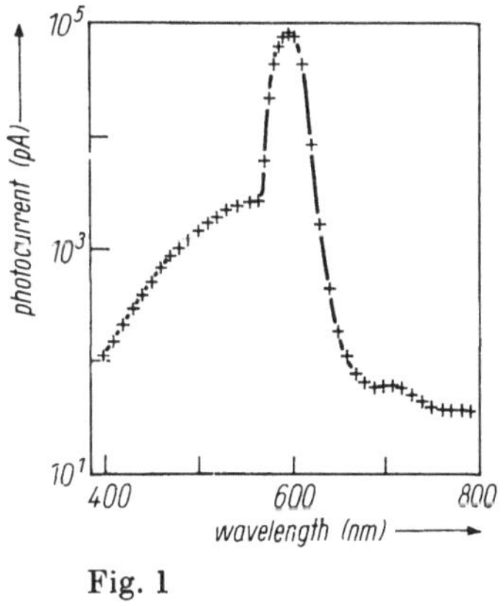

Fig. 1

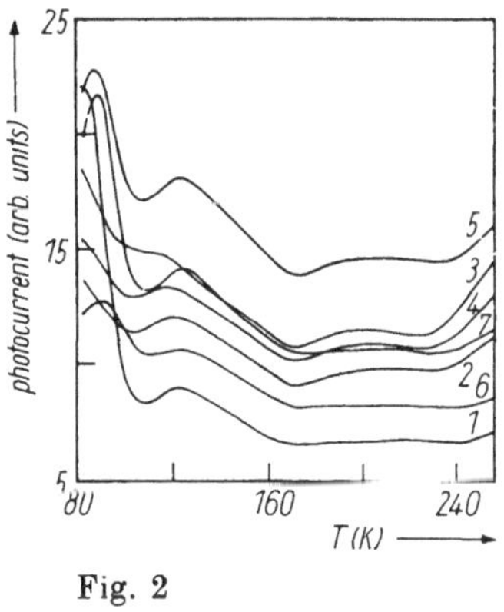

Fig. 2

Fig. 1. Photoconductivity spectra

Fig. 2. Photocurrent vs. temperature (for seven intensities). 1 arb. unit means: 25 nA for curves (1), (6); 12.5 nA for (7); 5 nA for (2), (5); 2.5 nA for 4; 1.25 nA for (3)

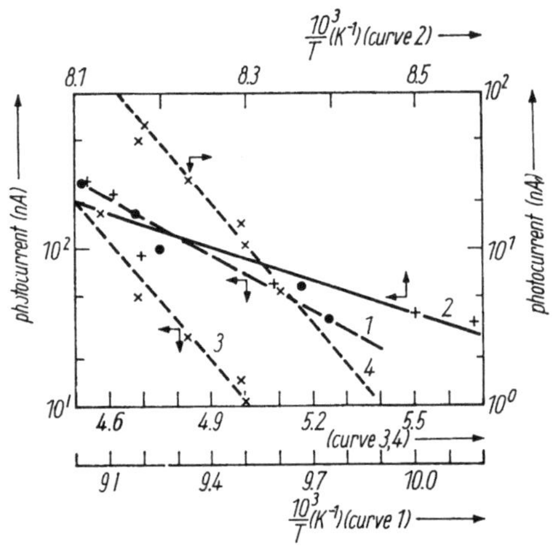

Fig. 3. Straight lines ln I vs. $1/T$ (from the three thermal quenchings of Fig. 2) (1) minimum of the current observed around 110 K, (2) maximum of the current situated around 120 K, (3), (4) maximum of the current situated around 200 K

[39]; s is the capture cross-section of the hole trapping level E, supposed to be the the same for trapping holes and electrons by the level E_r situated very near the midgap; s' is the capture cross-section of the level E for electrons; N is the concentration of centres E and N_r that of centres E_r.

Using (1) at the maxima of the photocurrent situated around 80, 120, and 200 K (Fig. 3) and (2) at the minimum of the current observed around 110 K, the three levels of activation energy E, the ratio of the capture cross-sections s'/s, and the ratio of the concentration N_r/N are estimated and tabulated (Table 1).

Table 1

T (K)	E (eV)	s/s'	N_r/N
200	0.5	7	
120	0.36	2.5×10^4	
80	0.21	50	
110			0.14

It must be mentioned that the experimental points giving the maxima of current around 120 K are quite dispersed. Also the straight line giving the activation energy corresponding to this peak could only be obtained as an average with about 10% deviation. The decrease of current after this maximum is also extended over a wide region (from 120 to 170 K). We think that this is due to the presence of a series of closely spaced levels and the activation energy calculated (0.36 eV) is only an average value. It should also be noted that the maximum of current around 80 K is rejected for temperatures below 80 K as these temperatures are inaccessible with our experimental set-up and for weak intensities of illumination. We could thus obtain it only for three intensities.

4.3 Optical quenching of photoconductivity

At 80 K a superposition of a secondary source of light of increasing energy reduces the primary current, due to the (primary) source of light with energy corresponding to the band gap, from 0.4 eV (Fig. 4). This reduction is followed by a second one from 0.48 eV. This quenching of primary current, continuous only for secondary excitation

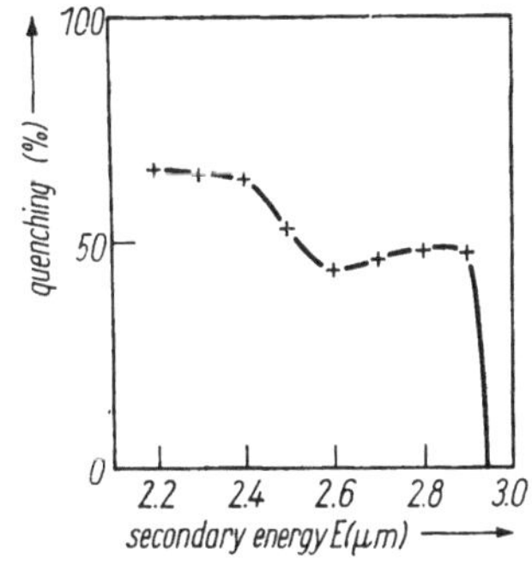

Fig. 4

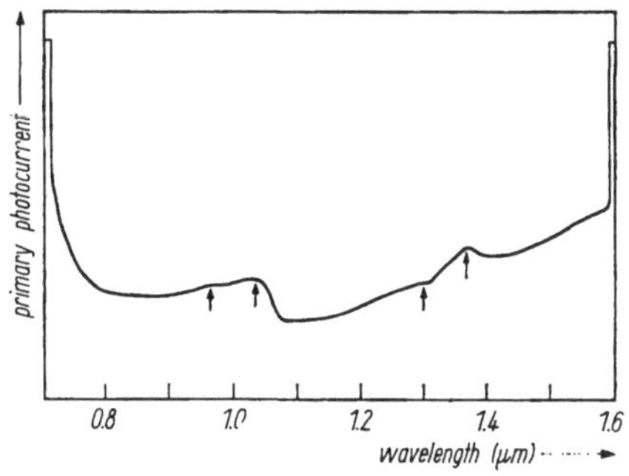

Fig. 5

Fig. 4. Percentage optical quenching at 2.9 μm (0.4 eV) and at 2.6 μm (0.48 eV)

Fig. 5. Optical quenching from 0.7 μm

of energy up to 1.77 eV (Fig. 5). Fig. 5 shows us also the existence of photocurrent peaks at 1.28, 0.95, and 0.91 eV. This type of quenching of photocurrent can be explained by introducing hole trapping levels at 0.4 eV. The quenching terminates when the excitation energy reaches 1.77 eV around which the process of excitation from these levels (0.4 and 0.48 eV) directly to the conduction band competes with the process responsible for the quenching. This experiment has permitted us to show, in addition, the existence of deep levels located near the mid-gap at 1.28, 0.95, and 0.91 eV.

4.4 Thermally stimulated currents (TSC)

The TSC spectra show eight peaks (Fig. 6). The corresponding activation energies and capture cross-sections, calculated by the method of Bube (position of the Fermi level) for both cases, electron and hole traps, are presented in Table 2. This calculation is made after separating the peaks by the numerical method mentioned earlier. Table 2 shows also, for comparison, the energies of the four levels reported by Bube [36] assuming them to be electron traps. One can see the excellent agreement between the two results. We give in the same table the capture cross-sections as well, which can be accepted at least for their order of magnitude, considering the difficulty or near impossibility (according to some workers) of obtaining these values from a TSC spectrum.

Fig. 7a to c show the TSC spectra obtained on the samples from LETI (solution grown), EG & G (grown from vapour phase), and from Strasbourg (also grown from vapour phase), respectively. It should be mentioned that there is remarkable similarity

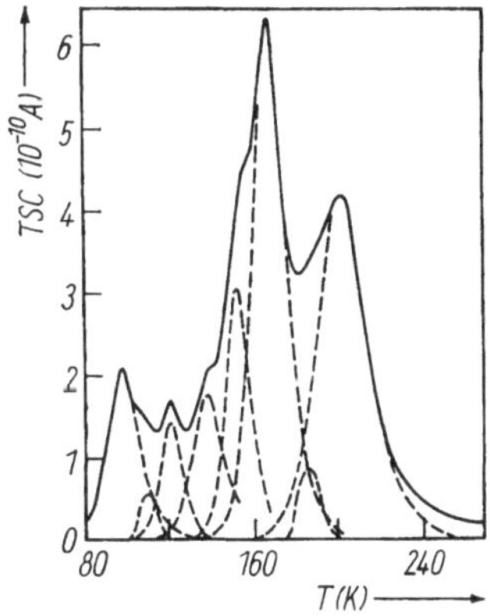

Fig. 6. Thermally stimulated current (TSC): eight peaks

Table 2

		T_{max} (K)	99	109	121	138	153	168	186	203
present results	electron traps	E (eV)	0.24	0.27	0.32	0.39	0.40	0.43	0.51	0.53
		S (cm²)	1.9×10^{-13}	3×10^{-12}	1.3×10^{-13}	1.8×10^{-12}	4.6×10^{-16}	1.8×10^{-14}	1.17×10^{-13}	1.8×10^{-14}
	hole traps	E (eV)	0.23	0.29	0.31	0.36	0.38	0.41	0.49	0.50
		S (cm²)	2.8×10^{-14}	3.6×10^{-14}	2.2×10^{-14}	1.1×10^{-14}	4.6×10^{-15}	1.7×10^{-15}	9.7×10^{-15}	1.3×10^{-15}
Bube	electron traps	T_{max} (K)			129	142		171	196	
		E (eV)			0.34	0.38		0.45	0.51	

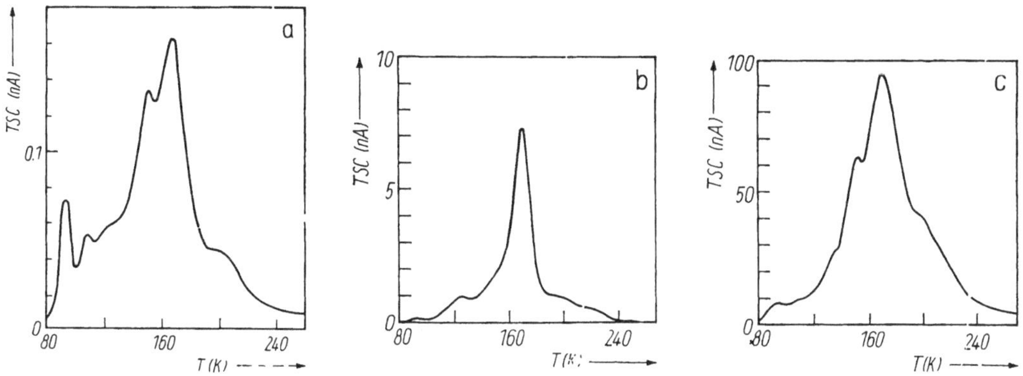

Fig. 7. TSC spectrum of a) LETI (solution grown) samples, b) EG&G (vapour phase grown) samples, c) Strasbourg (vapour phase grown) samples

Table 3

E		S_p	S_p/S_n	N/N_r
E_1	0.21	2.83×10^{-14}	50	10
E_5	0.29	6.62×10^{-14}	2.5×10^4	
	0.31	2.2×10^{-14}		
	0.35	1.13×10^{-14}		
	0.38	4.6×10^{-15}		
E_4	0.41	1.71×10^{-15}	7	
	0.49	9.74×10^{-15}		
	0.50	1.37×10^{-15}		

among the spectra having the same peaks, all of them centred around the peak at 170 K. Thus such a spectrum can be taken to be typical for the material HgI_2 in the present state of technology.

4.5 Probable energy level diagram

The results of thermal quenching experiment showed a hole trap at 0.5 eV, those of optical quenching indicated the existence of two levels at 0.4 and 0.48 eV, and finally those of TSC showed three levels at 0.41, 0.49, and 0.5 eV. We can therefore assume the presence of three hole trapping levels between 0.4 and 0.5 eV. Another level is also observed in the TSC and thermal quenching experiments in the region 0.21 to 0.23 eV. From an analysis by the latter method, it is found to be a hole trap. Finally four more levels between 0.3 and 0.4 eV are found from the TSC experiments. These closely spaced levels could not be resolved by thermal quenching. We have, however, observed a decrease in photocurrent very much extended in the temperatrue range (120 to 170 K), which may be due to these hole trapping levels being close to each other. The decrease in current has given an activation energy of 0.36 eV which is the average of the four levels in the range 0.3 to 0.4 eV. To these hole trapping levels one could add the deep levels near the mid-gap which are evidenced by the optical quenching experiment. We therefore propose an energy level scheme consisting of recombination centres at the mid-gap, four hole trapping levels in the region 0.3 to 0.4 eV, one level at 0.21 eV, and three levels in the region 0.4 to 0.5 eV. The characteristics of this scheme are given in Table 3 and are shown in Fig. 8.

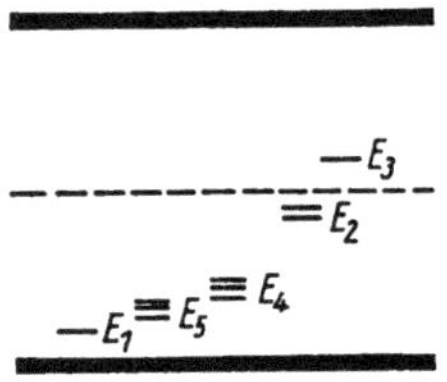

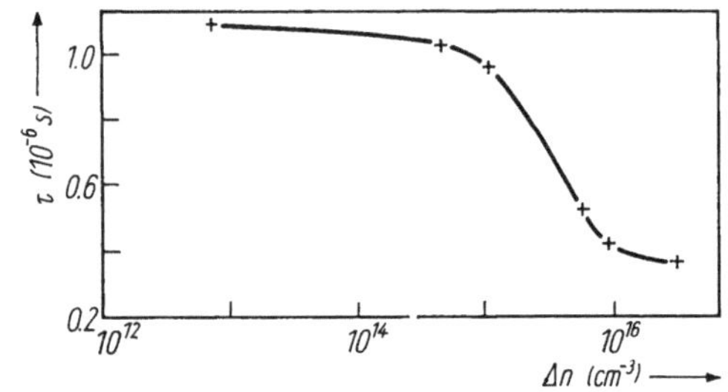

Fig. 8 Fig. 9

Fig. 8. Energy level scheme

Fig. 9. Lifetime τ vs. excess electron concentration Δn

5. Correlation between the Proposed Model and the Lifetime Measurements at Room Temperature

Fig. 9 shows the variation of the lifetime of electrons with the concentration of electron hole pairs Δ created by a pulsed laser excitation. We notice a reduction of lifetime as Δ increases. The model we proposed does not permit any levels other than hole traps. When the concentration Δ increases, the capture of holes at these levels may become more probable than their liberation at a certain value of Δ. In such a situation the concentration of free holes may be reduced. The capture of the free holes at the recombination levels E_2 and E_3 reduces and therefore capture of electrons at these levels also is decreased in the same way. Thus the concentration and the lifetime of the electrons increase. But we observe a decreasing lifetime for these electrons. Thus we have a model describing all the experimentally observed results except those concerning the lifetime. The proposed model is based on the fact that the capture cross-section for holes at the hole trap levels is much higher than that for electrons, such that the levels cannot act as recombination centres. We introduce a modification assuming that the ratio of capture cross-section for holes and electrons for the hole trap levels is not very high, such that the lifetime of electrons permits a contribution due to the recombination at the levels in addition to that of recombination centres E_2, E_3. When Δ increases, the capture of holes and therefore that of electrons by the hole trap levels increases. In that case the recombination of electrons at these levels increase while that at E_2 and E_3 levels always decreases. Now let us suppose in addition that the concentration of hole traps is larger than that of the recombination centres E_2, E_3, so that these centres get saturated above a certain value of Δ. The recombination at the centres E_2, E_3 therefore remains constant while that at the trap levels increases. Then we have an effective reduction in the lifetime of electrons. In the proposed model, the level E_1 (0.21 eV) combines the two conditions such that its interference leads to a decrease of the lifetime of the electrons. the ratio of capture cross-sections of holes and electrons is small (50) and the concentration of recombination centres is less than that of the E_1 levels (ratio 0.1). We can then verify by calculation the assumption that the interference of E_1 leads to a reduction of the electron lifetime. For such a calculation we take the following energy scheme:

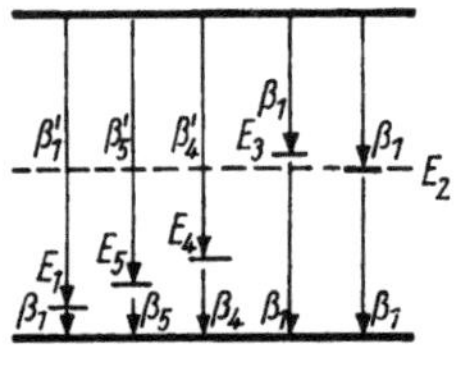

The levels E_1, E_2, E_4, and E_5 are assumed to be filled with electrons in the dark and the level E_3 vacant.

The characteristics of the levels are:

— energies (determined by different experiments)

$$E_1 = 0.21 \text{ eV}, \qquad E_4 = 0.50 \text{ eV}, \qquad E_5 = 0.36 \text{ eV},$$

$$E_2 \text{ and } E_3 = \text{mid-gap};$$

— concentrations

level E_1: $N_1 = 5 \times 10^{14}\ \text{cm}^{-3}$,

levels E_2 and E_4: $N_2 = N_4 = 2.3 \times 10^{12}\ \text{cm}^{-3}$,

level E_3: $N_3 = 4.6 \times 10^{12}\ \text{cm}^{-3}$,

level E_5: $N_5 = 2 \times 10^{13}\ \text{cm}^{-3}$,

— capture cross-sections for holes (TSC)

by E_1: $Sp_1 = 2.8 \times 10^{-14}\ \text{cm}^2$,

by E_4: $Sp_4 = 1.4 \times 10^{-15}\ \text{cm}^2$,

by E_5: $Sp_5 = 2.8 \times 10^{-14}\ \text{cm}^2$,

the capture cross-sections for electrons and holes by E_2 and E_3 are the same and equal to Sp_1;

— ratio of capture cross-sections for holes Sp and electrons Sn

$Sp_4/Sn_4 = 7$,

$Sp_1/Sn_1 = 50$,

$Sp_5/Sn_5 = 10^4$;

— probabilities for hole liberation

$$P = N_v \beta \exp\left(\frac{-E}{kT}\right),$$

where $\beta = Sp V_{th}$ (product of capture cross-section and thermal velocity).

The equations giving the concentration kinetics are

$$\frac{dn}{dt} = -n \sum_2^3 \beta_i(N_i - n_i) - n\beta_1'(N_1 - n_1) - n\beta_4'(N_4 - n_4) - n\beta_5'(N_5 - n_5) + F, \tag{3}$$

$$\frac{dp}{dt} = -\beta_1 p \sum_1^3 n_i - p \sum_4^5 \beta_i n_i + (N_1 - n_1) P_1 + (N_4 - n_4) P_4 + (N_5 - n_5) P_5 + F, \tag{4}$$

$$\frac{dn_j}{dt} = (N_j - n_j) P_j - \beta_j n_j p + \beta_j' n(N_j - n_j) \qquad j = 1, 4, 5, \tag{5, 6, 7}$$

$$\frac{dn_j}{dt} = \beta_1 n(N_j - n_j) - \beta_1 n_j p \qquad j = 2, 3. \tag{8, 9}$$

We have in addition the equation of the neutrality condition,

$$n + n_3 = p + \sum_{\substack{i=1 \\ i\neq 3}}^{5} (N_i - n_i) , \tag{10}$$

n_i concentration of centres E_i occupied by electrons, n and p concentration of free electrons and free holes, respectively, F number of electron–hole pairs created per second in unit volume. We neglect the band-to-band recombination.

If we assume that the equilibrium is attained at the maximum of the current response to laser excitation, (5) to (10) give

$$n + N_3 \frac{n}{n+p} = p + \sum_{\substack{i=1 \\ i\neq 2,3}}^{5} N_i \frac{P}{p + P_i/\beta_i + \beta_i'/\beta_i \cdot n} + N_2 \frac{P}{p+n}. \tag{11}$$

The concentrations of free electrons n is given experimentally by the amplitude of the peak in the current response. Equation (11) gives p and then one can calculate n_1, n_2, n_3, n_4, and n_5. The lifetime τ is then obtained by the equation

$$\tau = \frac{1}{\beta_1 \sum_{2}^{3} (N_i - n_i) + \sum_{\substack{i=1 \\ i\neq 2,3}}^{5} \beta_i' (N_i - n_i)} . \tag{12}$$

Table 4 and Fig. 10 show the good correlation obtained between the values of lifetime calculated and those obtained experimentally. The calculation has confirmed therefore the assumption that the decrease of lifetime is due to the interference with the level E_1. The model described well all the experimental results presented here including the variation of lifetime if we assume that the recombination can take place via the intermediate trap level E_1.

Table 4

E/E_0	Δn (cm^{-3})	τ_{exp} (μs)	τ_{calc} (μs)
2×10^{-3}	6.6×10^{12}	1.08	1.08
1×10^{-2}	4.3×10^{14}	1.02	0.94
2×10^{-2}	1×10^{15}	0.95	0.84
5×10^{-2}	5.8×10^{15}	0.52	0.52
2×10^{-1}	9.1×10^{15}	0.41	0.44
3×10^{-1}	3.1×10^{16}	0.36	0.33

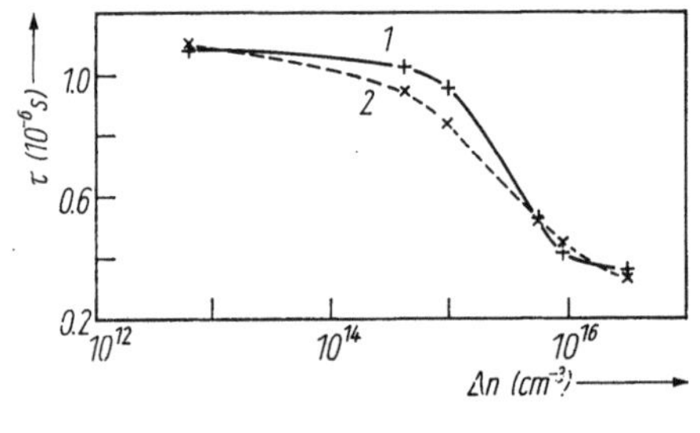

Fig. 10. Correlation between experimental (1) and calculated (2) lifetimes

6. Conclusion

In this paper we have brought out information about some levels appearing commonly in all the HgI_2 crystals presently available. This led us to propose a model of deep levels existing in mercuric iodide samples in their present state of technology. The fact that these levels are common in all the crystals tends to give credit to the hypothesis that they are due to the intrinsic defects of the crystal. The presence of such defects is not surprising if we take into account the layered structure of the material which permits a considerable relative movement between the layers.

References

[1] G. Athanasiu, C.R. Acad. Sci. (France) **176**, 1389 (1923).
[2] F. C. Nix, Phys. Rev. **47**, 71 (1935).
[3] S. Nikitine and M. Sieskind, C.R. Acad. Sci. (France) **240**, 1324 (1955).
[4] M. Sieskind, Rev. Opt. théor. instrument. **39**, 239 (1960).
[5] E. F. Gross and B. V. Novikov, J. Phys. Chem. Solids **22**, 87 (1961).
[6] W. R. Willig, Nuclear Instrum. and Methods **96**, 615 (1971).
[7] W. R. Willig and S. Roth, Nuclear Instrum. and Methods **101**, 23 (1972).
[8] H. L. Malm, IEEE Trans. Nuclear Sci. **19**, 263 (1972).
[9] H. L. Malm, T. W. Raudorf, P. M. Martini, and K. R. Zanio, IEEE Trans. Nuclear Sci. **29**, 500 (1973).
[10] S. P. Swierkowski, G. A. Armantrout, and R. Wichner, Appl. Phys. Letters **23**, 281 (1973).
[11] S. P. Swierkowski, G. A. Armantrout, and R. Wichner, IEEE Trans. Nuclear Sci. **21**, 395 (1974).
[12] G. M. Martin, P. Bach, and P. Guetin, Appl. Phys. Letters **25**, 286 (1974).
[13] Moon Kyu Chung and Song Mon Cho, IEEE Trans. Nuclear. Sci. **23**, 112 (1976).
[14] M. Schiefer, Nuclear Instrum. and Methods **144**, 469 (1977).
[15] J. Dabrowski and G. C. Huth, IEEE Trans. Nuclear Sci. **25**, 205 (1978).
[16] M. Schieber, I. Beinglass, G. Dishon, A. Holzer, and G. Yaron, IEEE Trans. Nuclear Sci. **25**, 644 (1978).
[17] M. Slapa, G. C. Huth, W. Seibt, M. Schieber, and P. T. Randtke, IEEE Trans. Nuclear Sci. **23**, 192 (1976).
[18] J. Llacer, M. M. K. Watt, M. Schieber, and R. Carlston, IEEE Trans. Nuclear Sci. **21**, 305 (1974).
[19] J. P. Ponpon, R. Stuck, P. Siffert, and C. Schwab, Nuclear Instrum. and Methods **119**, 197 (1974).
[20] A. Friant, J. Mellet, C. Saliou, and T. Mohammed-Brahim, IEEE Trans. Nuclear Sci. **27**, 282 (1980).
[21] L. van Denberg and R. C. Whited, IEEE Trans. Nuclear Sci. **25**, 395 (1978).
[22] R. C. Whited and L. van Denberg, IEEE Trans. Nuclear Sci. **24**, 165 (1977).
[23] P. T. Randtke and C. Ortale, IEEE Trans. Nuclear Sci. **24**, 129 (1977).
[24] M. Slapa, P. A. Tove, and G. Boberg, Nuclear Instrum. and Methods **150**, 55 (1978).
[25] G. A. Armantrout, S. P. Swierkowski, J. W. Sheroman, and J. H. Yee, IEEE Trans. Nuclear Sci. **24**, 21 (1977).
[26] R. Stuck, J. C. Muller, J. P. Ponpon, C. Scharager, C. Schawb, and P. Siffert, J. appl. Phys. **47**, 1545 (1976).
[27] R. C. Whited, M. Schieber, and P. T. Randtke, J. appl. Phys. **47**, 2230 (1976).
[28] M. Schieber, I. Beinglass, G. Dishon, A. Holzer, and G. Yaron, Nuclear Instrum. and Methods **150**, 71 (1978).
[29] I. Beinglass and M. Schieber, Nuclear Instrum. and Methods **167**, 423 (1979).
[30] I. Beinglass, G. Dishon, A. Holzer, S. Ofer, and M. Schieber, Appl. Phys. Letters **30**, 611 (1977).
[31] W. Seibt, M. Slapa, and G. C. Huth, Nuclear Instrum. and Methods **135**, 573 (1976).
[32] I. F. Nicolau and J. P. Joly, J. Crystal Growth **48**, 61 (1980).
[33] M. Schieber, W. F. Schnepple, and L. van Denberg, J. Crystal Growth **33**, 125 (1976).
[34] Topics in Appl. Phys., Vol. 37, Ed. P. Braunlich, Springer-Verlag, Berlin 1979.

[35] T. Mohammed-Brahim, phys. stat. sol. (a) **65**, K1 (1981).
[36] R. H. Bube, Phys. Rev. **106**, 703 (1957).
[37] A. Rose, R.C.A. Rev. **12**, 362 (1951).
[38] R. H. Bube, J. Phys. Chem. Solids **1**, 234 (1957).
[39] J. H. Yee, J. W. Sheroman, and G. A. Armantrout, IEEE Trans. Nuclear Sci. **23**, 117 (1976).

(Received March 2, 1983)

phys. stat. sol. (a) 79, 83 (1983)

Subject classification: 14.3.1; 2; 13.4; 16; 19; 22.1.2

Institute of Semiconductor Physics, Academy of Sciences of the USSR, Siberian Branch, Novosibirsk[1] *(a),*
and Sektion Physik der Humboldt-Universität Berlin, Berlin[2] *(b)*

Amorphous Silicon–Phosphorus Layers

By

A. V. Dvurechenskii (a), I. A. Ryazantsev (a), L. S. Smirnov (a), H. Klose (b), and M. Rieth (b)

EPR, photo- and dark conductivity are investigated for amorphous silicon films prepared by vacuum deposition of the silicon–phosphorus vapour mixture. It is shown that as the phosphorus vapour pressure increases ($p_P \leqq 10^{-3}$ Pa) the conductivity decreases and paramagnetic centres with $g = 2.0055$ disappear. The temperature dependence of the dark conductivity and EPR data permit the conclusion that the phosphorus doping by vacuum evaporation technique enables one to decrease the localized state density in a-Si down to 10^{17} cm^{-3} eV^{-1}. With a further increase of p_P ($> 10^{-3}$ Pa) the n-type conductivity appears. The maximum value of σ is 10^{-3} (Ω cm)$^{-1}$, though the phosphorus concentration in a-Si:P ranges from 10^{19} to 3×10^{22} cm^{-3}. It is found that the a-Si:P layers are characterized by high values of photoconductivity and thermally stable when heated up to 500 °C.

Исследованы злектрофизические свойства аморфного кремния, полученного методом вакуумного испарения с одновременным введением в вакуумный объем паров фосфора. Показано, что при увеличении давления паров фосфора ($p_P \leqq 10^{-3}$ Pa) имеет место уменъшение проводимости и исчезновение парамагнитных центров с $g = 2{,}0055$. Анализ хода температурной зависимости проводимости и данных ЭПР позволил сделать вывод, что введение фосфора методом вакуумного испарения дает возможность снизить плотность локализованных состояний в a-Si до 10^{17} cm^{-3} eV^{-1}. При дальнейшем увеличении ($p_P > 10^{-3}$ Pa) наблюдается рост проводимости электронного типа. Максималъная величина σ составила $\approx 10^{-3}$ (Ω cm)$^{-1}$, хотя концентрация фосфора в a- Si:P менялась в широких пределах: от 10^{19} до 3×10^{22} cm^{-3}. Установлено, что слои a- Si:P характеризуются высоким значением фотопроводимости и термостабильны при нагреве до 500 °C.

1. Introduction

The great attention to studies of amorphous silicon is paid to hydrogenated material that is, as expected, most promising for producing solar cells [1]. A serious shortcoming of hydrogenated amorphous silicon (a-Si:H) is its low thermal stability. At $T \geqq 300$ °C hydrogen begins going out of a-Si:H and the material properties change seriously [2]. Hydrogen in a-Si promotes the decrease of the localized state density N_F in the mobility gap down to 10^{17} cm^{-3} eV^{-1} and the increase of the photoconductivity [3]. The decrease of N_F was observed in a-Si with doped impurities, e.g. phosphorus [4]. The microscopic mechanism of this process was proposed in [5, 6].

The purpose of the present paper is an investigation of the possibilities to decrease the density of localized states in a-Si by impurity doping, a study of properties of a-Si formed by vacuum evaporation techniques, and their comparison with the parameters of a-Si:H.

2. Experimental

The films of amorphous Si were prepared by electron-beam evaporation of crystalline silicon in vacuum at a pressure $p \leqq 10^{-4}$ Pa. The deposition rate was 3 to 5 Å/s on

[1]) Academician Lavrentev prospekt 13, 630090 Novosibirsk 90, USSR.

[2]) Invalidenstr. 42, DDR-1040 Berlin, GDR.

polished quartz substrates at $T = 350$ °C. During evacuation of the system and source training the substrate was protected by a shutter. A quartz container with crystalline (red) phosphorus was placed into the vacuum volume for introducing the doping impurity. When the container walls are heated from to 200 to 300 °C the phosphorus vapours arrive at the zone of silicon vapour transfer near the substrate. To saturate the a-Si layers with hydrogen (hydrogenation) the deposition was performed in some experiments in hydrogen atmosphere. The structure of the films was controlled by the RHEED technique.

The EPR method was used to measure the concentration (N_s) of paramagnetic centres ($g = 2.0055$). In the measurements of dark conductivity (σ), photoconductivity (σ_{ph}), and sign of thermopower usable contacts were made of aluminium, indium, and an In–Ga mixture. The photoconductivity is measured at illumination by a defocused radiation from the LG-126 laser with a wavelength of 0,63 μm and a flow of 10^{15} cm^{-2}s^{-1}. The ohmic conductivity in the samples under investigation was retained in the field range from 10 to 10^3 V/cm. For additional saturation of defects with hydrogen the layers were annealed in a glow discharge in H_2 atmosphere at different temperature [7].

The chemical depth distribution of phosphorus atoms was determined by secondary ion mass spectrometry (SIMS). The instrument used was a CAMECA ion microscope. The samples were bombarded with positive oxygen ions with an energy of 10.5 keV. The primary ion beam diameter was 40 μm, the beam current 0.5 μA. To obtain flat crater bottoms the primary beam was rastered. The size of the rastered area was a square of 250×250 μm^2, the diameter of the analyzed area was 70 μm. To prevent charging a thin gold film was deposited on the high-resistant a-Si layers before measurement. Additionally the surface was bombarded with electrons. During the measurement process the $^{31}P^+$ and $^{30}Si^+$ signals were detected. The silicon signal served as reference signal to compare the results with slightly different apparatus parameters. For depth calibration the crater depth was determined by TALYSTEP (Taylor-Hobson) after the SIMS measurement. The absolute value of phosphorus concentration in a-Si was determined by calibration with standards. These standards were prepared by ion implantation in the dose range between 5×10^{15} and 10^{17} cm^{-2}.

The results of the measurements of these standards indicated that the SIMS signal varies proportionally to the phosphorus concentration up to the maximum of 10^{22} atoms/cm^3.

3. Results and Discusison

The undoped a-Si layers had p-type conductivity $\sigma = (2$ to $5) \times 10^{-5}$ $(\Omega^{-1} \cdot \text{cm})^{-1}$ and $N_s = (4$ to $6) \times 10^{19}$ cm^{-3}. Doping with phosphorus during the silicon deposition results in a decrease of σ and a complete disappearance of paramagnetic centres even at a pressure $p_p = 8 \times 10^{-4}$ Pa (Fig. 1). The temperature dependence of conductivity on these samples is indicative of a reduction of the localized state contribution to the low-temperature conductivity with growing p_p (Fig. 2a). As with $p_p = 10^{-3}$ Pa we could not fix the conductivity via defect states, we estimated an upper limit of the density of these states on the basis of a comparison of the temperature dependences in a-Si:P and a-Si:H, prepared by rf-glow discharge decomposition of silane. This comparison with allowance for the data of [8] permits the conclusion that the a-Si:P layers prepared have a density of localized states not exceeding 10^{17} cm^{-3} eV^{-1}. An estimation of the EPR data also provides the similar value of N_F in a-Si:P. The spin density N_s is related with N_F through the relation $N_F = N_s/\Delta E$, where ΔE is the energy gap occupied by states with non-zero spin. The quantity ΔE may be estimated by taking into account that the density of states in undoped a-Si is about 10^{20} cm^{-3} eV^{-1}

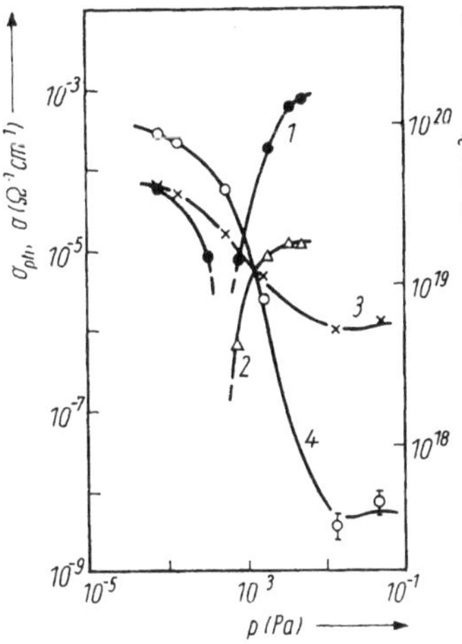

Fig. 1. Dark conductivity σ (1, 3), photoconductivity σ_{ph} (2), and spin concentration N_s (4) vs. the pressures of phosphorus vapour (1, 2) and hydrogen (3, 4)

[9] and $N_s \approx 5 \times 10^{19}$ cm^{-3}. Hence $\Delta E \approx 0.5$ eV. The minimum spin density recorded by the EPR in our case was approximately 10^{17} cm^{-3} (spectrometer sensitivity 10^{11} spin/G, linewidth approximately 10 G, thickness and area of the a-Si film 2×10^{-5} cm and 0.5 cm^2, respectively). Consequently, the minimum number of N_F recorded in our a-Si films by EPR was 2×10^{17} cm^{-3} eV^{-1}. As within the limits of sensitivity the EPR signal in a-Si:P vanishes, $N_F < 2 \times 10^{17}$ cm^{-3} eV^{-1}.

An additional argument for the low density of states in a-Si:P is provided also by experiments on the hydrogenation of these layers in a hydrogen plasma at 500 °C. We could not find any change of conductivity in a-Si:P after treatment in hydrogen plasma though this method gave a reduction of the state density in a-Si down to 10^{17} cm^{-3} [10] according to the EPR data.

A further rise of the phosphorus vapour pressure during the growth of a-Si:P results in a change of the conductivity type (from p- to n-type) and an increase of σ up to 10^{-3} Ω^{-1} cm^{-1} (Fig. 1). In this case the Fermi level approaches the conductivity band and is fixed in the position $E_c - 0.2$ eV (Fig. 2b). The obtained value of conductivity and the energy difference $E_c - E_F$ are somewhat worse than similar values observed during the doping of a-Si:H from the gaseous phase [1] or during ion implantation [5], though the concentration of phosphorus trapped in a-Si ranges within rather wide limits: from 10^{19} cm^{-3} at $p_P = 10^{-4}$ Pa to approximately 3×10^{22} cm^{-3} at 10^{-2} Pa (Fig. 3). Nevertheless the a-Si:P layers described are of practical interest. For instance, due to the low density of states these layers have a photoconductivity $\sigma_{ph} \approx (5$ to $6) \times 10^{-6}$ Ω^{-1} cm^{-1} (at $p_P \approx 10^{-3}$ Pa, Fig. 1) which corresponds to σ_{ph} measured in a-Si:H [11] at laser illumination with the same parameters as in our case. In

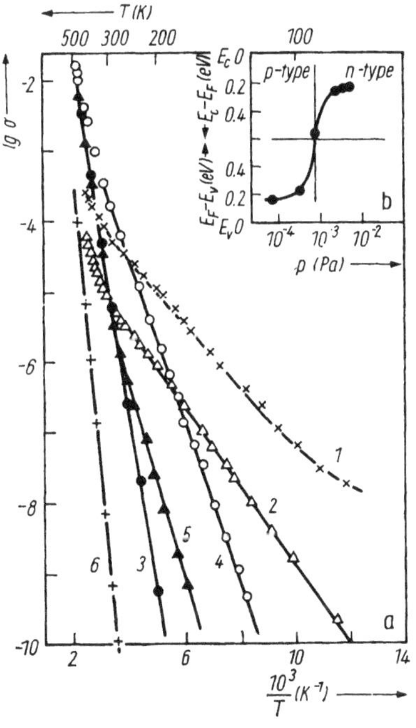

Fig. 2. Temperature dependence of conductivity in undoped a-Si (1) at a vacuum of 7×10^{-5} Pa and in a-Si deposited at different pressures of phosphorus vapour (p_P) and hydrogen (p_H): (2) $p_P = 3 \times 10^{-4}$, (3) 8×10^{-4}, and (4) 2×10^{-3} Pa; (5) $p_H = 1.8 \times$ $\times\ 10^{-3}$ Pa. (6) a-Si:H, prepared with rf glow discharge decomposition of silane at the substrate temperature 250 °C

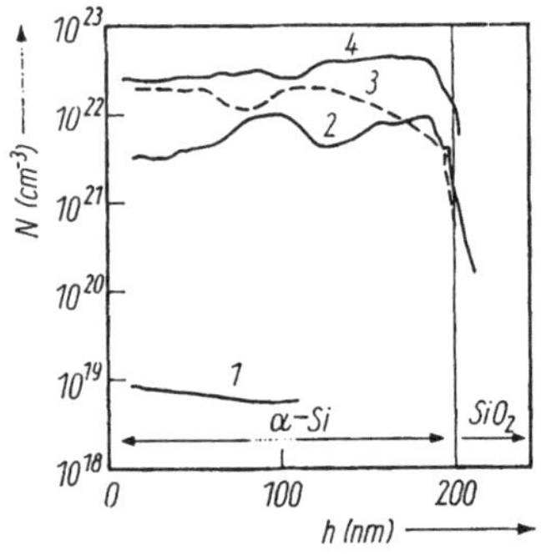

Fig. 3. Phosphorus depth distribution in a-Si:P. (1) $p_P = 10^{-4}$, (2) 3×10^{-3}, (3) 5×10^{-3}, and (4) 8×10^{-3} Pa

addition, the a-Si:P layers have a considerably better thermal stability: the value of σ does not change during heating up to 500 °C.

As has already been noted an ion implantation of P^+ in a-Si enables one to obtain layers with a higher value of σ than that in the present technique, i.e. the efficiency of impurity incorporation into substitutional sites with tetrahedral environment of atoms turns out to be higher at ion implantation [5]. This permits a conclusion on the possibility to increase σ (and utilization coefficient) in this method by implantation of low-energy ions in the process of a-Si film deposition. A decrease in energy will permit a reduction of the density of defects usually revealed in a-Si after ion implantation [5].

We have also made experiments on hydrogenation of a-Si by filling the vacuum chamber with H_2 gas during a-Si deposition. With increasing hydrogen pressure p_H the conductivity and density of paramagnetic defects are decreased (Fig. 1). However, even at the highest values of p_H (10^{-1} Pa) we could not completely passivate paramagnetic defects (Fig. 1) and, hence, suppress their contribution to the conductivity in the region of low temperatures (Fig. 2). The value of photoconductivity was also low: $10^{-6}\ \Omega^{-1}\ cm^{-1}$.

The deposition of a-Si films with simultaneous introduction of phosphorus vapour and hydrogen did not result in a considerable change of conductivity though a certain increase of σ (up to $3 \times 10^{-3}\ \Omega^{-1}\ cm^{-1}$) was observed.

4. Conclusions

Amorphous silicon–phosphorus films have been prepared by vacuum evaporation with different phosphorus contents. EPR, dark and photoconductivity investigations allow us to make the conclusion that the density of localized states and the photoconductivity in a-Si:P films obtained by vacuum evaporation proved to be almost the same as the best similar parameters of a-Si:H grown by rf glow discharge decomposition of silane. The thermal stability of a-Si:P is better than that of a-Si:H. The conductivity of a-Si:P films does not change at annealing up to 500 °C.

References

[1] Amorphous Semiconductors, Ed. M. H. Brodsky, Springer-Verlag, Berlin/Heidelberg/New York 1979.
[2] H. Fritzsche, Proc. 7th Internat. Conf. Amorphous and Liquid Semiconductors, Edinburgh 1977 (p. 3).
[3] D. A. Anderson and W. E. Spear, Phil. Mag. **36**, 695 (1978).
[5] A. V. Dvurechenskii and I. A. Ryazantsev, phys. stat. sol. (a) **69**, K117 (1982).
[6] A. V. Dvurechenskii, I. A. Ryazantsev, and V. A. Dravin, phys. stat. sol. (a) **71**, K133 (1982).
[7] A. V. Dvurechenskii, I. A. Ryazantsev, and L. S. Smirnov, Fiz. Tekh. Poluprov. **16**, 621 (1982).
[8] A. Madan, P. G. Le Comber, and W. E. Spear, J. non-crystall. Solids **20**, 239 (1976).
[9] W. E. Spear, Adv. Phys. **26**, 811 (1977).
[10] D. Kaplan, N. Sol, and G. Velasco, Appl. Phys. Letters **33**, 440 (1978).
[11] A. Madan, S. R. Ovshinsky, and W. Czubatyj, J. electronic Mater. **9**, 385 (1980).

(Received April 14, 1983)

phys. stat. sol. (a) 79, 87 (1983)

Subject classification: 14.1; 20.1; 21; 21.1

Institute of Metal Physics, Academy of Sciences of the USSR, Ural Scientific Centre, Sverdlovsk[1])

Anomalies in the Optical and Electrical Properties of Cr–Al Alloys

By

L. V. NOMEROVANNAYA and V. A. RASSOKHIN

The frequency dependences of the real and imaginary parts of the complex dielectric constant in the spectral range of 0.069 to 5 eV and the temperature dependences of the resistance of the $Cr_{100-x}Al_x$ ($x \leqq 32$ at% Al) alloy system are investigated. In alloys having 16 to 28 at% of Al, a negative temperature coefficient of resistance, and a minimum metallic conductivity $\sigma(0) < < 1000\ \Omega^{-1}\ cm^{-1}$ a dielectrization of the optical properties is found characterized by strong anomalous concentration and frequency dependences of the dielectric function in the infrared frequency range. From the analysis of the behaviour of the optical $\sigma(\omega)$ and the static $\sigma(0)$ conductivities it is concluded that the anomalies of optical and electrical properties result from the localization of free carriers near the Fermi level.

Исследованы частотные зависимочти действительной и мнимой частей комплексной диэлектрической проницаемости в области спектра 0,069 до 5 eV и температурные зависимости электросопротивления сплавов системы $Cr_{100-x}Al_x$ ($x \leqq 32$ at% Al). Для сплавов с 16 до 28 at% Al, имеющих отрицательный температурный коэффициент сопротивления и минимальную металлическую проводимость $\sigma(0) < 1000\ \Omega^{-1}cm^{-1}$, обнаружена диэлектризация оптических свойств, выразившаяся в резко аномальной концентрационной и частотной зависимостях диэлектрических функций в инфракрасном диапазоне частот. Из анализа поведения оптической $\sigma(\omega)$ и статической $\sigma(0)$ проводимостей делается заключение, что аномалии оптических и электрических свойств вызваны локализацией свободный носителей вблизи уровня Ферми.

1. Introduction

The experimental investigation of the electrical properties [1] of antiferromagnetic (AFM) alloys of the $Cr_{100-x}Al_x$ (x denotes at% of Al) system has shown that in the range of concentrations $x = 15$ to 25 at% Al these alloys exhibit a negative temperature coefficient of resistivity, thus indicating a nonmetallic behaviour. In the same composition range the concentration dependence of electrical conductivity and of the low-temperature electronic specific heat coefficient shows a deep minimum, and the alloys have rather large negative Hall coefficient which vary strongly with temperature. No satisfactory explanation of these anomalies exists as yet, although the authors of [1] have suggested that they may be attributed to the appearance of an energy gap between the d-subbands of Cr opening up and closing in accordance with the concentration of admixtures of Al, but which is not related with the AFM energy gap.

If the energy gap is formed in the electron band spectrum $E(\boldsymbol{k})$ at the Fermi level (E_F), its dielectric character may manifest itself in the optical absorption spectrum [2]. It seemed worthwhile to carry out an investigation of the optical properties of alloys

[1]) ul. S. Kovalevskoi 18, 620219 Sverdlovsk GSP-170, USSR.

of this system and to compare them with the temperature dependence of the resistance. The results of such an investigation of alloys of the Cr–Al system in the concentration range $x = 1$ to 32 at% Al are presented below.

2. Experiment and Results

The alloys were prepared by melting in an induction furnace from electrolytic chromium of 99.99% purity and high-purity aluminium (99.999%) in a magnetic suspension followed by quenching on a water-cooled copper bottom.

From the ingots strips of $15 \times 1 \times 1\ mm^3$ were cut out for resistance measurements and plates of $10 \times 10 \times 2\ mm^3$ from which the specimens for optical measurements were selected. The samples for resistance measurements underwent electropolishing in a mixture of perchloric acid and acetic anhydride. Surface for optical measurements were prepared by mechanical polishing.

The electrical resistivity of alloys with $x = 3, 5, 8, 12, 16, 18, 22, 24, 25, 30$, and 32 at% Al was determined by the four-electrode method in the temperature range from 4.2 to 300 K. The temperature was measured by means of germanium and platinum thermometers to within 0.01 K. The relative error of the resistance measurement amounted to 0.005%. The absolute error of resistivity measurements was within 5%, being due mainly to the error in the determination of the geometry of the sample.

The measurements of the refractive index n and of the absorption index k of alloys with $x = 1, 3, 5, 8, 23$, and 32 at% Al were carried out at room temperature in the

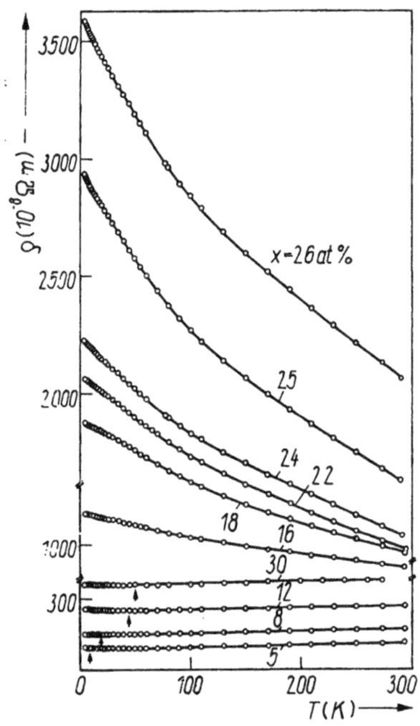

Fig. 1

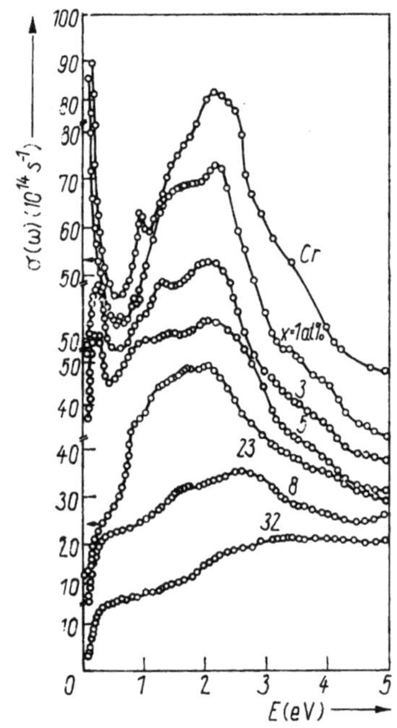

Fig. 2

Fig. 1. Temperature dependence of resistivity of $Cr_{100-x}Al_x$ alloys. Arrows indicate the positions of minima on $\varrho(T)$ curves

Fig. 2. High-frequency optical conductivity $\sigma(\omega)$ of $Cr_{100-x}Al_x$ alloys spectral range $\hbar\omega = 0.069$ to 5.0 eV ($\lambda = 18$ to 0.25 μm) by the polarimetric method with an error of 2 to 4%.

The temperature dependence of the electrical resistivity of the investigated alloys is presented in Fig. 1. The high-frequency conductivity $\sigma(\omega) = nk\omega/2\pi$ (ω is the cyclic frequency of the light wave) is shown in Fig. 2.

3. The Crystal Structure of Samples

The crystal structure of Cr–Al alloys was studied by various methods. Thus, in [3] X-ray and metallographic investigations in conjunction with the available state phase diagram have shown that chromium-based alloys with up to 25.8 at% Al are disordered solid solutions. The same result was obtained with X-ray investigations in [4] for samples subjected to long-time annealing and then to a slow cooling. However, electron diffraction patterns revealed two systems of additional reflections for alloys with 15 to 33 at% Al. One of them pertains to the already known ordered b.c.c. tetragonal β-phase Cr_2Al (26 to 33 at% Al). The other system of diffuse intensity contours is associated with a certain new phase (15 to 29 at% Al) called by the authors of [4] X-phase. For the description of the structure of the X-phase which occupies a substantial volume fraction of the samples and features domains measuring ≈ 3 nm three crystallographically equivalent models have been proposed. Each of these may be used with certain assumptions. Nevertheless, none of them explains all the special features of the observed complex diffraction pattern. The X-phase in [4] is described by a rhombohedral superlattice of Cr_3Al or Cr_5Al_3 type, or by a concentration modulation of the α-phase, or by a periodic antiphase boundary structure of the β-phase. Structure investigations carried out on our quenched samples [5] have shown that the X-phase structure may be described with the help of the known noncommensurate ω-phase occurring at the $\beta \rightarrow \omega$ transition in titanium alloys.

It can be deduced from the above that the crystal structure of Cr–Al alloys is a complex nonequilibrium state.

4. Discussion of Results

4.1 Electrical properties

The anomalies in the behaviour of the transport properties of Cr–Al system alloys distinguish the latter sharply from the metallic solid solutions (Fig. 1). Thus, with varying concentration a transition is observed from typically metallic properties to nonmetallic behaviour. Characteristic is the existence of a certain critical concentration $x = 16$ at which the temperature coefficient of resistivity changes sign from positive to negative. Furthermore, unusual for solid solutions is a strongly pronounced concentration dependence of residual resistivity ϱ_0 (Fig. 3). At maximum, ϱ_0 attains the large value of $\approx 3500 \times 10^{-8}\ \Omega$ m (alloy with $x = 26$), corresponding to a conductivity of 285 Ω^{-1} cm^{-1}. It can be seen that for the alloy with $x = 26$ at 4.2 K the value of the static conductivity, termed by Mott as a minimum metallic conductivity σ_{min} and according to his estimates equal to 300 to 1000 Ω^{-1} cm^{-1} [6, 7], has been attained. In Fig. 3 the horizontal is the upper borderline of this estimate (1000 Ω^{-1} cm^{-1}). Alloys with $x = 16$ to 28 have a conductivity less than the upper borderline.

For alloys with $2 < x < 16$, featuring a metallic type of concentration, a low-temperature minimum in the $\varrho(T)$ dependence is evident (Fig. 1). It is notable that the concentration $x = 2$ is outstanding as regards magnetic properties also. The investigation of the temperature dependence of magnetic susceptibility [8] has shown that in the concentration range $x < 2$ the alloys behave, similarly to pure chromium, as band

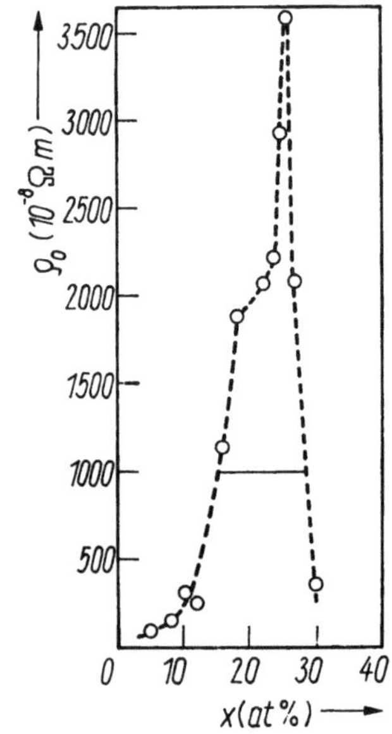

Fig. 3. Concentration dependence of residual resistivity of $Cr_{100-x}Al_x$ alloys. The horizontal line is the estimate of the upper limit of σ_{min} according to Mott and Davis [6]

antiferromagnets. It seems that only alloys with $x \leqq 2$ at% Al are chromium-base solid solutions. The temperature dependences of magnetic susceptibility and resistivity for alloys with $x = 2$ to 4 are more complex, reflecting the coexistence of noncompensated (as in the case of pure chromium) and compensated AFM structures, which have been confirmed by neutron diffraction analysis data [9]. In the case of alloys with $4 < x < 24$ the Néel temperature T_N, the atomic magnetic moment, and lattice parameter increase rapidly with x, this being attributed at present to defects and to the destruction of Cr–Cr interaction as Al is added, the latter reducing the degree of mobility of d-electrons. The low-temperature minimum in $\varrho(T)$ for alloys with $x = 2$ to 15 (Fig. 1) is possibly associated with the complex nonequilibrium crystal structure. For alloys with $x = 30$ to 32 the temperature coefficient of resistance becomes positive again, while the values of ϱ_0 drop (Fig. 1 and 3). Such a behaviour may be attributed to the formation of the metallic β-phase Cr_2Al.

4.2 Optical properties

To analyze the optical properties it is convenient to divide the alloys into two groups: one with a low content of aluminium ($x = 1, 3, 5$), the second with $x = 8, 23, 32$. Let us begin with examining the alloys of the first group. Similarly to pure chromium [2], the alloy with $x = 1$ exhibits a maximum of the main absorption band at 2.2 eV, but the band is narrower (Fig. 2). The fine structures in the curve $\sigma(\omega)$ in the infrared region are at 0.41, 0.56, and 0.83 eV (in pure chromium the energy of optical structures are 0.40, 0.50, and 0.65 eV). The optical conductivity $\sigma(\omega)$ of alloys with $x = 3$ and 5 manifests qualitative and quantitative changes: a drop in intensity of the main absorption band, its further splitting into two subbands (in the alloy with $x = 3$ into 2.0 and 1.2 eV, in the alloy with $x = 5$ into 2.0 and 1.3 eV), and a shift to the infrared region. The shift of the main absorption band towards lower energies, as calculations show [10], is caused by the increase in the scattering of electrons in alloys, infringement of selection rules, and growth of the percentage of indirect transitions. With the addition of aluminium to chromium the main absorption band blurs also. A sharp distinction between $\sigma(\omega)$ curves for pure Cr and the alloy with $x = 1$ and those for alloys with $x = 3$ and $x = 5$ is introduced by the appearance of a new absorption band in the infrared region of the spectrum with a maximum at 0.16 eV.

Recent theoretical work [11, 12] has shown that the addition of a small admixture to the matrix leads to the "induction" of indirect transitions between critical points of the main band structure. This gives new bands in the interband absorption spectrum. It is possible that in alloys with a small admixture of aluminium ($x = 3, 5$) transitions at low energies are induced in the band spectrum via gaps which in pure chromium are forbidden by the selection rules.

The analysis of infrared absorption shows that the optical properties of Cr and alloys with a small content of aluminium in the range $\hbar\omega < 1$ eV are determined by the mechanism of intraband acceleration of electrons and interband transitions: $\sigma(\omega)_{\text{exp}} = \sigma_{\text{intra}}(\omega) + \sigma_{\text{inter}}(\omega)$. The change in the optical absorption spectrum in alloys with $x = 1, 3, 5$ proceeds also as a result of increasing scattering of electrons on admixtures. The total collision frequency γ of electrons determined from the dispersion of the real $\varepsilon_1 = n^2 - k^2$ and imaginary $\varepsilon_2 = 2nk$ parts of the complex dielectric constant has increased in the alloy with $x = 5$ by more than four times as compared with pure chromium. Note that the resistance at $T = 293$ K has increased from 13.3×10^{-8} Ω m for Cr to 112×10^{-8} Ω m for the alloy with $x = 5$.

The theoretical behaviour of the intraband optical conductivity $\sigma_{\text{intra}}(\omega)$ in a binary disordered substitutional alloy was examined in the coherent-potential approximation by [13]. The authors calculated numerically the frequency dependence $\sigma_{\text{intra}}(\omega)$ for various parameters describing the alloy and for different positions of E_F in the band. The anomalous behaviour of $\sigma_{\text{intra}}(\omega)$ (deviation from a Drude-like dependence) was observed as the scattering of electrons in the alloy increased and E_F reached the minimum density of states with changing aluminium content. The maximum deviation occurred with alloys having minimum static conductivity.

Thus, in quenched alloys with small admixtures of Al in Cr ($x = 3, 5$) the appearance of an intensive absorption band at 0.16 eV may be due to two competing mechanisms which lead to the reduction of $\sigma(\omega)$ as the frequency drops. These are interband transition and deviation of $\sigma_{\text{intra}}(\omega)$ from the Drude dependence as the electron scattering on admixtures in the alloy grows. A similar behaviour of $\sigma(\omega)$ in Ti–V alloys was observed by [14]. However, it should be noted that in Cr–Al alloys with $x = 3$ and $x = 5$ the temperature coefficient of resistance remains positive.

With aluminium contents of 8, 23, and 32 at% a typical feature in the frequency dependence $\sigma(\omega)$ is a growing deviation from the Drude dependence in the infrared part of the spectrum, a further reduction in the intensity of the band, and its lack of structure (Fig. 2). An anomalously sharp drop in $\sigma(\omega)$ with lowering frequency is observed in all alloys.

The calculations of density of states $N(E)$ of alloys in the coherent-potential approximation show that the increase of admixture concentration and, consequently, scattering on admixtures, leads to a substantial distortion of $N(E)$ of the alloy especially near E_F. With the formation of the alloy the d-band undergoes the most significant change. The stronger electron–impurity scattering in disordered alloys as compared to electron–phonon scattering makes the description of the intraband conduction by a Drude-like dependence useless.

Note the similarity between the optical conduction curves $\sigma(\omega)$ for alloys with $x = 8$ and 32, having a positive temperature resistance coefficient, and the deviation from $\sigma(\omega)$ for the alloy with $x = 23$ having a negative temperature resistance coefficient. In the case of the alloy with $x = 23$ the curve $\sigma(\omega)$ is higher, while the drop with diminishing frequency is sharper.

It is seen from the plot of the real part ε_1 of the dielectric constant versus concentration at low frequencies $\hbar\omega = 0.177$ and 0.069 eV (Fig. 4) that for alloys with $x = 3$ and 5 $\varepsilon_1 < 0$ and has a large absolute value in the entire infrared region, characterizing the alloys as metals. With further increase of Al concentration $\varepsilon_1(\omega)$ rises sharply and at $x = 23$ becomes positive. The change in sign of ε_1 at low energies from negative to positive for $x = 23$ coincides with the change in the character of the dispersion (slope)

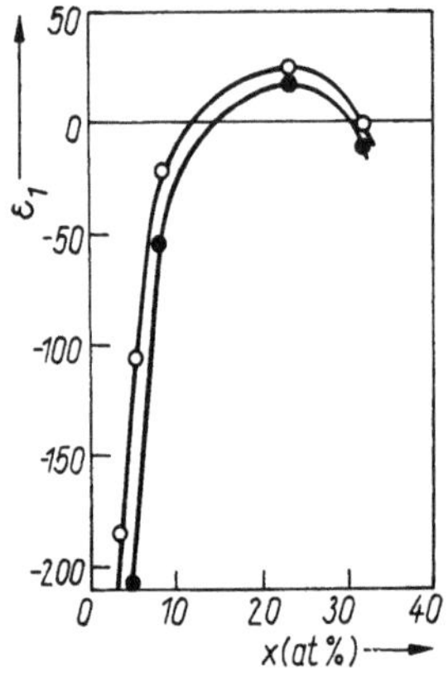

Fig. 4

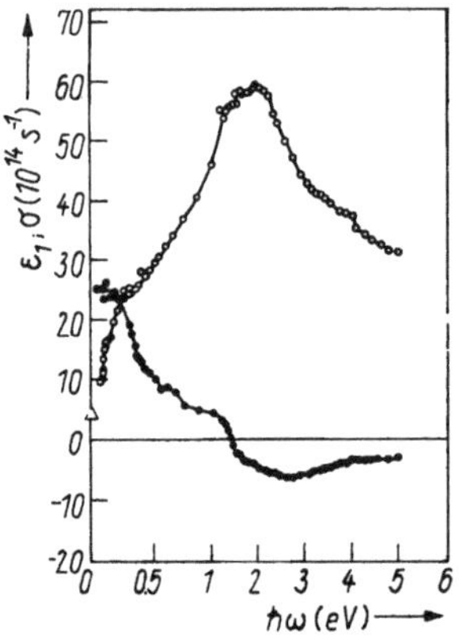

Fig. 5

Fig. 4. Concentration dependence of the real part of the complex dielectric constant ε_1 for $\hbar\omega = 0.177$ (○) and 0.069 eV (●)

Fig. 5. Variance $\varepsilon_1(\omega)$ (●) and $\sigma(\omega)$ (○) of the alloy Cr–23 at% Al. The value of static conductivity $\sigma(0)$ is indicated by ▲

of $\sigma(\omega)$ as compared with that for alloys with $x = 8$ and $x = 32$. It is evident that in the Cr–Al system the real part ε_1 of the complex dielectric constant experiences an anomely in the infrared region of the spectrum as the Al concentration varies.

Let us examine the frequency dependences of $\varepsilon_1(\omega)$ and $\sigma(\omega)$ of the alloy with $x = 23$ (Fig. 5). The positive values and the high absolute magnitude of ε_1 in the energy range $\hbar\omega < 1$ eV are unusual for metallic alloys. The positive values of ε_1 enable one to assert that the contribution to the optical conductivity $\sigma(\omega)$ by intraband transitions of conduction electrons $\sigma_{intra}(\omega)$ disappears completely and the optical properties are determined by quantum interband transitions $\sigma_{inter}(\omega)$ only. Note that throughout the entire frequency range investigated the optical conduction $\sigma(\omega)$ is higher than the static $\sigma(0)$. In Fig. 5 $\sigma(0)$ is indicated by a triangle on the $\sigma(\omega)$ axis. Apparently, electrons participating in the transitions from states below the Fermi level E_F to states above E_F are more mobile than electrons at E_F. A similar behaviour of optical conductivity was found experimentally for liquid tellurium, glasses [6], and amorphous Mg–Bi alloys [15].

As the Al content in the Cr–Al system grows, the Fermi level apparently crosses the mobility edge defined as the energy value separating the localized states from the nonlocalized ones. The negative temperature coefficient of resistance, the minimum metallic conductivity $\sigma_{min}(0) < 1000\ \Omega^{-1}\ cm^{-1}$, and the positive values of ε_1 at low energies for alloys with 16 to 28 at% of Al lead us to the conclusion that E_F for these alloys is found to be in the region of localized electron states.

Thus, in a nonequilibrium Cr–Al alloy system with 16 to 28 at% of Al a dielectric character of optical properties has been revealed: sharply anomalous concentration and frequency dependences of the real and imaginary parts of the complex dielectric constant in the infrared region. The analysis of the behaviour of the optical $\sigma(\omega)$ and static $\sigma(0)$ conductivities allows one to assume that the anomalies of optical and electrical properties are caused by the localization of free carriers near E_F as a result of a reduction in the degree of mobility.

Acknowledgements

We wish to express our gratitude to B. A. Bolotin, M. M. Kirillova, and A. F. Prekul for helpful discussion of results.

References

[1] D. J. Chakrabarty and P. A. Beck, J. Phys. Chem. Solids **32**, 1609 (1971).
[2] M. M. Kirillova and L. V. Nomerovannaya, II. Internat. Symp. Electron Structure of Transition Metals, Their Alloys and Compounds, Izd. Naukova Dumka Kiev 1979 (p. 57 to 61) (in Russian).
[3] W. Köster, E. Wachtel, and K. Grube, Z. Metallk. **54**, 393 (1963).
[4] G. van Tendeloo, F. J. A. den Broeder, A. Amelinckx, R. de Ridder, J. van Landuyt, and H. J. van Daal, phys. stat. sol. (a) **67**, 217 (1981).
[5] S. V. Sudareva, V. A. Rassokhin, and A. F. Prekul, phys. stat. sol. (a) **76**, 101 (1983).
[6] N. Mott and E. Davis, Electron Processes in Non-Crystalline Substances Vol. 1, Izd. Mir, Moscow 1982 (Russian translation).
[7] A. L. Efros, Uspekhi fiz. Nauk **126**, 41 (1978).
[8] J. Pop, D. Dadariat, T. Petrison, and A. Ginrgin, J. Phys. Chem. Solids **42**, 927 (1981).
[9] A. Kallel and F. de Bergevin, Solid State Commun. **5**, 955 (1967).
[10] I. N. Anohina, V. E. Egorushkin, S. E. Kulkova, and V. P. Fadin, Fiz. tverd. Tela **23**, 2971 (1981).
[11] B. Velicky and K. Levin, Phys. Rev. B **2**, 938 (1970).
[12] J. C. Parlebas and D. L. Mills, Phys. Rev. B **18**, 3988 (1978).
[13] K. J. Wysokinsky, E. Taranko, and R. Taranko, J. Phys. **13**, 6659 (1980).
[14] I. I. Sasovskaya, S. V. Sudareva, and S. V. Yartsev, Fiz. Metallov i Metallovedenie **52**, 86 (1981).
[15] K. M. Sutton, Solid State Commun. **16**, 327 (1975).

(Received April 18, 1983)

phys. stat. sol. (a) 79, 95 (1983)

Subject classification: 20.3; 12.2; 22.5.2

Department of Physics, Government Science College, Raipur[1])

Pressure and Temperature Dependence of the Mechanoluminescence of γ-Irradiated Alkali Halide Crystals

By

B. P. Chandra, M. Elyas, and B. Majumdar

The intensity of mechanoluminescence (ML) produced during application and release of uniaxial pressure, in γ-irradiated KBr, KCl, KI, LiF, and NaCl crystals decreases with the number of applications and releases of the pressure. The ML intensity of γ-irradiated KBr, KCl, LiF, and NaCl crystals increases with the temperature, however, the ML intensity of γ-irradiated KI crystals decreases with the temperature. The ratio of the ML intensity during the relase of the pressure to the ML intensity during the corresponding number of applications of the pressure, decreases with the temperature of the crystals. Assuming that the movement of dislocations is responsible for the ML excitation, the pressure and temperature dependence of the ML is discussed. The activation energy related to the increase in the number of dislocations with the temperature is estimated from the ML measurement and it is found to be 0.24, 0.25, 0.24, and 0.32 eV for γ-irradiated KBr, KCl, LiF, and NaCl crystals respectively.

Die Intensität der Mechanolumineszenz (ML), die während des Anlegens eines uniaxialen Drucks und während der Entspannung in γ-bestrahltem KBr-, KCl-, KJ-, LiF-, und NaCl-Kristallen nimmt mit der Zahl der Druck-Zyklen zu. Die ML-Intensität der γ-bestrahlten KBr-, KCl-, LiF- und NaCl-Kritalle nimmt mit der Temperatur zu, jedoch die ML-Intensität der γ-bestrahlten KJ-Kristalle mit der Temperatur ab. Das Verhältnis der ML-Intensität während der Druckentspannung zur ML-Intensität der entsprechenden Zahl von Druckanlegungen nimmt mit der Temperatur des Kristalls ab. Unter der Annahme, daß die Versetzungswanderung für die ML-Anregung verantwortlich ist, wird die Druck- und Temperatur-Abhängigkeit der ML diskutiert. Die mit dem Anwachsen der Zahl der Versetzungen mit der Temperatur verknüpfte Aktivierungsenergie wird aus den ML-Messungen zu 0,24; 0,25; 0,24 und 0,32 eV für γ-bestrahlte KBr-, KCl-, LiF- und NaCl-Kristalle bestimmt.

1. Introduction

The luminescence produced during mechanical deformation of solids is known as mechanoluminescence (ML) or triboluminescence. It is known that the γ -or X-irradiated alkali halide crystals exhibit ML in their elastic, plastic, and fracture regions [1 to 5]. However, the further details of the ML in the coloured alkali halide crystals have not been investigated. We have reported earlier that the X- or γ-irradiated alkali halide crystals exhibit ML during the application as well as release of a uniaxial pressure [6, 7]. The present paper reports the pressure and temperature dependence of ML of γ-irradiated KBr, KCl, KI, LiF, and NaCl crystals.

2. Experimental Procedure

The single crystals of KBr, KCl, KI, and NaCl used in the present investigation were supplied by the National Physical Laboratory, New Delhi. The crystals of LiF-TLD 100 were supplied by the Harshaw Chemical Company. The size of the specimens used for the ML measurements was $4.18 \times 4 \times 2.85$, $4 \times 3.80 \times 2.50$, $4.20 \times 3.9 \times 2.5$,

[1]) 492002 Raipur, India.

2.90 × 2.8 × 2.3, and 4.5 × 4.2 × 2.9 mm^3 for KBr, KCl, KI, LiF, and NaCl, respectively. The small-size crystals were annealed at 450 °C for 2 h and cooled very slowly. The γ-irradiation was carried out for 2 h from a ^{60}Co source. The irradiated specimen wrapped in aluminium foil was kept in dark for about an hour to allow the after-glow to decay to a value well below that expected in the ML measurements. Care was taken to avoid the exposure of the irradiated crystals to stray light during the ML measurements.

A uniaxial pressure was applied to the crystal by placing heavy loads statically using the device described previously [8]. The crystal was kept pressed for 30 s and then the pressure was withdrawn by releasing the lever. The ML intensities both at the time of pressing and release were measured in terms of the deflection of a ballistic galvanometer which was connected to the amplifier coupled to an IP 28 photomultiplier tube. The cycles of pressing and release were repeated for a number of times, till the intensity of ML becomes small. All the measurements were made by applying and releasing the pressure along (100) direction of the crystals.

For the measurements of the ML at different temperatures, a ceramic cylinder was placed on the glass platform of the ML measurement device described in an earlier investigation [8]. A heater coil was wound round the cylinder for heating the crystal, and by changing the voltage, the crystals could be heated to any desired temperature. The ML measurement was carried out when the device has attained a steady state temperature. The temperature of the crystals was measured by a copper–constantan thermocouple. The effect of temperature on the ML of crystals was studied by applying and releasing a load of 12.5 kp. A piston of 5 cm length attached with a plane mirror at the bottom, was connected to the bottom of the hanger of the device described previously [8]. Such arrangement was needed to deform the crystal placed at the bottom of a cylinder of 4 cm length. To avoid the heating of the photomultiplier tube, a thick rubber sheet with a hole at its centre was placed between the glass plate and the photomultiplier housing. Four crystals were studied at each temperature. The standard error in the ML measurement was found to be ±6%.

3. Results

Fig. 1 shows that the intensity of ML produced during application and release of the pressure, that is, I_n^p and I_n^r of γ-irradiated KBr crystals decreases with the successive number n_p and n_r of applications and releases of the pressure and follows the relations

$$I_n^p = I_1 \exp\left[-\beta(n_p - 1)\right], \tag{1}$$

$$I_n^r = I_1^r \exp\left[-\beta_1(n_r - 1)\right], \tag{2}$$

where β and β_1 are constants and I_1^p and I_1^r are the ML intensity during the first application and first release of the pressure, respectively. The values of I_1^p and I_1^r increase with the applied pressure, however, the values of β and β_1 decrease with the applied pressure. For a given value of the applied pressure β is always greater than β_1.

Fig. 2 shows that the dependence of the ML intensity (during application and release of the pressure) on the successive number of applications and releases of the pressure at different temperatures also follows the above equations (1) and (2).

The values of β and β_1 increase with temperature. However, for a given temperature, the value of β_1 is always less than the value of β.

The effects of pressure and temperature on the ML intensity of γ-irradiated KCl, LiF, and NaCl crystals were similar to that of γ-irradiated KBr crystals. Fig. 3 shows that the dependence of the ML intensity during the application and release of the pres-

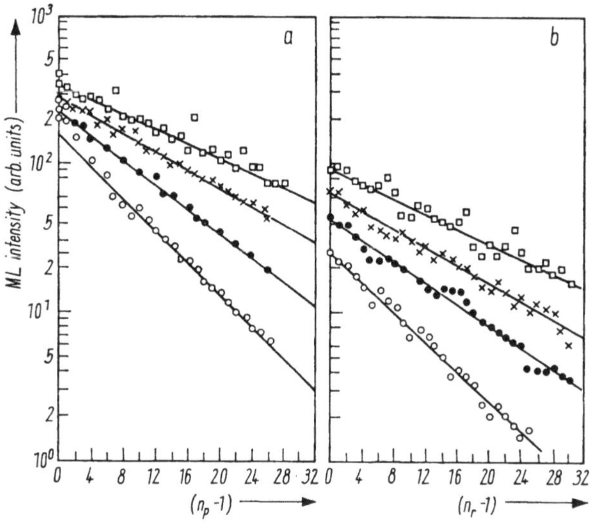

Fig. 1. a) Plot of the ML intensity during pressing vs. $(n_p - 1)$ in γ-irradiated KBr crystals for different values of the applied stress: ○ 4.5, ● 8.5, × 12.5, □ 16.5 kg b) Plot of the ML intensity during release vs. $(n_r - 1)$ in γ-irradiated KBr crystals for different values of pressure: ○ 4.5, ● 6.5, × 12.5, □ 16.5 kp

sure in the γ-irradiated KI crystals on the number of applications and releases of pressure follows (1) and (2). However, I_1^p and I_1^r decrease with the temperature of the crystals. The ratio of the ML intensity during the release of the pressure to the ML intensity during the corresponding number of applications of the pressure, decreases with increasing temperature of the crystals.

Fig. 4 reveals that the total ML intensity, that is, the sum of the areas below the I_n^p versus n_p and I_n^r versus n_r curves of γ-irradiated KBr, KCl, LiF, and NaCl crystals

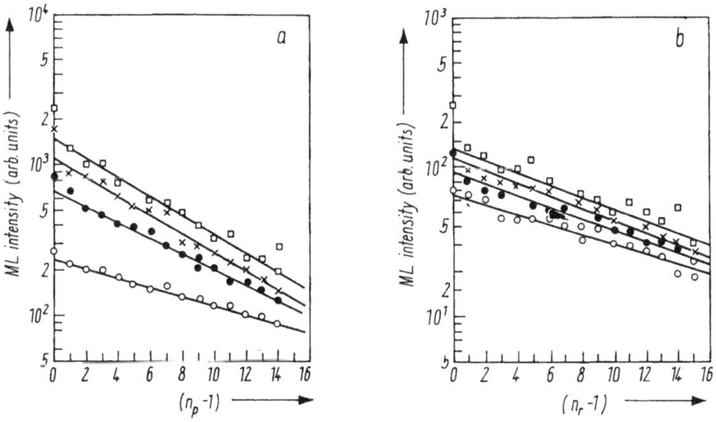

Fig. 2. a) Plot of the ML intensity during pressing vs. $(n_p - 1)$ in γ-irradiated KBr crystal for different values of temperature: ○ 40, ● 60, × 80, □ 100 °C b) Plot of the ML intensity during release vs. $(n_r - 1)$ in γ-irradiated KBr crystals for different values of temperature: ○ 40, ● 60, × 80, □ 100 °C

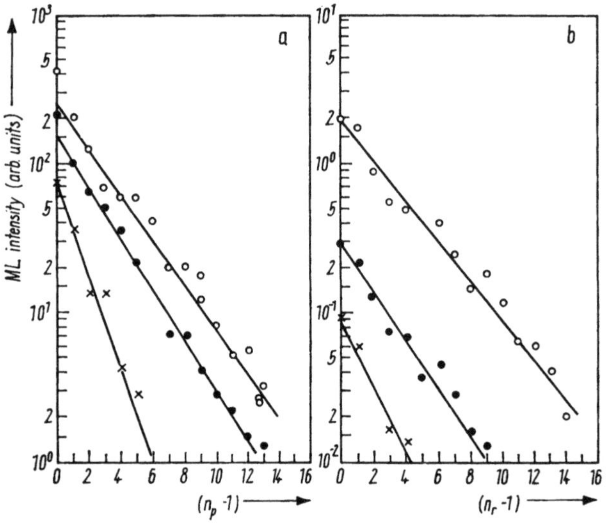

Fig. 3. a) Plot of the ML intensity during pressing vs. $(n_p - 1)$ in γ-irradiated KI crystal for different values of temperature: ○ 37, ● 60, × 80 °C. b) Plot of the ML intensity during release vs. $(n_r - 1)$ in γ-irradiated KI crystals for different values of temperature: ○ 37, ● 60, × 80 °C

increases with temperature. However, the total intensity of the ML of γ-irradiated KI crystals decreases with temperature.

For the study of temperature effects on the ML of the crystals, the crystals were placed on the glass plate above the photomultiplier tube which was in a steady state. The time of annealing at this temperature has no considerable effect on the ML intensity of γ-irradiated KBr, KCl, NaCl, and LiF crystals. The ML intensity of γ-irradiated KI crystals decreases with the time of annealing.

The ML intensity is directly related to the density of the colour centres [2, 3], and the density of the colour centres is directly related to the area below the thermoluminescence glow curves of the crystals [9, 10]. Hence, the ML intensity can be normalized for the decrease in the density of the colour centres with the temperature of the crystals using the thermoluminescence glow curves. For normalizing the ML intensity, the total ML intensity I_T at a given temperature T was multiplied by $A_T/(A_T - \int_{T_0}^{T} I_{TL}\, dT)$

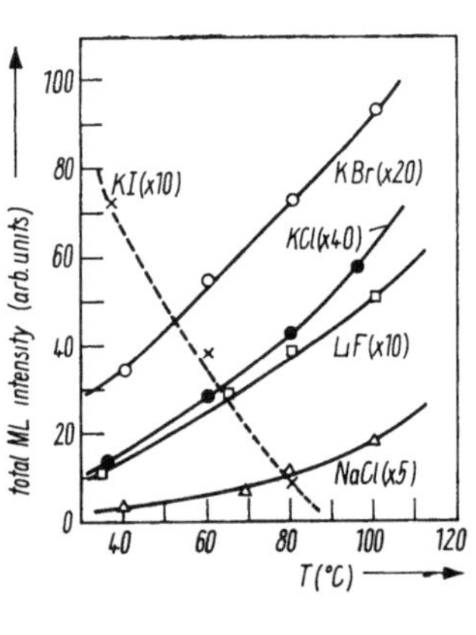

Fig. 4. Effect of temperature on the total ML intensity I_T of γ-irradiated KBr, KCl, KI, LiF, and NaCl crystals

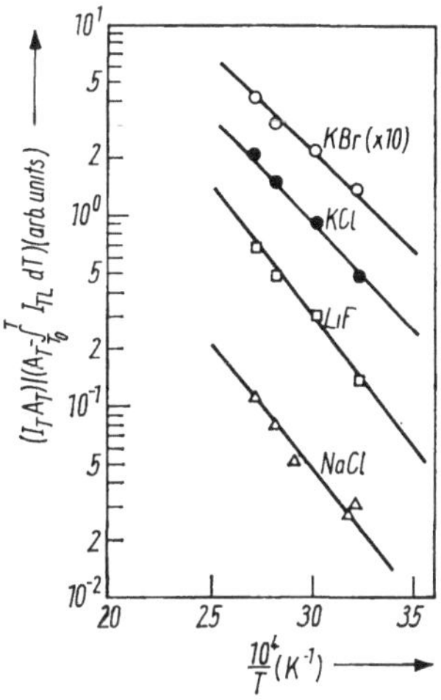

Fig. 5. Plot of $\lg I_T A_T/(A_T - \int_{T_0}^{T} I_{TL}\, dT)$ vs. $1/T$

where I_{TL} is the TL intensity, A_T the total area below the glow curve, and T_0 the room temperature. The value of $\int_{T_0}^{T} I_{TL}\, dT$ was estimated from the area below the glow curve, from room temperature to the temperature T, because the TL measurements were started from room temperature. The TL glow curves of these crystals have been reported in our earlier investigation [11]. Fig. 5 illustrates that the plot of $\lg I_T \times A_T/(A_T - \int_{T_0}^{T} I_{TL}\, dT)$ versus $1/T$ is a straight line with a negative slope for γ-irraidated KBr, KCl, LiF, and NaCl crystals. This result suggests the relation

$$\frac{I_T A_T}{A_T - \int_{T_0}^{T} I_{TL}\, dT} = A_0 \exp\left(-\frac{E_0}{kT}\right), \tag{3}$$

where A_0 is a constant, k the Boltzmann constant, and E_0 the activation energy. The value of E_0 estimated from Fig. 5 is found to be 0.24, 0.25, 0.27, and 0.32 eV for γ-irradiated KBr, KCl, LiF, and NaCl crystals, respectively. The annealing time strongly decreases the density of colour centres in KI crystals. The decrease in the ML intensity of γ-irradiated KI crystals with temperature may chiefly be due to the decrease in the density of colour centres with annealing time. Since the annealing time decreases the density of colour centres in γ-irradiated KI crystals, it is not possible from the ML measurements to determine the value of the activation energy E_0.

It is seen from Fig. 1 and 3 that the difference between the extrapolated and the experimental values of I_1^p decreases with increasing temperature of the crystals. This fact suggests that the higher experimental values of I_1^p may be due to the presence of shallow traps, which disappear during the initial application of the pressure.

4. Discussion

The ML appears during the elastic and plastic deformation of γ-irradiated alkali halide crystals [1 to 5]. This fact suggests that the mobile dislocations are responsible for the ML excitation. It will be interesting to discuss the dependence of ML intensity on the successive number of applications and release of a uniaxial pressure in γ-irradiated alkali halide crystals, on the basis of the movement of the newly created dislocations.

Let $N_1, N_2, N_3, \ldots, N_n$ be the numbers of dislocations created during 1st, 2nd, 3rd, and n-th application of the pressure. The total number N_T of newly created dislocations up to the n-th pressing may be given by

$$N_{T,n} = N_1 + N_2 + N_3 + \ldots + N_n\,. \tag{4}$$

Assuming that the number of newly created dislocations decreases exponentially with increasing number of applications of the pressure, the above equation may be written as

$$N_{T,n} = N_1 + N_1\,e^{-\alpha} + N_1\,e^{-2\alpha} + \dots + N_1\,e^{-(n-1)\alpha}$$

or

$$N_{T,n} = \frac{N_1(1 - e^{-n\alpha})}{(1 - e^{-\alpha})}$$

or

$$N_{T,n} = N_0(1 - e^{-n\alpha})\,, \tag{5}$$

where

$$N_0 = \frac{N_1}{(1 - e^{-\alpha})}$$

and α is a constant.

The number N_n^p of the newly created dislocations during the n-th pressing may be given by

$$N_n^p = N_1^p\,e^{-(n_p-1)\alpha}\,, \tag{6}$$

where N_1^p is the number of newly created dislocations during the first pressing.

An independent investigation [12] suggests that the plot of N_n^p versus $\lg(n_p - 1)$ is a straight line with a negative slope, which supports equation (6).

When pressure is applied to a crystal, the dislocation sources become operative and the mobile dislocations are produced. Some of the mobile dislocations disappear due to the annihilation. The other mobile dislocations relax and become stationary. Thus, the density of dislocations increases in the crystal with the application of pressure. The ML intensity measured in terms of the deflection of the ballistic galvanometer is linearly related to the number N of the newly created dislocations. This may be expressed as

$$\int I\,dt = CN$$

or

$$\int \frac{dn}{dt}\,dt = n_l = CN\,, \tag{7}$$

where n_l is the number of excited luminescence centres and C a constant. The linear relation between the ML intensity (measured by an XY recorder) and the strain rate [1] suggests that the rate of excitation of the luminescence centres is directly related to the rate of creation of the dislocation. This is in accord with (7).

If the dislocation annihilation model is the major process for the ML excitation in X- and γ-irradiated alkali halide crystals, then (7) indicates that the rate of annihilation (which is responsible for dn/dt) should be directly proportional to the rate of creation of the mobile dislocation. Such a result may be expected because the number of dislocations responsible for the annihilations may be a fraction of the number of mobile dislocations. Moreover, the number of ML centres excited by the unpinning of dislocations, stripping of interstitial halide ions, and by the interaction of dislocations may also be directly related to the number of newly created dislocations. Thus, the intensity I_n^p of ML (monitored by the deflection of the ballistic galvanometer) produced during n-th pressing may be expressed as

$$I_n^p = \eta N_1\,e^{-(n_p-1)\alpha}\,\varrho\,, \tag{8}$$

where η is a factor related to the ML efficiency of the crystal and ϱ the density of the colour centres.

The density ϱ of colour centres decreases with the mechanical deformation of crystals [2, 5]. For the simplification of the problem, let us assume that the density of colour centres decreases exponentially with the number of pressings, that is, it holds the relation

$$\varrho_n^{\mathrm{p}} = \varrho_0 \exp(-\alpha_1 n_{\mathrm{p}}) , \tag{9}$$

where ϱ_0 and ϱ_n^{p} are the density of the colour centres without pressings and after the n-th pressing of the crystal, respectively, and α_1 is a constant.

From (8) and (9), I_n^{p} may be written as

$$I_n^{\mathrm{p}} = \eta N_1 \exp[-\alpha(n_{\mathrm{p}} - 1)]\, \varrho_0 \exp[-\alpha_1(n_{\mathrm{p}} - 1)]$$

or

$$I_n^{\mathrm{p}} = \eta N_1 \varrho_0 \exp[-(\alpha + \alpha_1)(n_{\mathrm{p}} - 1)]$$

or

$$I_n^{\mathrm{p}} = I_1^{\mathrm{p}} \exp[-\beta(n_{\mathrm{p}} - 1)] , \tag{10}$$

where $I_1^{\mathrm{p}} = \eta N_1 \varrho_0$ is the ML intensity in the first pressing.

It is observed that the plot of lg I_n^{p} versus $(n_{\mathrm{p}} - 1)$ is a straight line with a negative slope, which supports (10). The value of β estimated from the plot of lg I_n^{p} versus $(n_{\mathrm{p}} - 1)$ is nearly equal to the value of the slope α estimated from the plot of lg N_n^{p} versus $(n_{\mathrm{p}} - 1)$ (for the same stress) [12]. This fact indicates that α_1 is much less than α. It seems that the decrease in the ML intensity with the number of pressings is primarily due to the decreased creation of new dislocations with the successive number of pressings. The decrease in the density of colour centres with the number of pressings is less responsible for the decrease in the ML intensity with the number of pressings.

The increase in the ML intensity with increasing values of applied pressure and with irradiation time can be understood from (10). I_1^{p} is directly related to N_1^{p}, the number of newly created dislocations in the first pressing. It will be higher for higher values of the applied pressure [13, 14]. Thus, the ML intensity will be higher for the higher values of the applied pressure. I_1^{p} is also related directly to the density of colour centres in the crystals. As such, the ML intensity may increase with the increase in the irradiation time of the crystals. It will get saturated for a longer time of irradiation, because the density of colour centres will attain a saturation value.

If it is assumed that the ML produced during the release of the pressure is related to the ML produced during the corresponding number of applications of the pressure, then, from (10), the ML intensity during the n-th release of the pressure may be given by

$$I_n^{\mathrm{r}} = I_1^{\mathrm{r}} \exp[-\beta_1(n_{\mathrm{r}} - 1)] , \tag{11}$$

where β_1 is a constant. For a given pressure, β is always greater than β_1. The ML emission during the release of the pressure is related to the number of the new traps produced during the deformation of crystals. The number of shallow traps may increase with the deformation of the crystals, which may increase the probability of the ML emission during the release of the pressure. Thus, β may be higher as compared to β_1.

It was found that the value of β decreases with increasing values of the pressure. This may be due to the plastic deformation in different stress–strain regions of the crystals, where the plastic behaviour may be different.

Since the probability for exciting an electron from the colour centre to the conduction band depends on the thermal spike rather than on the average temperature of the crystal, the less variation of the ML intensity with temperature is expected. It has been described earlier that the ML intensity of γ-irradiated KBr, KCl, LiF, and NaCl crystals increases with their temperature, and follows the relation

$$\frac{I_{\mathrm{T}} A_{\mathrm{TL}}}{A_{\mathrm{TL}} - \int_{T_0}^{T} I_{\mathrm{TL}}\, \mathrm{d}T} = A_0 \exp\left(-\frac{E_0}{kT}\right). \tag{12}$$

The values of E_0 estimated from the ML measurements are found to be 0.24, 0.25, 0.27, and 0.32 eV for γ-irradiated KBr, KCl, LiF, and NaCl crystals, respectively.

It is known that the number of newly created dislocations for a given value of the applied stress increases with the temperature of the crystal [13, 14]. It seems that the increase in the ML intensity with temperature (after normalization for the decrease in the density of the colour centres with temperature) should be due to the increase in the number of the newly created dislocations. Thus, the factor E_0 should be related to the activation energy for the increase in the number of dislocations with the temperature of the crystal.

It has already been described that the ML intensity of γ-irradiated KBr, KCl, LiF, and NaCl crystals increases; however, the ML intensity of γ-irradiated KI crystals decreases with temperature. The density of colour centres in γ-irradiated KI crystals decreases strongly with annealing time. The annealing time does not affect considerably the density of colour centres in γ-irradiated KBr, KCl, LiF, and NaCl crystals. Thus, the decrease in the ML intensity of γ-irradiated KI crystals with temperature may chiefly be due to the decrease in the density of the colour centres with annealing time. Since the TL glow curves were recorded during the fast heating of the crystals, the normalization for the decrease in the density of colour centres using the TL glow curves may not be correct for the γ-irradiated KI crystals.

The ratio of I_n^r/I_n^p decreases with increasing temperature of the crystals. The number of retrapped electrons may decrease with increasing temperature of the crystals. Since I_n^r is chiefly related to the number of retrapped electrons, its decraese with temperature is expected. A decrease in the ratio I_n^r/I_n^p is also expected, if the temperature affects differently the number of dislocations moved during the release of the pressure and the number of dislocations moved during the application of the pressure.

The factor β is related to the rate constant of the decrease in the number of newly created dislocations with the number of application of pressure. The increase in the values of β and β_1 with increasing temperature may be due to the change in the stress–strain behaviour with the temperature of the crystals.

Acknowledgements

The authors wish to thank Dr. P. R. Khandekar and Prof. P. V. Ratnam for their constant encouragement throughout the work. One of the authors (B. M.) is also thankful to the University Grants Commission, New Delhi, for the award of a Junior Research Fellowship.

References

[1] G. ALZETTA, I. CHUDÁČEK, and R. SCARMOZZINO, phys. stat. sol. (a) **1**, 775 (1970).
[2] C. G. BUTLER, Phys. Rev. **141**, 150 (1966).
[3] F. I. METZ, R. N. SCHWEIGER, H. R. LEIDER, and L. A. GIRIFALCO, J. phys. Chem. **61**, 86 (1957).
[4] A. S. KRUGLOV, I. A. EL-SHANSHOURY, and M. K. MATTA, J. Phys. Soc. Japan **21**, 2147 (1966).
[5] F. D. SENCHUKOV and S. Z. SHMURAK, Fiz. tverd. Tela **12**, 9 (1970).
[6] B. P. CHANDRA and M. ELYAS, Kristall und Technik **13**, 1341 (1978).
[7] M. ELYAS, B. P. CHANDRA, and S. P. KATHURIA, phys. stat. sol. (a) **66**, K21 (1981).
[8] B. P. CHANDRA and M. ELYAS, Indian J. pure appl. Phys. **15**, 744 (1977).
[9] V. AUSIN and J. L. ALVAREZ RIVAS, Phys. Rev. B **6**, 4828 (1972).
[10] S. C. JAIN and P. C. MEHENDRU, Phys. Rev. A **140**, 957 (1965).
[11] M. ELYAS, B. P. CHANDRA, and S. P. KATHURIA, Radiat. Eff. **33**, 1399 (1982).
[12] B. P. CHANDRA, M. ELYAS, and B. MAJUMDAR, Solid State Commun. **42**, 753 (1982).
[13] N. S. AKULOV, Dislocation and Plasticity, Raj Kamal Prakashan Private Ltd. Delhi, 1964 (p. 39).
[14] G. SCHOECK, Adv. appl. Mech. **4**, 229 (1956).

(Received April 20, 1983)

phys. stat. sol. (a) 78, 103 (1983)

Subject classification: 14.3; 22.7

Department of Physics, Sadar Patel University, Vallabh Vidyanagar[1])

Electrical Studies on (Mo/W)Se_2 Single Crystals

III. Anisotropy in Resistivity[2])

By

M. K. AGARWAL, P. D. PATEL, and O. VIJAYAN

The anisotropy in resistivity along and across the layers of the layered single crystals in the series $Mo_xW_{1-x}Se_2$ ($0 \leqq x \leqq 1$) are studied in the temperature range from 300 to about 600 K. The crystals are grown by a direct vapour transport technique. The anisotropy parameters of these crystals have values $\leqq 10^6$. The band mass anisotropy (m_a/m_c) is found to be ≈ 10 near the WSe_2-end and $\approx 10^4$ near the $MoSe_2$-end of the system.

Die Anisotropie des Widerstands in Richtung und senkrecht zur Schicht von geschichteten Einkristallen werden von 300 bis etwa 600 K untersucht. Die Kristalle werden durch direkte Dampftransportmethoden gezogen. Die Anisotropieparameter dieser Kristalle zeigen Werte $\leqq 10^6$. Die Bandmassenanisotropie (m_a/m_c) beträgt ≈ 10 am WSe_2-Ende und $\approx 10^4$ am $MoSe_2$-Ende des Systems.

1. Introduction

The layered crystals of $Mo_xW_{1-x}Se_2$ ($0 \leqq x \leqq 1$) have been grown by a direct transport technique [1]. The optical properties [2] and electrical properties [3, 4] of these single crystals have been studied. This paper reports the study of the anisotropy in the electrical resistivity along and across the layers of these crystals. The group VIA–VIB layered crystals show markedly easy cleavage along the layer planes (i.e., perpendicular to the *c*-axis). The crystals have a hexagonal structure with each layer consisting of a sandwich of a metal layer (Mo or W) within two Se layers. The intralayer bonding is partly ionic and partly covalent and the interlayer bonding is of the weak van der Waal's type. Because of this strong anisotropy in structure these crystals show strong anisotropy in the electrical and optical properties, too. The anisotropy in the optical absorption of WS_2, WSe_2, $MoTe_2$, $NbSe_2$, and NbS_2 crystals has been studied by Liang [5]. Studies on the anisotropy in the electrical properties of SnS_2 [6], $SnSe_2$ [7], GaSe [8], [9], InSe [10], and MoS_2 [11] are available from literature, but no studies seem to have been made on the anisotropy in the resistivity of a mixed system like (Mo/W)Se_2, nor on the temperature variation of the anisotropy in resistivity of $MoSe_2$ or WSe_2 crystals.

2. Experimental

The four point probe method described by Schanbel [12] has been employed for the study. This method makes use of a crystal sandwiched between two pairs of diametrically opposite fine points, with the condition that the contact diameter is much smaller than the crystal thickness. As the crystals were very thin (≈ 0.001 cm) the fulfilment of this condition becomes nearly impossible with evaporated contacts. To overcome this, a jig was fabricated with two stainless steel pins having pointed tips of diameter

[1]) Vallabh Vidyanagar 388120, India.

[2]) Part I and II see phys. stat. sol. (a) 78, 133 and 469 (1983), respectively.

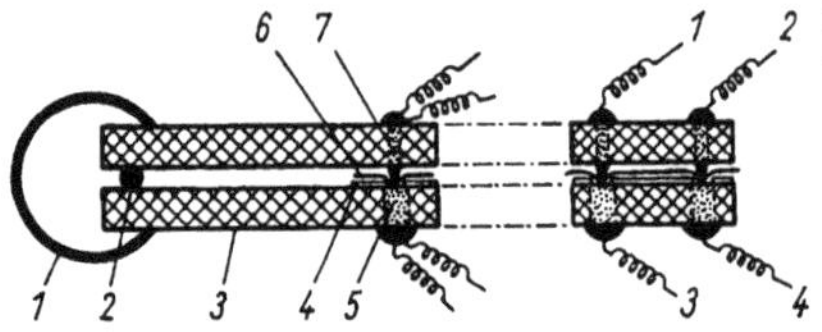

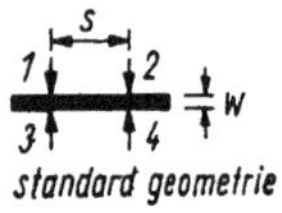

Fig. 1. Details of the fig. 1 spring, 2 pivot, 3 ebonite bar, 4 mica sheet, 5 brass rod, 6 crystal, 7 stainless steel pins

of about 0.0006 cm fixed with a separation of 3 mm between the tips, on an ebonite strip of $60 \times 15 \times 5\ \text{mm}^3$ dimensions. Another ebonite strip of the same dimensions was hinged at one end to this strip and the two strips were pressed together by a steel spring. The tips of the two pins rested on the flattened ends of two brass rods embedded in the lower ebonite strip, with a pressure of about 5g. A thin mica sheet (≈ 0.01 cm thick) was glued to the lower ebonite strip with two holes (of diameter ≈ 0.5 mm) over the brass rods at the centres of these holes (Fig. 1). Wheln the crystal is placed over the mica sheet and the pins allowed to rest on the crystal surface, the crystal bends at the points of contact of the pins and establish contacts with the lower brass rods with contact areas essentially the same as those of the pin tips. This set-up was kept inside the heater assembly (described in a previous communication [3]), which was kept at a vacuum of $\approx 10^{-5}$ Torr.

The contacts were often found noisy. Further, the resistance of a sample with a given pair of contacts was found to be different for different regions of the sample. This is most probably due to the non-uniformity of the crystal thickness. Although a number of crystals in each composition were studied, the consistency obtained was not very satisfactory. Another difficulty experienced while measuring the resistivity anisotropy was that of switching in these crystals [13]. When the applied voltage across the contacts exceeds a value U_{TH}, the threshold voltage for switching, the resistivity of the crystal falls rapidly and a heavy current passes through the sample. Consequently, the voltage developed across the other terminals increases. Besides, at higher temperatures the contacts become too noisy. Considering all these effects, the results obtained have errors of 20 to 30%.

3. Results and Discussion

The anisotropy ratio, ν, varies with temperature as

$$\nu = A\, e^{E/kT}\,.$$

Fig. 2 shows the plots of $\lg \nu$ versus $1/T$ for the crystals studied. The activation energies calculated from the slopes of these plots are plotted versus crystal composition in Fig. 3. The anisotropy parameter is defined as

$$\nu = \sigma_a/\sigma_c\,,$$

where σ_a and σ_c are the conductivities along the a-axis (i.e., parallel to the layers) and along the c-axis (i.e. perpendicular to the layers), respectively. Now,

$$\sigma_a = \sigma_{a_0}\, e^{-E_a/kT}\,,$$

$$\sigma_c = \sigma_{c_0}\, e^{-E_c/kT}\,,$$

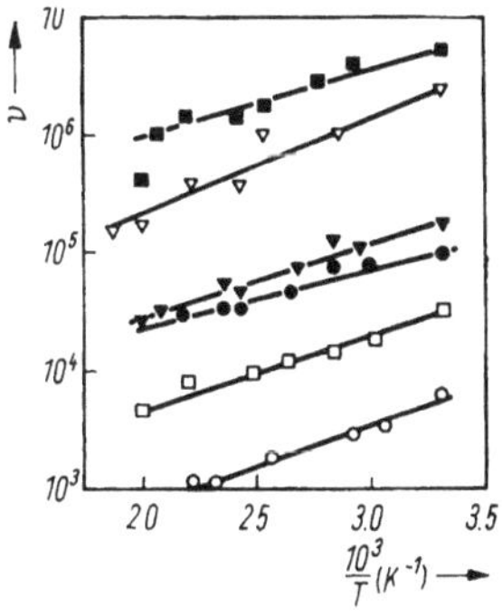

Fig. 2

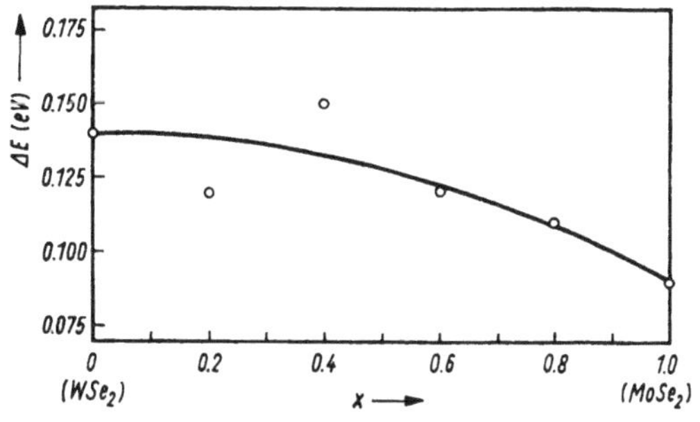

Fig. 3

Fig. 2. Plots of lg ν vs. $1/T$. ○ $x = 0$, □ 0.2, ▽ 0.4, ▼ 0.6, ■ 0.8, ● 1.0

Fig. 3. Plot of activation energy vs. crystal composition

where E_a and E_c are the corresponding thermal activation energies. Thus,

$$\nu = A\, e^{(E_c - E_a)/kT}$$

or

$$E_c = E_a + \Delta E\,,$$

i.e., the conductivity along the c-axis differs from that along the a-axis by an activation energy ΔE. The room temperature anisotropy ratios have been evaluated for 25 °C by extrapolating the ln ν versus $1/T$ plots. The $\nu_{25\,°C}$-values have been plotted versus crystal composition in Fig. 4. The room temperature resistivities along the c-axis have been plotted versus crystal composition in Fig. 4. The room temperature resistivities along the c-axis have been evaluated from the anisotropy parameters and the resistivity data and have been plotted versus crystal composition in Fig. 5. Fig. 6 shows plots of E_c versus x. The pre-exponential factors have also been evaluated and plotted versus crystal composition in Fig. 7.

It is seen that the room temperature anisotropies are in excess of 10^3, as has been pointed out by Wilson and Yoffe [14] and reaches values as high as 10^6 towards the $MoSe_2$ end. The room temperature anisotropy parameters for layered crystals reported are 4.21×10^4 to 4.96×10^4 for SnS_2 [6]; ≈ 10 [15], 13.3 [16], and 32.7 [17] for TaS_2; ≈ 30 [18] and ≈ 1200 [11] for MoS_2.

ΔE-values show a smooth nonlinear variation with crystal composition. The thermal activation energies for vertical conduction (i.e. along the c-axis), E_c also show a similar

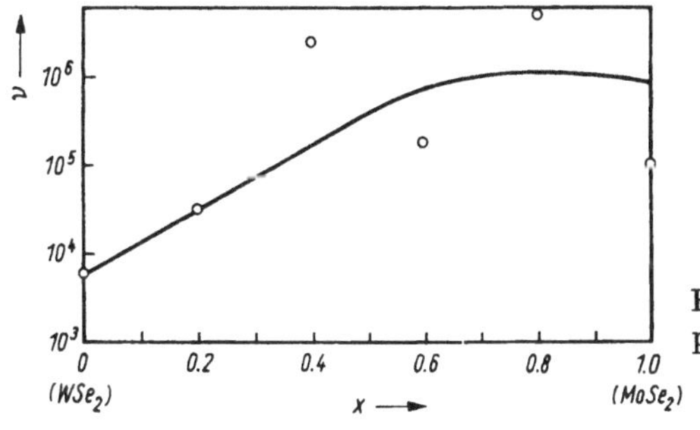

Fig. 4. Plot of room temperature anisotropy parameter vs. crystal composition

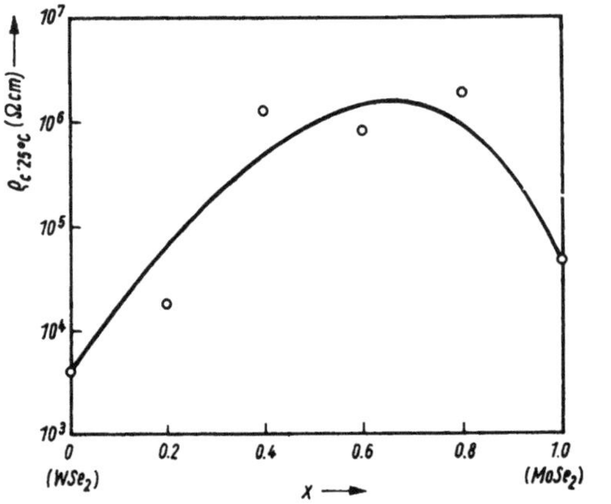

Fig. 5

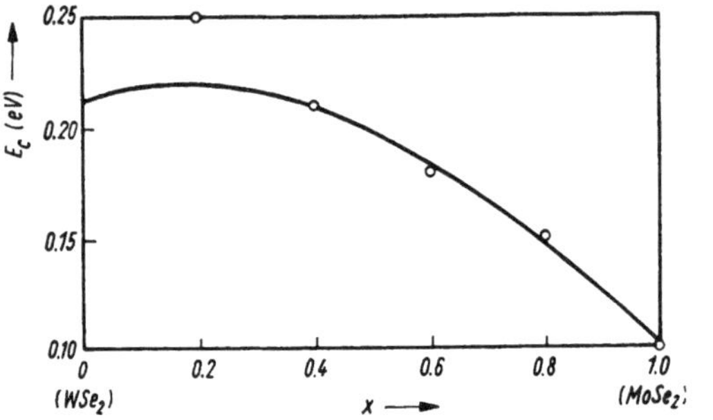

Fig. 6

Fig. 5. Plot of room temperature resistivity along the c-axis vs. crystal composition

Fig. 6. Plot of activation energy for conduction along the c-axis vs. crystal composition

variation, decreasing towards the $MoSe_2$ end. ΔE-values are below 150 meV as exhibited by most of the layered crystals [19]. Tredgold and Clark [8] suggested that the ΔE-values can be strongly increased due to plastic deformation of the samples. The high values of ΔE (compared with $\Delta E = 0.04$ to 0.08 eV given by Souder and Brodie [11] for MoS_2) obtained in the present study, is evidently a result of the mechanical handling of the crystals. However, no such structural dependence of E_a was observed [3] suggesting a strong dependence of the vertical conductivity of structural imperfections.

The pre-exponential factors reach a value in excess of 10^4 towards the $MoSe_2$ end. The similarity between the A versus x (Fig. 7) and the ν versus x (Fig. 4) curves is a result of the nearly linear variation of ΔE with x (Fig. 3). The pre-exponential factors can be approximated to be equal to the ratio of band masses [20]; i.e.,

$$A \sim \frac{m_c}{m_a},$$

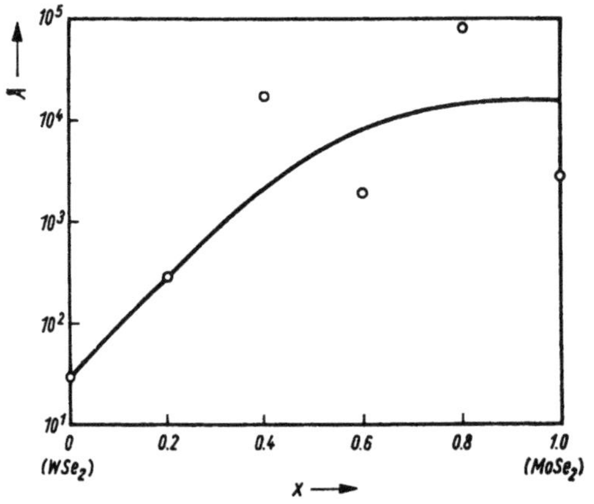

Fig. 7. Plots of the pre-exponential factors vs. crystal composition

Thus the charge carrier effective mass along the layers is greater than that across the layers. The values of A obtained seem to be unrealistically high to be explained by a description of the band mass anisotropy of a perfect layered structure.

Layered crystals often show predominance of stacking faults [21]. Stacking faults are favoured in layered crystals by their weak interlayer bonding. Stacking faults do not affect the crystalline symmetry along the layers and hence do not affect the properties along the layers. But the existence of a large number of planar defects make the crystal a one-dimensional disordered solid when viewed along the c-axis. The exponential tails of the localized states overlap through many layers and the vertical conduction no longer remains a hopping process, except at very low temperatures [19]. The solid is then to be described in terms of a mobility gap rather than by an energy band gap. That is, the vertical conduction is to be explained by a quasi-amorphous solid model. Conan et al. [22] explained the anomalous results obtained from the thermoelectric and Hall effect studies of $MoTe_{2-x}$ and $MoSe_{2-x}$ crystals on the basis of such a model. Thus, the higher values of the band mass anisotropy obtained seem to be due to the continuous nature of the energy bands along the c-axis, introduced by structural defects, the band structure along the a-axis remaining unaffected by such defects.

Another supplementary evidence of the quasi-amorphous nature of these crystals came from the switching phenomenon exhibited by them. (Mo/W)Se_2 crystals are found to show electrical threshold switching [13]. Switching has been observed only across the layers. Switching and the nonlinear conductivity behaviour prior to switching are characteristics of amorphous materials [23].

Acknowledgement

One of the authors (O. V.) wishes to thank the UGC (New Delhi) for financial assistance.

References

[1] M. K. Agarwal and P. A. Wani, Mater. Res. Bull. **14**, 825 (1979).
[2] M. K. Agarwal, P. D. Patel, and O. Vijayan, Bull. Mater. Sci. **4**, 467 (1982).
[3] M. K. Agarwal, P. D. Patel, and O. Vijayan, phys. stat. sol. (a) **78**, 133 (1983).
[4] M. K. Agarwal, P. D. Patel, and O. Vijayan, phys. stat. sol. (a) **78**, 469 (1983).
[5] W. Y. Liang, J. Phys. C **6**, 551 (1973).
[6] S. G. Patil and R. H. Tredgold, J. Phys. D **4**, 718 (1971).
[7] A. I. Likhter, E. G. Pel, and S. I. Prysyazhnuk, phys. stat. sol. (a) **14**, 265 (1972).
[8] R. H. Tredgold and A. Clark, Solid State Commun. **7**, 1519 (1969).
[9] Ph. Schmid and E. Mooser, Helv. phys. Acta **45**, 870 (1972).
[10] S. M. Atakishiev and G. A. Akhundov, phys. stat. sol. **32**, K33 (1969).
[11] A. Souder and D. E. Brodie, Canad. J. Phys. **49**, 2565 (1971).
[12] P. Schnabel, Philips Res. Rep. **19**, 43 (1964).
[13] M. K. Agarwal, P. D. Patel, and O. Vijayan, Silver Jubilee Phys. Symp. BARC (Bombay) **24c**, 499 (1982).
[14] J. A. Wilson and A. D. Yoffe, Adv. Phys. **18**, 193 (1969).
[15] F. J. Di Salvo, B. G. Bagley, J. M. Voorhoeve, and J. V. Waszczak, J. Phys. Chem. Solids **34**, 1357 (1973).
[16] A. H. Thompson and F. R. Gamble, Phys. Rev. B **5**, 2811 (1972).
[17] J. P. Tidman, O. Singh, A. E. Curzon, and R. F. Frindt, Phil. Mag. **30**, 1191 (1974).
[18] J. Edwards and F. R. Frindt, J. Phys. Chem. Solids **32**, 2217 (1971).

[19] R. FIVAZ and PH. SCHMIDT, Physics and Chemistry of Materials with Layered Structures, Vol. 4, D. Riedel, Dordrecht (Holland) 1976 (p. 343).
[20] K. MASCHKE and PH. SCHMID, Phys. Rev. B **12**, 4312 (1975).
[21] G. C. TRIGUNAYAT, phys. stat. sol. (a) **4**, 281 (1971).
[22] A. CONAN, G. GOUREAUX, and M. ZOATER, J. Phys. Chem. Solids **36**, 315 (1975).
[23] D. ADLER, H. K. HENISCH, and N. F. MOTT, Rev. mod. Phys. **50**, 209 (1978).

(Received May 2, 1983)

phys. stat. sol. (a) 79, 109 (1983)

Subject classification: 18.2; 2; 10.1; 21.1; 21.1.1

Institut für Theoretische und Angewandte Physik der Universität Stuttgart[1])
and Max-Planck-Institut für Metallforschung, Institut für Physik, Stuttgart[2])

On the Coercive Field of the Nearly Non-Magnetostrictive Amorphous Ferromagnetic Alloy $Co_{58}Ni_{10}Fe_5Si_{11}B_{16}$

By

W. BECK and H. KRONMÜLLER

The coercive field of nearly non-magnetostrictive amorphous alloys is predominantly determined by quenching conditions, relaxation processes, and the surface roughness. Taking into account surface effects and magnetostrictive contributions to the coercive field it is possible to separate different coercive field contributions and to predict the value and the conditions for obtaining a minimum coercive field. In this paper the effects of the quenching rate on spin quenched ribbons are measured.

Die Koerzitivfeldstärke schwachmagnetostriktiver amorpher Legierungen wird vorwiegend durch die Abschreckbedingungen, Relaxationsprozesse und die Oberflächenrauhigkeit bestimmt. Unter Berücksichtigung von Oberflächeneffekten und magnetostriktiven Beiträgen ist eine Trennung der verschiedenen Anteile der Koerzitivfeldstärke und eine Vorhersage des kleinsten Koerzitivfeldes möglich. In der vorliegenden Arbeit wird der Einfluß der Abschreckbedingungen auf Bänder untersucht, welche mit dem Melt-Spin-Verfahren hergestellt wurden.

1. Introduction

The coercive field, H_c, as well as the initial susceptibility, χ_0, of amorphous ferromagnetic alloys is determined by the pinning forces acting on domain walls [1]. In crystals the pinning centres for domain wall motion are known to be dislocations, grain boundaries, and precipitates. In amorphous ferromagnetic alloys these kinds of pinning centres are not present and consequently these materials are expected to supply excellent soft magnetic materials. Nevertheless the smallest coercive fields of amorphous alloys are a factor of two to five larger than that of crystalline permalloy materials. Therefore, we have to assume that also in amorphous alloys defect structures exist which act as pinning centres for domain wall motions.

2. Experimental Procedure

In order to keep magnetostrictive contributions to the coercive field as small as possible the nearly non-magnetostrictive alloy $Co_{58}Ni_{10}Fe_5Si_{11}B_{16}$ ($\lambda_s \approx -10^{-7}$) was chosen for the following investigations. Ribbons of 3 to 5 m length were produced by the common spin-quenching technique with a 12 cm diameter copper–2% beryllium wheel. This alloy has the advantage of a very large critical flow stress in comparison to pure copper. Therefore, rotation frequencies of maximal 8000 rotations min^{-1} were possible.

The coercive field was measured with an earth-field-compensated Förster coercimeter (the earth field compensation had to be adjusted very carefully). The samples investigated were 5 cm in length and 3 mm wide and were measured in three different

[1]) Pfaffenwaldring 57, D-7000 Stuttgart 80, FRG.
[2]) Heisenbergstr. 1, D-7000 Stuttgart 80, FRG.

runs. The received results were reproducible within ± 1 mOe in almost all cases. Domain patterns were recorded by the method of the magnetooptical Kerr effect [2]. Annealing treatments in general were performed for 2 h in earth field and longitudinal fields of 410 Oe.

3. Experimental Results

3.1 Quenching conditions

The homogeneity of the ribbon thickness over the tape length was ± 0.5 to 1 μm at an average thickness between 15 and 40 μm. The reproducibility of chemical composition was within 0.1 at% for the metallic components and 0.6 at% for the non-metallic components.

The contact time τ_k of the ribbon on the wheel during the quenching procedure could be determined from the optically determined contact length and the rotation frequency Ω (see Fig. 1). In order to obtain ribbons of good quality, the temperature

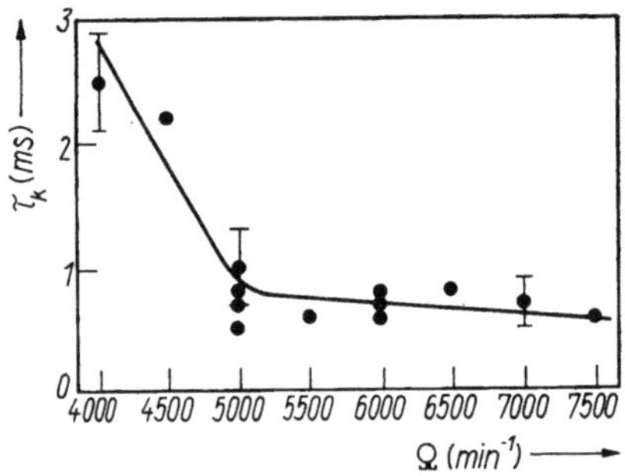

Fig. 1. The contact time, τ_k, as a function of rotation frequency

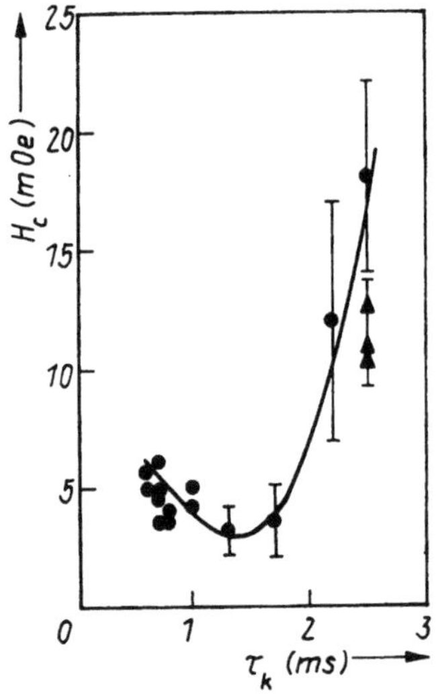

Fig. 2

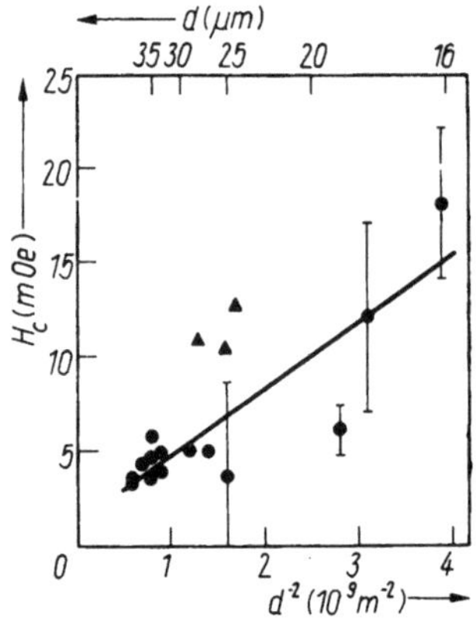

Fig. 3

Fig. 2. The coercive field as a function of the contact time, τ_k

Fig. 3. The coervice field as a function of the reciprocal square of the ribbon thickness of the melt could only be varied within a small temperature interval of about ±50 K at an average melt temperature of 1400 °C. Contrary to the temperature of the melt the rotation frequency could bc varied in a wide range between 4000 to 7500 cycles min^{-1}.

Fig. 2 and 3 show the coercive field in dependence of τ_k and of the reciprocal square of the ribbon thickness, $1/d^2$, respectively (printed tolerances belong to tapes for which the H_c distribution over tape length exceeds the above-mentioned 1 to 2 mOe, caused, e.g., by bad tape quality). Similar results were reported by other authors [3, 4].

3.2 Domain patterns and annealing treatments

In Fig. 4 domain patterns for increasing contact time, τ_k, are shown.

For lower τ_k's (Fig. 4 a, b) preferentially large transversal laminar domains are observed, embedding isolated regions of fine domains. Around the regions of fine domains the transversal domains are markedly deformed.

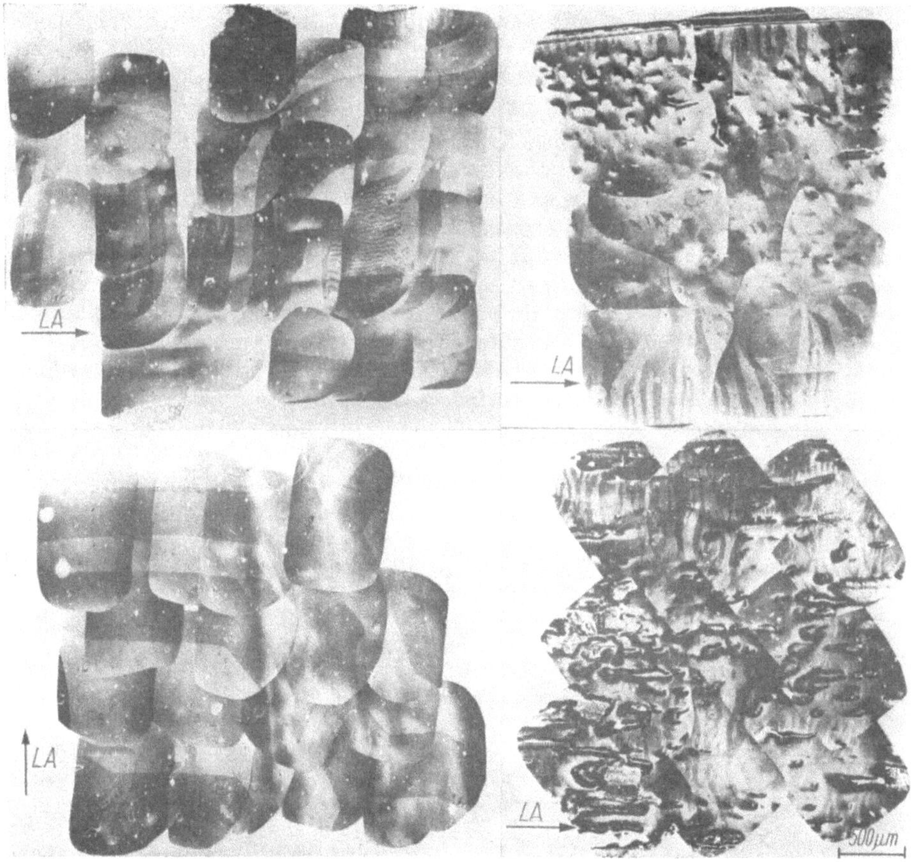

a c b d

Fig. 4. Domain patterns of ribbons produced with different contact times: a) $\tau_k = 0.6$, b) 1.3, c) 2.2, d) 2.5 ms

Tapes produced with higher contact time, τ_k, show smaller transversal domains at the edges of the ribbons (Fig. 4 c, d). In the central regions of the ribbons the domain structure is very irregular. Regions with fine-structure domains are found also. Their number increases for larger τ_k's.

4. Discussion

4.1 Influence of quenching rate and annealing treatment on H_c

To describe glass formation, nucleation of the crystalline phase and crystallite growth must be considered [5 to 7]. Both processes are governed by the atomic mobility described by the viscosity η in the liquid phase. A phase diagram for the liquid and solid phase in dependence of temperature and quenching time is shown in Fig. 5 (after [5]). This diagram is known as the time–temperature transition diagram (TTTD). The phase boundary curve separates the liquid from the crystalline region and reflects nucleation (upper part) and crystal growth (lower part). To obtain an amorphous state, the cooling curve should lie completely out of the crystalline phase boundary in the temperature range between the melting temperature, T_m, down to the glass forming temperature, T_g.

Depending on the contact time τ_k, the lift-off of the ribbon from the wheel takes place at high temperatures, about T_g (short τ_k), or at lower temperatures of about room temperature (large τ_k). After the take-off of the ribbon the quenching rate in the surrounding atmosphere is lowered, thus corresponding to an annealing process. During quenching, the T_g-isotherm moves through the ribbon from wheel to top side. Therefore, we get a second-order heat transport and the quenching rate $\dot{T}$ is proportional to the square of the reciprocal ribbon thickness [7]. In Fig. 3 we see the expected rise of H_c with the quenching rate. Higher quenching rates, i.e., larger contact times, τ_k, and thinner ribbons result in larger internal stresses leading to larger coercive fields. The τ_k-dependence of H_c shows a minimum at $\tau_k = 1.5$ ms. For higher τ_k's quenching is complete, for lower τ_k annealing effects eventually together with a possible nucleation of small crystalline regions are superimposed leading to higher coercive fields.

Annealing treatments in the earth field show a H_c-maximum at about 175 to 200 °C (see Fig. 6). Since this effect is not observed after a longitudinal field annealing (see Fig. 7) it is supposed that this effect is due to the formation of induced anisotropies (similar to the well-known maxima in FeNiB alloys [8]) and not only due to changes in magnetostriction [9].

As is well known domain patterns represent the distribution of long-range internal stresses [1]. Depending on the sign of the magnetostriction constant, λ_s, large- and fine-structured domains represent regions with compressive and tensile stresses [1, 11], respectively. Because of the larger amount of quenched-in free volume in the case of

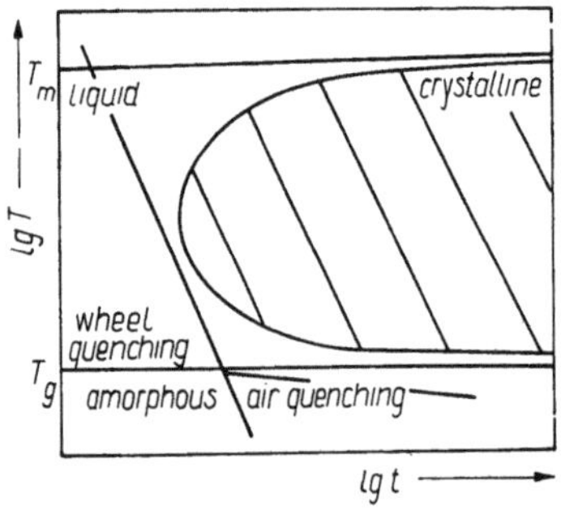

Fig. 5. Time–temperature transition (TTT) diagram (modified after [5])

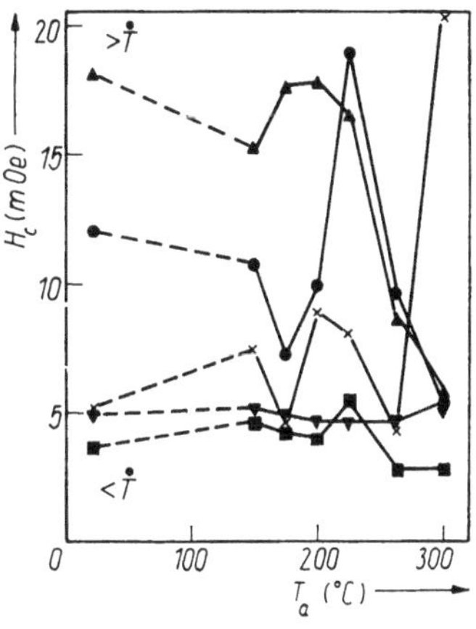

Fig. 6

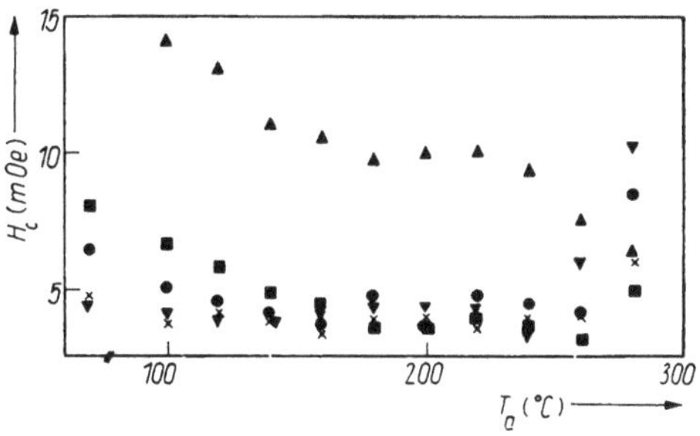

Fig. 7

Fig. 6. The coercive field after various annealing treatments (2 h, earth field)

Fig. 7. The coercive field after various annealing treatments (2 h, longitudinal field of 410 G)

higher quenching rates, larger internal stresses and anisotropies as well as more fine-structured domains are found in these ribbons. The sizes of the remaining transverse domain regions are smaller than in slowly cooled ribbons.

4.2 Contributions to the coercive field in the nearly non-magnetostrictive alloy $Co_{58}Ni_{10}Fe_5Si_{11}B_{16}$

In amorphous alloys the following pinning mechanisms of domain walls contribute to the coercive field [10]:

a) Natural intrinsic fluctuations of exchange and crystal anisotropy energy.

b) Chemical short-range order changes locally the intrinsic magnetic material parameters.

c) Defect structures of the amorphous matrix, caused by a local increase or decrease of the atomic number density. These regions produce internal stresses which cause additional anisotropies by the magnetoelastic coupling energy. Careful investigations were made by [11] for this H_c-contribution. For FeNiB alloys it was found that H_c is a function of the square root of the quantities which characterize the defect structures, whereas H_c depends linearly on saturation magnetostriction constant λ_s.

d) Surface roughness causes changes of the domain wall area resulting in a spatial variation of the total domain wall energy.

e) Thermally activated relaxation processes assisted by the free volume alter the orientation of metallic atom pairs within domain walls. If the energetically favoured atom pair directions lie parallel to the spontaneous magnetization, a directional ordering of the atom pairs takes place within the domain walls, leading to a stabilization potential of the domain walls. The additional contribution to the coercive field becomes time-dependent due to this relaxation process and may be written as [10]

$$H_c^{rel} \sim \frac{1 - e^{-t/\tau}}{kT},$$

where τ denotes the relaxation time of the thermally activated reorientation process of pair axes.

The surface contribution to the coercive field depends linearly on the surface roughness of the ribbon [10],

$$H_c^{\text{surf}} \sim \frac{\Delta d}{\langle d \rangle \, l},$$

where $\Delta d/\langle d \rangle$ denotes the relative surface roughness (thickness fluctuations divided by average ribbon thickness) and l the repetition length of the surface irregularities. Pure surface contributions to H_c turned out to be about 3 mOe [12]. By separation of the surface contributions, there remains a bulk coercive field which depends on the quenching rate. The above-mentioned relaxation processes due to pair ordering induce an additional coercive field of about 3 to 5 mOe in the as-quenched state [13].

5. Conclusions

By producing ribbons of $Co_{58}Ni_{10}Fe_5Si_{11}B_{16}$ with different quenching rates and by analysing the annealing behaviour of H_c as well as of the surface and relaxation effects, the following contributions to the coercive field of as-quenched ribbons have been separated:

1. Internal stresses:	1 mOe
2. Pair ordering, induced anisotropy, and intrinsic fluctuations:	3 mOe
3. Surface roughness:	3 mOe

Here the first two contributions are strongly dependent on the quenching rate $\dot{T}$. Assuming that neither anisotropy nor stresses contribute to H_c if we deal with the minimum in the coercive field as a function of quenching rate we may determine approximately the minimum possible H_c to be about 2 to 3 mOe, and which is caused exclusively by the surface roughness.

Acknowledgements

The authors are grateful to Mr. X. Z. Dong (on leave of Institute for Steel and Iron, Shaanxi, China) for many helpful discussions and the help of Mrs. T. Jendrysik at preparing Kerr effect photographs. This work was supported by the "Deutsche Forschungsgemeinschaft".

References

[1] H. KRONMÜLLER, M. FÄHNLE, M. DOMANN, H. GRIMM, R. GRIMM, and B. GRÖGER, J. Magnetism magnetic Mater. **13**, 53 (1975).
[2] J. KRANZ and A. HUBERT, Z. angew. Phys. **15**, 220 (1963).
[3] S. TAKAYAMA and T. OI, J. appl. Phys. **50**, 1595 (1979).
[4] L. NOVÁK, L. POTOCKÝ, A. LOVAS, É. KISDI-KOSZÓ, and J. TAKÁCS, J. Magnetism magnetic Mater. **19**, 149 (1980).
[5] H. A. DAVIES, in: Proc. 3rd Internat. Conf. Rapidly Quenched Metals, Vol. 1, Ed. B. CANTOR, The Metals Society, London 1978 (p. 1).
[6] D. R. UHLMANN, J. non-crystall. Solids **7**, 337 (1972).
[7] H. H. LIEBERMANN, IEEE Trans. Magnetics **15**, 1393 (1979).
[8] H. MEHRER, G. FLIK, J. HORVÁTH, and H. KRONMÜLLER, phys. stat. sol. (a) **72**, 215 (1982).
[9] X. Z. DONG, W. FERNENGEL, and H. KRONMÜLLER, Appl. Phys. **A28**, 103 (1982).
[10] H. KRONMÜLLER, J. Magnetism magnetic Mater. **24**, 159 (1981).
[11] B. GRÖGER and H. KRONMÜLLER, Appl. Phys. **24**, 287 (1981).
[12] W. BECK, Diploma Work, Universität Stuttgart, 1982.
[13] B. GRÖGER, W. BECK, X. Z. DONG, N. MOSER, and H. KRONMÜLLER, J. Magnetism magnetic Mater. **26**, 264 (1982).

(Received June 23, 1983)

phys. stat. sol. (a) 79, 115 (1983)

Subject classification: 10.1 and 12.1; 21

The Institute of Physical and Chemical Research, Saitama[1])

High-Temperature Internal Friction Peaks of Pure Aluminium

By

K. IWASAKI

The effects of deformation and successive annealings on the high temperature internal friction peak (so-called grain boundary peak) of pure aluminum are investigated. The peak is divided into two components, a low and a high temperature component, with the use of differentiated vibration period. The fomer and the latter components are considered to be related with lattice dislocations and grain boundary dislocations, respectively.

Der Einfluß einer Verformung und anschließender Anlaßbehandlungen auf das Hochtemperaturdämpfungsmaximum (sogenanntes Korngrenzmaximum) des reinen Aluminums wird untersucht. Die Methode der differenzierten Schwingungsperiode zeigt, daß das Maximum aus zwei Komponenten, der Tief- und Hochtemperaturkomponente, besteht. Die erste scheint durch Versetzungen in den Körnern und die zweite durch Korngrenzenversetzungen verursacht zu sein.

1. Introduction

In 1947 Kê [1, 2] discovered a large internal friction peak at about 285 °C in poly crystalline pure aluminum but not in single, actually bamboo, crystals. These experimental results led him quite naturally to interpret that the peak was related with grain boundaries. Since then similar peaks have been observed in various kinds of metals and several models have been proposed for the grain boundary relaxation mechanisms [3].

Kê's first interpretation given above had been believed conceptionally doubtless until a few groups [4, 5] pointed out that the peak might not be related with grain boundaries but with dislocations inside the grains. Their arguments are based on the following observations [6]. i) The peak is observed also in single crystals [4, 5]. ii) It has a clear correlation with stacking fault energy [7]. iii) It is strain amplitude dependent [8]. iv) It is very sensitive to the mechanical conditions of specimens [9].

In opposition to them Sun and Kê [10] recently proposed a continuous grain boundary dislocation model and gave the following counterevidence. As the relaxation behavior of high-angle grain boundaries is similar to that of slip bands, similar internal friction peaks are expected to be observed. These peaks, however, may be different in their peak temperatures and annealing behavior. The peak observed in deformed single crystals is therefore quite different from that in annealed polycrystals in their origins [11].

The arguments of both groups, however, are based on the experimental results taken in different specimen conditions such as purity and dislocation density. If a polycrystalline specimen with stable grain distribution is deformed in an appropriate manner, both the effects of lattice dislocations and those of grain boundary dislocations are expected to be observed at the same time in one specimen. Although the

[1]) 2-1 Hirosawa, Wako, Saitama 351, Japan.

internal structure of the specimen is not simple in this case, careful experiments will make it possible to separate the two effects.

In the present work the effects of deformation and successive annealings are investigated in detail for polycrystalline pure aluminum with stable bamboo structure. The separation of an internal friction peak into some components is made with the use of differentiated vibration period which is known to give rise to a sharper peak than the corresponding internal friction peak by a factor of about 0.67 [12]. Activation properties of the peak are also determined in this way [13, 14].

2. Experimental Procedure

Pure aluminum (99.999%) specimen wires of 1 mm in diameter and 80 mm in effective length were prepared by wire drawing at room temperature. They are chemically polished at room temperature with a solution of 67% HNO_3 + 29% HCl + 4% HF in order to remove contaminated surface layers and to observe the grain distribution.

Internal friction and the vibration period are measured as functions of temperature with an inverted torsion pendulum. Fully automatic measurements and real-time data processing are carried out with a micro-computer [15, 16]. The maximum surface shear strain is 5×10^{-6} and the vibrational frequency is about 1.5 Hz. The counterweight of the pendulum can be varied from outside without breaking vacuum, which makes it possible to deform a specimen by elongation in situ at any temperature [17]. The specimen elongation is detected with a differential transformer.

The normalized period of vibration is differentiated with respect to temperature, in which three adjacent data points are used. The obtained values are used for the separation of an internal friction peak into some components and for the determination of activation properties [12 to 14].

3. Experimental Results

Isochronal and isothermal annealing behavior of an as-drawn specimen is shown in Fig. 1, where the normalized vibration period (t), the differentiated vibration period (t'), and internal friction (Q^{-1}) are plotted against temperature (T). In Fig. 1a, for example, symbol □ indicates that the specimen was annealed at 400 °C for 1 h in the measuring apparatus free from stress, that is, no excess counterweight was added to the balancing one during the annealing. A large internal friction peak observed in the temperature range between 200 and 300 °C is called "main peak" or simply "peak" in the present paper.

In the as-drawn state the main peak is associated with several unstable subpeaks due to annealing effects. After annealing at 400 °C for 1 h these subpeaks disappear and the main peak becomes well-shaped. As the isochronal (1 h) annealing temperature is increased from 400 to 600 °C, both the peak height and the peak temperature increase. Quite a clear double peak is observed in the t' curve after annealing at 600 °C for 1 h, which means that the main peak consists of at least two components, a low-temperature component (LTC) and a high-temperature component (HTC). It is to be noted, however, that the accuracy of the t' values is rather poor because they are calculated by taking small differences between large values. Small serrations in the t' curve, therefore, should not be taken too seriously. The main peak becomes widest in this annealed state as will be described later (Table 1). When the specimen is isothermally annealed at 600 °C for a longer time, the peak height decreases, while the peak temperature increases (Fig. 1b).

The annealing behavior is summarized in Fig. 2, where the data which are not shown in Fig. 1 are also included. Fig. 2a and b correspond to isochronal and isothermal annealing, respectively.

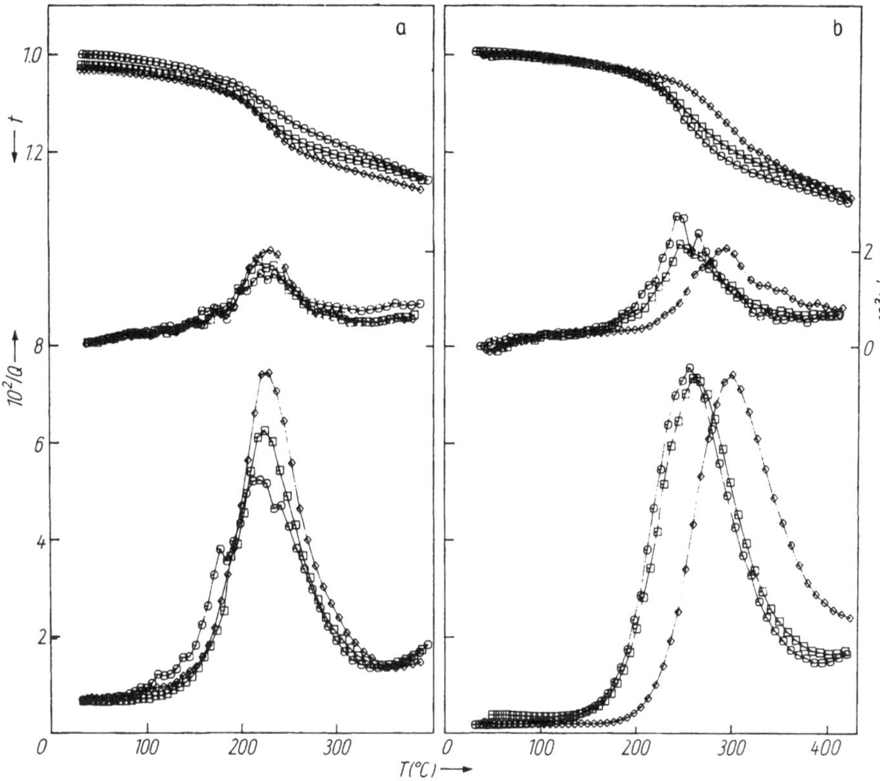

Fig. 1. Annealing behavior of an as-drawn specimen. a) ○ as drawn; □ 400 °C, 1 h, 0 gA; ◇ 500 °C, 1h, 0 gA; ○ 600 °C, 1 h, 0 gA; □ 600 °C, 10 h, 0 gA; ◇ 600 °C, 141 h, 0gA

The surface observations were made after the same annealing treatments as were performed above and the results are shown in Fig. 3. In roder to avoid the nonreproducible effects of specimen handling on the internal friction measurements another specimen than that used in Fig. 1 and 2 is used here. The magnification factor of the photographs is apparent from the specimen diameter of 1 mm. Almost all the small grains that exist in the as-drawn state disappear after annealing at 400 °C for 1 h.

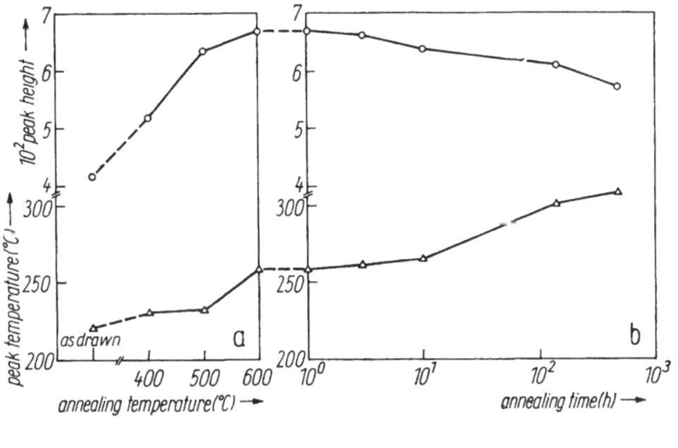

Fig. 2. Summary of annealing behavior of an as-drawn specimen. a) isochronal (1 h) annealing; b) isothermal (600 °C) annealing

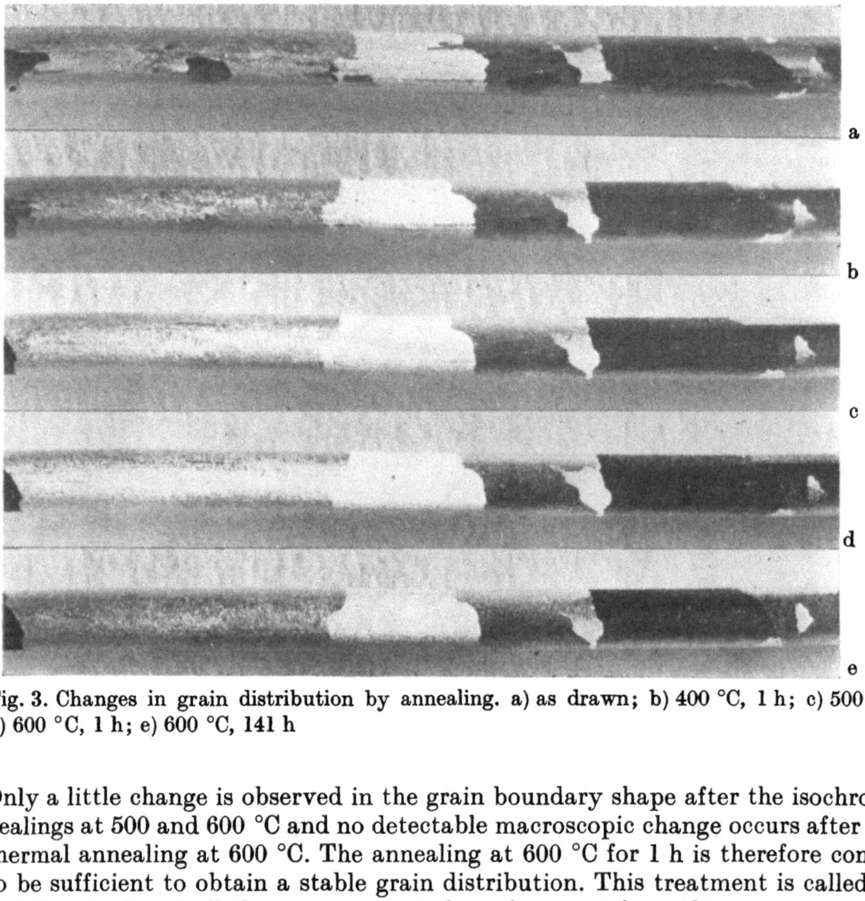

Fig. 3. Changes in grain distribution by annealing. a) as drawn; b) 400 °C, 1 h; c) 500 °C, 1 h; d) 600 °C, 1 h; e) 600 °C, 141 h

Only a little change is observed in the grain boundary shape after the isochronal annealings at 500 and 600 °C and no detectable macroscopic change occurs after the isothermal annealing at 600 °C. The annealing at 600 °C for 1 h is therefore considered to be sufficient to obtain a stable grain distribution. This treatment is called "grain stabilization" and all the measurements hereafter start from this state.

The effects of a tensile stress imposed on a specimen during annealing are shown in Fig. 4, where a stabilized secpimen was annealed in situ with an additional counterweight of 100 g (indicated as "100 gA" in the figure caption). During the stressed annealing the specimen is deformed slowly and the total specimen elongation at the end of each annealing is given in per cent in the figure caption. The double t' peaks are observed in all the t' curves in this case. The main peak shifts to lower temperature as the stressed annealing proceeds. Corresponding to this peak shift the increase in the LTC and the decrease in the HTC are clearly seen. This behaviour is just opposite to that shown in Fig. 1b where no stress was applied during the annealing.

A similar behavior is observed also when a specimen is deformed rapidly. An example is shown in Fig. 5, where "450 °C, 3.9%, 200 gD", for example, indicates that the specimen was deformed at 450 °C in situ with an additional counterweight of 200 g to the total elongation of 3.9%. The deformation finishes in a few seconds in this case. As the degree of deformation increases, both the peak height and the peak temperature decrease and the background damping increases. In addition to the main peak there appears a small peak at about 70 °C due to the deformation. A peak of this type is often observed around 100 °C when specimens are deformed rapidly at low temperature.

The effects of annealing after the rapid deformation on the main peak are shown in Fig. 6, where a stabilized specimen was deformed at 430 °C with an additional counter-

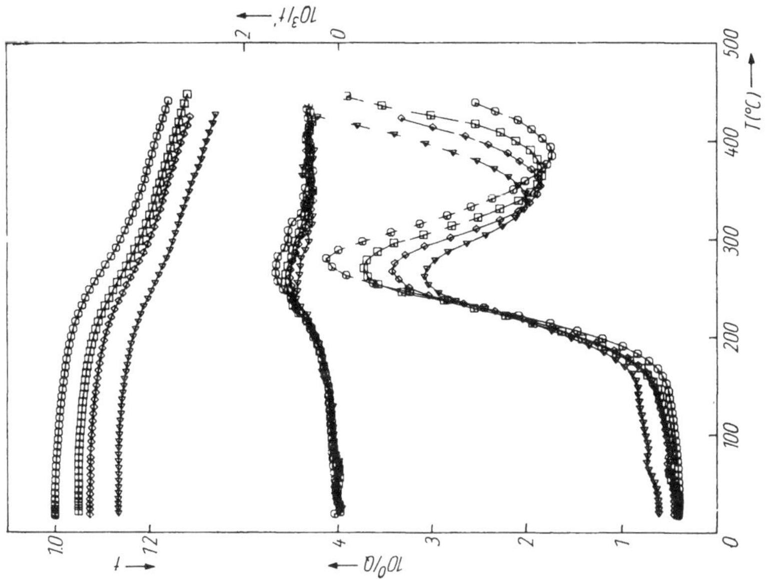

Fig. 5

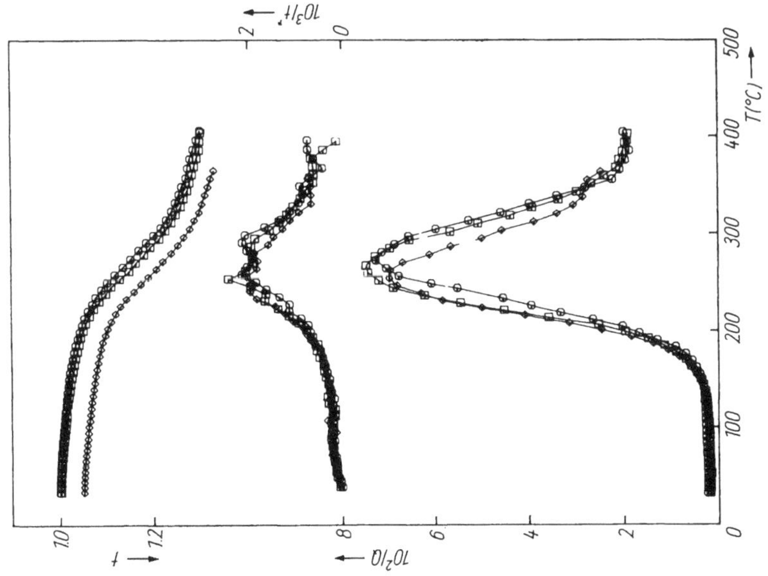

Fig. 4

Fig. 4. Effects of stressed annealing. ○ 600 °C, 1 h, 0 gA; □ +600 °C, 1 h, 100 gA, 1.9%; ◇ +600 °C, 1h, 100 gA, 4.8%

Fig. 5. Effects of high-temperature deformation. ○ 600 °C, 1 h, 0 gA; □ 450 °C, 1.5%, 200 gD; ◇ 450° C, 3.9%, 200 gD; △ 450 °C, 7.7%, 200 gD

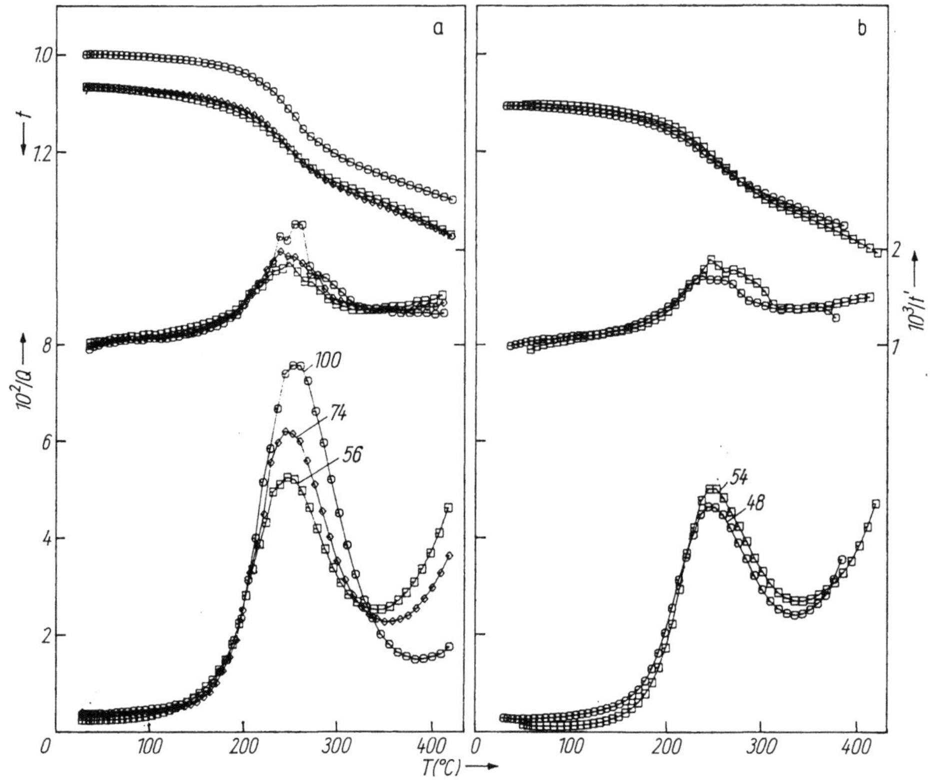

Fig. 6. Effects of deformation and successive annealing. a) ○ 600 °C, 1 h, 0 gA; □ 430 °C, 4.6%, 210 gD; ◇ +600 °C, 1 h, 0 gA; b) ○ 430 °C, 8.9%, 210 gD; □ +600 °C, 1 h, 0 gA

weight of 210 g and then annealed at 600 °C without stress. A clear double t' peak is observed again in the grain stabilized state (600 °C, 1 h, 0 gA). The main peak is reduced by the deformation and recovers partially to its original state by the successive annealing at 600 °C. The changes in the peak height are shown with the normalized values in the figure. As the degree of deformation is increased, the amount of recovery by the annealing decreases. The same behavior is observed independent of deformation temperature.

The specimen surface was observed after the rapid deformation at 430 °C and the results are shown in Fig. 7, where a specimen different from that used in Fig. 6 was used by the same reason as given before. Fine slip bands are observed after deformation by 5%, while double slip occurs when a specimen is deformed by 10%. During the deformation the grain boundary seems to remain unchanged, that is, the deformation takes place mainly in the grains. When the specimen is deformed further by more than 10%, a neck appears in the grain. The experimental values obtained with such a specimen does not reflect the whole specimen equivalently because the vibrational stress is concentrated in the necked part.

The dependence of the peak height on the grain size is shown in Fig. 8, where the experimental data for specimens annealed at 600 °C for more than 2 h and for those deformed at temperatures between 430 and 600 °C are shown with different symbols. As almost all the grains except those in the as-drawn state are larger than the specimen

Fig. 7. Changes in specimen surface by deformation. a) 600 °C, 1 h; b) 5%, c) 10%

diameter (so-called bamboo structure) as shown in Fig. 3 and 7, the number of grains per unit length along the wire direction is used as an index for the grain size here. The annealed specimens show a clear correlation between the peak hiehgt and the number of grains as shown by a broken line, while the correlation is very poor for the deformed ones.

The activation energy (E) is determined using the differentiated vibration period for each measuring run [13]. The obtained values are listed in Table 1 together with other activation properties calculated from E such as frequency factor (ν_0), with ratio (α), and peak temperature (T_p) at 1 Hz. All the values of the present work are the averages of several different specimens in nearly the same experimental conditions. The standard deviation is given with $\pm$ sign after each value. As the isochronal annealing temperature is increased from the as-drawn state, all the values have a tendency to increase as listed in the first four rows in the table. When the specimens are annealed isothermally at 600 °C for more than 100 h, E, ν_0, and α decrease, while T_p increases. On the

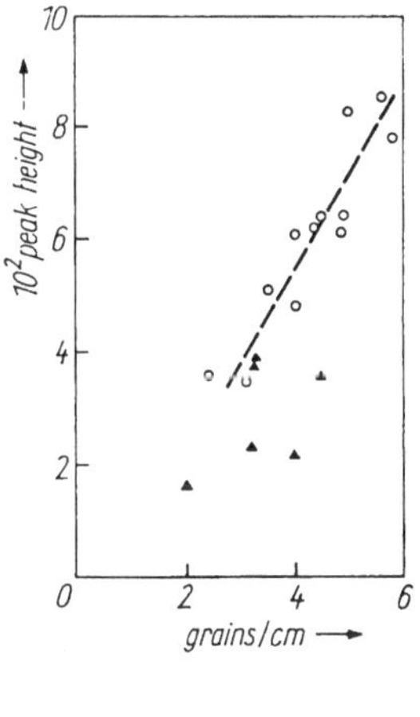

Fig. 8. Grain size dependence of the peak height. ○ annealed, ▲ deformed

Table 1
List of activation properties

specimen condition	E (eV)	ν_0	α	T_p at 1 Hz (°C)	authors
as-drawn	1.09 ± 0.09	9.9 ± 0.4	1.82 ± 0.15	215.0 ± 2.0	
400°C × 1 H	1.23 ± 0.13	12.4 ± 0.7	1.48 ± 0.07	226.9 ± 5.1	
500°C × 1 H	1.22 ± 0.12	12.3 ± 1.1	1.53 ± 0.19	228.0 ± 3.4	Iwasaki
600°C × 1 H	1.62 ± 0.05	15.4 ± 0.4	2.15 ± 0.10	260.3 ± 6.9	(present work)
600°C × (> 100) H	1.58 ± 0.08	13.1 ± 0.4	1.88 ± 0.06	297.7 ± 8.3	
Deformation at (430—600) C	1.49 ± 0.08	14.3 ± 0.7	1.95 ± 0.14	254.3 ± 8.2	
450°C × 5 H	1.39	14		285	Kê [1, 2]
325 C × 2 H	1.50	14.5		300	Cordea and Spretnak [18]
400°C × 16 H + (410 to 615) °C× 2 H	1.65 ± 0.13	13.5		295	Williams and Leak [19]
230°C × 1 H	1.05 ± 0.1	11.6 ± 0.1		180	Esnouf and Fantozzi [20]
deformed single or macro-	1.3 ± 0.1	14 ± 0.5	1.1 ± 0.2	220	Bonetti et al. [5, 21]
crystals	1.7 ± 0.1	16 ± 0.5	1.2 ± 0.2	270	
deformed single and	0.95	12.5		130	
polycrystals	1.02	9.3		330	Riviere et al. [22]
	0.98	10		410	

other hand, when the stabilized specimens are deformed at high temperature, all the values decrease. It is to be noted here that the peak becomes widest in the stabilized state and that a clear double t' peak is observed correspondingly.

The results on pure aluminum by other authors are also listed in the lower part of Table 1. All the results except those by the last two groups are made on polycrystals. As far as the peak which appears at temperatures between 200 and 300 °C is concerned, the present results agree with the existing data comparatively well.

4. Discussion

After the grain stabilization (annealing at 600 °C for 1 h) the main peak becomes widest (Table 1) and splits into at least two components, LTC and HTC (Fig. 1b, 4, 5, and 6a). Both components seem to be nearly equal-weighted in the stabilized state. When the stabilized specimen is annealed further at 600 °C, the peak moves to higher temperature becoming a little sharaer (Fig. 1b and Table 1). These results show that the HTC is predominant over the LTC in the well-annealed state. On the other hand, if the stabilized specimen is deformed, the peak moves to lower temperature becoming again a little sharper (Fig. 4, 5, and 6 and Table 1). The LTC is now predominant over the HTC in the deformed state. It is clear from the grain size dependence shown in Fig. 8 that only the HTC has a correlation between the peak height and the number of grains. The surface observations show that the macroscopic grain boundary shape does not change by long-time annealing and that the deformation takes place mainly in the grains (Fig. 3 and 7).

From all these experimental results LTC and HTC are considered to be related with the lattice dislocations and the grain boundary dislocations, respectively. The situation, however, may be much more complicated because both types of dislocations are interrelated with each other and they cannot be treated independently. For example, some of the lattice dislocations introduced by the deformation may pile up in the grain boundaries. They may rearrange or disappear there during the successive annealing. These processes change the configuration of the grain boundary dislocations [23], though the macroscopic shape of the boundaries looks unchanged. The lattice dislocations, on the other hand, are also influenced by the existence of the grain boundaries. The density of the lattice dislocations, for example, depends largely on the grain size [21]. They also have a possibility to gather together inside the grains to form new low-angle grain boundaries or subboundaries [11].

Although the main peak was divided into two components here, each component seems to split further into a few components. Further decomposition, however, is very difficult at the present stage because of the rather poor accuracy of the t' values as stated before.

The amount of recovery by the successive annealing after rapid deformation decreases as the degree of deformation increases (Fig. 6). This may be due to the irrecoverable steps (slip bands) formed by the deformation. The height and the number of steps increase with the increase in the degree of deformation. Once the steps are formed, they do not disappear by the annealing. As the specimen diameter of the slipped region is apparently a little smaller than that of the non-slipped one, the vibrational stress is concentrated in the former region. The contribution of the irrecoverable slipped region to the total internal friction is expected to increase as the degree of deformation increases.

The small peak often observed around 100 °C seems to correspond to the low temperature peak reported by Esnouf et al. [20] and Riviere et al. [22]. Recent preliminary measurements on single and polycrystalline sheet specimens with a newly developed inverted flexure pendulum [17] also exhibit a large internal friction peak in this temperature range [24]. This peak is different from the LTC, but may be also related with the lattice dislocations, though the details have not yet been known.

The conflict about the origin of the large high-temperature peak seems to be solved now. Two different peaks seem to have been discussed as an identical one. The peak observed in deformed single crystals and that in well-annealed polycrystals may correspond to LTC and HTC, respectively.

References

[1] T. S. Kê, Phys. Rev. 71, 533 (1947).
[2] T. S. Kê, Phys. Rev. 72, 41 (1947).
[3] C. C. Smith and G. M. Leak, Nuovo Cimento **33B**, 388 (1976).
[4] J. Woirgard, J. P. Amirault, and J. de Fouquet, Internal Friction and Ultrasonic Attenuation in Solids, Vol. I, Springer-Verlag, 1975 (p. 392).
[5] E. Bonetti, E. Evangelista, P. Gondi and R. Tognato, Nuovo Cimento **33B**, 408 (1976).
[6] J. Woirgard, A. Riviere, and J. de Fouquet, J. Physique **C5**, 407 (1981).
[7] J. T. A. Roberts, J. Inst. Metals **98**, 381 (1970).
[8] C. C. Smith and G. M. Leak, Internal Friction and Ultrasonic Attenuation in Solids, Vol. I, Springer-Verlag, 1975 (p. 383).
[9] J. Woirgard, J. P. Amirault, and J. de Fouquet, Acta metall. **22**, 1003 (1974).
[10] Z. Q. Sun and T. S. Kê, J. Physique **C5**, 451 (1981).
[11] T. S. Kê, J. Physique **C5**, 421 (1981).
[12] K. Iwasaki, K. Lücke, and G. Sokolowski, Acta metall. **28**, 855 (1980).
[13] K. Iwasaki, J. Phys. Soc. Japan **49**, 271 (1980).
[14] K. Iwasaki, phys. stat. sol. (a) **63**, K137 (1981).
[15] K. Iwasaki, Bull. Japan. Inst. Metals **15**, 43 (1976).
[16] K. Iwasaki, Bull. Japan. Inst. Metals **21**, 715 (1982).
[17] K. Iwasaki, J. Phys. E **16**, 421 (1983).
[18] J. N. Cordea and J. W. Spretnak, Trans. AIME **236**, 1685 (1966).
[19] T. M. Williams and G. M. Leak, Acta metall. **15**, 1111 (1967).
[20] C. Esnouf and G. Fantozzi, J. Physique **C5**, 445 (1981).
[21] E. Bonetti, E. Evangelista, P. Gondi, and R. Tognato, phys. stat. sol. (a) **39**, 661 (1977).
[22] A. Riviere, J. P. Amirault, and J. Woirgard, J. Physique **C5**, 439 (1981).
[23] R. A. Berrisford, C. C. Smith, and G. M. Leak, Internal Friction and Ultrasonic Attenuation in Solids University Tokyo Press, 1977 (p. 307).
[24] K. Iwasaki, to be published.

(Received May 17, 1983)

phys. stat. sol. (a) **79**, 125 (1983)

Subject classification: 1.3 and 5; 14.1; 18.2; 21.1.1

Institute of Physical Metallurgy, Czechoslovak Academy of Sciences, Brno[1]) *(a)*
and Zentralinstitut für Festkörperphysik und Werkstofforschung
der Akademie der Wissenschaften der DDR, Dresden[2]) *(b)*

Atomic Structure and Magnetic Properties of the Pseudobinary Alloys Fe_3(Al, Si)

By

O. SCHNEEWEISS (a), T. ZEMČÍK (a), T. ŽÁK (a), and S. MAGER (b)

The influence of substitution of Al by Si on the atomic order, saturation magnetic polarization, and electrical resistivity is studied in $Fe_3Al_{1-x}Si_x$, $0 \leqq x \leqq 1$, alloys. The changes in the hyperfine parameters of the ^{57}Fe Mössbauer spectra indicate that the degree of DO_3 order after order anneal is rised, accompanied by a separation of the Fe_3Si and Fe_3Al types of local surroundings. The most significant changes of isomer shifts, magnetic hyperfine fields, saturation magnetic polarization, and electrical resistivity due to atomic order are observed near the Fe_3Al and $Fe_3Al_{0.25}Si_{0.75}$ compositions.

Der Einfluß der Substitution von Al durch Si auf die Atomordnung, die magnetische Sättigungspolarisation und den spezifischen elektrischen Widerstand wird in $Fe_3Al_{1-x}Si_x$ $(0 \leqq x \leqq 1)$ untersucht. Die Änderungen der Hyperfein-Parameter der ^{57}Fe-Mössbauer-Spektren zeigen, daß der Grad der DO_3-Ordnung mit der Wärmebehandlung ansteigt und gleichzeitig eine Auftrennung in die Fe_3Si- bzw. Fe_3Al-Typen der lokalen Umgebung stattfindet. Die bedeutendsten Änderungen der isomeren Verschiebungen, der magnetischen Hyperfeinfelder, der magnetischen Sättigungspolarisation und des elektrischen Widerstandes infolge atomarer Ordnung werden nahe den Zusammensetzungen Fe_3Al und $Fe_3Al_{0,25}Si_{0,75}$ beobachtet.

1. Introduction

Ternary Fe–Al–Si alloys are attractive not only because of their physical properties, but also for their technical applicability as soft magnetic materials (Sendust, Alsifer). For the latter reason, it is the range $Fe_3Al_{1-x}Si_x$ which is most studied.

Until now, the main effort was concentrated on understanding the phase diagram [1]. In the whole concentration range, these alloys form a pseudobinary solid solution Fe_3(Al, Si), of which both the structure and the order types are identical with those of Fe_3Al. X-ray measurements complemented with microhardness, saturation magnetization, and Curie temperature indicate the existence of the ternary superstructure Fe_6AlSi [1]. This was suggested in [2] on the basis of the ordering study of these alloys by the Debye X-ray method and the temperature dependence of the electrical resistivity. Besides that, a superstructure $Fe_{75}Al_{18}Si_7$ is considered.

The study of the order–disorder phenomena in these alloys is the subject of a whole series of more recent works. In [3], from the critical temperatures $T_1(DO_3-B2)$ and $T_2(B2-A2)$, the energy characteristics of atomic ordering were determined. The rise of T_1 and T_2 with increasing Si content prohibited the evaluation of the whole concentration range, because the B2 phase extends up to the solidus for Si content above 12.5 at%. Thus, the ordering energy obtained for the first two neighbouring shells, v_1 and v_2, rises with increasing Si content, the ratio v_1/v_2 remaining independent of composition.

[1]) Žižkova 22, 616 62 Brno, Czechoslovakia.
[2]) Helmholtzstraße 20, DDR-8027 Dresden, GDR.

The influence of the substitution of Si by Al atoms is shown in [4]. Namely, the increasing Al content reduces the degree of the DO_3 order towards B2 and brings about more remarkable changes in the atomic order due to various heat treatments. Changes of the lattice spacings and deviations from Vegard's law were followed in [5]. The aim of this work is to find the atomic order as determined from the Mössbauer spectra and some other physical quantities, such as the average isomer shift and average hyperfine field at the iron nucleus, saturation magnetic polarization, and electrical resistivity.

2. Experimental

The alloys of the composition $Fe_3Al_{1-x}Si_x$ were prepared by melting high purity Fe, Al, and Si (99.99). The samples have been cast in the form of discs (30 mm in diameter and 3 mm thick) or ground and chemically polished from the ingots down to 50 μm foils. Sample analyses are given in Table 1. Heat treatment of the samples was performed in two ways: (i) quenching from 950 °C, (ii) annealing at 700 °C for 1 h and cooling down to 300 °C at the rate 3.5 K/h. In the first way a DO_3 order as low as possible should be prepared, the second way, on the contrary, was directed towards the highest order. The disc samples, however, could not be quenched, because of many macroscopic cracks arising.

All the measurements were performed at room temperature. Mössbauer spectra of 14.4 keV γ-rays were taken with the foils in transmission arrangement and with the discs by scattering technique using ^{57}Co in Rh source. Spectra analyses were done by a standard least-squares fit to sets of Zeeman sextets of Lorentzian lines. This procedure yielded relative intensities of the partial components and corresponding hyperfine structure (hfs) parameters, i.e. ground state splitting g_0 and isomer shifts IS. Saturation magnetic polarization J_s was measured at samples of ≈ 10 mg by the magnetic balance.

Electrical resistivity was obtained by the van der Pauw four-point method on the discs, except in the state after quenching where the measurements were prohibited due to macroscopic defects.

Table 1

Composition of the samples. Relative contents of Si and Al in the samples are denoted c_{Si} and c_{Al}, respectively

sample No.	1	2	3	4	5	6	7	8
c_{Si}	—	0.046	0.090	0.116	0.173	0.161	0.193	0.228
c_{Al}	0.240	0.202	0.183	0.125	0.110	0.086	0.068	—

3. Results

A typical Mössbauer spectrum for $x = 11.6$ at% Si is shown in Fig. 1. An example of the discrete distribution of hyperfine fields is depicted in Fig. 2 in the form of a bar diagram. The average internal magnetic field at ^{57}Fe was calculated by the formula

$$\bar{B}_{int} = k \sum_{i=1}^{n} I_i g_{0i} ;$$

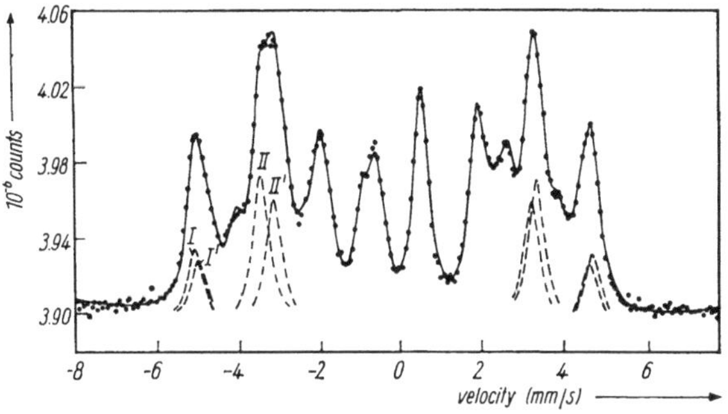

Fig. 1

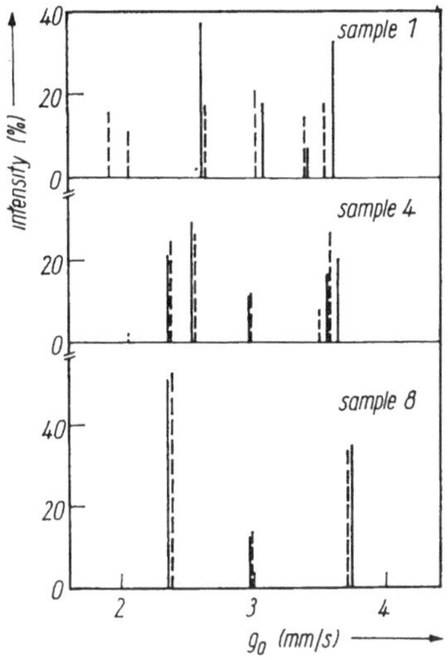

Fig. 2

Fig. 1. Mössbauer spectrum of the sample $Fe_{75.9}Al_{12.5}Si_{11.6}$

Fig. 2. Hyperfine field distribution in some samples (compositions see Table 1). The bar heights and positions, respectively, represent the intensities and splittings g_0 of the spectral components, respectively, — — — as-cast state, ——— annealed state

I_i being the relative intensity of i-th Zeeman component with g_{0i}, n the number of components, and $k = 8.428$ T/mm s^{-1} transforms the splittings g_0 (in mm s^{-1}) to fields B_{int} (in T) [6]. The average isomer shift $\overline{IS}$ was calculated analogously. Concentration dependences of $\overline{IS}$ and $\overline{H}_{eff}$ are shown in Fig. 3 and 4a, respectively. Similarly, the results for the macroscopic quantities J_s and ϱ are given in Fig. 4b and 5.

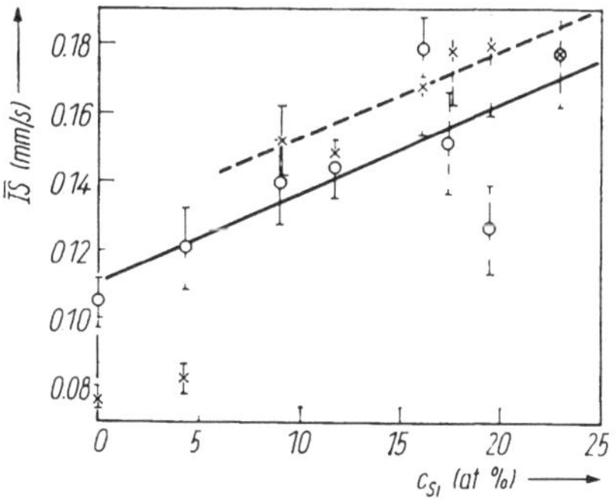

Fig. 3

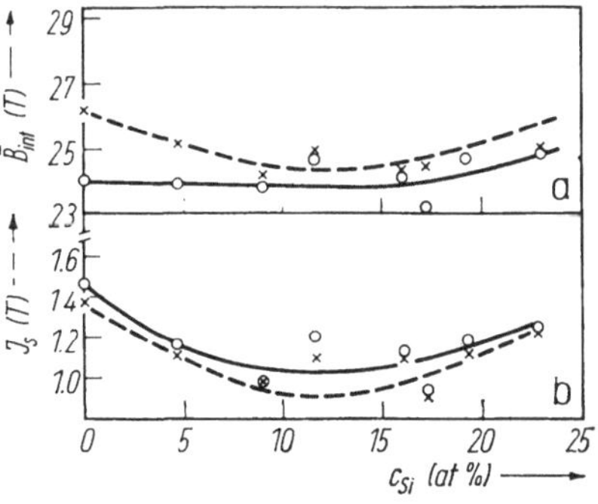

Fig. 4

Fig. 3. Composition dependence of a verage isomer shift $\overline{IS}$ (×, — — — as-cast state, ○, ——— annealed state with corresponding linear regressions)

Fig. 4. Dependence of average hyperfine magnetic field $\overline{H}_{eff}$ and saturation magnetic polarization on the composition. Symbols see Fig. 3

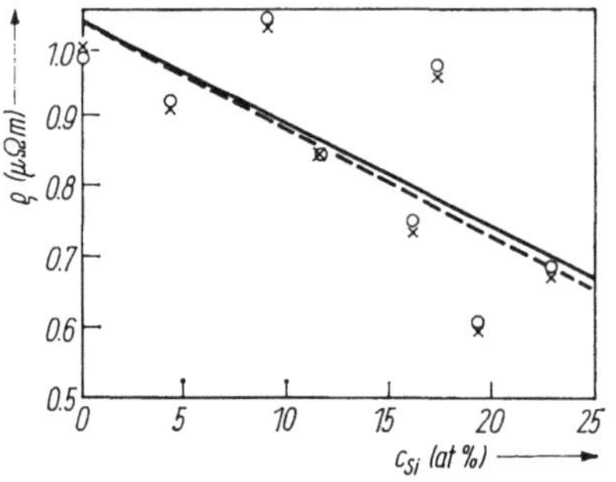

Fig. 5. Dependence of the electric resistivity ϱ on the composition. Symbols see Fig. 3

4. Discussion

The variation of the local order parameter at the transition between the DO_3 and B2 types of order for the considered range of the ternary system was shown in [4]. More recent results generally agree with those already published. A detailed analysis of the hyperfine field distribution leads to the conclusion that in the annealed/slowly cooled samples a rearrangement of Si and Al atoms probably takes place. From the point of view of the short-range order a separation of the Fe_3Si and Fe_3Al type of surroundings occurs because of the repulsive interaction between the Si and Al atoms. An analogous effect was observed in the $Fe_3(Al, Ge)$ alloys [7], where it led to the destabilization of the DO_3 structure.

In our spectra, the separation of both surroundings caused the decomposition of the dominant spectral components into two parts whose hfs parameters agree with those of pure Fe_3Si and Fe_3Al. This situation is illustrated with the components I, I′ and II, II′ in Fig. 1. The proportion of sub-components changed according to the Si–Al substitution, their separation being more pronounced at higher Si contents. It is, nevertheless, difficult to decide from these results whether a simultaneous formation of the above-mentioned ternary superstructures occurs.

The average isomer shift $\overline{IS}$ increases (relative to pure iron) with the substitution of Al by Si. Except the Al side, the as-cast samples give somewhat lower values of $\overline{IS}$ than the annealed and slowly cooled ones. However, in Al-rich alloys ordering annealing causes a decrease of $\overline{IS}$, probably due to a high density of structure defects in cast samples.

The dependence $\overline{H}_{eff}$ versus silicon content shows that in the as-cast samples (nearly as if quenched), the average field is almost composition independent despite the remarkably lower degree of DO_3 order in the vicinity of Fe_3Al. On the other hand, after annealing/slow cooling, i.e. increasing the DO_3 order, $\overline{H}_{eff}$ rises due to the positive influence of Al and, secondly, becomes anomalous around 18 at% Si (Fig. 4a).

On the assumption of localized spins and interactions in the alloys, a simple relation between $\overline{H}_{eff}$ and the average magnetic moment of iron $\overline{\mu}$ can be considered of the type

$$\overline{B}_{int} = a\overline{\mu} ;$$

a being the hyperfine coupling constant given by core polarization [8]. In the case of constant iron content, $\overline{\mu}$ is proportional to J_s. Really, the J_s versus x dependence has features similar to $\overline{B}_{int}$ versus x, being highest for Fe_3Al and diminishing by substitution with a flat minimum around Fe_6AlSi. However, in contrast to $\overline{B}_{int}$, the as-cast and quenched states exhibit higher J_s than the annealed one. This discrepancy suggests that the relation between $\overline{B}_{int}$ and J_s is not so simple. Higher values of J_s in the disordered state are in contradiction not only to the published values for Fe_3Si and Fe_3Al [9, 10], but also to the above results for $\overline{B}_{int}$. Hence the explanation is to be sought probably in factors other than the changes of J_s. It seems to be a change during the order anneal that made the magnetic saturation of tiny samples more difficult. The magnetic balance gives a magnetic induction of about 0.8 T. This is usually enough to saturate samples, but not in case of Si-rich Fe_3(Al, Si) alloys, where the sample volume corresponded to a single grain.

In contrast to the above results, no remarkable changes in the electrical resistivity were observed resulting from the various heat treatments of the samples. This might be due to the local character of the above order variations which influence only slightly the value of ϱ reflecting rather the state of long-range order, and due to deviations of iron content in the samples, which should be 75 at%. In this way, the exceptional deviations from the linear decrease from Fe_3Al to Fe_3Si occurring at about 12.5 at% Si can be explained by the existence of a ternary superstructure Fe_6AlSi.

In conclusion we can state that we have observed by means of Mössbauer spectroscopy an important phenomenon in the atomic structure of the alloys: separation of Fe_3Al and Fe_3Si surroundings. The macroscopic magnetic and electric properties and mean values of hfs, however, do not react on these changes in the atomic ordering. The annealing, which should increase the degree of DO_3 order in the samples, has only weak influence on the mentioned physical properties. The dominant role is played by the composition of the samples.

Acknowledgement

The authors are very grateful to Prof. U. Gonser for stimulated discussion of the preprint of this paper.

References

[1] F. Lihl, R. Burger, F. Sturm, and H. Ebel, Arch. Eisenhüttenwesen **39**, 877 (1968).
[2] S. V. Arutjunyan, Izv. Akad. Nauk Arm. SSR, Ser. Tech. Nauk **24**, 54 (1971).
[3] A. A. Katznelzon and V. E. Polishchuk, Fiz. Metallov i Metallovedenie **36**, 321 (1973).
[4] T. Zemčík and O. Schneeweiss, Proc. Conf. Soft Magn. Mater. 3, Bratislava 1977, Vol. 2 (p. 748).
[5] S. J. Cowdery and F. X. Kayser, Mater. Res. Bull. **14**, 91 (1979).
[6] J. G. Stevens and V. E. Stevens, Ed., Mössbauer Effect Data Index, IFI Plenum, New York 1976 (p. 5).
[7] G. Athanassiadis, G. Le Caer, J. Foct, and L. Rimlinger, phys. stat. sol. (a) **40**, 425 (1977).
[8] F. van der Woude and I. Vincze, J. Physique **40**, C1-151 (1980).
[9] K. Numakura, A. Tsugawa, M. Sugano, and Y. Sato, Sci. Rep. Res. Inst. Tôhokù Univ. **24A**, 48 (1972).
[10] A. Arrott and H. Sato, Phys. Rev. **114**, 1420 (1959).

(Received July 8, 1983)

phys. stat. sol. (a) 79, 131 (1983)

Subject classification: 20.3; 22.8.1

Fachgebiet Werkstoffe der Elektrotechnik der Universität Duisburg[1])

Micron-Scaled Spectral-Resolved Cathodoluminescence of Grains in Bariumtitanate Ceramics

By

G. KOSCHEK and E. KUBALEK

The grain cathodoluminescence (CL) properties of bariumtitanate ceramics, sintered at different oxygen partial pressures are systematically investigated. The spectra are found to be composed of different bands, the intensities of which depends strongly on the specimen temperature and on the oxygen partial pressure of the ambient atmosphere during the sintering process. By fitting the obtained spectra with superposed Gaussian curves it is shown that the involved recombination processes can be satisfactorily explained within the configurational coordinate model. Furthermore it is possible to trace back the observed electronic transitions to vacancy luminescence, corroborating existing theories on band schemes, and on dependences of vacancy concentrations on oxygen partial pressure during the sintering. In this way the observed bands may be assigned to transitions in the energy gap of $BaTiO_3$.

Die Kathodolumineszenzeigenschaften von Kristalliten in Bariumtitanat-Keramiken, die unter verschiedenen Sauerstoffpartialdrücken gesintert werden, werden systematisch untersucht. Es kann gezeigt werden, daß die Spektren aus verschiedenen Banden bestehen, deren Intensitäten starke Abhängigkeit von der Probentemperatur und vom Sauerstoffpartialdruck der Sinteratmosphäre aufweisen. Durch das Anpassen der erhaltenen Spektren mit superponierten Gaußkurven wird gezeigt, daß die beteiligten Rekombinationsprozesse mit dem Konfigurationskoordinaten-Modell erklärt werden können. Ferner ist es möglich, die beobachteten elektronischen Übergänge auf Leerstellenlumineszenz zurückzuführen und bestehende Theorien über Bandschemata und Abhängigkeiten von Leerstellenkonzentrationen vom Sauerstoffpartialdruck während des Sinterns zu untermauern. Auf diese Weise kann eine Zuordnung der beobachteten Bänder zu Übergängen in der verbotenen Zone von $BaTiO_3$ erfolgen.

1. Introduction

Devices of n-doped bariumtitanate ceramics (BTC) are of increasing importance, as they open up a large variety of applications and are cheap and reliable, too. In these devices not only the dielectric but also the cold conductor properties of BTC are exploited (high dielectricity (HD) capacitors and positive temperature coefficient (PTC) resistors, respectively).

Both properties are traced back to the existence of depletion layers at the grain boundaries of the $BaTiO_3$ crystallites [1 to 3]. It is assumed that the depletion regions are caused by barium vacancies acting as acceptors. Thermodynamical calculations predict a strong dependence of the vacancy concentrations on the oxygen partial pressure of the sintering atmosphere. A higher value of the oxygen partial pressure causes a higher concentration of the Ba vacancies [9].

Up to now no experimental verification neither of the existence of these acceptors nor of their mentioned dependence on oxygen partial pressure had been made. In this paper an experimental proof of these theoretical predictions is reported, using spectral-resolved cathodoluminescence (CL) measurement techniques of high spatial resolution for a systematic analysis of the electronic structure in BTC grains.

[1]) Kommandantenstraße 60, D-4100 Duisburg 1, FRG.

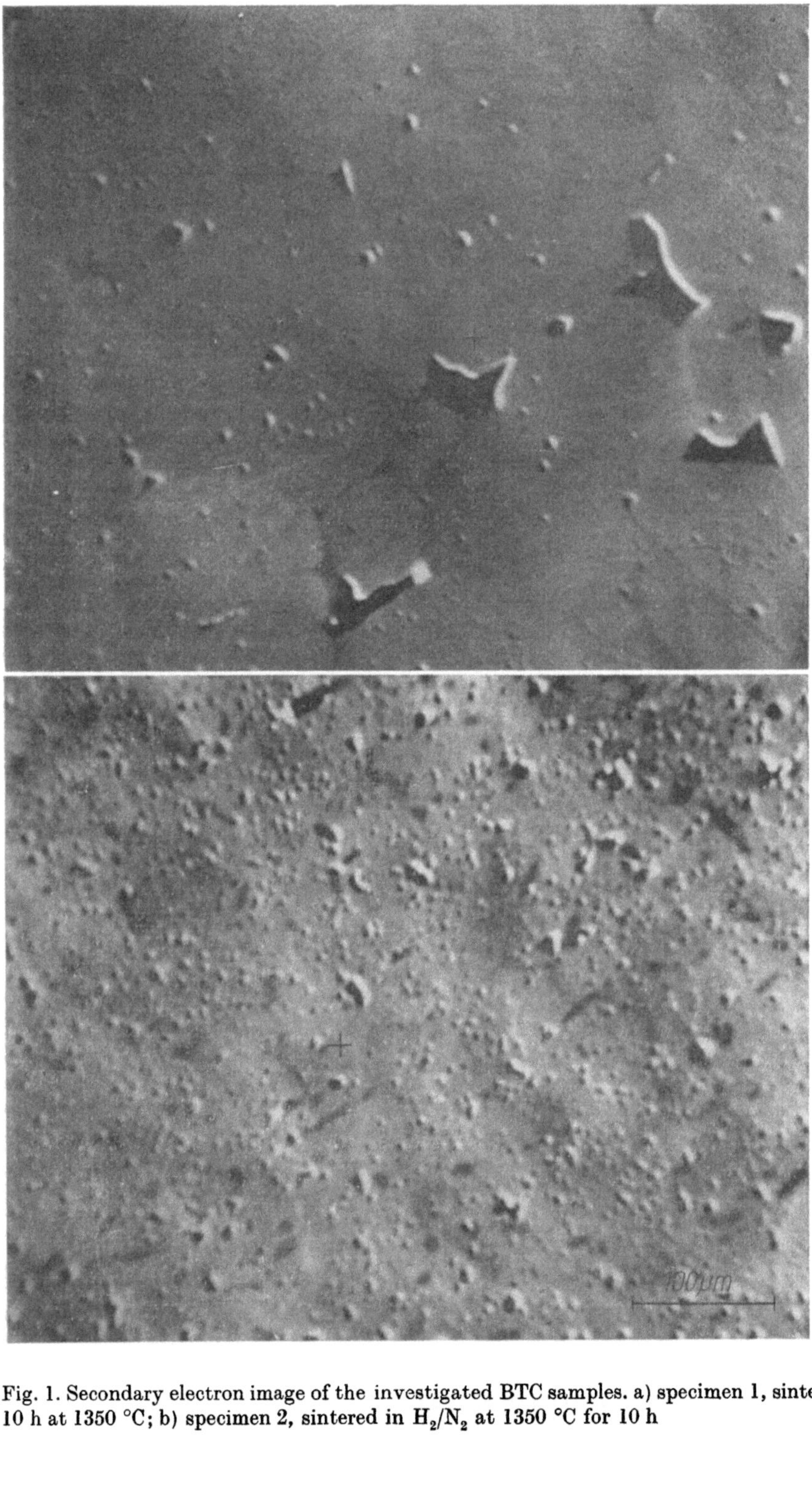

a

b

Fig. 1. Secondary electron image of the investigated BTC samples. a) specimen 1, sintered in air for 10 h at 1350 °C; b) specimen 2, sintered in H_2/N_2 at 1350 °C for 10 h

2. Preparation

The samples were prepared from $BaCO_3$ (Merck 1714) and TiO_2 (Kronos) with an excess of 3 mol% TiO_2 to obtain Ti-rich second phases which are considered to be sintering aids during the sintering process. The n-doping was realized by adding of 0.2 mol% Y_2O_3. After a milling process in water with agate balls the material was dried and then calcined in air at 1050 °C for 2 h in order to obtain the solid-state reaction $BaCO_3 + TiO_2 \rightarrow BaTiO_3 + CO_{2\uparrow}$. After renewed milling and drying the substance was granulated with polyviol and then pressed into disks with a diameter of 15 mm and a thickness of about 1.3 mm. These disks were sintered at various temperatures between 1200 and 1400 °C in steps of 30 K for 10 or 20 h, respectively. The sinterings were carried out in a H_2/N_2 steam (reducing), or in air (oxidizing), respectively. Every sample was cooled down with 5 K min^{-1} to room temperature. Afterwards the samples were polished with Syton W 30.

3. Experimental

The micron-scaled CL measurements were carried out in a modified scanning electron microscope (SEM; Cambridge S4-10) [4]. These modifications resulted in essential improvements of the CL mode of the SEM, enabling uniquely a correlation of topographic, chemical, and electronic structures of the material. In detail by installation of an efficient photon collection system, adaptation of a vacuum grating monochromator (GCA/Mc Pherson 218), and of a cooled GaAs photomultiplier to the SEM, spectral CL analysis in the wavelength range from 300 to 900 nm is enabled. For the specimens a cryostat is adapted, making possible CL investigations down to liquid helium temperatures. The obtained CL spectra were corrected with the spectral efficiency characteristic of the GaAs photomultiplier.

4. Results

Fig. 1 a, b are backscattered electron images of typical BTC samples (1 a: oxidizing atmosphere (air), 1350 °C for 10 h, specimen 1; 1 b: reducing atmosphere (H_2/N_2), 1350 °C for 10 h, specimen 2). At the cross-marked spots in Fig. 1 a, b CL spectra were measured at various specimen temperatures. The normalized spectra are shown in Fig. 2 a, b. The CL spectra of specimens, sintered in oxidizing atmosphere (Fig. 2 a),

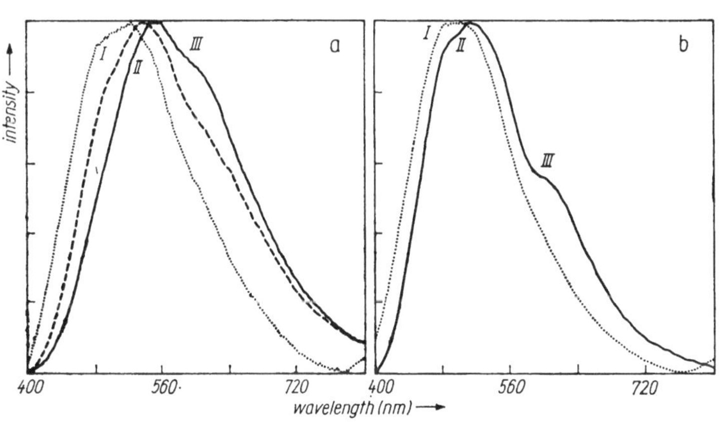

Fig. 2. Normalized CL spectra of a) specimen 1 (·········· 77 K; – – – 130 K, ——— 293 K; (I) to (III) CL maxima); b) specimen 2 (·········· 77 K, ——— 293 K; (I) to (III) CL maxima)

exhibit for every temperature a broad luminescence of about 185 nm full width at half maximum (FWHM). For 77 K the spectrum shows an emission maximum at 550 nm (II) and a shoulder at 610 nm (III). For 130 K the intensity of (III) compared to (II) has decreased, at 480 nm another shoulder (I) has appeared. (II) has shifted to 530 nm. For 293 K (III) has almost disappeared, (II) again has shifted by 20 nm to lower wavelengths, (I) now appears to be a marked maximum.

For specimens, sintered in reducing atmosphere, the CL spectrum (Fig. 2b) also shows a broad luminescence of about 185 nm FWHM for every temperature. For 77 K an emission maximum at 520 nm can be observed (II), additionally two shoulders at 480 nm (I) and at 610 nm (III) appear. For 293 K the relative intensities of (II) and (III) have decreased, but increased for (I). (II) has shifted by 20 nm to lower wavelengths. Additional CL measurements on both specimens at different specimen temperatures confirm the observed behaviour: The observed broad luminescence represents a superposition of three different, clearly distinguishable, temperature-dependent emission bands. Compared to the intensity of (II), for increasing temperatures the relative intensity of (III) decreases and increases continuously for (I). Measurements at other specimen positions inside the BTC grains showed the same spectral behaviour. Except for the obvious influences of the oxygen partial pressure on the spectra, no dependence on other sintering parameters such as sintering temperature or sintering time or the choice of the starting raw materials could be observed.

The grain luminescence intensity of samples sintered in reducing atmosphere is about three orders of magnitude lower than that of oxidized sintered samples. In samples which were sintered at reducing atmosphere (II) seems to be shifted by 30 nm to lower wavelengths compared to samples which were sintered at oxidizing conditions. For a given temperature the relative intensity of (III) compared to (II) is much lower than under oxidizing conditions and (I) already at 77 K is clearly observable.

5. Discussion

The appearance of particular band in the measured spectra implies that this emission behaviour is not due to a pure band–band transition as proposed by Ihrig et al. [5]. In agreement with their opinion, the great FWHM of the luminescence indeed gives rise to the conclusion that the recombination results from both radiative and non-radiative processes. They supposed that the recombination can be theoretically described with the configurational coordinate model (CC model) (cf. [6]). This assumption is corroborated in the following.

In the CC model, the shape of each spectral band is expected to be Gaussian. Therefore, the validity of the CC model is proved, if the experimentally obtained spectra can be fitted with superimposed Gaussian curves of the form

$$I(\lambda) = \frac{1}{\lambda^2} \sum_{i=1}^{z} I_{0i} \exp\left[-P_i\left(\frac{1}{\lambda} - \frac{1}{\lambda_{0i}}\right)^2\right], \tag{1}$$

where I is the emission intensity, λ the wavelength, z the number of the superimposed Gaussian curves, I_{0i} the intensity in the emission maxima, P_i are factors inversely proportional to FWHM; the λ_{0i} are connected to the band extremum wavelengths λ_{Ei} as $1/\lambda_{0i} = -\lambda_{Ei}/P_i + 1/\lambda_{Ei}$; the I_{Ei} are connected with the I_{0i} as $I_{Ei} = I_{0i}/\lambda_{Ei}^2$).

Corresponding to the experimental results we superimposed three Gaussian curves. In Table 1 the data of these curves are given. In Fig. 3a, b for two temperature values measured (solid) and fitted (dotted) spectra are shown together. For wavelengths lower than 620 nm they are in a very good agreement, for higher wavelengths, however,

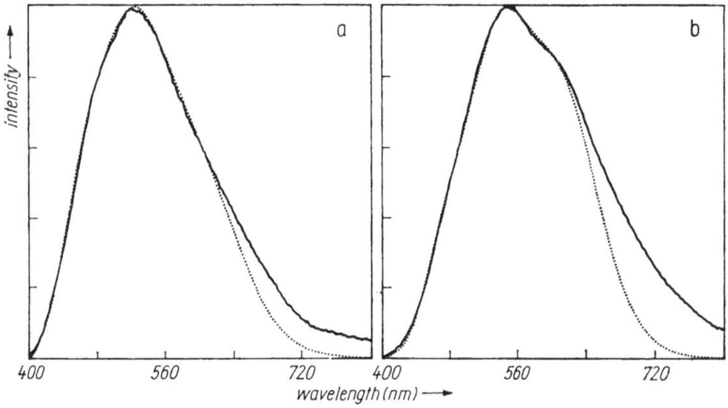

Fig. 3. Measured CL spectra (solid) a) specimen 1, 190 K and b) specimen 1, 77 K fitted by superposition of three Gaussians (dotted)

the fit curves show distinctly lower values of intensity compared to the measured spectra.

A reason for this discrepancy in the higher-wavelength range could be a fault in the luminescence model itself or a systematic, wavelength-dependent error. Furthermore, however, the existence of a fourth CL band could be taken for explanation. The first possibility cannot be completely excluded, because it is known that for example ZnS band spectra show low- and high-energy exponential tails [12], though exhibiting Gaussian shapes around the emission maximum, as predicted by the CC model for the whole wavelength range of the particular band. In the case of BTC grain luminescence, however, the deviation from Gaussian behaviour exclusively occurs in the long, but not in the short-wavelength range, where the validity of the CC model is thus demonstrated.

Table 1

Fitparameters (equation (1), $z = 3$)

specimen 1	λ_{Ei} (nm)			relative intensities I_{Ei} (arb. un.)			P_i (10^{-6} nm^2)		
temperature (K)	$i = 1$	2	3	$i = 1$	2	3	$i = 1$	2	3
190	623	549	489	0.53	1	1.11	39	31	22
77	630	557	492	0.69	1	0.47	51	34	29

The second possibility, a systematic measuring error, is to be excluded, because experiments showed that the signal detection and processing unit does not cause a wavelength-dependent background.

To prove the effect of a fourth CL band, the measured curves now were fitted with four Gaussian curves, adding to the three significant bands another band of longer wavelength with small intensity. In Table 2 the fit parameters for different temperatures and for the investigated specimens (1, 2) are listed. Fig. 4a, b show the results of this fit for the two CL spectra of Fig. 3a, b and demonstrate an excellent agreement

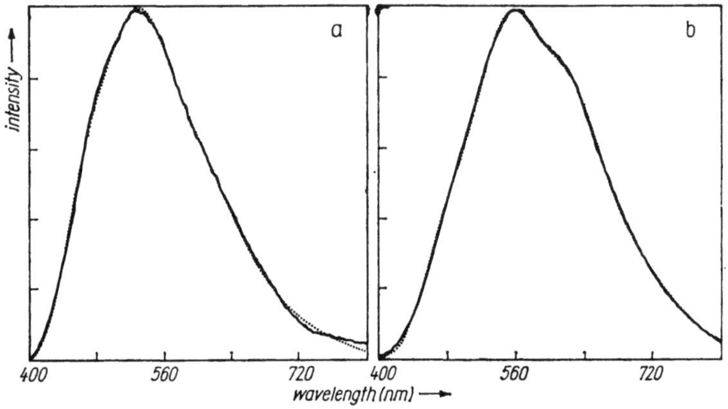

Fig. 4. Measured CL spectra (solid) a) as in Fig. 3a and b) as in Fig. 3b fitted by superposition of four Gaussians (dotted)

in the whole wavelength range. In Fig. 5 the fitted curve of Fig. 4a is shown with its four Gaussian components.

To explain the nature of these transitions, the band scheme, theoretically calculated by Daniels and Härdtl [9], can be taken as a basis (Fig. 6). Transitions like $V_{Ba}^{--} \rightarrow V_{Ba}^{-}$, $V_{Ba}^{-} \rightarrow V$, $C \rightarrow V_{O}^{+}$, $C \rightarrow Y_{+}$ (arrows indicate the transition, V valence band, C conduction band) could not be detected in any case, because the photomultiplier response is not sufficient in this wavelength range. A transition $V_{O}^{++} \rightarrow V$ (wavelength 775 nm (1.6 eV)) is not probable because of the trap character of the doubly ionized oxygen vacancies. The transition $V_{Ba}^{--} \rightarrow V$ that would cause an emission at about the same wavelength is to be excluded for the same reason, too. An influence of $Y_{+} \rightarrow V$ also may not have any importance, because X-ray excited luminescence spectra of undoped bariumtitanate single crystals show spectral comparability [10].

Table 2

Fitparameters (equation (1), $z = 4$)

specimen	λ_{Ei} (nm)				relative intensities I_{Ei} (arb.un.)				P_i (10^{-6} nm^2)			
temperature (K)	$i = 1$	2	3	4	$i = 1$	2	3	4	$i = 1$	2	3	4
$\frac{1}{300}$	667	602	540.5	482	0.2	0.7	1	1.9	18	35	41	15
$\frac{1}{190}$	676	606.5	537	479	0.2	0.5	1	0.9	22	39	33	21
$\frac{1}{77}$	687	615.5	545.5	488	0.32	0.61	1	0.47	30	51	36	28
$\frac{2}{77}$	687	615.5	545.5	488	0.16	0.51	1	0.71	30	51	36	28

The significant relative and absolute dependence of the luminescence intensity on oxygen partial pressure of the sintering atmosphere indicates a connection to vacancies, because just their concentration should decisively depend on oxygen partial pressure

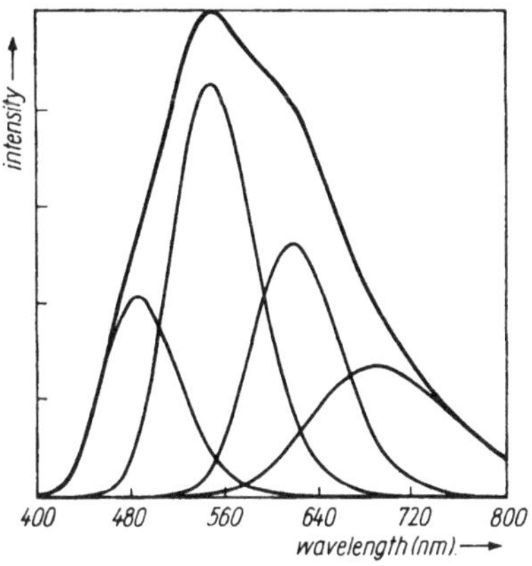

Fig. 5. Fit curve of Fig. 4b with its four Gaussian components. $T = 77$ K

according to the mass action law $[V_O] \times [V_{Ba}] = K$ ([11]; $[V_O]$, $[V_{Ba}]$ concentration of the oxygen and barium vacancies, respectively; K a constant depending on temperature and oxygen partial pressure).

In the case of sintering at reducing atmospheres the luminescence intensity decreases four orders of magnitude compared to sinterings at oxidizing conditions. This gives rise to the conclusion that the influence of the sintering atmosphere causes the insertion of a "killer" term between the present terms, thus quenching the radiative transitions. A similar luminescence behaviour is observed at ZnS phosphors if smallest amounts of iron are added [13]. In the case of BTC the oxygen vacancy concentration should take very high values if reducing sintering is performed (10^{19} cm^{-3} at $p_{O_2} \leqq 10^{-10}$ atmospheres [9]). Thus the observed dependence of the luminescence behaviour on the sintering atmosphere indicates a strong quenching influence of oxygen vacancy terms, especially of the doubly ionized oxygen vacancies. The suppositions that these terms because of their energetic position in the band gap either act as centres of infrared emission or cause an intensified non-radiative emission, could not be experimentally verified up to now.

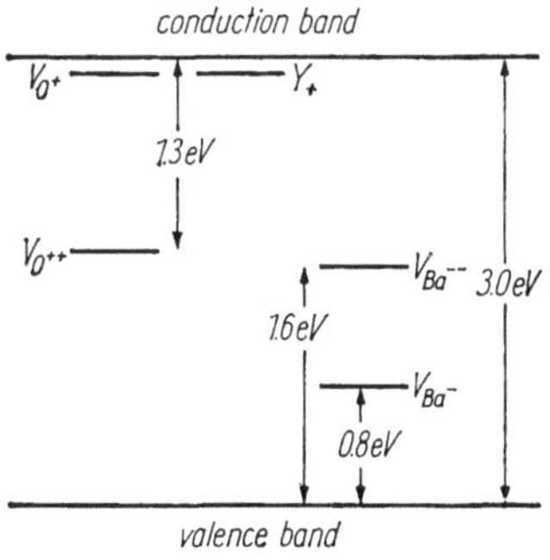

Fig. 6

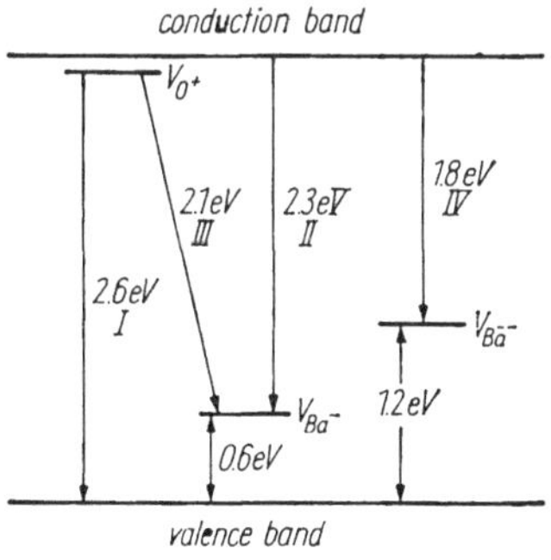

Fig. 7

Fig. 6. Band scheme of $BaTiO_3$ after Daniels and Härdtl [9] (V_{O^+}, $V_{O^{++}}$, V_{Ba^-}, $V_{Ba^{--}}$ single or doubly ionized oxygen or barium vacancies, respectively; Y_+ ionized n-dopant; V_{O^+} and Y_+ lie about 100 meV below the conduction band edge)

Fig. 7. Recombination processes in bariumtitanate grains (observed transitions marked with (I) to (IV))

Concerning the relative intensities of the spectral components, it must be considered that in the case of samples sintered at reducing atmospheres the intensity of every luminescence process involving oxygen vacancies should increase because of the high concentration of these defects. On the contrary, the concentration of the Ba vacancies should be very low for reducing sinterings (10^{16} cm^{-3} at $p_{O_2} \leqq 10^{10}$ atmospheres). So the number of recombination processes at these vacancies should be lowered compared to oxidized material, leading to lower intensities of recombination processes with participating Ba vacancies.

Consequently the observed bands can be associated to the following transitions (Fig. 7; C conduction band, V valence band)

(I) (2.6 eV at RT) $V_{O^+} \rightarrow V$,

(II) (2.3 eV at RT) $C \rightarrow V_{Ba^-}$,

(III) (2.1 eV at RT) $V_{O^+} \rightarrow V_{Ba^-}$,

(IV) (1.8 eV at RT) $C \rightarrow V_{Ba^{--}}$.

At low temperatures the electron in the V_{Ba^-} term will recombine with the hole in the valence band after excitation by the electron beam. Then the radiative transitions $V_{O^+} \rightarrow V_{Ba^-}$ and $C \rightarrow V_{Ba^-}$ as well as $C \rightarrow V_{Ba^{--}}$ will occur (II–IV), the latter supposed to be followed by a nonradiative transition to V_{Ba^-}. At higher temperatures, however, the V_{Ba^-} term may not be emptied completely. Thus for increasing temperatures less electrons will recombine, and therefore the luminescence intensity of (II to IV) will decrease. This gives rise to augmented recombination by other luminescence channels (here $V_{O^+} \rightarrow V$, (I)) which causes an increasing relative intensity of this particular band.

The observed energies imply that the V_{Ba^-} term is to position 200 meV, the $V_{Ba^{--}}$ term 400 meV closer to the valence band edge as predicted by the theoretical calculation of Daniels and Härdtl [9]. The Stokes shift which can be attributed to every transition within the CC model is distinctly appearing only at $V_{O^+} \rightarrow V$ and amounts to about 400 meV.

Summarizing the validity of the CC model for the CL spectra of BTC grains is shown and the nature of the observed recombination processes is presented. Unsatisfactory is still that only the unsymmetric shape of the spectra indicates the postulated transition (IV). Additionally performed experiments to measure (IV) as a CL maximum or a shoulder did not succeed neither by using other photomultiplier types nor by performing the measurements at lower specimen temperatures.

Furthermore it is true that with regard to the great number of parameters the fit values include uncertainties which cannot be estimated, but nevertheless the following tendencies are evident, being in agreement with the experimental results. The FWHM of the bands increases with higher temperatures. This observation is expected because higher energetic phonons take part at the recombination process which causes a broadening within the CC model. Moreover, the luminescence maxima weakly shift to higher energies for increasing temperature which is in accordance with the self-activated centre luminescence, known since many years, e.g. from ZnS phosphors or GaAs (7, 8).

For a fixed temperature the fit curves for specimens, sintered at reducing atmosphere, result from the ones for samples sintered at oxidizing conditions, if only the intensities I_{Ei} are varied, but all other fit parameters are retained. Especially the apparent oxygen partial pressure-dependent shift of (II) can be explained in this way, because it is solely the result of the superposition of the same band with different intensities.

Acknowledgements

The authors wish to thank the Siemens AG, Munich for technological assistance and the Deutsche Forschungsgemeinschaft for financial support.

References

[1] W. Heywang, Solid State Electronics **3**, 51 (1961).
[2] J. Daniels and R. Wernicke, Philips Res. Rep. **31**, 544 (1976).
[3] L. Hanke, Siemens Forsch.- u. Entwickl.-Ber. **8**, 209 (1979).
[4] K. Löhnert, M. Hastenrath, L. J. Balk, and E. Kubalek, Inst. Phys. Conf. Ser. **60**, 179 (1981).
[5] H. Ihrig, J. H. T. Hengst, and M. Klerk, Z. Phys. B **40**, 301 (1981).
[6] C. C. Klick and J. H. Schulman, Solid State Phys. **5**, 97 (1957).
[7] D. Curie and J. S. Prener in: Physics and Chemistry of the II–VI Compounds, Ed. M. Aven and J. S. Prener, Amsterdam 1967.
[8] E. W. Williams, Phys. Rev. **168**, 922 (1967).
[9] J. Daniels and K. H. Härdtl, Philips Res. Rep. **31**, 489 (1976).
[10] M. Aguilar, C. Gonzalo, and G. Godefroy, Ferroelectrics **25**, 467 (1980).
[11] R. Wernicke, Philips Res. Rep. **31**, 526 (1976).
[12] S. Datta, B. G. Yacobi, and D. B. Holt, J. Lum. **21**, 53 (1979).
[13] N. Riehl, Ed., Einführung in die Lumineszenz, Verlag Karl Thiemig, München 1971.

(Received June 10, 1983)

phys. stat. sol. (a) 79, 141 (1983)

Subject classification: 1.3 and 10.1; 21; 21.1

Sektion Mathematik/Physik der Pädagogischen Hochschule Halle (Saale)[1])

On the Size–Shape Relation of Guinier-Preston Zones and α'_R-Precipitates in an Al–6.8 at% Zn Alloy

By

R. RAMLAU and H. LÖFFLER

Ellipsoidal Guinier-Preston zones and lenticular precipitates of the α'_R-phase grown at 120 °C in an Al–6.8 Zn at% alloy are examined by means of transmission electron microscopy. For each particle two characteristic lengths are determined and the corresponding two-dimensional histograms are established. It can be shown that the precipitates monotonically flatten with increasing size. Hence, the size–shape relation did not provide any indication of the phase transition.

Elliptische Guinier-Preston-Zonen und lentikular Präzipitate der α'_R-Phase, die bei 120 °C in einer Al–6,8 At% Zn-Legierung wachsen, werden mittels Transmissionselektronenmikroskopie untersucht. Für jedes Teilchen werden zwei charakteristische Längen bestimmt und die entsprechenden zweidimensionalen Histogramme aufgestellt. Es kann gezeigt werden, daß sich die Präzipitate mit zunehmender Größe monoton abflachen. Somit liefert die Beziehung, Größe–Form, kein Anzeichen für einen Phasenübergang.

1. Introduction

According to Merz and Gerold [1] the metastable precipitates occurring in the Al–Zn system are spherical and ellipsoidal Guinier-Preston (G.P.) zones, precipitates of the rhombohedrally distorted f.c.c. α'_R-phase, and eventually precipitate of the f.c.c. α'_m-phase. Both the ellipsoidal G.P. zones and the precipitates of the α'_R-phase are in the shape of an oblate spheroid whose polar axis is parallel to one of the four ⟨111⟩ directions of the f.c.c. matrix (in this paper we disregard that the shape of the considered α'_R-precipitates is more precisely described as lenticular). The lengths of the major and the minor semi-axis of the spheroids are referred to as a and c, respectively.

Criteria of distinction between ellipsoidal G.P. zones and α'_R-precipitates were, up till now, controversially discussed (see e.g. [2 to 4]). Besides investigations on the internal structure [4, 5] and the degree of coherency of the respective precipitate–matrix interface [6], examinations of the shape of these precipitates seem to us a suitable means to elucidate this open problem.

Hitherto, the course of the ratio $v = c/a$ in dependence on the size of the precipitates had been determined only by X-ray investigations [4, 5, 7]. It is the purpose of this work to examine the alteration of $v = c/a$ during the growth of the concerned precipitates by means of transmission electron microscopy (TEM).

2. Experimental

Samples of an Al–6.8 at% Zn alloy were carefully homogenized and, after intermediate annealing, rolled in several steps to sheets of a thickness of about 0.15 mm. Afterwards these sheets were again homogenized at 400 °C for 1 h and then exposed to three different heat treatments (HT):

[1]) Kröllwitzer Str. 44, DDR-4020 Halle (Saale), GDR.

HT A: Direct quenching (DQ) from 400 °C into an oil bath of 120 °C and storing there for 68 h, or in brief:

400 °C/DQ to 120 °C (oil)/68 h at 120 °C.

HT B: 400 °C/DQ to 20 °C (oil)/19 d at 20 °C/11 d at 120 °C.

HT C: 400 °C/DQ to 20 °C (oil)/1 min at 20 °C/69 h at 120 °C.

For the TEM investigations thin foils were prepared by electropolishing (window technique) at about −15 °C. The samples were examined in an electron microscope of type JEM-100C operating at 100 kV.

The foils prepared by use of the window technique seldom (particularly in the sufficiently thin parts) offer larger regions of equal orientation. Hence a variety of samples had to be investigated in order to obtain the necessary stock of information. For each sample the electron-optical magnification was calibrated by means of lattice fringe images reducing in this manner the maximum error of the magnification in the finally evaluated positives to ±5%. The foil orientation and the image conditions were chosen in such a manner that at least one of the ⟨111⟩ directions of the matrix was exactly perpendicular to the primary beam and that strain contrast did not appear. The electron-optical magnification was about 70000 times and the photographic post-magnification twice. The particle size was determined from the positives by use of a precision magnifier. In this way the error of the characteristic lengths amounted to ±0.5 mm, this corresponds to an error of $\Delta a = \Delta c = \pm 0.36$ nm in the real scale. Thus the intervals in the histograms were selected to 0.5 nm.

Doubtlessly it would have been more suitable to refer the statistical analysis to the lengths originally measured on the positives, i.e. to use intervals of 0.05 mm and to transform only results of this analysis (inclusive the end points of the intervals) into the real scale. But since the electron-optical magnifications varied in the order of ±15% for the individual samples and, on the other hand, only the totality of the samples delivered a sufficient ensemble of particles this could not be carried out.

The reliability of our method was confirmed by the fact that the a–c histogram established in the real scale for a total ensemble belonging to the same heat treatment did not qualitatively differ from those established in the non-transformed scale for each single sample.

Comparative investigations were made by applying small angle X-ray scattering (SAXS). Details concerning the experimental procedure and the special way of evaluation are reported in [8].

3. Results

3.1 Heat treatment A

The precipitates nucleated and grown during this heat treatment (400 °C/DQ to 120 °C/68 h at 120 °C) are shown in Fig. 1. From this and the related micrographs the values of a and c were determined for about 650 particles. The relative frequencies for the various intervals of the a–c histogram are presented in Table 1. The mean values of the semi-axes are $\bar{a} = 13.6$ nm and $\bar{c} = 3.5$ nm, respectively.

From the individual a–c pairs the mean radius (r_m) was calculated for each precipitate, applying the formula $r_m = (a^2c)^{1/3}$, i.e. r_m is the radius of a sphere possessing the same volume as the considered spheroid. We found $\bar{r}_m = 8.7$ nm.

From the r_m-histogram (see Fig. 2a) a fairly narrow size distribution of the precipitates can be concluded. Consequently this system of precipitates seems to be appropriate for SAXS investigations based upon the supposition of a monodisperse system. The evaluation of the SAXS experiment frequently results in an approximate value for the mean radius of gyration $\bar{r}^*$ (Guinier's approximation etc.) [9]. Actually, these

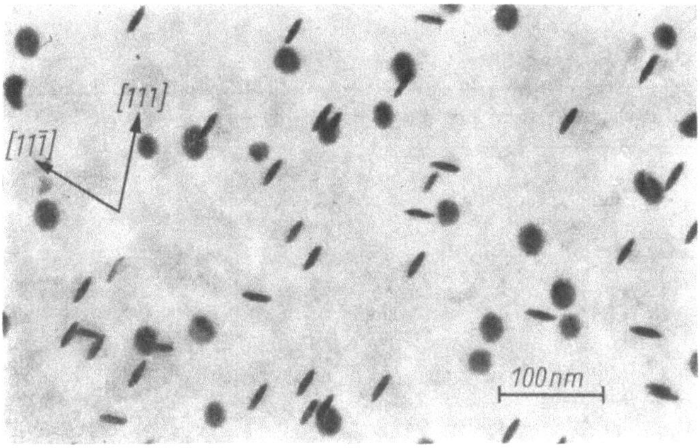

Fig. 1. TEM image of the alloy after heat treatment A. The foil orientation is [1$\bar{1}$0]

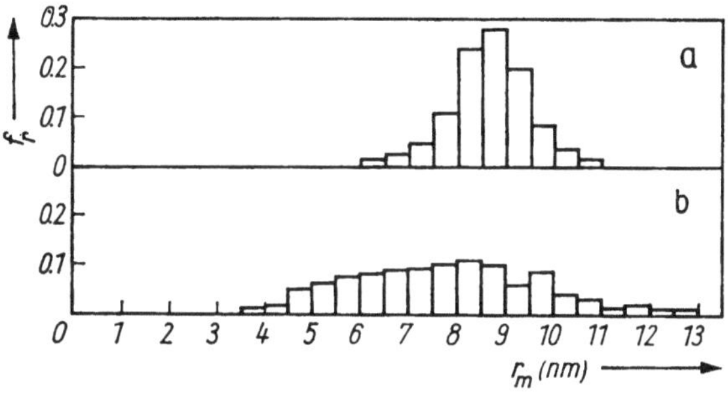

Fig. 2. Relative frequencies (f_r) of the mean radius r_m; a) heat treatment A, b) heat treatment B

approximate values are certain mean values, but not in the arithmetic sense. Only in cases of monodisperse, dilute systems the result of the SAXS experiment, for instance the Guinier radius (r_G), should be equal to the gyration radius of the single particle and thus equal to $\bar{r}^*$.

For an oblate spheroid $r^* = [(2a^2 + c^2)/5]^{1/2}$. By use of this relation we found from the TEM investigations $\bar{r}^* = 8.8$ nm, whereas the SAXS experiments resulted in $\bar{r}^*_{SAXS} = 8.0$ nm [8]. The mean value of $v = c/a$ was determined to $\bar{v} = 0.26$. Assuming that the scattering system consists of equisized ellipsoidal particles the SAXS investigations delivered $\bar{v}_{SAXS} = 0.25$ [8]. It is easily seen that the corresponding results confirm each other within the range of experimental error.

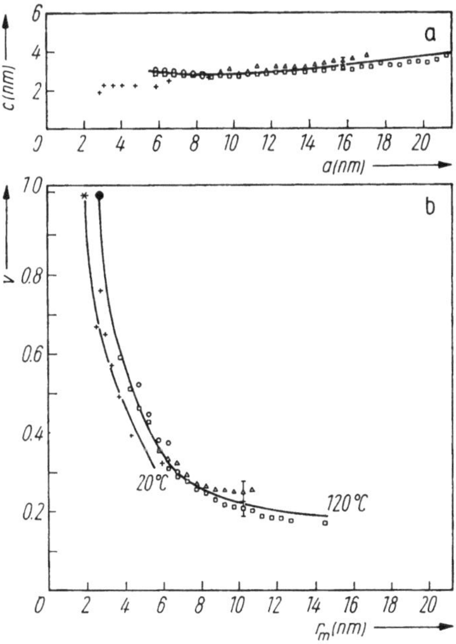

Fig. 3. a) Course of c versus a for the precipitates grown at 120 °C (△ heat treatment A, □ heat treatment B, ○ heat treatment C), + values taken from the work of Fumeron et al. [4]. b) Course of $v = c/a$ in dependence on the mean radius r_m for the precipitates grown at 120 and 20 °C, respectively. (● Deguercy et al. [5]; × Ramlau and Löffler [10]; other symbols same as in a)) Fig. 4

Table 1

Relative frequencies for the intervals of the a–c histogram in case of heat treatment A (400 °C/DQ to 120 °C/68 h at 120 °C); a and c are the lengths of the major and the minor semi-axis, respectively, of the ellipsoidal precipitates

a (nm)	c (nm)				
	2.5	3.0	3.5	4.0	4.5
7.0					
7.5		0.0016			
8.0					
8.5		0.0016			
9.0		0.0031			
9.5	0.0031	0.0047	0.0016		
10.0	0.0016	0.0047	0.0031		
10.5	0.0031	0.0031			
11.0	0.0031	0.0250	0.0016		
11.5		0.0078	0.0125		
12.0	0.0016	0.0313	0.0034		
12.5		0.0250	0.0344		
13.0	0.0094	0.0767	0.0751	0.0016	
13.5	0.0047	0.0391	0.0407		
14.0	0.0016	0.0297	0.0329		
14.5		0.0736	0.2019	0.0031	
15.0		0.0125	0.0501		
15.5		0.0125	0.0610		
16.0		0.0031	0.0344	0.0063	
16.5		0.0031	0.0172	0.0063	
17.0			0.0031	0.0031	
17.5			0.0031		
18.0				0.0047	
18.5					
19.0					
19.5					
20.0			0.0016	0.0016	
20.5			0.0016	0.0016	

The course of c versus a and that of v versus r_m is shown in Fig. 3 (open triangles). For each interval of a and r_m the mean values of all c and v, respectively, were calculated.

3.2 Heat treatment B

In Fig. 4 the kind of precipitates nucleated and grown during this heat treatment (400 °C/DQ to 20 °C/19 d at 20 °C/11 d at 120 °C) is shown. Obviously the intermediate storing at 20 °C leads to a considerable broadening of the size distribution. The a–c histogram, based upon the information gained from about 600 particles, is presented in Table 2. One finds $\bar{a} = 12.4$ nm and $\bar{c} = 3.1$ nm, i.e. values which resemble those obtained by heat treatment A. The histogram of r_m is given in Fig. 2b and the courses of $c(a)$ and $v(r_m)$ in Fig. 3 (open squares).

3.3 Heat treatment C

This heat treatment (400 °C/DQ to 20 °C/1 min at 20 °C/69 h at 120 °C) was chosen to get smaller mean sizes of the precipitates in comparison with the heat treatments A and B. Fig. 5 is representative of the precipitates obtained after performing this heat

Table 2

Relative frequencies for the intervals of the *a*–*c* histogram in case of heat treatment B (400 °C/DQ to 20 °C/19 d at 20 °C/11 d at 120 °C)

a (nm)	*c* (nm) 2.0 2.5	2.5 3.0	3.0 3.5	3.5 4.0	4.0 4.5
4.0					
4.5		0.0034			
5.0	0.0034	0.0017	0.0017		
5.5			0.0017		
6.0	0.0051	0.0135	0.0068		
6.5	0.0017	0.0118	0.0051	0.0017	
7.0	0.0084	0.0202	0.0101		
7.5		0.0169	0.0034		
8.0	0.0017	0.0135	0.0084		
8.5	0.0017	0.0304	0.0135		
9.0	0.0034	0.0118			
9.5	0.0051	0.0304	0.0152		
10.0	0.0017	0.0253	0.0084		
10.5	0.0068	0.0371	0.0169		
11.0	0.0017	0.0169	0.0118		
11.5	0.0034	0.0287	0.0219		
12.0		0.0253	0.0204		
12.5		0.0169	0.0354		
13.0		0.0169	0.0455		
13.5		0.0101	0.0152		
14.0	0.0017	0.0118	0.0253	0.0017	
14.5		0.0067	0.0422		
15.0	0.0017	0.0084	0.0354		
15.5		0.0034	0.0253		
16.0			0.0304	0.0034	
16.5	0.0017	0.0034	0.0320		
17.0		0.0017	0.0118	0.0051	
17.5		0.0017	0.0320	0.0118	
18.0			0.0017	0.0068	
18.5		0.0017	0.0068	0.0084	
19.0			0.0051	0.0051	
19.5			0.0017	0.0101	
20.0			0.0017	0.0068	
20.5				0.0033	
21.0				0.0068	
21.5				0.0017	0.0017
22.0			0.0017		0.0017
22.5			0.0017		0.0017
23.0				0.0017	0.0017
23.5					
24.0					
24.5					
25.0			0.0017		
25.5					
26.0					
26.5					0.0017

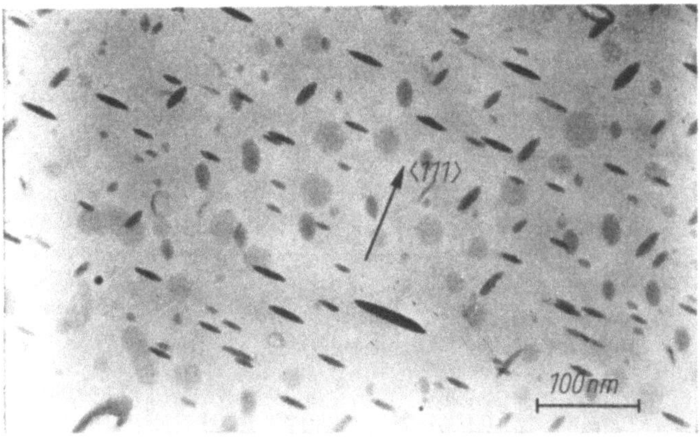

Fig. 4. TEM image of the alloy after heat treatment B. The foil normal is perpendicular to one of the ⟨111⟩ directions

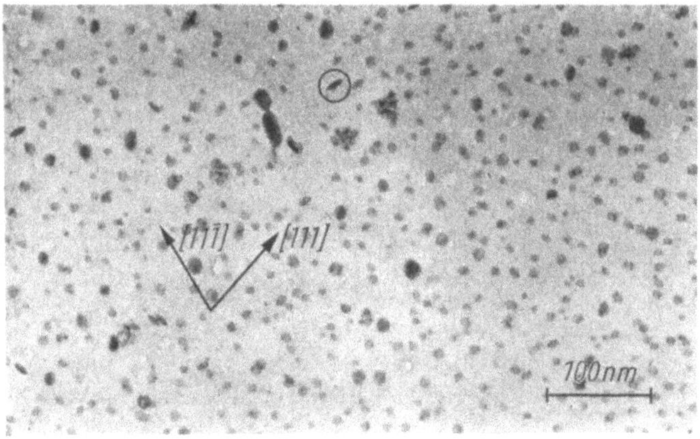

Fig. 5. TEM image of the alloy after heat treatment C (400 °C/DQ to 20 °C/1 min at 20 °C /69 h at 120 °C). The foil orientation is [1̄10]. Only precipitates having the size of the encircled particle were analysed

treatment. From this figure it can be realized that it is the more difficult to identify and measure the semi-axes precisely the smaller the precipit ates are. For this reason we did not evaluate the whole set of precipitates and selected only about 100 particles of sizes marked in Fig. 5 by an open circle. The resulting courses of $c(a)$ and $v(r_m)$ are presented in Fig. 3 (open circles).

4. Discussion

As already mentioned above, both the ellipsoidal G.P. zones and the α'_R-precipitates are in the shape of an oblate spheroid, with its polar axis parallel to one of the four ⟨111⟩ directions of the matrix. It is obvious that only for those precipitates whose polar axis is perpendicular to the incident electron beam can a and c be directly measured. This is in the most favourable case fulfilled for two of the four "kinds" of precipitates, namely when the incident beam is exactly in a ⟨110⟩ direction.

From Fig. 1 it is apparent that at least in special areas of the foil these four "kinds" of precipitates do not occur equivalently. Since the results gained by SAXS, which characterize the whole polycrystal, agree fairly well with those obtained by the selective TEM investigations, however, we conclude that the above-mentioned phenomenon does not yield additional uncertainties.

The deviations in the courses of the three $c(a)$- and $v(r_m)$-curves (see Fig. 3) are attributed to the systematic error made in the calibration of the magnification and not to the influence of the different heat treatments. From our point of view, the holding times at 120 °C were sufficiently long to attain the metastable equilibrium of the zinc content valid at 120 °C in spite of the different prehistories of the samples.

In the following particular emphasis is laid on the discussion of the size–shape relation. Fumeron et al. [4] investigated the size–shape relation of the precipitates nucleated and grown in an Al–6.8 at% Zn alloy during the ageing at 20 °C by means of X-ray investigations (see the crosses in Fig. 3). By use of high resolution TEM the authors of this paper already determined the critical radius (r_c) for the transition of spherical into ellipsoidal zones in an Al–3 at% Zn alloy aged at 20 °C and found $r_c \approx \approx 2$ nm [10]. This value was also included in Fig. 3b (asterisk).

This transition is accompanied with a rhombohedral distorsion of the f.c.c. lattice inside the ellipsoidal G.P. zones, which results in a certain increase of the chemical part of the free enthalpy but obviously diminishes to a higher degree the stress term of the free enthalpy as compared to the respective still spherically assumed zone.

From Fig. 3 it becomes transparent that the courses of $c(a)$ and $v(r_m)$ determined by Fumeron et al. [4] in the alloy aged at 20 °C are shifted, at least for $a < 7$ nm and $r_m < 5$ nm, towards somewhat lower c- and v-values compared with our results. This tendency is in full agreement with the results reported by Deguercy et al. [5], who found that r_c is increasing with the rise of the ageing temperature. In our opinion this behaviour can be interpreted taking into account the metastable phase diagram [11], according to which the zinc content of the zones decreases when increasing the ageing temperature as well as the diminution of the rigidity of the alloy with the rise of the ageing temperature. From the course of r_c in dependence on temperature, presented in the paper of Deguercy et al. [5], for 120 °C $r_c \approx 2.8$ nm can be extrapolated. This value fits well the results presented in this work (see the full circle in Fig. 3b). Degischer et al. [12] found a maximum thickness of the "plate-like" α_R'-precipitates grown at 20 °C of $d = 4$ nm ($\triangleq 2c$). After ageing at 120 °C, however, we observed lenticular precipitates with $2c = 7.5$ nm. These differences indicate that the ageing temperature will not only influence the value of r_c but also the courses of $c(a)$ and $v(r_m)$.

From Fig. 3a a slight downward tendency of $c(a)$ for small a-values (6 nm $< a <$ $<$ 9 nm) can be supposed. Taking this rather non-distinct decrease of $c(a)$ for granted, the G.P. zones after becoming ellipsoidal ought, to a certain degree, shrink in the direction of their polar axis.

By means of hardness and SAXS investigations performed with several Al–Zn alloys ranging in zinc content between 12 and 18 at% Löffler et al. [7] found that the transition from spherical into ellipsoidal G.P. zones takes place at a Guinier radius of $r_G \approx 2$ nm and that fluctuations of both the hardness number and the v-value occur immediately after the transition. A reliable proof, whether in fact such fluctuations of the shape of the precipitates occur, can, unfortunately, hardly be delivered by TEM, due to the difficulties already mentioned in Section 3.3.

According to the literature the transition from ellipsoidal G.P. zones into precipitates of the α_R'-phase takes place at $r_G = 5$ nm [7] or $2c = 6$ nm [13]. From the course of v in dependence on r_m as well as from the relevant histograms it can be concluded that the precipitates monotonically flatten when increasing in size. That means, our investigations concerning the size–shape relation do not provide any indication of this phase transition.

Eventually we would like to emphasize that from Table 1 it could be concluded that not only the mean value $\bar{c}$ of the minor semiaxes of the oblate precipitates but also their individual values do not considerably exceed 3 nm, which is assumed to be the critical value where the transition of ellipsoidal G.P. zones into precipitates of the α_R'-phase takes place [13] by forming an extra {111} plane perpendicular to the polar axis of the oblate precipitate. It is of particular interest that even for the largest precipitates observed in this investigation, namely about $a = 25$ nm, the corresponding c-values do not exceed $c = 4.5$ nm. Obviously the increase of the stress energy related with an assumed growth of the thickness of the precipitate becomes so high that the formation of a second extra plane (parallel to the first one) cannot happen. This problem will be treated more carefully in another paper of the authors [14].

5. Final Conclusions

The critical radius for the transition of the spherical G.P. zones into ellipsoidal ones is increasing with the rise of the ageing temperature.

The ratio $v = c/a$ of the ellipsoidal precipitates monotonically diminishes with increasing size. Even for larger values of r_m, the course of $v(r_m)$ might slightly depend on the ageing temperature. Due to the monotonous decrease of $v(r_m)$, the transition of ellipsoidal G.P. zones into α'_R-precipitates cannot be proved by investigations on the size-shape relation. Since investigations on the internal crystal structure of the precipitates [4, 5] did not produce evidence for the transition, either, the differences between ellipsoidal G.P. zones and precipitates of the α'_R-phase can only be sought in the degree of coherency of the matrix-precipitate interface. This matter will be discussed in another paper of the authors [14].

References

[1] W. MERZ and V. GEROLD, Z. Metallk. 57, 607 (1966).
[2] M. SIMERSKÁ and V. SYNEČEK, Acta metall. **15**, 223 (1967).
[3] G. LAZLAZ and P. GUYOT, Acta metall. **25**, 277 (1977).
[4] M. FUMERON, J. P. GUILLOT, A. DAUGER, and J. CAISSO, Scripta metall. **14**, 189 (1980).
[5] J. DEGUERCY, M. F. DENANOT, M. FUMERON, J. P. GUILLOT, and J. CAISSO, Acta metall. **30**, 1921 (1982).
[6] R. RAMLAU and H. LÖFFLER, IV. Konf. Metallkundliche Probleme der Werkstoffentwicklung, Freiberg (Sa.) 1983.
[7] H. LÖFFLER, O. KABISCH, and R. KROGGEL, phys. stat. sol. (a) **49**, 299 (1978).
[8] W. GILLE, Thesis, Pädagogische Hochschule Halle, 1983.
[9] A. GUINIER and G. FOURNET, Small-Angle Scattering of X-Rays, John Wiley & Sons, New York 1955.
[10] R. RAMLAU and H. LÖFFLER, phys. stat. sol. (a) **68**, 531 (1981).
[11] A. KRUPKOWSKI, Bull. Acad. Polon. Sci., Ser. Sci. Techn. **17**, 87 (1969).
[12] P. DEGISCHER, CHR. Y. ZAHRA, and A. ZAHRA, Z. Metallk. **73**, 635 (1982).
[13] V. SYNEČEK, private communication.
[14] R. RAMLAU and H. LÖFFLER, in preparation.

(Received July 13, 1983)

phys. stat. sol. (a) 79, 149 (1983)

Subject classification: 11

Sektion Physik der Humboldt-Universität zu Berlin,
Bereich Angewandte Massenspektrometrie[1])

Energy Spectra of Backscattered Electrons for Solid Films and Double Layers by Use of Werner's Analytic Model

By

S. Rogaschewski

Expressions for the energy spectrum of electrons backscattered from solid films and double layers as a function of incident electron energy, angle of incidence, film thickness, and atomic number of target are derived from Werner's analytic model. Values calculated from these expressions agree well with existing experimental and Monte Carlo data.

Aus dem analytischen Modell nach Werner werden Ausdrücke für das Energiespektrum von Rückstreuelektronen aus dünnen Festkörperschichten und Doppelschichten als Funktion der Energie der einfallenden Elektronen, des Einfallswinkels, der Schichtdicke und der Ordnungszahl des Targets hergeleitet. Aus diesen Ausdrücken berechnete Werte stimmen gut mit vorliegenden experimentellen und Daten aus Monte-Carlo-Rechnungen überein.

1. Introduction

Scanning electron microscopy, electron probe microanalysis, and electron beam technologies have stimulated an interest in experimental and theoretical studies on the interaction of intermediate energy electrons with a solid target. As an energetic electron penetrates into a target, the probability that it will emerge as a backscattered electron is the result of the simultaneous and interacting processes of scattering (change of direction) and energy loss. Attempts to describe the very complex transport process by models, based on the alternative assumptions that backscattering is due to a single large-angle scattering or to electron diffusion, have been made by Bothe [1] and Everhart [2] (single scattering) and by Bothe [1], Bethe et al. [3], Archard [4], Kanaya and Okayama [5], Thümmel [6] (p. 189) (diffusion). The real scattering process includes a combination of small-angle scattering and occasional large-angle scattering events. Recently simplified analytic approaches have been suggested which allow for both kinds of events (Werner [7], Werner et al. [8], Darlinski [9], Niedrig [10]).

In this paper the model developed in [7, 8] is briefly described and then it is used to derive expressions for the energy spectra of electrons backscattered at oblique incidence from solid films and targets consisting of a film of one element on a thick substrate of another element (double layer). Values calculated from these expressions are compared with existing experimental data for thick films (bulk targets), double layers, and with existing results of the Monte Carlo method to show generally good agreement.

---

[1]) Invalidenstr. 42, DDR-1040 Berlin, GDR.

2. The Model

Werner's non-relativistic model of electron backscattering for the energy range of 10 to 100 keV and normal electron incidence is based on the following assumptions:

(i) The incident electrons suffer energy losses due to inelastic collisions with atomic electrons according to Bethe's continuous-slowing-down model [11]. Their mean path length s travelled until the initial energy E_0 is reduced to the mean energy E is approximately obtained from

$$s = \frac{4\pi\varepsilon_0^2 A}{e_0^4 N_A \varrho Z} \frac{E_0^2 - E^2}{\ln (1.166 E_0/J)}, \tag{1}$$

where ε_0 is the dielectric constant, A the relative atomic mass, e_0 the electronic charge, N_A is Avogadro's number, ϱ the density, and Z the atomic number. J is the atomic mean excitation energy. An approximation to the experimental values of Bakker and Segré [12] is used: $J/Z = 10.5$ eV.

According to the suggestion of Knecht and Bothe [13], the penetration depth is related to s as

$$x = \frac{2s}{1 + \sqrt{3}}. \tag{2}$$

If $E = 0$ (1) yields the mean total path length s_0 (Bethe range). Dividing by s_0 in (1) one obtains

$$t = 1 - \varepsilon^2 , \tag{3}$$

where $t = s/s_0$ is the fractional mean path length and $\varepsilon = E/E_0$ is the normalized mean electron energy.

For future convenience the quantity τ measuring the travelled path length is introduced,

$$\tau = -\ln (1 - t) . \tag{4}$$

(ii) The electron scattering is caused by elastic collisions with the atomic nuclei. Small-angle multiple scattering and single large-angle scattering are taken into account simultaneously by means of a simple differential equation system.

As the consequence of the small-angle scattering the incident electrons reach the state of isotropic diffusion continuously along their path. The density of the transition probability into this stage is given as an approximation to the density function of Bothe's diffusion theory [1]. In contrast to the latter the approximation allows for diffusion also for very small path lengths and takes into account approximately plural scattering through medium angles. According to Archard's point-source-diffusion model [4], the isotropically diffused electrons move without further deflection in such a way that their overall paths equal the Bethe range s_0, as proposed by Thümmel [6] (p. 189) in his modification of Bothe's theory.

The single large-angle scattering contribution is determined from the well-known Everhart model [2]. Everhart assumed that the electrons penetrate into the target without deflection before being scattered by a single-Rutherford-scattering event into angles greater than $\pi/2$. A singly scattered electron moves without further deflection through the residual path length.

The scattering and slowing-down process is characterized by two dimensionless parameters

$$a_T = \frac{8}{5} \frac{s_0}{\lambda_D(E_0)}, \qquad a_S = \frac{s_0}{\lambda_S(E_0)}, \tag{5}$$

where the subscripts T and S indicate small-angle and single scattering, respectively, s_0 is the Bethe range, and λ_D and λ_S are the transport mean free paths for diffusion and single scattering, respectively. From the approximations involved in the treatment it follows that the transport mean free paths are assumed to decrease linearly with increasing fractional path length.

λ_D is adopted from Brown and Ogilvie [14] and λ_S is calculated from the Rutherford cross section according to Everhart's model (see Appendix). From these mean free paths, equation (1), and $J/Z = 10.5$ eV, (5) can be written as

$$a_T = \frac{Z \ln (41 E_0/Z^{2/3})}{5 \ln (111 E_0/Z)}, \qquad a_S = \frac{Z}{16 \ln (111 E_0/Z)}, \tag{6}$$

where E_0 is in keV. Equations (6) can be roughly approximated by $a_T = Z/5$ and $a_S = Z/80$.

A generalization to scattering in mixed-element targets is given in the Appendix.

3. Solid Films[2])

For the case of oblique electron incidence Fig. 1 shows the geometry of the model outlined above. The incident electrons of energy E_0, including an angle α with the surface normal, travel a distance t, measured by τ (see (4)), diffuse isotropically or undergo a single scattering through an angle $\theta' > \pi/2$, move through the residual path length, and emerge with energy E.

Introducing by the transformation of the coordinate system θ' and φ' into the expression given by Werner et al. [8] (equation (14)) the number of electrons to be scattered along $d\tau$ into the elemental solid angle $d\Omega' = \sin \theta' \, d\theta' \, d\varphi'$ by both scattering processes included may be written as

$$dn(\alpha, \tau, \theta', \varphi') = dn_T + dn_S, \tag{7}$$

where

$$dn_T = a_T^2 n(\alpha, \tau) \frac{\tau \, d\tau \, d\Omega'}{4\pi(1 + a_T \tau)},$$

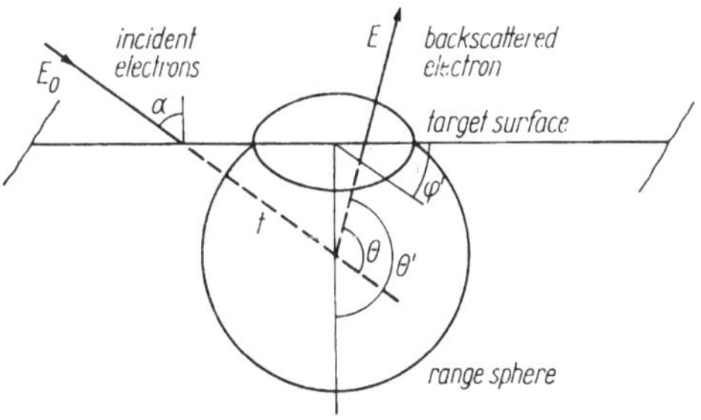

Fig. 1. Transport and escape geometry of Werner's model for oblique electron incidence

[2]) Numerous authors have calculated within Everhart's model the backscattering coefficient at oblique incidence for both solid films (Nakhodkin et al. [15], Shimizu and Shinoda [16]) and double layers (Nakhodkin et al.) and the backscattered electron energy spectrum at normal incidence for solid films (Nakhodkin et al. [15], McAfee [17], Iafrate et al. [18], Sogard [19]) and double layers (Iafrate et al.). The single-scattering fraction according to Everhart's model can be gained from the results presented here and in the next section, setting $a_T = 0$, replacing a_S by the commonly used adjusted Everhart coefficient, $a_E = 0.045Z$, and replacing t_x by $y = x/(R_{TW} \cos \alpha)$, where x is the target thickness and R_{TW} is the Thomson-Whiddington range.

and

$$\mathrm{d}n_{\mathrm{S}} = a_{\mathrm{S}} n(\alpha, \tau) \frac{\mathrm{d}\tau \, \mathrm{d}\Omega'}{\pi(1 - \cos\theta' \cos\alpha - \sin\theta' \cos\varphi' \sin\alpha)^2} .$$

$n(\alpha, \tau)$ is the number of electrons which have travelled the fractional path length t without deflection. It is obtained from

$$n(\alpha, \tau) = n_0 - \int_0^{\tau} \int_0^{\pi} \int_0^{2\pi} \mathrm{d}n_{\mathrm{T}} - \int_0^{\tau} \int_{\pi/2}^{\pi} \int_0^{2\pi} \mathrm{d}n_{\mathrm{S}} , \tag{8}$$

where the second and the third term on the right-hand side give the number of electrons lost from the incident ones by small-angle and single scattering, respectively. n_0 is the number of incident electrons. Equation (8) is solved by

$$n(\alpha, \tau) = n_0(1 + a_{\mathrm{T}}\tau) \, \mathrm{e}^{-a^*\tau} , \tag{9}$$

where $a^* = a_{\mathrm{S}}^* + a_{\mathrm{T}}$ and $a_{\mathrm{S}}^* = a_{\mathrm{S}}/\cos^2\alpha$.

According to (3) for a backscattered electron of normalized energy ε from a depth corresponding to τ the angle of emerging is given by

$$\cos\theta' = \frac{1 - \mathrm{e}^{-\tau}}{\varepsilon^2 - \mathrm{e}^{-\tau}} \cos\alpha . \tag{10}$$

Consider a target of thickness x corresponding via (2) to the fractional path length

$$t_x = \frac{1 + \sqrt{3}}{2 s_0 \cos\alpha} x \tag{11}$$

and to the quantity $\tau_x = -\ln(1 - t_x)$. Since the path length travelled by a backscattered electron must be less then s_0, the maximum t-value consistent with the model is

$$t_{x,\max} = (1 + \cos\alpha)^{-1}$$

which gives $\tau_{x,\max} = \ln(1 + \sec\alpha)$. For thicknesses corresponding to values of or greater than $t_{x,\max}$, the target is considered to be a bulk target.

The integrated energy spectrum, i.e., the fraction of backscattered electrons having energy ranging from εE_0 to E_0 can be obtained from

$$\eta(\alpha, \tau_x, \varepsilon) = \frac{1}{n_0} \int_{\varphi}^{\tau_0} \int_{\theta'_{\min}}^{\pi} \int_0^{2\pi} \mathrm{d}n(\alpha, \tau, \theta', \varphi') , \tag{12}$$

where $\theta'_{\min}$ and τ_0 are determined by (10). With $\theta' = \theta'_{\min}$ one has from (10) the lower limit of the θ'-integration. Letting $\theta' = \pi$ one has again from (10)

$$\tau_\varepsilon = \ln\left(\frac{1 + \cos\alpha}{\cos\alpha + \varepsilon^2}\right)$$

and the upper limit of the τ-integration is then

$$\tau_0 = \begin{cases} \tau_x & \text{for} \quad \tau_\varepsilon > \tau_x \quad \text{or} \quad 0 \leqq \varepsilon < \varepsilon_x \\ \tau_\varepsilon & \text{for} \quad \tau_\varepsilon \leqq \tau_x \quad \text{or} \quad \varepsilon_x \leqq \varepsilon \leqq 1 , \end{cases} \tag{13}$$

where ε_x is the normalized energy of an electron backscattered through $\theta' = \pi$ from the maximum penetration depth. ε_x can be obtained from

$$\varepsilon_x^2 = (1 + \cos\alpha) \, \mathrm{e}^{-\tau_x} - \cos\alpha ,$$

where $\varepsilon_x = 0$ for a bulk target.

By integrating (12) one gets

$$\eta(\alpha, \tau_x, \varepsilon) = \frac{a_T^2}{2} \int_0^{\tau_0} \left(1 + \frac{1 - e^{-\tau}}{\varepsilon^2 - e^{-\tau}} \cos\alpha\right) \tau\, e^{-a^*\tau}\, d\tau + \\ + \frac{2a_S^*}{1 - \varepsilon^2}\left[F(a^* + 1, a_T) - \frac{1 + \varepsilon^2}{2} F(a^*, a_T)\right] + \\ + \frac{a_S^* \sin^2\alpha}{(1 - \varepsilon^2)^2} \{F(a^*, a_T) - 2F(a^* + 1, a_T) + F(a^* + 2, a_T)\}, \quad (14)$$

where

$$F(p, q) = \left[\left(\frac{p + q}{p^2} + \frac{a_T}{p}\tau\right) e^{-p\tau}\right]_{\tau_0}^{0} \quad (15)$$

and τ_0 is given by the relations (13).

Taking the first derivate of (14) with respect to ε the differential energy spectrum is obtained as

$$\frac{d\eta(\alpha, \tau_x, \varepsilon)}{-d\varepsilon} = a_T^2 \varepsilon \cos\alpha \int_0^{\tau_0} \frac{1 - e^{-\tau}}{(\varepsilon^2 - e^{-\tau})^2} \tau\, e^{-a^*\tau}\, d\tau + \\ + \frac{4a_S^*\varepsilon}{(1 - \varepsilon^2)^2}\Big\{F(a^*, a_T) - F(a^* + 1, a_T) + \\ + \frac{\sin^2\alpha}{1 - \varepsilon^2}[2F(a^* + 1, a_T) - F(a^*, a_T) - F(a^* + 2, a_T)]\Big\} \quad (16)$$

with the limit

$$\left(\frac{d\eta}{-d\varepsilon}\right)_{\varepsilon=1} = \frac{2(1 + 2\cos\alpha)}{3(1 + \cos\alpha)^2} a_S^*,$$

where $F(p, q)$ and τ_0 are given again by (15) and the relations (13), respectively.

Setting $\varepsilon = 0$ and taking into account $\varepsilon_x = 0$ for a bulk target, the backscattering coefficient for arbitrary target thickness is obtained from (14) as

$$\eta(\alpha, \tau_x) = \frac{a_T}{2}[F(a^*, -a_S^*)(1 + \cos\alpha) - F(a^* - 1, -a_S^* + 1)\cos\alpha] + \\ + 2a_S^*\Big\{[F(a^* + 1, a_T) - \frac{1}{2} F(a^*, a_T)]\cos^2\alpha + \\ + \frac{1}{2} F(a^* + 2, a_T)\sin^2\alpha\Big\}. \quad (17)$$

The theoretical predictions for the dependence of η on target thickness and atomic number have been compared by Werner [7] and Werner et al. [8] with experiments for normal electron incidence and several energies. The comparison yields good agreement. As shown in Fig. 2, the same good agreement yields for bulk targets at various angles of electron incidence. At oblique incidence electron scattering takes place closer to the target surface and therefore the probability for backscattering is enhanced. For normal incidence the model suggests the predominance of small-angle over single scattering for a wide range of atomic number whereas at large angles of incidence the single

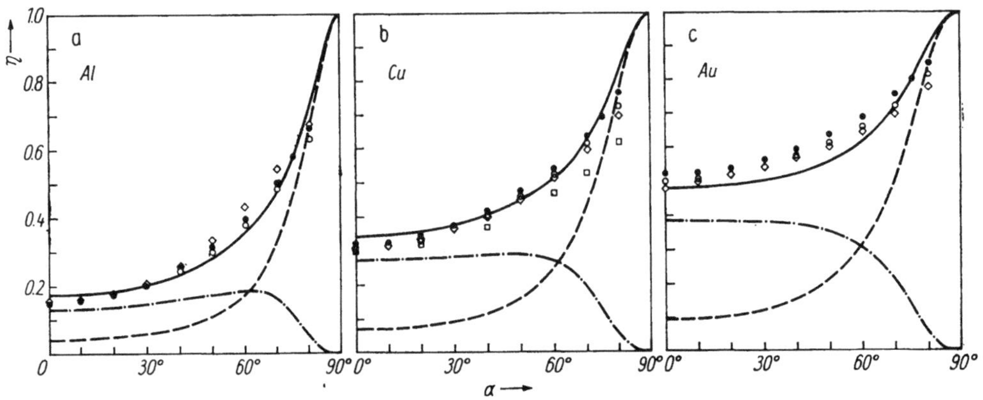

Fig. 2. Bulk backscattering coefficient at oblique incidence for a) Al, b) Cu, and c) Au. Experimental values: ◇ Kanter [20] for 10 to 70 keV, ○ Drescher et al. [21] for 25 keV, □ Radzimski [22] for 10 keV, ● Neubert and Rogaschewski [23] for 40 keV; ——— calculated from (17) for $a_T = Z/5$ and $a_S = Z/80$, – – – and –·–·–·– calculated single and small-angle scattering reaction, respectively

scattering fraction, which is calculated from the second term on the right-hand side of (17), shows the expected strong increase.

The model describes also the striking features of the backscattered electron energy spectra. As has been shown for the case of bulk targets at normal incidence, the calculated differential spectra for elements with low and medium atomic numbers agree well with experiments [7, 8]. For targets with high atomic number a somewhat too small yield at the low-energy end of the spectra is obtained. This can, of course, also be seen from the integrated energy spectra as calculated from (14). The lack of agreement between theory and experiment for high-Z elements which is explained to be due to the simplified scattering model, results in a somewhat to high calculated mean energy $\bar{\varepsilon}$. A comparison of values obtained by numerical integration of (14) with experimental results is shown in Fig. 3.

For the energy spectra from bulk targets at oblique incidence also a considerable degree of agreement with experiments is found (see Fig. 4). In accordance with the experiments, with increase in α the spectra are peaked nearer to the initial energy and the yield at the initial energy is enhanced. In general, the predictions of the analytic model describe the experimental results as well as the results of the Monte-Carlo simulation of Matsukawa et al. [25].

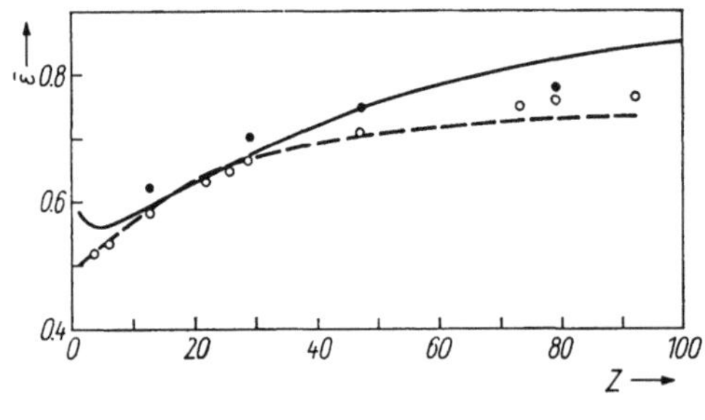

Fig. 3. Mean energy at normal incidence and 20 keV as a function of the atomic number of bulk targets. Experimental values: ● Darlington [24], – – – Matsukawa et al. [25]; ○ values obtained from $\bar{\varepsilon} = 0.5(1 + \eta)$ according to Fitting [26], where η is from Neubert and Rogaschewski [23]; ——— calculated from (14)

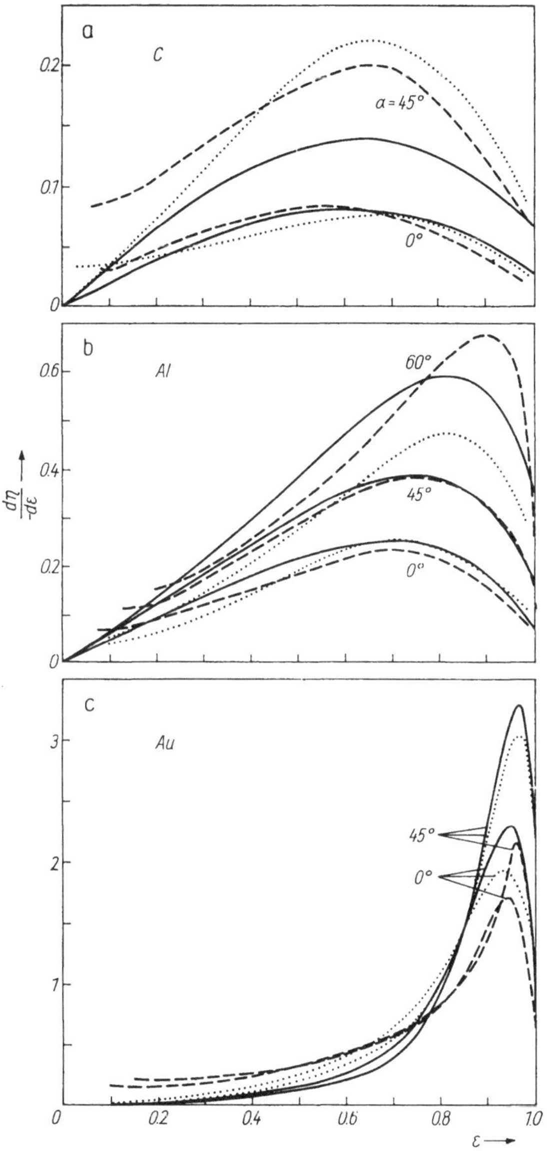

Fig. 4. Differential energy spectra for bulk targets at different angles of incidence and 20 keV for a) C, b) Al, and c) Au. — — — experiments and ·········· Monte-Carlo simulation from Matsukawa et al. [25], ——— calculated from (16)

4. Double Layers

Let the target now be composed of a film with atomic number Z_1 and thickness x on top of a thick substrate with atomic number Z_2. The derivation of the integrated energy spectrum requires then to determine the contribution of the substrate backscattering, η_2, to the total backscattering which is given by

$$\eta_{12} = \eta_1 + \eta_2 ,$$

where η_1 is the film contribution according to the previous section.

To calculate η_2, the fractional path length travelled by an incident electron is now written as

$$t = t_x + t_2 \,, \tag{18}$$

where $t_2 = s_2/s_{02}$. Following the derivations given above, the number of electrons which have travelled within the substrate through t without deflection is

$$n_2(\alpha, \tau) = n_{02} - \int_{\tau_x}^{\tau} \int_0^{\pi} \int_0^{2\pi} \mathrm{d}n_{\mathrm{T}_2} - \int_{\tau_x}^{\tau} \int_{\pi/2}^{\pi} \int_0^{2\pi} \mathrm{d}n_{\mathrm{S}_2} \,, \tag{19}$$

where, according to (9),

$$n_{02} = n_0(1 + a_{\mathrm{T}_1}\tau_x)\, \mathrm{e}^{-a_1^*\tau_x} \,.$$

Equation (19) is solved by

$$n_2(\alpha, \tau) = n_0 Q(\alpha, \tau_x)\,(1 + a_{\mathrm{T}_2}\tau)\, \mathrm{e}^{-a_2^*\tau},$$

where

$$Q(\alpha, \tau_x) = \frac{1 + a_{\mathrm{T}1}\tau_x}{1 + a_{\mathrm{T}2}\tau_x}\, \mathrm{e}^{-(a_1^* - a_2^*)\tau_x} \,.$$

Regarding that t is given by (18), the angle of emerging for an electron backscattered from the substrate with normalized energy ε can be found again from (10).

Thus the total integrated energy spectrum from a double layer can be obtained from

$$\eta_{12}(\alpha, \tau_x, \varepsilon) = \begin{cases} \eta_1(\alpha, \tau_x, \varepsilon) + Q(\alpha, \tau_x)\, \eta_2'(\alpha, \tau_x, \varepsilon) & \text{for} \quad 0 \leqq \varepsilon < \varepsilon_x \\ \eta_1(\alpha, \tau_x, \varepsilon) \quad \text{for} \quad \varepsilon_x \leqq \varepsilon \leqq 1 \,, \end{cases} \tag{20}$$

where η_1 is given by (14) and $\eta_2'(\alpha, \tau_x, \varepsilon)$ can also be calculated from (14) inserting $a_{\mathrm{T}_2}(E_0, Z_2)$, $a_{\mathrm{S}2}^*(E_0, Z_2)$, and the appropriate limits of τ-integration which are τ_x instead of zero and $\tau_0 = \tau_\varepsilon$. The restriction of ε is caused by the film covering the substrate.

Taking the ε-derivate of (20) the differential energy spectrum is obtained. $\mathrm{d}\eta_1/\mathrm{d}\varepsilon$ is given by (16) and $\mathrm{d}\eta_2'/\mathrm{d}\varepsilon$ can also be calculated from (16) taking into account the same changes as for the case of $\eta_2'(\alpha, \tau_x, \varepsilon)$.

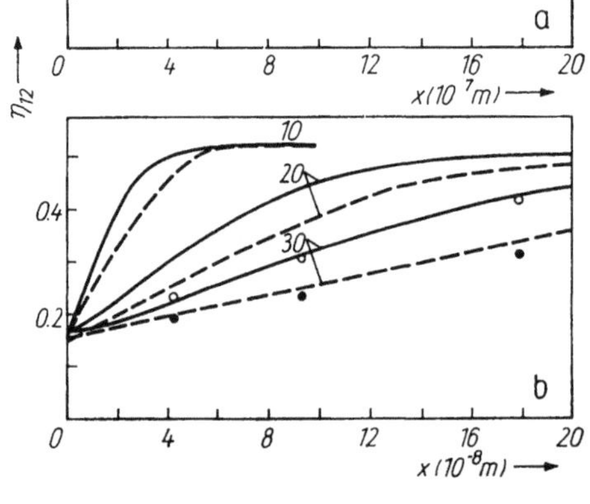

Fig. 5. Backscattering coefficient at normal incidence as a function of film thickness for a) Al films on Au substrate, b) Au films on Al substrate; – – – Monte Carlo simulation from Kyser and Murata [27], ○ experimental values for 20 keV and ● 30 keV from Hohn et al. [28], ——— calculated from (21)

The backscattering coefficient may be written as

$$\eta_{12}(\alpha, \tau_x) = \eta_1(\alpha, \tau_x) + Q(\alpha, \tau_x)\,[\eta_2(\alpha) - \eta_2(\alpha, \tau_x)]\,, \tag{21}$$

where all η included can be calculated from (17) using the appropriate scattering coefficients. $\eta_2(\alpha)$ is the bulk backscattering coefficient for the substrate.

Equation (21) is now used to calculate the backscattering coefficient for aluminium films on a gold substrate and gold films on an aluminium substrate for normal incidence at different incident electron energies. In Fig. 5 the result of these calculations are compared with those of Monte-Carlo simulations of Kyser and Murata [27] and recent experiments of Hohn et al. [28]. For the case of aluminium films on a gold substrate the calculated curves agree with each other fairly well, whereas for the case of gold. films on an aluminium substrate the quantitative agreement is less close. Additional experimental results are needed for a comparison of the calculations with experiment Both the analytic model and the Monte-Carlo simulations do not predict correctly the energy dependence for the gold substrate, which is experimentally found to increase slowly with increasing energy. A more exact description of the scattering and slowing-down process should yield an improvement of the theoretical values. Recently Kotera et al. [29] concluded that mainly the use of the Mott equation for the elastic scattering cross-section in their Monte-Carlo model gives the expected variation of the back-scattering coefficient for a bulk gold target in the energy range of 1 to 10 keV.

The results of further calculations show that a linear relation between η_{12} and η_1 as suggested earlier by De Nee [30], is a good approximation as long as the film atomic number Z_1 is not too large in comparison with the substrate atomic number Z_2. This allows simply to remove the contribution of the substrate to the backscattering coefficient of a suitable double layer to get the backscattering coefficient of the self-supporting film [31].

Experimental values for energy spectra from double layers seem not to be available in the literature. Therefore, the differential energy spectra for aluminium films of various thickness on a gold substrate at normal incidence, shown in Fig. 6, are calculated with the approximated expressions for a_T and a_S. With increasing film thickness the peak position of the spectra is shifted from that of gold ($t_x = 0$) to lower energies than the most probable energy for the bulk aluminium target ($t_x = 0.5$). The sharp rise of the yield, occurring at the beginning of the combined film–substrate backscattering, is due to the model which reflects the average behaviour of the electrons, disregarding straggling and using a unique relation between penetration depth and travelled path length. A smooth variation with energy should be expected for experimental spectra.

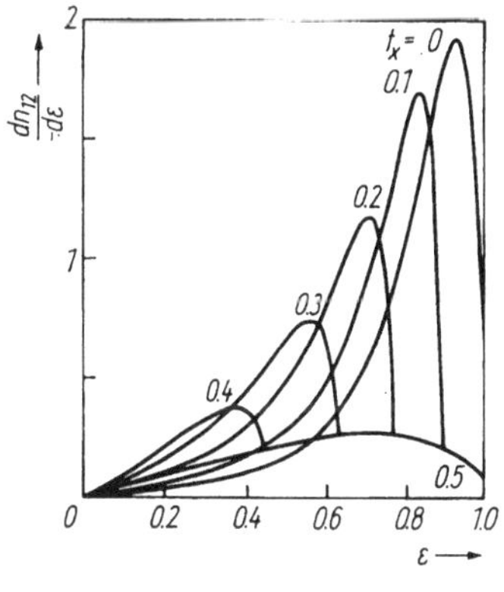

Fig. 6. Differential energy spectra for Al films on Au substrate at normal incidence calculated from the ε-derivate of (20) for $a_T = Z/5$ and $a_S = Z/80$. t_x is the fractional path length according to (11)

5. Conclusions

Werner's analytic model has been used to derive expressions for the energy spectra of electrons backscattered from solid films and double layers. The good agreement of the values calculated from these expressions with experimental and Monte-Carlo data confirms that the main features of electron backscattering at intermediate energies can be described by a simple analytic model which combines small-angle and single large-angle scattering, where both scattering processes are characterized by the quotient of electron range and transport mean free path.

Acknowledgements

Dr. U. Werner and G. Neubert are gratefully acknowledged for reading the manuscript of this paper and helpful remarks.

Appendix

The parameters a_{T} and a_{S} can be generalized to scattering in mixed-element targets using the additivity rules (see [6], p. 253)

$$\frac{\mathrm{d}E}{\varrho\,\mathrm{d}s} = \sum_i c_i \frac{\mathrm{d}E}{\mathrm{d}(\varrho s)_i} \tag{A1}$$

and

$$\frac{1}{\varrho\lambda} = \sum_i c_i \frac{1}{(\varrho\lambda)_i}, \tag{A2}$$

where c_i is the mass concentration of the i-th element. From (A1) and (1) one has

$$\varrho s_0 = \frac{4\pi\varepsilon_0^2}{e_0^4 N_{\mathrm{A}}} \frac{E_0^2}{\sum_i c_i(Z_i/A_i) \ln (k_1 E_0/Z_i)}, \tag{A3}$$

where $k_1 = 111\ \mathrm{keV}^{-1}$.

The transport mean free paths are

$$(\varrho\lambda)_{\mathrm{D}}^{-1} = \frac{e_0^4 N_{\mathrm{A}}}{32\pi\varepsilon_0^2} \frac{Z^2}{AE^2} \ln\left(\frac{k_2 E}{Z^{2/3}}\right), \tag{A4}$$

where $k_2 = 41\ \mathrm{keV}^{-1}$ and

$$(\varrho\lambda)_{\mathrm{S}}^{-1} = \frac{e_0^4 N_{\mathrm{A}}}{64\pi\varepsilon_0^2} \frac{Z^2}{AE^2}. \tag{A5}$$

The definition of λ_{D} differs by a factor of two from that used by Brown and Ogilvie [14].

Putting (A4) and (A5) into (A2) one has from (5) and (A3)

$$a_{\mathrm{T}} = \frac{\sum_i c_i(Z_i^2/A_i) \ln (41 E_0/Z_i^{2/3})}{5 \sum_i c_i(Z_i/A_i) \ln (111 E_0/Z_i)},$$

$$a_{\mathrm{S}} = \frac{\sum_i c_i(Z_i^2/A_i)}{16 \sum_i c_i(Z_i/A_i) \ln (111 E_0/Z_i)},$$

where E_0 is in keV.

References

[1] W. Bothe, Ann. Phys. (Leipzig) **6**, 44 (1949).
[2] T. E. Everhart, J. appl. Phys. **31**, 1483 (1960).
[3] H. A. Bethe, M. E. Rose, and L. P. Smith, Proc. Amer. Phil. Soc. **78**, 573 (1938).
[4] G. D. Archard, J. appl. Phys. **32**, 1505 (1961).
[5] K. Kanaya and S. Okayama, J. Phys. D **5**, 43 (1972).
[6] H.-W. Thümmel, Durchgang von Elektronen- und Betastrahlung durch Materieschichten, Akademie-Verlag, Berlin 1974
[7] U. Werner, Doctoral Thesis, Martin-Luther-University Halle/Wittenberg, Halle (Saale) 1978.
[8] U. Werner, H. Bethge, and J. Heydenreich, Ultramicroscopy **8**, 417 (1982).
[9] A. Darliński, Ph. D. Thesis, Tech. Univ. Wrocław (Poland), 1979; phys. stat. sol. (a) **63**, 663 (1981).
[10] H. Niedrig, Beitr. elektronenmikroskop. Direktabb. Oberfl. **14**, 291 (1981); Scanning Electron Microscopy 1981, Vol. I, SEM Inc., AMF O'Hare (Chicago) (p. 29).
[11] H. Bethe, Hdb. Phys. Vol. 24, Julius Springer, Berlin 1933 (p. 491).
[12] C. J. Bakker and E. Segré, Phys. Rev. **81**, 489 (1951).
[13] O. Knecht and W. Bothe, Z. Naturf. **8a**, 805 (1953).
[14] D. B. Brown and R. E. Ogilvie, J. appl. Phys. **37**, 4429 (1966).
[15] N. G. Nakhodkin, A. A. Ostroukhov, and V. A. Romanovskii, Fiz. tverd Tela **4**, 1514 (1962).
[16] R. Shimizu and G. Shinoda, in: X-Ray Optics and X-Ray Microanalysis, Ed. H. H. Pattee, V. E. Cosslett, and A. Engstöm, Academic Press, New York/London 1963 (p. 419).
[17] W. S. McAfee, J. appl. Phys. **47**, 1179 (1976).
[18] G. J. Iafrate, W. S. McAfee, and A. Ballato, J. Vacuum Sci. Technol. **13**, 843 (1976).
[19] M. R. Sogard, J. appl. Phys. **51**, 4412 (1980).
[20] H. Kanter, Ann. Phys. (Leipzig) **20**, 144 (1957).
[21] H. Drescher, L. Reimer, and H. Seidel, Z. angew. Phys. **29**, 331 (1970).
[22] Z. Radzimski, Acta phys. Polon. **A53**, 783 (1978).
[23] G. Neubert and S. Rogaschewski, phys. stat. sol. (a) **59**, 35 (1980).
[24] E. H. Darlington, J. Phys. D **8**, 85 (1975).
[25] T. Matsukawa, R. Shimizu, and H. Hashimoto, J. Phys. D **7**, 695 (1974).
[26] H. J. Fitting, J. Phys. D **8**, 1480 (1975).
[27] D. F. Kyser and K. Murata, Nat. Bur. Standards spec. Publ. **460**, 129 (1976).
[28] F. J. Hohn, M. Kundt, H. Niedrig, and B. Stuht, Optik **46**, 491 (1976).
[29] M. Kotera, K. Murata, and K. Nagami, J. appl. Phys. **52**, 997 (1981).
[30] P. B. De Nee, Scanning Electron Microscopy 1978, Vol. I, SEM Inc., AMF O'Hare (Chicago) (p. 741).
[31] G. Neubert and S. Rogaschewski, phys. stat. sol. (a) **75**, K191 (1983).

(Received April 12, 1983)

phys. stat. sol. (a) 79, 161 (1983)

Subject classification: 2 and 20.3; 22.1.2

Forschungslaboratorien der Siemens AG, München[1]) *(a)*
and Institut für Physik, Fachbereich Elektrotechnik, HSBw München[2]) *(b)*

Photoluminescence Decay in Amorphous Silicon Alloys

By

B. RAUSCHER[3]) (a) and B. BULLEMER (b)

The photoluminescence (PL) decay of undoped a-Si:H, a-Si:O:H, and a-Si:C:H alloys during the first microsecond after excitation is investigated. In all alloys the fastest decay is observed for the short-wavelength PL photons, this is attributed to charge carriers trapped in shallow localized states tunneling to deeper states. This tunneling process competes with the radiative recombination. The nonradiative recombination at nonzero temperatures is interpreted as a thermal activation of the trapped charge carriers to the mobility edges followed by diffusion to nonradiative recombination centres. Whereas the initial decay is thereby accelerated, the radiative decay time in the long-term decay is observed to be prolonged if the temperature rises from 30 to 75 K.

Das Abklingen der Photolumineszenz (PL) von undotierten a-Si:H-, a-Si:O:H- und a-Si:C:H-Mischsystemen wird während der ersten Mikrosekunde nach der Anregung untersucht. In allen Mischsystemen beobachtet man für die kurzwelligen PL Photonen das schnellste Abklingverhalten, das durch Tunneln der in flachen lokalisierten Zuständen getrappten Ladungsträger in tiefere Zustände erklärt wird. Dieser Tunnelprozeß konkurriert mit der strahlenden Rekombination. Die nichtstrahlende Rekombination bei höheren Temperaturen wird durch eine thermische Aktivierung der getrappten Ladungsträger zu den Beweglichkeitskanten mit anschließender Diffusion zu nichtstrahlenden Rekombinationszentren interpretiert. Während das anfängliche Abklingen dadurch schneller wird, beobachtet man im Langzeit-Abklingen eine Zunahme der strahlenden Zerfallszeit, wenn die Temperatur von 30 auf 75 K erhöht wird.

1. Introduction

Photoluminescence (PL) in a-Si:H at low temperatures (about 50 K) is generally considered to stem from the recombination of electron–hole pairs in localized tail states ([1] and references therein). Characteristically the decay of the PL after excitation by a short pulse is found to be nonexponential up to 10^{-2} s after excitation. As a possible way to analyze the experimental data, a decay-time distribution has been used [2] that is attributed to the radiative tunneling of electron–hole pairs exhibiting a wide spread in their pair separation. This spread results from the diffusion of charge carriers away from each other after photogeneration. The Xerox group has recently published a quantitative description of the diffusion and recombination process [3] based on the Onsager theory [4]. Their model which neglects the occupation of the tail states and the electron–phonon interaction, yields the gross features of the PL decay at $\hbar\omega_{PL} = 1.4$ eV for times greater than 1 μs after excitation. This model fails, however, to describe the wavelength dependence of the PL decay and the temperature response of stationary spectra.

The present report investigates the influence of the density of localized states in the mobility gap on the PL decay. The charge carrier pairs are assumed to be trapped

[1]) Otto-Hahn-Ring 6, D-8000 München 83, FRG.
[2]) Werner-Heisenberg-Weg 39, D-8014 Neubiberg, FRG.
[3]) Presentaddress: Siemens Corp., 105 College Road East, Princeton, New Jersey 08540, USA.

in the tail states, from where they tunnel to deeper states or, in a parallel process, recombine radiatively with each other. Since the probability of tunneling to states deeper in the mobility gap decreases with increasing trapping depth due to the rapidly declining density of states, the decay rate of the charge carriers in the various tail states differs appreciably. The decay rate of the short-wavelength PL ($\hbar\omega_{PL} > 1.3$ eV) is consequently accelerated, which is shown to be generally observable in amorphous silicon alloys, such as a-Si:H, a-Si:O:H, and a-Si:C:H. In addition, the PL decay of these a-Si alloys is investigated at nonzero temperatures. The results are discussed in terms of a recently proposed model for PL in a-Si:H [5].

2. Experimental

The samples used were prepared by plasma deposition of undiluted SiH_4 using low rf power (0.04 W/cm^2 at 13.56 MHz). The density of states in the gap has been estimated from capacitance measurements to be about 10^{16} (cm^3 eV)$^{-1}$ [6]. Details on the deposition process are given in [7], while the electrical parameters of the films are surveyed in [8]. The time-resolved luminescence measurements were performed by time-correlated single-photon counting (time resolution about 2 ns) using a computer-controlled multichannel analyzer. A cavity-dumped laser (514.5 nm) with typical pulse widths ranging from 8 to 15 ns and a pulse power of about 10^{-9} J was used for excitation. The PL was dispersed in a monochromator with a spectral width of 25 nm and monitored with an S1 photomultiplier. The sample was mounted on a closed-loop helium cryostat that can be set to temperatures down to 30 K. The accuracy of the measurement of the PL decay time was continuously monitored using a special $GaAs_{0.35}P_{0.65}$:N LED with a known decay time of 75 ns.

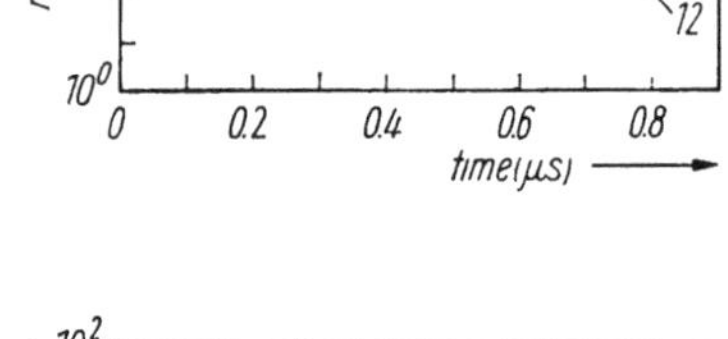

Fig. 1. Dependence of the PL decay on the excitation pulse length t_{exc}. $T = 30$ K, $\hbar\omega_{PL} = 1.3$ eV, $\lambda_{exc} = 514.5$ nm

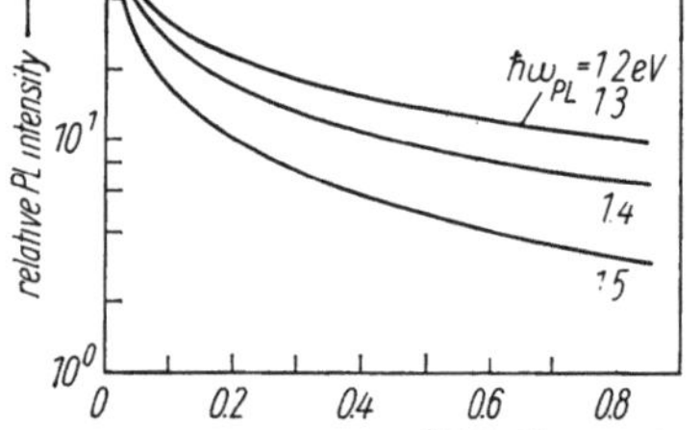

Fig. 2. PL decay at various PL photon energies $\hbar\omega_{PL}$. $t_{exc} = 10$ ns, $\lambda_{exc} = 514,5$ nm, $T = 30$ K

3. Low-Temperature PL Decay

Comparison of the results of PL decay experiments reported by various authors is hardly possible on account of the differences in time resolution, in exciting laser pulse width, and in the chosen PL wavelengths. Moreover, to observe the effects of the thermalization of the carriers in the tail states, i.e. the fast decay of the PL, a minimum time resolution of less than 10 ns is required [9] and the width of the pulses used for excitation must be of the same order. Larger pulse widths will increase the number of long-lived electron–hole pairs, thereby decreasing the rate of the initial decay, as shown in Fig. 1 for a boron-doped sample. This feature is characteristic of the existence of a decay time distribution, which in a-Si:H is reported to range from 10^{-8} to 10^{-2} s [3].

Let us first consider the dependence of the PL decay on the luminescence energy at low temperature as shown in Fig. 2. Below 1.3 eV the decay does not depend on the luminescence energy, but above 1.3 eV the rate of decay increases. This can be interpreted with the aid of our recently introduced model that assumes geminate recombination and explicitly takes account of the tunneling of the photogenerated charge carriers from shallow to deeper tail states [5, 10]. Assume a charge carrier, say an electron, to be in a localized state at an energy ε below the conduction band ($\varepsilon = E_c - E$). Neglecting nonradiative recombination, an electron in a localized state can recombine radiatively with a hole in the valence-band tail or, alternatively, it can tunnel to a more localized deeper state and relax by emitting several phonons. Thus the total decay rate p_{tot} in the state characterized by ε is determined by the radiative decay rate p_r and the tunneling rate p_t,

$$p_{tot}(\varepsilon) = p_r + p_t(\varepsilon)\,. \tag{1}$$

The rate $p_t(\varepsilon)$ is given by

$$p_t(\varepsilon) = \nu_a \exp\left[-2\alpha r(\varepsilon)\right], \tag{2}$$

where ν_a is a typical vibronic frequency of 10^{12} Hz, α the localization parameter of the carrier wave function, and r the mean distance to the next unoccupied state deeper in the gap. From calculations of the density of states in the mobility gap [11] it is obvious that $r(\varepsilon)$ increases drastically with increasing energy ε, whence $p_t(\varepsilon)$ in (2) goes to zero if ε is greater than the tail width. As radiative recombination, which emits low-energy photons ($\hbar\omega_{PL} \leqq 1.3$ eV), occurs mainly between carriers localized deep in the mobility gap, it is unaffected by competetive tunneling. The high-energyPL ($\hbar\omega_{PL} > 1.3$ eV), however, stems from charge carriers in shallow traps, and thus pro-

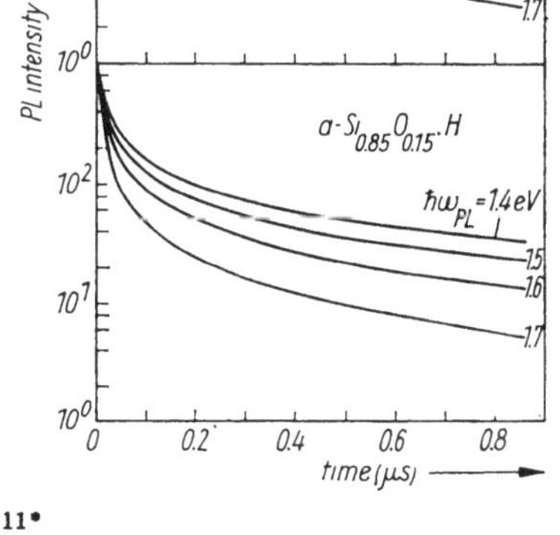

Fig. 3. Dependence of the decay of an a-$Si_{0.78}C_{0.22}$:H sample and of an a-$Si_{0.85}O_{0.15}$:H sample on the PL photon energy $\hbar\omega_{PL}$. $t_{exc} = 10$ ns, $\lambda_{exc} = 514.5$ nm, $T = 30$ K

nounced tunneling accelerates the PL decay. The wavelength dependence of the PL decay is a common feature of "pure" a-Si:H and of alloys, as shown in Fig. 3 for a-$Si_{0.78}C_{0.22}$:H and a-$Si_{0.63}O_{0.15}$:H. Compared to a-Si:H, the photoluminescence in these alloys generally decays faster due to the increased incorporation of nonradiative defect centres, which reduces the stationary PL quantum efficiencies.

4. Temperature Dependence of the PL Decay

Section 3 considered only the low-temperature case, where the nonradiative recombination can be neglected. It is well known that the quantum efficiency and the decay time of the PL in a-Si:H decrease at higher temperatures (Fig. 4) due to the competitive nonradiative recombination. In our model the charge carriers are located in the tail states, from where they recombine radiatively. Nonradiative recombination is possible by two parallel mechanisms: the electrons or holes tunnel in a hopping process directly to a nonradiative recombination centre, or the carriers are thermally activated to the mobility edges and then diffuse to such a centre. For a low density of defect states, the hopping distances are very large, whence the thermal activation will determine the recombination rate. For an electron trapped with an energy ε below E_c an activation energy of ε plus the pair binding energy E_B is necessary to escape the Coulomb attraction of the hole and to diffuse away from it. Thus the nonradiative recombination rate $p_{nr}(\varepsilon)$ of a charge carrier in a state is given by [5] as

$$p_{nr}(\varepsilon) = \nu_a \exp\left[-\frac{(E_B + \varepsilon)}{kT}\right], \tag{3}$$

where ν_a is the attempt frequency (10^{12} Hz) and k the Boltzmann constant. It should be stressed that the binding energy E_B in (3) is small relative to the mean trap depth energy ε. Our recent model [5] yielded a mean trap depth energy ε of 0.13 to 0.19 eV

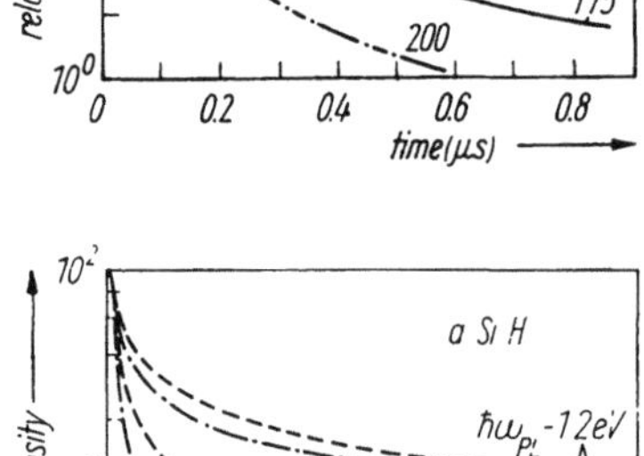

Fig. 4. PL decay in a-Si:H ($\hbar\omega_{LP}$ = 1.2 eV) at various sample temperatures. t_{exc} = 10 ns, λ_{exc} = 514.5 nm

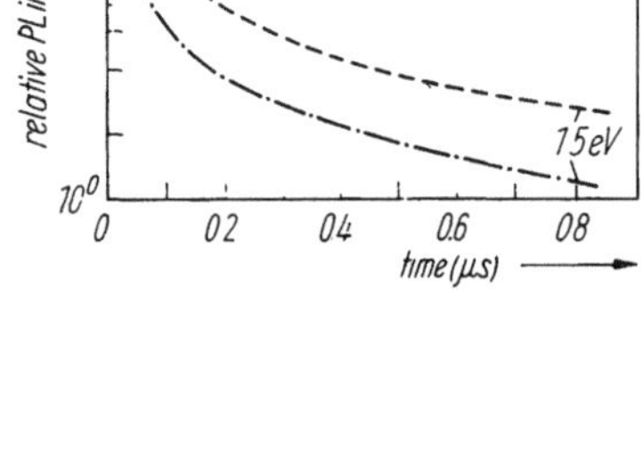

Fig. 5. Difference in influence of the temperature on the PL decay of a-Si:H at two PL energies $\hbar\omega_{PL}$ = 1.2 eV and $\hbar\omega_{PL}$ = 1.5 eV. ——— T = 50 K, — · — 100 K, t_{exc} = 10ns, λ_{exc} = 514.5 nm

for electrons and of 0.29 to 0.35 eV for holes, which agrees closely with the activation energy for the drift mobility [12]. The Coulomb binding energy E_B, however, is 25 meV calculated on the basis of the mean electron–hole separation of 5 nm estimated by the Xerox group [2] from the onset of bimolecular recombination observed in the excitation-power dependence of the PL.

An as yet unreported feature of the PL decay is the difference in the influence of the temperature on the decay in the high- and low-energy spectral regions of the PL (Fig. 5). A temperature rise from 50 to 100 K will influence the PL decay only slightly at 1.2 eV but markedly at 1.5 eV. Together with the general wavelength dependence of the PL decay in Fig. 2, these effects contradict an explanation of the PL in terms of a phonon interaction as proposed earlier in [13]. If, however, the PL is regarded as a tail–tail luminescence paralleled by nonradiative recombination, the interpretation of Fig. 5 is quite natural. The high-energy luminescence stems from electron–hole pairs in localized states closer to the band edges than the low-energy luminescence. Thus the activation energy ε in (3) is smaller for the PL of 1.5 eV than for the PL of 1.2 eV, whence the nonradiative recombination rate for the PL of 1.5 eV is higher. A corresponding behaviour is also observed in stationary PL experiments, as the high-energy region of the PL band decreases instantly with rising temperature, but the low-energy region remains almost constant up to a temperature of 100 K [5].

A closer study of Fig. 4 reveals still another interesting feature. The decay curves for the temperatures 30 and 100 K, which are normalized at $t = 0$, intersect at about $t = 0.8$ μs. Fig. 6 and 7 depict this situation in greater detail. The decay time up to $t = 0.1$ μs decreases steadily with rising temperature due to the competitive nonradiative recombination. At higher temperatures, however, the slope of the curves between $t = 0.5$ and 0.8 μs decreases indicating an increase of radiative decay time with temperature.

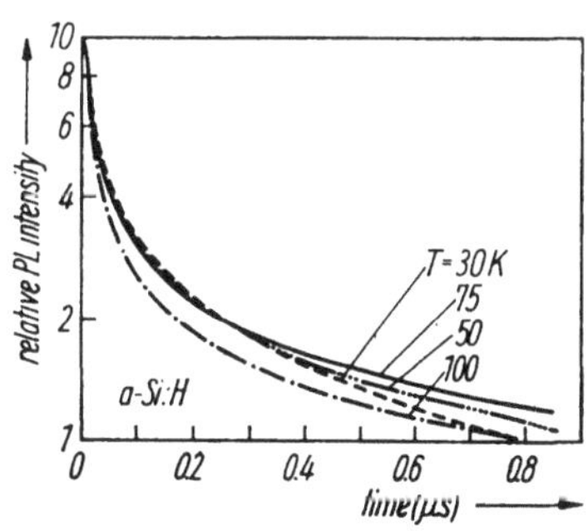

Fig. 6

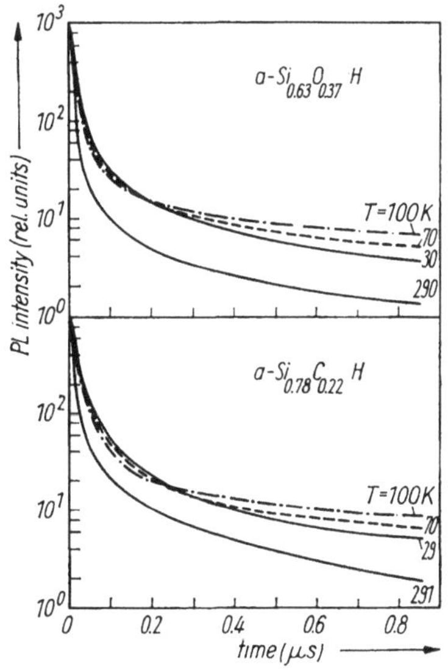

Fig. 7

Fig. 6. PL decay ($\hbar\omega_{PL} = 1.2$ eV) in the temperature region 30 K $\leqq T \leqq$ 100 K (curves normalized at $t = 0$). $t_{exc} = 10$ ns, $\lambda_{exc} = 514.5$ nm

Fig. 7. Temperature dependence of the PL decay of an a-$Si_{0.63}O_{0.37}$:H sample ($\hbar\omega_{PL} = 1.7$ eV) and an a-$Si_{0.78}C_{0.22}$:H sample ($\hbar\omega_{PL} = 1.5$ eV). $t_{exc} = 10$ ns, $\lambda_{exc} = 514.5$ nm

The rise in temperature causes the electron–hole pairs to diffuse farther apart as discussed in [5]. This increase in pair separation may either occur during thermalization to the mobility edges after excitation, or during a diffusion of the pairs in shallow localized states. As the radiative decay rate is given [2] as

$$p_r = p_0 \exp(-2\alpha r) , \tag{4}$$

where α is the localization parameter of the wave function ($\alpha^{-1} = 1.1$ nm) and $p_0 = 10^8\ s^{-1}$, a reduction of the electron–hole wave function overlap results in a lower decay rate p_r. At higher temperatures (> 100 K) this reduction in p_r is overcompensated by the predominant nonradiative recombination, which is regarded as a competitive process. Thus the slope of the decay curves in Fig. 8 again increases for temperatures T above 100 K. It should be stressed that this reduction in p_r was also observed in a-Si:O:H and a-Si:C:H samples (Fig. 7). These experimental results convincingly demonstrate that the geminate pair recombination model holds for all amorphous silicon alloys at low temperatures.

Acknowledgements

The authors gratefully acknowledge the preparation of the a-Si:H samples by M. Möller, W. Stetter, and W. Juergens. Prof. Dr. Kukimoto and Dr. E. Holzenkämpfer kindly supplied the a-Si:C and a-Si:O samples. Dr. H. Pfleiderer took part in many stimulating discussions.

References

[1] R. A. STREET, Adv. Phy. **30**, 593 (1981).
[2] C. TSANG and R. A. STREET, Phys. Rev. B **19**, 3027 (1979).
[3] K. M. HONG, J. NOOLANDI, and R. A. STREET, Phys. Rev. B **23**, 2967 (1981).
[4] L. ONSAGER, Phys. Rev. **54**, 554 (1938).
[5] B. RAUSCHER, H. PFLEIDERER, and B. BULLEMER, phys. stat. sol. (a) **78**, 623 (1983).
[6] H. PFLEIDERER and B. RAUSCHER, phys. stat. sol. (a) **75**, 537 (1983).
[7] M. MÖLLER, R. PLÄTTNER, W. STETTER, and B. RAUSCHER, Proc. III EC Photovoltaic Solar Energy Conf., Cannes, Ed. W. PATZ, Reidel Publ. Co., Dordrecht 1980 (p. 322).
[8] W. KRÜHLER, M. MÖLLER, H. PFLEIDERER, R. PLÄTTNER, and B. RAUSCHER, Proc. IV EC Photovoltaic Solar Energy Conf., Stresa, Ed. W. H. BLOSS and G. GRASSI, Reidel Publ. Co., Dordrecht 1982 (p. 754).
[9] W. REHM and R. FISCHER, phys. stat. sol. (b) **94**, 595 (1979).
[10] B. RAUSCHER, Thesis, Federal Armed Forces University Munich, 1982.
[11] M. HIROSE, T. SUZUKI, and G. H. DÖHLER, Appl. Phys. Letters **34**, 234 (1979).
[12] A. R. MOORE, Appl. Phys. Letters **31**, 762 (1977).
[13] R. A. STREET, Phil. Mag. **B37**, 35 (1978).

(Received March 9, 1983)

phys. stat. sol. (a) 79, 167 (1983)

Subject classification: 10.2 and 20.3; 22.5.2

Department of Physics, Indian Institute of Technology, Kharagpur[1]) *(a)*
and Department of Physics Manipur University, Imphal[2]) *(b)*

Effect of Post-Irradiation Deformation on Thermoluminescence Spectra of KCl:Ca and NaCl:Ca Crystals

By

J. MANAM (a), R. K. GARTIA (b), and V. V. RATNAM (a)

The effect of post-irradiation deformation on the thermoluminescence spectra of KCl:Ca and NaCl:Ca crystals is studied. Deformation does not cause any appreciable change in the recombination mechanisms and sites as revealed by the occurrence of the peaks in TL emission in the same spectral regions. The deformation causes only a preferential redistribution of the traps as revealed by the changes in the relative intensities of the emissions occurring in almost the same regions. The results in KCl:Ca are in confirmity with observations of Arsenovici and Townsend on the redistribution of F-centres by F-light bleaching, whereas in NaCl:Ca they are in disagreement

Der Einfluß von Deformation und Bestrahlung auf die Thermolumineszenzspektren von KCl:Ca- und NaCl:Ca-Kristallen wird untersucht. Die Deformation verursacht keine wesentliche Änderung im Rekombinationsmechanismus und der Rekombinationsplätze, wie durch das Auftreten der Maxima der TL-Emission in den gleichen spektralen Bereichen nachgewiesen wird. Die Deformation verursacht nur eine bevorzugte Umverteilung der Haftstellen, die sich in Änderungen der relativen Intensitäten der Emission, die in fast denselben Bereichen auftreten, bemerkbar macht. Die Ergebnisse in KCl:Ca sind in Übereinstimmung mit Beobachtungen von Arsenovici und Townsend über die Umverteilung von F-Zentren durch F-Lichtbleichung, während sie für NaCl:Ca nicht übereinstimmen.

1. Introduction

The phenomenon of deformation luminescence, namely the emission of light during post-irradiation deformation of irradiated alkali halides is well-known. This effect has been investigated by various workers reference to which can be found in the work of Butler [1] who has recorded the RT deformation luminescence spectra of KCl, KBr, and NaCl. Similar effects have been shown to occur below RT in systems like KCl:Cu and KBr:Cu by Markova et al. [2]. Deformation luminescence is believed to arise due to the fact that moving dislocations (produced during deformation) can directly ionize F centres as a result of the sensitivity of their electronic energy to the ionic configuration. The ionized F centre electron can recombine with the emission centre giving rise to DL or it may be retrapped in some other traps giving rise to new thermoluminescence (TL) peaks [3]. Recently attempts have been made to understand the effect of post-irradiation deformation on the TL of Ca-doped NaCl [4]. However, no attempt has been made to study the effect of post-irradiation deformation on TL spectra of alkali halides.

In the present work, we have studied the effect of post-irradiation deformation on TL spectra of KCl:Ca and NaCl:Ca. Furthermore, in these systems it is known that TL occurs by the recombination of electrons and holes [5 to 7] and DL occurs at RT [1] in these irradiated crystals.

---

[1]) Kharagpur 721302, India.
[2]) Imphal 795003, India.

2. Experimental

The samples used were single crystals. In all cases before irradiation the samples were annealed at 400 °C for about 10 min and quenched to room temperature (RT) (≈ 30 °C). The samples were irradiated by X-rays obtained from the MO target of a Machelett tube operated at 30 kV, 10 mA. The TL curves are recorded under identical conditions, the sample being heated in vacuum at a uniform rate of (30 ± 0.5) K per min. The TL emission was detected by an RCA 1P28B photomultiplier tube and recorded on a strip-chart recorder. The TL emission spectra were recorded by scanning the emission near the peak temperature with the help of a Jarrel-Ash 0.25 m grating monochromator. The irradiated samples were deformed by compressing them between the two jaws of a laboratory screw gauge during which process visible DL was also observed. Care was taken to see that the stress applied did not cause fracture.

3. Results and Discussion

3.1 Effect of post-irradiation deformation on TL spectra of KCl:Ca crystals

TL of KCl:Ca crystals X-irradiated at RT for 2 min shows a prominent TL peak around 200 °C and two overlapping weak peaks around 110 and 135 °C (curve 1, Fig. 1). This result is more or less similar to earlier results [8]. When the irradiated crystal is deformed and TL is recorded, it is found that low-temperature TL peaks increase in intensity, whereas that of the high-temperature TL peak decreases (curve 2, Fig. 1). Since the 135 and 200 °C TL peaks are assigned to the first/early and late/second stage F centres, respectively, the present results may be interpreted as transfer of F centre electrons from the second stage to the first during deformation. A similar behaviour has been reported recently in NaCl:Ca [4]. However, it is to be noted that in the work of Gartia et al. [4] deformation was done by grinding the micro crystals which naturally includes fracture as well, whereas in the present study care was taken that fracture does not occur during deformation of the single crystals. Therefore, the

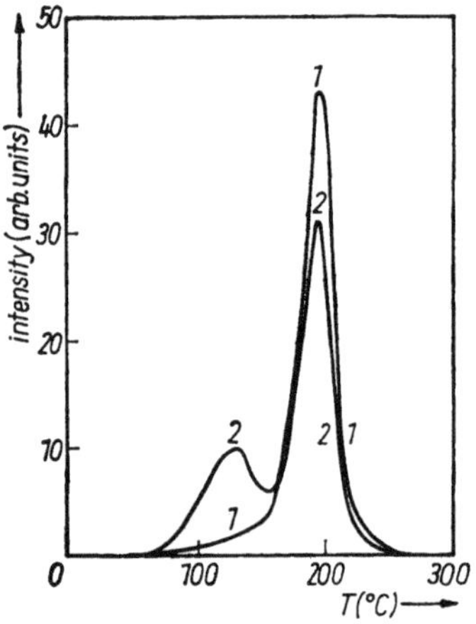

Fig. 1

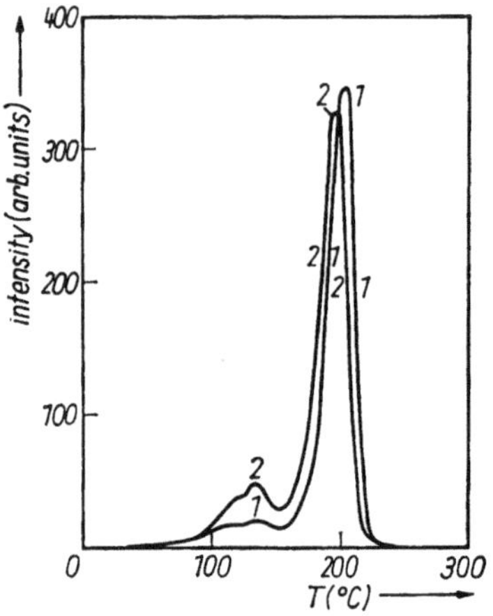

Fig. 2

Fig. 1. Thermoluminescence glow curves of a KCl:Ca (250 ppm) crystal irradiated by X-rays for 2 min at RT (≈ 30 °C): (1) as-irradiated, (2) after deformation at RT

Fig. 2. Thermoluminescence glow curves of a KCl:Ca (250 ppm) crystal irradiated by X-rays for 2 min at RT: (1) TL recorded after thermally cleaned to 165 °C and cooled to RT, (2) same as (1) but after deformation at RT present results show that if the deformation is sufficient to allow the movement of dislocations (which can be checked by observing whether DL is produced or not during deformation), one can observe the deformation stimulated thermoluminescence. In an attempt to understand the above results the following experiment was carried out.

The crystal was irradiated for 2 min and heated to 165 °C to thermally clean the low-temperature TL peaks. The sample was then cooled down to RT and TL is recorded. As expected the TL curve shows an isolated TL peak around 195 °C (curve 1, Fig. 2). When the crystal is irradiated at RT, and the low-temperature TL peaks are thermally cleaned (as done in the case of curve 1, Fig. 2) and then deformed at RT and TL is recorded, the post-irradiation deformation stimulates a TL peak around 130 °C (curve 2, Fig. 2). A substantial decrease in intensity of the 195 °C TL peak is also observed. This result clearly establishes the transfer of F centre electrons from the second stage to the first.

Table 1
Deformation luminescence and thermoluminescence spectra of KCl and NaCl

sample	emission (nm)	
KCl	430 ± 10, 520 ± 20	DL spectra [1]
KCl:Ca	430 to 435, 510 to 520	TL spectra (present work)
NaCl	282 ± 10, 318 ± 10, 362 ± 5, 610 ± 20	DL spectra [1]
NaCl:Ca	365 to 370, 450 to 470, 510 to 520, 590 to 600	TL spectra (present work)

3.2 Effect of post-irradiation deformation on TL spectra of KCl:Ca

TL spectra of the 135 and 200 °C TL peaks were recorded in the range 300 to 700 nm both in undeformed and deformed crystals. The results are shown in Fig. 3. In both cases two prominent bands are observed in the regions of 430 nm and 510 to 520 nm. In addition to these two emissions there is an indication of an emission around 620 nm. These two main emission bands occur in the same regions as those reported by Butler [1] in RT DL (Table 1). At the very outset it should be noted that no new emission bands are observed in TL indicating that no new recombination centres are formed by deforma-

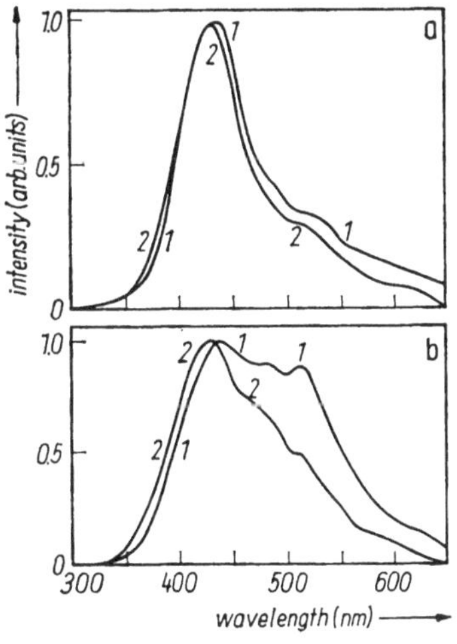

Fig. 3. Effect of deformation on TL spectra of KCl:Ca irradiated by X-rays at RT: a) TL spectra at 135 °C (1) as-irradiated, (2) after deformation at RT, (b) TL spectra at 200 °C (1) as-irradiated, (2) after deformation at RT; curves normalized for 435 nm emission

tion. It is clear that the recombination processes giving luminescence are identical both for DL and TL in the KCl system. However, one point which is worth noting is that the relative intensities of the 430 and 520 nm emissions are not the same in undeformed and deformed crystals. The effect is more pronounced in the case of the high-temperature (205 °C) TL peak. In the undeformed crystal both the TL emissions are comparable, whereas in the deformed one, the 510 to 520 nm emission is suppressed to a greater extent compared to the undeformed one. This fact leads us to infer that during DL preferential radiative recombination takes place at sites responsible for the 435 to 440 nm emission and that more of the emission centres responsible for the 510 to 520 nm emission are destroyed by deformation. Remembering that the 520 nm emission in KCl is due to the radiative decay of the self-trapped excitons (STE) in the unperturbed regions of the crystal, that is to say, due to the radiative decay of STEs formed when moving V_K centres (thermal dissociation products of the radiation stabilized V_3 and similar trihalogen centres) approach the free isolated F centres. These F centres are naturally the randomly distributed F centres. The 430 to 440 nm emission is known to be due to the radiative decay of the similarly formed STEs next to cation vacancy sites. Comparing the relative intensities of the two emission bands in TL before and after deformation we are led to the conclusion that deformation, apart from facilitating the radiative annihilation of the radiation induced electron and the hole centres, redistributes the traps. To be specific, deformation, so to say, pushes the F centres from random location sites to sites near impurities, vacancies, etc. That is why the emission in the 435 nm band is relatively stronger in the deformed samples than in the undeformed samples. We can attribute the centres responsible for the 520 nm emission to the second stage of F centres and those responsible for the 435 nm emission to the first stage in the light of the observations of Arsenovici and Townsend [9]. It is to be further noted that they have reported that even the randomly distributed second stage F centres move towards dislocations and get localized by light bleaching. Our observations and inference are in confirmity with the above findings.

3.3 Effect of post-irradiation deformation on TL spectra of NaCl:Ca

TL spectra of the 135 and 180 °C TL peaks of NaCl:Ca were recorded in the range 300 to 700 nm both in undeformed and deformed crystals. The results are shown in Fig. 4. In both cases four emissions around 367, 450, 521, and 594 nm are observed. Unlike the case of KCl where similarity between TL and DL spectra has been observed, in the case of NaCl there is a considerabel difference (Table 1). Though some of the emissions like the 367 and 594 nm emissions are common to both DL and TL spectra, there are dissimilarities. Therefore, unlike the KCl system, in the NaCl system the recombinations during DL and TL do not seem to take place under idnetical conditions. To visualize this the

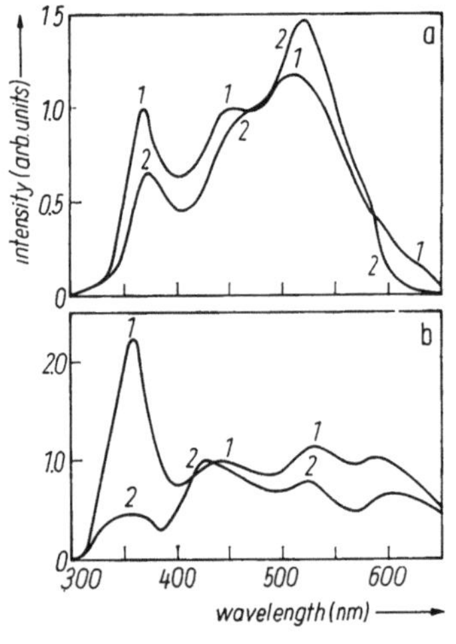

Fig. 4. Effect of deformation on TL spectra of NaCl:Ca irradiated by X-rays for 15 min at RT: (a) TL spectra at 135 °C (1) as-irradiated, (2) after deformation at RT; (b) TL spectra at 180 °C (1) as-irradiated, (2) after deformation at RT; curves normalized for 450 nm emission

TL spectra for the two main glow peaks before and after deformation are shown in Fig. 4a and b where the spectra normalized for the 450 nm region emission are plotted. From these spectra it is evident that in the TL spectra of the 180 °C glow peak, compared to the 430 nm region emission, emission in the 365 nm region is reduced to a relatively greater extent than the emissions in 530 nm and 600 nm regions. In the TL spectra of the 135 °C glow peak, while the 365 to 375 nm region emission is reduced, the emission in the 520 nm region increases relative to the 450 to 470 nm region emission, though due to deformation there is an overall decrease in intensities of all the emissions. This result is clearly different from that for KCl:Ca. Another difference in the behaviour of NaCl:Ca at deformation is that the peak positions of the emission bands shift to higher wavelengths in this system, whereas in the KCl:Ca system the deformation shifts the peak positions to lower wavelengths. A shift to lower wavelengths can be attributed to a greater perturbation at the emitting sites caused during redistribution of the defect centres. Then the shift of the emission peaks to longer wavelengths and the increase in the emission at the sites of isolated unperturbed F centres can be understood in this way. During deformation the interacting dislocations do not affect the randomly distributed F centres to the same extent as in KCl.

Acknowledgements

One of the authors (R.K.G.) is grateful to the authorities of UGC, New Delhi, Manipur University, Imphal and I.I.T., Kharagpur for making the collaboration possible under the National Association Programme.

References

[1] C. T. Butler, Phys. Rev. **141**, 750 (1966).
[2] I. A. Markova, O. Sorgina, and S. Z. Shmurak, phys. stat. sol. (b) **74**, 753 (1976).
[3] E. Panizza, Phys. Letters, **10**, 37 (1964).
[4] R. K. Gartia, R. Singh, and K. Y. Singh, phys. stat. sol. (b) **113**, K57 (1982).
[5] I. Katz, B. Chenfoux, and N. Kristianpoller, phys. stat. sol. (a) **12**, 307 (1972).
[6] P. Sridharan, R. K. Gartia, R. C. Bhuniya, and V. V. Ratnam, phys. stat. sol. (a) **64**, 127 (1981).
[7] R. K. Gartia, phys. stat. sol. (a) **37**, 571 (1976).
[8] S. C. Jain and P. C. Mehendru, Phys. Rev. **140**, A957 (1965).
[9] L. C. Arsenovici and P. D. Townsend, Phil. Mag. **25**, 381 (1972).

(Received April 15, 1983)

phys. stat. sol. (a) 79, 173 (1983)

Subject classification: 13.4 and 20.3; 22.1.2

The Research Institute for Iron, Steel, and Other Metals, Tôhoku University, Sendai[1])

Dependence of Photoluminescence on Temperature in Dislocated Silicon Crystals

By

M. SUEZAWA, Y. SASAKI, and K. SUMINO

The intensities of photoluminescence lines D1 through D4 and also of the background component that are characteristic in a dislocated silicon crystal are measured as a function of the temperature. All of them become weaker as the temperature is raised. Such behaviour is interpreted in a quantitative way with the model in which two different kinds of traps, one with a shallow level and the other with a deep level, participate in the radiative recombination process giving rise to each photoluminescence line. The shallower levels are assigned to the levels associated with hole traps that are drawn from the valence band by the deformation potential of a dislocation, and are in the range 4 to 12 meV above the top of the valence band. The deeper levels are proposed to be associated with electron traps due to the antibonding states of reconstructed bonds at the dislocation core, being located in the range 0.16 to 0.35 eV below the bottom of the conduction band.

Die Photolumineszenzintensitäten der Linien D1 bis D4 sowie der Untergrundkomponente, die charakteristisch ist für einen versetzten Siliziumkristall, werden in Abhängigkeit von der Temperatur gemessen. Alle werden schwächer, wenn die Temperatur erhöht wird. Ein solches Verhalten wird quantitativ mit einem Modell interpretiert, in dem zwei verschiedene Haftstellenarten, eins mit einem flachen Niveau und das andere mit tiefem Niveau, an dem strahlenden Rekombinationsprozeß teilnehmen und zu je einer Photolumineszenzlinie Anlaß geben. Die flacheren Niveaus werden den Niveaus zugeordnet, die mit Löchertraps verknüpft sind, die aus dem Valenzband durch das Deformationspotential einer Versetzung stammen und im Bereich 4 bis 12 meV oberhalb der Valenzbandkante liegen. Für die tieferen Niveaus wird angenommen, daß sie verknüpft sind mit Elektronenhaftstellen, hervorgerufen durch die antibindenden Zustände der rekonstruierenden Bindungen am Versetzungskern und im Bereich 0,16 bis 0,35 eV unterhalb der Leitungsbandkante lokalisiert sind.

1. Introduction

It has been known that characteristic photoluminescence lines are developed in silicon crystals upon plastic deformation [1 to 7]. In previous papers [3, 7], the present authors have obtained the strong evidence showing that the deformation-induced photoluminescence is related to dislocations introduced by deformation. The parameters that affect the photoluminescence spectrum have also been clarified. The important information which can be deduced from the photoluminescence spectrum is the position of electronic energy levels of the defects concerned with the phenomenon. However, such energy levels are not necessarily related directly to the spectrum itself, since some complicated processes for trapping of electrons and/or holes may be involved in the recombination process. In this paper, the dependence of the photoluminescence spectrum on the temperature is investigated in detail for plastically deformed silicon crystals. The energy levels for both electrons and holes participating in the luminescence are deduced on the basis of a theoretical analysis of the experimental results.

[1]) 2-1-1 Katahira, Sendai 980, Japan.

2. Experimental

Specimens were prepared from both floating-zone-grown silicon crystals (abbreviated as FZ-Si) and Czochralski-grown silicon crystals (CZ-Si). They were plastically deformed in compression along the [123] direction at 900 °C under a constant strain rate in a vacuum.

Two kinds of excitation light the wavelengths of which were 488 nm and 1 μm, respectively, were used to excite the photoluminescence. The source for the excitation light of 488 nm wavelength was an argon ion laser. The excitation power was chosen so as to reduce the effect of heating of the specimen. A halogen lamp and a grating monochromator were used for the excitation by the light of 1 μm wavelength, the excitation power being lower than 1 mW. As was reported in the previous paper [7], the light of 1 μm wavelength was suitable for the measurement of the dependence of the background component of the photoluminescence on temperature.

The specimen was set inside a double-walled cryostat filled with helium gas of a low pressure and having a heater. Since helium gas worked as a good heat exchanger, a good uniformity in the temperature distribution along the specimen was assured with this technique. The cryostat was surrounded by liquid helium. The temperature of the specimen was kept constant within an accuracy of ±0.1 K. The photoluminescent light was detected by a cooled germanium detector.

The dependence of the intensities of D3 and D4 lines on temperature was found not to be influenced by the kind of the excitation lights used even if the excitation powers were much different. This seems to show that the effect of local heating of the specimen due to the irradiation by the excitation light is negligibly small under the experimental conditions of the present work.

3. Experimental Results and Analysis of the Data

3.1 Dependence of the photoluminescent intensities on temperature

Fig. 1 shows the photoluminescence spectra at 4.2 and 77 K of a high-purity FZ-Si specimen of p-type which has been deformed to a shear strain of 2.1% under rather high excitation condition. The specimen in this case was immersed directly into liquid helium or liquid nitrogen and was excited with the light of 488 nm wavelength and 200 mW power. The heights of all the lines D1 through D4 are much smaller at 77 K than at 4.2 K, while the linewidths are larger at 77 K than at 4.2 K. The peak positions of D1 and D2 lines are almost independent of temperature. On the other hand, those of D3 and D4 lines shift to the higher-energy side by 2 meV as temperature is raised from 4.2 to 77 K. Such small dependences of the peak positions of the lu-

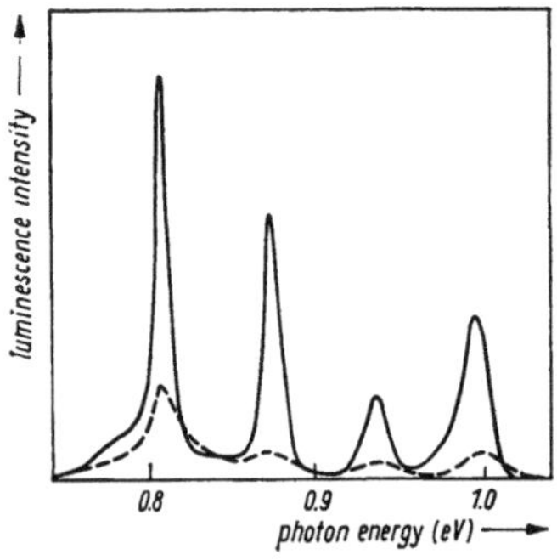

Fig. 1. Photoluminescence spectra of a lightly dislocated FZ-Si crystal at 4.2 (———) and 77 K (– – –) under rather high excitation with an argon-ion laser beam

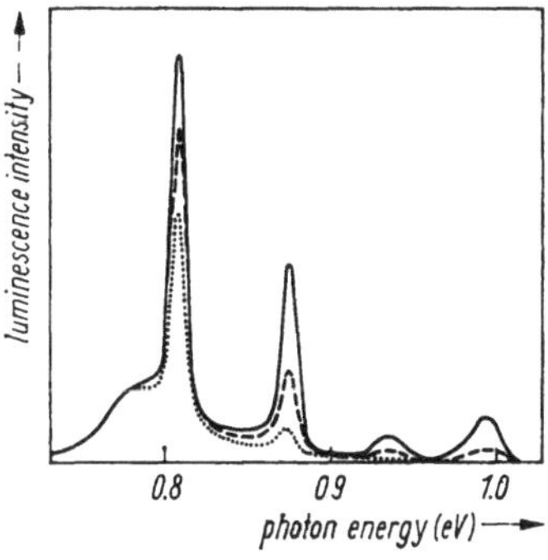

Fig. 2. Photoluminescence spectra of the ligthly dislocated FZ-Si crystal same as that in Fig. 1 at various temperatures under low excitation with anf agon-ion laser beam: ------ 8.2, ——— 13.8, ——— 23.7 K

minescence lines on temperature are thought to reflect a small dependence of the band-gap in silicon crystals on temperature in the low-temperature range. The increase in the linewidths at high temperature seen in Fig. 1 is interpreted to be due to the interaction of electrons and lattice vibrations. We are concerned only with the peak height of each line in the following section. The dependences of the heights of lines D1 through D4 on temperature are different from one another even in the same specimen. This seems to suggest that different energy levels are responsible for different luminescence lines. Almost the same behaviour of the photoluminescence spectrum as that seen above has been observed also for a CZ-Si specimen. However, the peak position of D1 in the CZ-Si specimen has been found to shift to the higher-energy side by about 9 meV when changing the temperature from 4.2 to 77 K.

Fig. 2 shows the change in the photoluminescence spectrum with increasing temperature for the FZ-Si specimen same as that in Fig. 1 under low excitation. The excitation light has 488 nm wavelength and 8 mW power. As temperature is raised, the intensities of all the lines D1 through D4 diminish drastically. The diminishing in the intensities of lines D3 and D4 is more pronounced than in those of lines D1 and D2.

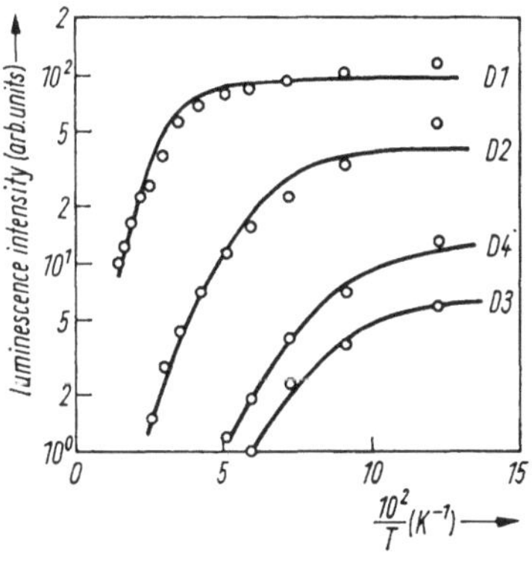

Fig. 3

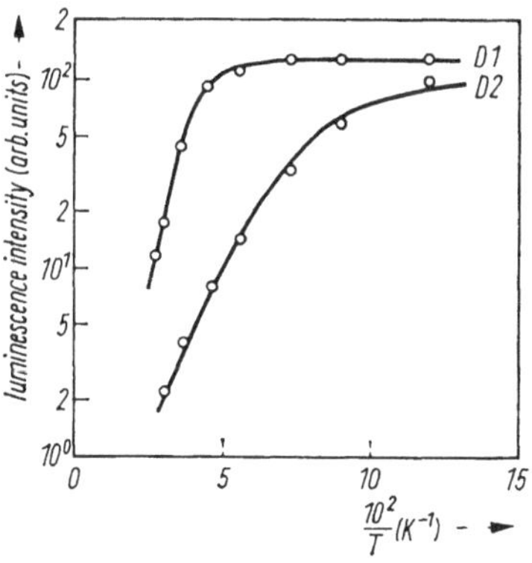

Fig. 4

Fig. 3. The dependence of the intensities of the lines D1 through D4 of the lightly dislocated FZ-Si crystal same as that in Fig. 1 on temperature at low excitation with an argon-ion laser beam. Full lines are those given by (1) with values of $C_1 = 60$, $E_1 = 7.0$ meV; $C_2 = 400$, $E_2 = 4.5$ meV; $C_3 = 1500$, $E_3 = 4.2$ meV; $C_4 = 1500$, $E_4 = 4.2$ meV, where the suffix stands for the line number

Fig. 4. The dependence of the intensities of D1 and D2 lines of a rather heavily dislocated CZ-Si crystal on temperature at low excitation with an argon-ion laser beam. Full curves are those given by (1) with values of $C_1 = 2000$, $E_1 = 12.0$ meV; $C_2 = 1000$, $E_2 = 4.0$ meV

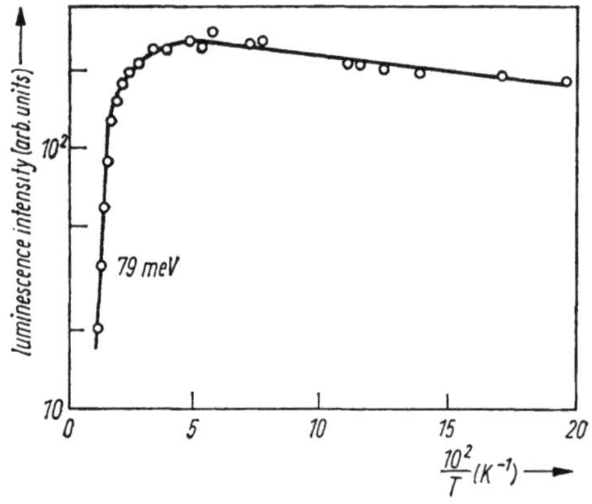

Fig. 5. The dependence of the integrated intensity of the background component in the FZ-Si crystal on temperature. The slope in the high-temperature range gives an energy of 79 meV for the level concerned

The D1 line survives distinctly up to a fairly high temperature. Such a behaviour of the line height is found to be reversible against the change in temperature. Therefore, the decrease in the luminescence intensities due to the rise of temperature is concluded not to be related to the annealing of the defects responsible for the luminescence. Essentially the same behaviour of the peak intensities against the temperature was observed also for the CZ-Si specimen. So, oxygen atoms both dissolved in the matrix crystal and segregated around dislocations are known to play no significant role in the dependence of the peak heights of the luminescence lines on temperature.

Fig. 3 and 4 show the dependence of the peak height of each luminescence line on temperature for the FZ-Si specimen and the CZ-Si specimen, respectively. The full lines in the figures show the curves derived from the following relation:

$$\frac{I(T)}{I(0)} = \frac{1}{1 + CT^{3/2} \exp(-E/kT)}, \tag{1}$$

where $I(T)/I(0)$ is the height of the luminescence line at temperature T (K) divided by that extrapolated to absolute zero temperature, and C and E are parameters. The magnitudes of C and E are chosen so as to give the best fits of the curves to the experimental data and are shown in the captions of the figures. Equation (1) has been proposed by Ashkinadze et al. [8] to describe the dependence of the photoluminescence intensity in GaP on temperature. However, no explanation has so far been given for the meaning of the equation. It is seen that the experimental data are well described by (1).

Fig. 5 shows the integrated intensity of the whole spectrum of the FZ-Si specimen excited with the 1 μm light as dependent on temperature. The contributions from D1 and D2 lines can be neglected for the excitation light of this wavelength as already shown in the previous paper [7]. Since D3 and D4 lines diminish at temperatures higher than 30 K, the ordinate in Fig. 5 in the high-temperature range may be regarded to represent the intensity of the background component of the spectrum.

3.2 *Model for the radiative recombination processes*

The marked dependence of the photoluminescence intensity on temperature seen in Fig. 3 to 5 suggests that at least one kind of traps participating in each radiative recombination process is likely to have a shallow level. If the luminescence intensity is assumed to be proportional to the density of carriers trapped on such a level, it is easy to understand why the intensity decreases with an increase in temperature, since carriers on the traps are thermally activated to the relevant band and cease to participate in the radiative recombination process concerned. The magnitude of the photon energy of the luminescence suggests that another level coupled with such a shallow level to give rise to the radiative recombination must be deep. Thus, the following model may be valid to interpret the radiative recombination processes for the deformation-induced photoluminescence. Namely, two energy levels, one being shallow and the other deep, are responsible for the recombination process. These two levels are occupied one by electrons and the other by holes. In the following, we deduce (1) on the basis of this model, assuming that electrons are trapped by the shallow level and holes by the deep level for the sake of convenience. Essentially the same type of equation is derived on the assumption that holes are trapped by the shallow level and electrons by the deep level.

The energy of the shallow level responsible for the recombination process giving rise to the photoluminescence line Di ($i = 1, 2, 3, 4$) is put E_D^i. The system under irradiation of light is not in thermal equilibrium so that the occupation statistics of the level is described with the use of the quasi-Fermi level. Since the magnitudes of the parameters such as the capture cross-section characterizing the traps may be different for different type of traps, the quasi-Fermi level is taken to be dependent on the species of the traps and is put E_F^i ($i = 1, 2, 3, 4$). The occupation ratio f_i of electrons for the level associated with the trap of the i-th kind is given by

$$f_i = \frac{1}{1 + \exp\left[(E_D^i - E_F^i)/kT\right]}, \tag{2}$$

and the local density of free electrons n_i in the vicinity of such a trap by

$$n_i = N_c \exp\left(\frac{E_F^i - E_c}{kT}\right), \tag{3}$$

where N_c is the effective density of states of the conduction band and E_c the energy of the bottom of the conduction band. Under the irradiation condition where n_i is primarily determined by the excitation power and influenced little by temperature, f_i as a function of the temperature turns out to be

$$f_i = \frac{1}{1 + (A/n_i)\, T^{3/2} \exp\left[(E_D^i - E_c)/kT\right]}, \tag{4}$$

where N_c is replaced by $AT^{3/2}$, A being a constant. Putting $E_D^i - E_c = -E_i$ and $A/n_i = C_i$, we obtain the relation

$$f_i = \frac{1}{1 + C_i T^{3/2} \exp\left(-E_i/kT\right)}, \tag{5}$$

where E_i is the depth of the level of the i-th kind measured from the bottom of the conduction band.

The occupation ratio of the deep level which couples with the above shallow level is little influenced by temperature in the temperature range concerned in the present experiment. Thus, with the assumption that the rate of occurrence of the radiative recombination is proportional to the product of the occupation ratios of the two levels, the dependence of the intensity of each luminescence line turns out to be given by (1).

Fig. 6 shows the behaviour of f_i against the temperature with C_i as a parameter. For a given magnitude of E_i, f_i at any temperature is smaller for a larger value of C_i. A larger value of C_i represents a lower quasi-Fermi level according to (3) which may be related to a smaller capture cross-section of the trap and leads to a lower efficiency in the occurrence of radiative recombination. A larger magnitude of E_i means that

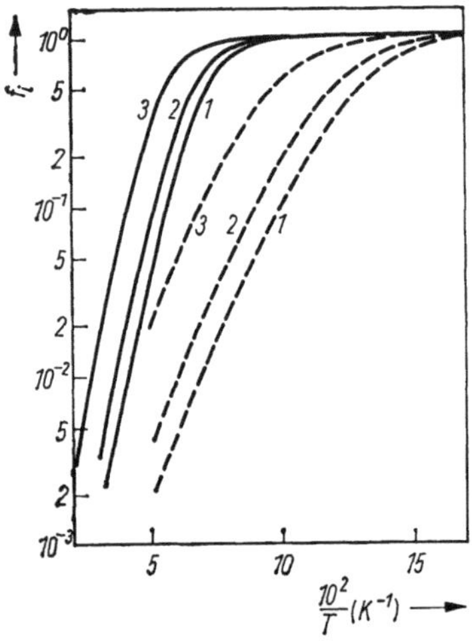

Fig. 6. The variation of f_i against the reciprocal temperature for various magnitudes of C_i and E_i. Full lines are for $E_i = 10$ meV and dashed lines for $E_i = 5$ meV. Curve numbers (1), (2), and (3) show that the magnitudes of C_i are 100, 50, and 10, respectively

the level of the shallow trap is deeper. The efficiency in the occurrence of radiative recombination remains high up to a relatively high temperature for a large value of E_i.

Fig. 1 and 3 show that the variation of the photoluminescence spectrum against temperature depends on the power of excitation light. This behaviour may be interpreted in the following way. Fig. 7a, b show the dependence of the intensities of the lines D1 through D4 on the excitation power at 4.2 and 77 K, respectively. It is seen that the intensities of all the lines depend weakly on the excitation power at 4.2 K. On the other hand, at 77 K the intensities of lines D1 and D2 depend weakly on the excitation power in the high-power range and rather strongly in the low-power range, while the intensities of lines D3 and D4 depend linearly on the excitation power over the whole range of excitation power. The behaviour of the spectra in Fig. 1 and 3 is just a reflection of the behaviour of the photoluminescence lines in Fig. 7 a, b. For a high excitation power, the quasi-Fermi level is high and, consequently, trapping of carriers at the recombination centres takes place very effectively. The magnitute of f_i in (5) is close to unity, and this leads to a weak dependence of the photoluminescence intensity on the excitation power. Such a situation is realized more readily at lower temperatures and also for traps with a larger value of E_i according to (5). The behaviour of the intensities of the lines D1 through D4 in Fig. 7a and that of lines D1 and

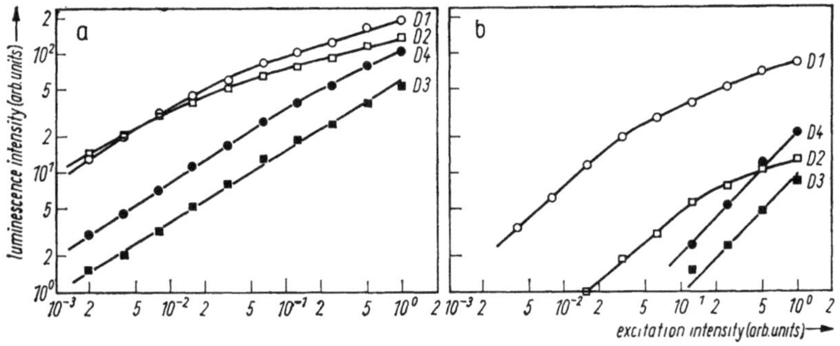

Fig. 7. The dependence of the intensities of the lines D1 through D4 on the excitation intensity. The abscissa shows the relative magnitude of excitation power, 10^0 corresponding to 250 mW of the 488 nm wavelength light and 2.5 nm beam diameter. a) 4.2 K, b) 77 K

D2 for high excitation powers in Fig. 7b may, thus, be interpreted with the above model. For low excitation powers at high temperatures the second term of the denominator in (5) may be dominant and f_i becomes apparently linear with respect to n_i. This is more pronounced for smaller values of E_i. The intensity of the photoluminescence in this case is thought to decrease linearly with decreasing excitation power, since n_i is thought to be proportional to the excitation power. The behaviour of the intensities of the luminescence lines in Fig. 7b is interpreted in this way.

Fig. 3 and 4 show that the experimentally observed dependence of the intensities of the lines D1 through D4 on temperature is well described with (5) by suitable choice of the values for C_i and E_i. As shown in the previous paper [7], the photon energy at the peak of line D1 in a CZ-Si crystal is higher than that in a FZ-Si crystal by about 7 meV. The magnitude of E_1 determined for a CZ-Si crystal is larger than that determined for a FZ-Si crystal by approximately the same amount.

The full line in the high-temperature range in Fig. 5 is the curve given by

$$I(T) \sim \exp\left(\frac{E_b}{kT}\right) \tag{6}$$

with $E_b = 79$ meV. No plausible explanation is available at present for the fact that the background intensity increases as temperature is raised in the low-temperature range. Since this type of behaviour has been reported for other kind of photoluminescence [9], it seems not to be an effect characteristic of dislocations.

3.3 Energy levels of radiative recombination centres

The energy levels which are responsible for the radiative recombination in a deformed silicon crystal are determined in this section. The bandgap energy E_g is given by

$$E_g = E_p^i + E_i + E_i' , \tag{7}$$

where E_p^i is the energy of photon at the peak position of line D*i*, and E_i and E_i' are the shallower and the deeper energy levels, respectively, coupled to give rise to line D*i*. The magnitudes of E_i have been determined in the preceding section. The dependence of the energy gap on temperature is very weak at low temperatures and may be ignored.

The positions of the shallower and deeper levels cannot be determined uniquely from the photoluminescence data alone, since the configurations of the levels in the bandgap which are in the relation of mirror image to each other with respect to the middle of the bandgap result in the same photoluminescence spectrum. In the previous paper [7], the present authors have shown that the photoluminescence is not related to dangling bonds in the crystal. Since the luminescence has been related to dislocations introduced by deformation, we may ascribe the deeper levels responsible for the luminescence to the reconstructed bonds at the dislocation core resulting from geometrical dangling bonds. Electrons are thought to be trapped in the antibonding state of the reconstructed bonds the levels of which are thought to be located in the upper half of the bandgap. The shallower levels may be attributed to the hole traps associated with the deformation potential of dislocations. The two types of levels which give rise to a photoluminescence line are attributed to the same dislocations, so that the recombination will occur readily between carriers trapped by these two levels. Different photoluminescence lines may be attributed to different types of dislocations. Fig. 8 shows the scheme of the energy levels in silicon crystals responsible for the deformation-induced photoluminescence which is deduced from the above discussion. Both the shallower and deeper levels for the background component are thought to have band structure. The data in Fig. 5 suggest the edges of the band of

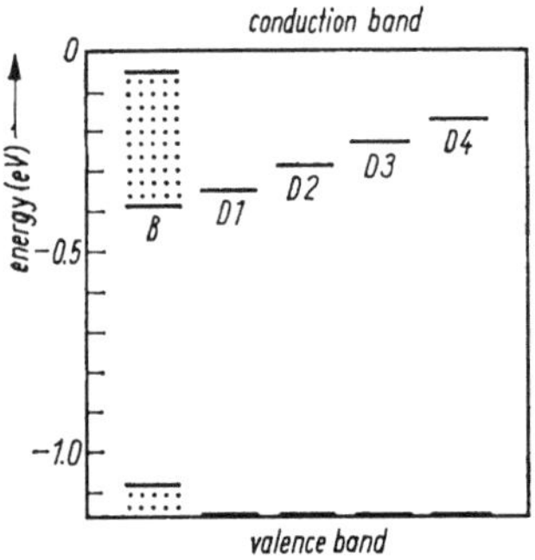

Fig. 8. The proposed energy levels for the traps responsible for the recombination processes giving rise to the lines D1 through D4 and also to the background component in a deformed silicon crystal

the shallower level to be located at approximately zero and 79 meV above the top of the valence band.

The energy levels in silicon crystals introduced by plastic deformation may be investigated by several kinds of techniques. The correlation between the levels obtained by the photoluminescence technique and those detected by other techniques is discussed here. The photo-EPR study [10] reveals the levels associated with dangling bonds residing at the so-called Si-K1 centres. Since the photoluminescence has been shown to have no relation with dangling bonds in the crystal [7], the levels measured by the photoluminescence and photo-EPR techniques are of different origins. It has indeed been verified that the level detected by photo-EPR is located approximately in the middle of the bandgap, while those measured by photoluminescence in the upper half of the bandgap, as seen in Fig. 8.

DLTS measurements have revealed a spectrum of energy levels in deformed silicon crystals [11 to 14]. The DLTS technique is useful only for the study of levels of traps having relatively low emission rates. On the other hand, the photoluminescence technique is free from such restrictions. It has been shown that the DLTS spectrum in p-type silicon crystals induced by deformation vanishes almost completely upon annealing the crystals at temperatures higher than about 800 °C even if a high density of dislocations remains unannealed. On the contrary, the photoluminescence lines survive even after annealing at temperatures around 1000 °C as long as dislocations remain in the crystals [7]. Thus, the energy levels so far determined with the DLTS technique are thought to originate from other types of traps than those giving rise to the photoluminescence. Possibly, different kinds of defects and/or different states of dislocations are responsible for the DLTS signal and the photoluminescence in deformed silicon crystals.

Measurements of the Hall effect are also useful for the determination of the deformation-induced levels [15 to 17]. It has been shown that the carrier concentration versus temperature relation in a deformed p-type silicon crystal obtained by Hall effect measurements is described well in terms of the levels determined by DLTS measurements [17]. Hence, the levels controlling the carrier concentration in the crystal are the same as those giving rise to the DLTS signal and are different from those responsible for the luminescence. The traps associated with the radiative recombination processes are, thus, concluded to be electrically inactive under thermal equilibrium conditions at least in p-type crystals. This agrees with the model which ascribes the traps concerned to the reconstructed bonds at the dislocation core.

References

[1] N. A. Drozdov, A. A. Patrin, and V. D. Tkachev, Soviet Phys. — J. exper. theor. Phys., Letters **23**, 579 (1976).

[2] N. A. Drozdov, A. A. Patrin, and V. D. Tkachev, phys. stat. sol. (b) **83**, K137 (1977).

[3] M. Suezawa, Y. Sasaki, Y. Nishina, and K. Sumino, Japan J. appl. Phys. **20**, L537 (1981)
[4] N. A. Drozdov, A. A. Patrin, and V. D. Tkachev, phys. stat. sol. (a) **64**, K63 (1981).
[5] D. Gwinner, and R. Labusch, phys. stat. sol. (a) **65**, K99 (1981).
[6] M. Suezawa, K. Sumino, and Y. Nishina, Japan J. appl. Phys. **21**, L518 (1982).
[7] M. Suezawa and K. Sumino, phys. stat. sol. (a) **78**, 639 (1983).
[8] B. M. Ashkinadze, I. P. Kretsu, S. L. Pyshkin, and S. M. Ryvkin, and I. D. Yaroshetskii, Soviet Phys. – Solid State **10**, 2921 (1969).
[9] H. Serizawa, Master Thesis, Tôhoku University, 1979.
[10] R. Erdmann and H. Alexander, phys. stat. sol. (a) **55**, 251 (1979).
[11] L. C. Kimerling and J. R. Patel, Appl. Phys. Letters **38**, 73 (1979).
[12] J. R. Patel and L. C. Kimerling, Crystal Res. and Tech. **16**, 187 (1981).
[13] V. V. Kveder, Y. A. Osipyan, W. Schröter, and G. Zoth, phys. stat. sol. (a) **72**, 701 (1982).
[14] H. Ono and K. Sumino, to be puplished.
H. Ono, Ph. D. Thesis, Tôhoku University, 1983.
[15] H. Weber, W. Schröter, and P. Haasen, Helv. phys. Acta **41**, 1255 (1968).
[16] V. A. Grazhulis, V. V. Kveder, and V. Yu. Mukhina, phys. stat. sol. (a) **43**, 407 (1977).
[17] H. Ono and K. Sumino, J. appl. Phys. **54**, (1983) in the press.

(Received May 20, 1983)

phys. stat. sol. (a) 79, 183 (1983)

Subject classification: 14.3.1; 22.2.3

Department of Physics, Sri Venkateswara University, Tirupati[1]) *(a)*
and Sri Satya Sai Institute of Higher Learning, Prashantinilayam[2]) *(b)*

Temperature and Thickness Dependence of Electrical Conductivity of Flash-Evaporated InSb Films

By

A. K. SHARMA (a) and P. JAYARAMA REDDY (b)

The electrical conductivity as a function of film temperature, substrate temperature, and thickness is reported on InSb films flash evaporated onto gold seal glass, freshly cleaved mica and cleaved NaCl substrates in a vacuum of 6.7×10^{-4} Pa. Post deposition annealing shows an interesting behaviour in the electrical conductivity. The structure of the films is found to improve with substrate temperature and single crystal growth is observed at $T_s = 573$ K. While the electrical conductivity shows an increasing behaviour with substrate temperature and thickness, the activation energy shows a decreasing trend.

Die elektrische Leitfähigkeit wird in Abhängigkeit von der Schichttemperatur, Substrattemperatur und Dicke an InSb-Schichten untersucht, die auf goldversiegeltem Glas, frisch gespaltenem Glimmer und im Vakuum von $6{,}7 \times 10^{-4}$ Pa gespaltenen NaCl-Substraten blitzaufgedampft werden. Eine Temperung nach der Aufbringung ergibt ein interessantes Verhalten der elektrischen Leitfähigkeit. Es wird gefunden, daß sich die Struktur der Schichten mit der Substrattemperatur verbessert und bei $T_s = 573$ K wird Einkristallwachstum beobachtet. Während die elektrische Leitfähigkeit mit der Substrattemperatur und Dicke ein steigendes Verhalten zeigt, hat die Aktivierungsenergie einen fallenden Trend.

1. Introduction

InSb films have attracted considerable attention in view of their potentialities in the preparation of many devices. Despite the large amount of work [1 to 10] the relation between the properties of the samples and the conditions for their formation have not been thoroughly studied. It is of interest to investigate the structure and electrical properties of these films as a function of deposition conditions. In this paper, we report investigations on the electrical conductivity of InSb films as a function of their thickness, temperature, substrate temperature and the effect of post-deposition heat treatment.

2. Experimental

Flash-evaporation was employed for the preparation of the films as it ensures stoichiometry of the deposited film. The powdered material was fed to a resistively heated tantalum boat, maintained at 1773 K, when instantaneous evaporation took place. Gold seal glass, freshly cleaved mica, and cleaved NaCl were used as substrates. The distance of the source to the substrates was 0.25 m. Pressures of the order of 6.7×10^{-4} Pa were obtained using a stainless steel bell-jar evacuated by means of an oil diffusion pump backed by a rotary pump. Substrate temperatures varied from 323 to 673 K and the thickness range was 100 to 300 nm. Before evaporation of the film

[1]) Tirupati 517502, India.
[2]) Prashantinilayam 515134, India.

the substrates were baked at temperatures higher than that required for film formation, which often resulted in dropping of the vacuum to 1.33×10^{-3} Pa. The temperature of the substrate was measured using a chromel–alumel thermocouple fixed onto the upper surface of the substrate. The thickness of the film was monitored by controlling the rate of the particle feed and was later verified by Tolansky's multiple beam interferometer.

Ohmic contacts to the film were obtained by predeposition of thick silver electrodes on the substrates. Resistivity measurements were made using standard four-probe technique. While the resistivity measurements above room temperature were made in-situ, at low temperatures, the films were transferred immediately into a cryostat evacuated to approximately 1.33×10^{-2} Pa. The resistivity measurements were carried down to liquid air temperature.

The structure of the films was determined by using a transmission electron microscope EM300 operated at 100 kV. The chemical composition was examined by polarography.

3. Results and Discussion

3.1 Structural studies

As already reported in our earlier publication [11], InSb films formed at and around room temperature were found to be amorphous. When the substrate temperature was

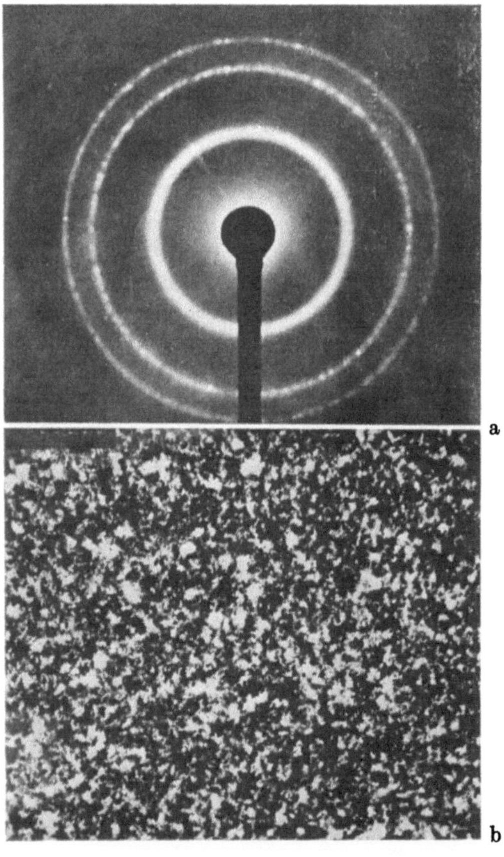

Fig. 1. a) Selected area diffraction (SAD) pattern and b) micrograph of 100 nm thick InSb film formed at $T_s = 473$ K

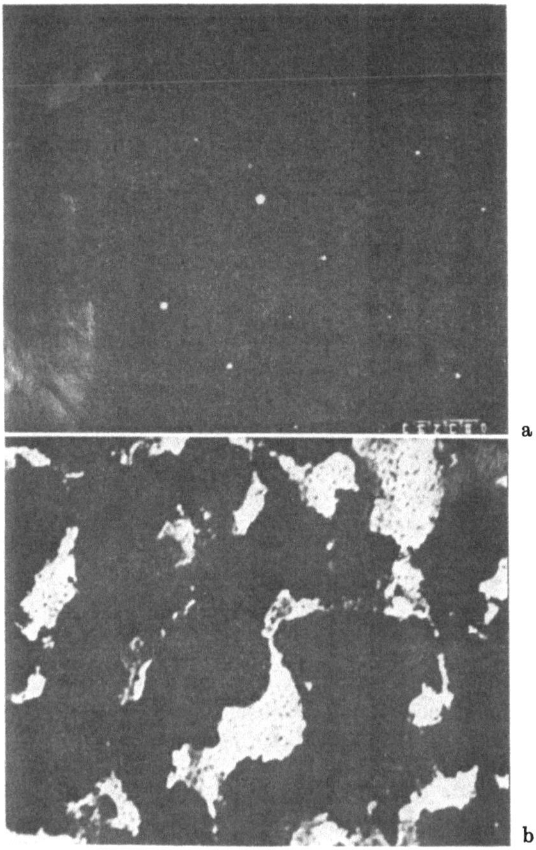

Fig. 2. a) Selected area diffraction pattern and b) micrograph of 100 nm thick InSb film formed at $T_s = 573$ K

ncreased, the crystalline content in the films increased. A perfectly polycrystalline film was obtained at 473 K (Fig. 1) and at $T_s = 573$ K (Fig. 2) preferred growth was observed, showing that the structure of the film improved from amorphous → polycrystalline → well oriented, with increasing substrate temperature. The increase in grain size with increasing substrate temperature is clearly evident from the micrographs (Fig. 1 b and 2 b).

3.2 Composition

Polarographic analysis of the composition showed that the films very nearly retained the composition of the starting material. The composition of films was found to be 36.76% In and 63.24% Sb, while that of the starting material was 36.78% In and 63.22% Sb, the accuracy of the analysis being ±0.01%.

3.3 Electrical conductivity

The electrical conductivity of the films was measured as a function of temperature. The heat-treatment procedure described elsewhere [11] was employed to measure the changes in the film conductivity during different stages of annealing. A uniform heating rate of about 2 K/min was employed throughout the whole range of investigations. Fig. 3 about the temperature dependence of electrical conductivity of 200 nm thick film formed at substrates temperatures 303, 373, 473, and 573 K, respectively.

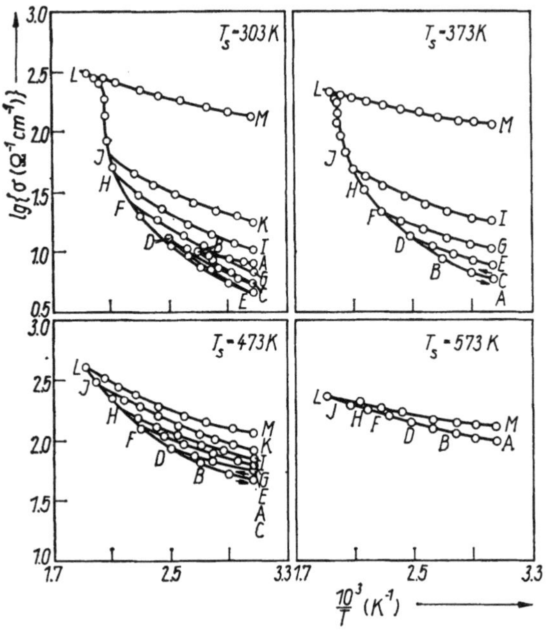

Fig. 3. Temperature dependence of electrical conductivity (during the heat treatment through various cycles) of InSb films, 200 nm thick, formed at different substrate temperatures. A represents the start of annealing. Heating cycles; 1st, ABC; 2nd CDE; 3rd EFG; 4th GFHI; 5th IHJK; 6th KJLM. Corresponding temperatures; B 373 K, D 403 K, F 440 K, H 470 K, J 490 K

For films formed at $T_s = 373$ K (Fig. 3b) it was seen that during the first cycle, which corresponded to 373 K, the electrical conductivity changed reversibly. But in the case of a film formed at room temperature (Fig. 3a) and subsequently heat treated to 373 K, no such reversible variation was noticed [11]. The comparison of these two suggests that in the films formed at $T_s = 373$ K, certain defects like vacancy sites are reduced.

On further heating beyond 373 K, the conductivity variation was found to be identical to that of the film formed at $T_s = 303$ K [11]. However, it was noticed that the transition from amorphous to crystalline phase took place at a lower temperature, i.e. 473 K as compared to 500 K in the films formed at $T_s = 303$ K. For thicknesses 100 and 300 nm, the transition temperatures were 478 and 471 K, respectively.

For films formed at $T_s = 473$ K (Fig. 3c) the room temperature electrical conductivity of the as-deposited film was found to be higher than that of the film formed at $T_s = 303$ K. The behaviour of the electrical conductivity with temperature is shown in the figure for a 200 nm thick film.

The conductivity behaviour exhibited an activation energy which was considerably less than that for films formed at $T_s = 303$ or 373 K. No steep and sudden increase in the electrical conductivity during the post-deposition heat treatment was observed. This was because the as-deposited films were already polycrystalline.

The temperature dependence of the electrical conductivity of films formed at $T_s = 573$ K is shown in Fig. 3d. The room temperature conductivity values of as-deposited films are nearly equal to those of the heat treated films formed at $T_s = 303$, 373 or 473 K. No significant change on heating was noticed in these films. The activation energies were smaller than those for the films formed at $T_s = 303$, 373 or 473 K. A similar behaviour was observed in all films irrespective of thickness.

The above data indicates that the electrical conductivity of a film of given thickness increases with increasing substrate temperature. This can be attributed to the improvement in the film structure with increasing substrate temperature. However, it

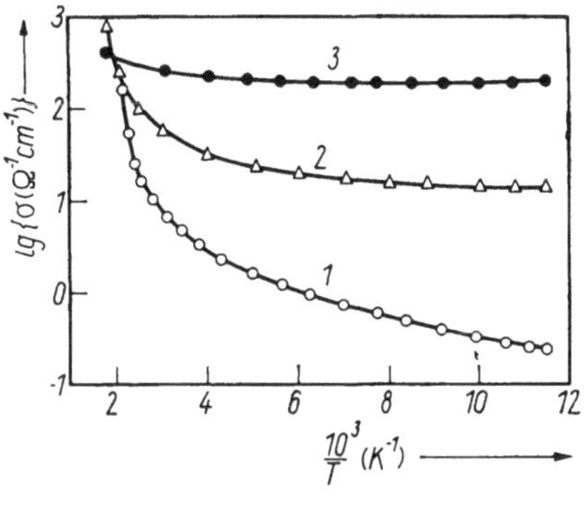

Fig. 4. Temperature dependence of the electrical conductivity of a 200 nm thick InSb film formed at different substrate temperatures: (1) $T_s = 373$, (2) 473, (3) 573 K

was interesting to note that the room temperature conductivity after heat treatment was found to be almost the same regardless of the temperature (T_s) at which the film was formed (Fig. 3).

3.4 Activation energies

The electrical conductivity of films formed at different substrate temperatures was measured from 90 to 550 K. Fig. 4 shows the typical variation for a 200 nm thick film formed at different substrate temperatures.

The activation energies were found to increase with increasing temperature. While the increase at higher temperatures was more pronounced, it was very weak at low temperatures. Also the values of activation energy at high temperatures were considerably higher than those at low temperatures (Table 1). A similar behaviour was observed for films formed at $T_s = 373$ and 573 K, independent of thickness.

In conclusion, it is seen that the activation energy decreases with increasing substrate temperature for a given thickness. However, it is found to increase with increasing temperature. It is also found to decrease with increasing film thickness. The values of activation energy (Table 1) indicate that while intrinsic conduction takes place at higher temperatures, extrinsic conduction takes place at low temperatures.

Table 1

Activation energies at different temperatures of InSb films formed at different substrate temperatures

T_s (K)	thickness (nm)	activation energy (eV)		
		400 K	300 K	100 K
373	100	0.201	0.184	0.032
	200	0.163	0.126	0.020
	300	0.102	0.074	0.011
473	100	0.124	0.065	0.008
	200	0.103	0.053	0.006
	300	0.080	0.041	0.005
573	100	0.024	0.008	0.003
	200	0.013	0.006	0.001
	300	0.007	0.004	0.001

4. Conclusions

The substrate temperature, thickness of the sample and post-deposition heat treatment play a very crucial role in determining the structure and electrical properties of thin InSb films. The structure of the films showed marked improvement with increasing substrate temperature. While amorphous films were obtained at and around room temperature, polycrystalline nature was seen at 473 K, and preferred orientation appeared at $T_s = 573$ K. The electrical conductivity was seen to increase with increasing substrate temperature, increasing thickness and upon heat treatment. However, for a given thickness the room temperature conductivities showed an almost constant value after heat treatment, regardless of the temperature at which they were prepared. The activation energy was found to decrease with increasing substrate temperature and increasing thickness, but increased with increasing temperature, the increase being stronger at higher temperatures. These studies help to identify clearly two types of conduction mechanisms taking place in the material, namely intrinsic at high temperatures and extrinsic at low temperatures.

Acknowledgement

One of the authors (A.K.S.) wishes to thank CSIR, New Delhi (India) for providing Junior Research Fellowship during the course of the present work.

References

[1] J. E. Greene and C. E. Wickersham, J. appl. Phys. 47, 3630 (1976).
[2] A. L. Dawar, K. V. Krishna, O. P. Taneja, and P. C. Mathur, phys. stat. sol. (a) 47, 375 (1978).
[3] J. Shigeta, N. Kotera, and T. Oi, J. appl. Phys. 47, 621 (1976).
[4] A. Zemel and J. R. Sites, Thin Solid Films 41, 297 (1977).
[5] M. le Countellec, J. Richard, and J. Henaff, Thin Solid Films 36, 151 (1976).
[6] X. M. Burvenich, Thin Solid Films 27, 129 (1975).
[7] M. Nagabhooshanam and V. Hari Babu, J. appl. Phys. 51, 3250 (1980).
[8] W. Hanus and M. Oszwaldowski, Thin Solid Films 61, 235 (1979).
[9] A. I. Solovev, T. A. Zeveke, V. A. Tolomasov, L. S. Eremina, N. V. Gudkova, and N. N. Kukoshkina, Inorg. Mater. 14, 1690 (1978).
[10] A. L. Dawar, B. K. Sachar, O. P. Taneja, and P. C. Mathur, Thin Solid Films 76, 387 (1981).
[11] A. K. Sharma and P. Jayarama Reddy, J. non-crystall. Solids. 41, 13 (1980).

(Received May 16, 1983)

phys. stat. sol. (a) 79, 189 (1983)

Subject classification: 1.5 and 10.1; 21.6

Institut für Festkörperphysik und Elektronenmikroskopie der Akademie der Wissenschaften der DDR, Halle (Saale)[1])

Influence of Substrate Deformations on Interface Structures

II. Experimental Investigations for Epitaxial Ag/Pd and Au/Pd Bicrystals[2])

By

E. PIPPEL and J. WOLTERSDORF

The influence of substrate relaxations on pseudomorphic growth in Ag/Pd island bicrystals and continuous Au/Pd films is investigated using transmission electron microscopy. For several substrate and deposit thicknesses the radii of curvature resulting from bending contours are measured. Further on the critical deposit thickness for complete pseudomorphism is deduced from the first appearance of misfit dislocations. Both quantities strongly depend on bending of the substrate layer. The experimental results are compared with theoretical values. Finally, some contrast problems of misfit dislocations and moiré lines are discussed as they arise in the case of epitaxial bicrystals.

Am Beispiel von Ag/Pd-Inseln und kontinuierlichen Au/Pd-Filmen wird der Einfluß von Substratrelaxationen auf das Pseudomorphieverhalten epitaktischer Bikristalle mit Hilfe der Transmissionselektronenmikroskopie untersucht. Für unterschiedliche Schichtdickenverhältnisse der Verwachsungspartner werden vermittels Biegekonturen die Krümmungsradien der Bikristalle gemessen, und aus dem ersten Auftreten von Anpassungsversetzungen wird auf die kritische Schichtdicke für vollständige Pseudomorphie geschlossen. Beide Größen sind wesentlich von der Verbiegung der Substratunterlage abhängig. Die erhaltenen experimentellen Ergebnisse werden mit entsprechenden theoretischen Werten verglichen. Abschließend werden noch einige Kontrastfragen von Anpassungsversetzungen und Moirélinien in epitaktischen Bikristallen diskutiert.

1. Introduction

As it was shown in the first part of this paper [1] (in the following referred to as Part I) substrate relaxations in epitaxial bicrystals may cause crystal bending and can even influence the misfit dislocation generation processes. This is a consequence of pseudomorphic strain. Based on this, the present Part II comprises experimental investigations with the help of transmission electron microscopy (TEM). Particularly, the radius of curvature of the bicrystals as well as the critical thickness for pseudomorphic growth were measured and compared with the corresponding theoretical values of Part I.

For examining the influence of elastic substrate relaxations on pseudomorphic growth such bicrystal systems are required which enable substrate relaxations already with the formation of the first deposit layers. In the present case bicrystals suitable for TEM investigation were prepared in two different ways to fulfil the above-mentioned condition: On the one hand, the substrates are separated monocrystalline silver islands condensed onto a rocksalt crystal; on the other hand, continuous gold thin films were used as substrates only supported by a copper grid (denominated as "un-

---

[1]) Weinberg 2, P.F. 250, DDR-4010 Halle (Saale), GDR.

[2]) Part I see phys. stat. sol. (a) 78, 475 (1983).

restrained films" in the following). In each case palladium was chosen as the deposit material.

Because of the weak interaction between the Ag islands and their rocksalt base (cf. [2]) the relaxation should be well possible. Moreover, the preparation of islands with a thickness gradient enables states of different degrees of relaxation to be illustrated in one and the same TEM image. A mutual influence can be excluded. As is well known (cf. [3]) the surface tension leads to lattice distortions in small particles. This effect could influence the effective misfit between Ag and Pd for islands smaller than 20 nm. With decreasing size the edges of the islands become more important: In this region the layer thickness decreases and the direction of the interface changes. Therefore, the state of plane stress as supposed in the model used in Part I no longer exists in the specimen. However, the Au films prepared to be unrestrained and of uniform thickness should better represent a state of plane stress. But after all, these films are bound also by the meshes of the grid and an unrestrained metal film will be more sensitive to mechanical influences.

Parts of the experimental results are based on the measurement of the distance of moiré fringes and of the density of misfit dislocations. Since it is difficult to distinguish both structures by TEM [4], finally some questions will be discussed concerning the contrast of misfit dislocations and moiré fringes in epitaxial bicrystals.

2. Pseudomorphic Growth of Ag/Pd Bicrystal Islands Influenced by Substrate Relaxations

2.1 Experimental

Substrate islands of different sizes were prepared on the (100) cleavage face of NaCl by vacuum deposition of Ag through a shutter in form of a grid. Epitaxial growth of 60 to 70 nm thick layers takes place at substrate temperatures of about 300 °C. The

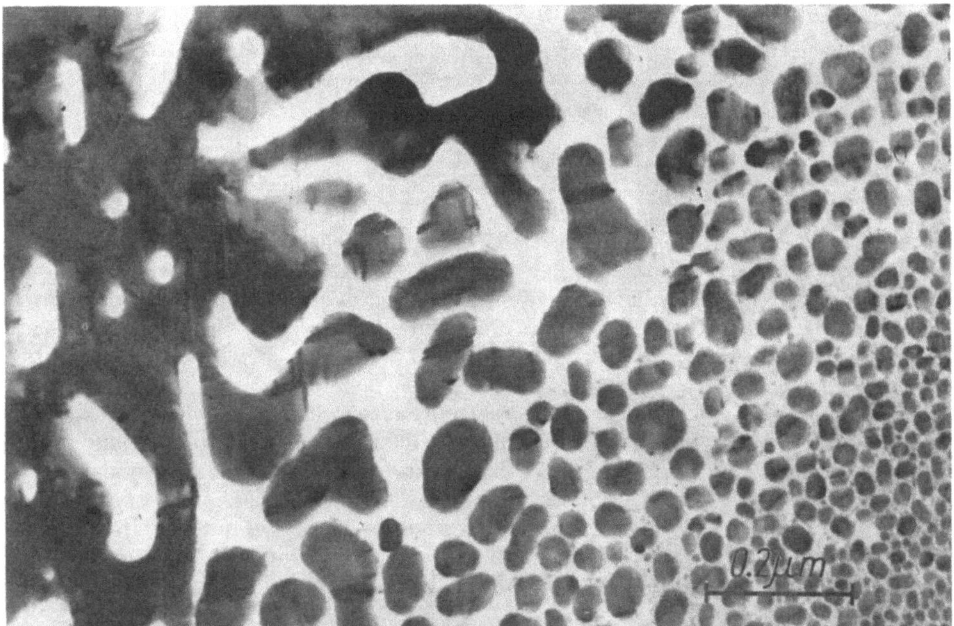

Fig. 1. Thickness gradient of Ag islands epitaxially grown on a NaCl cleavage face

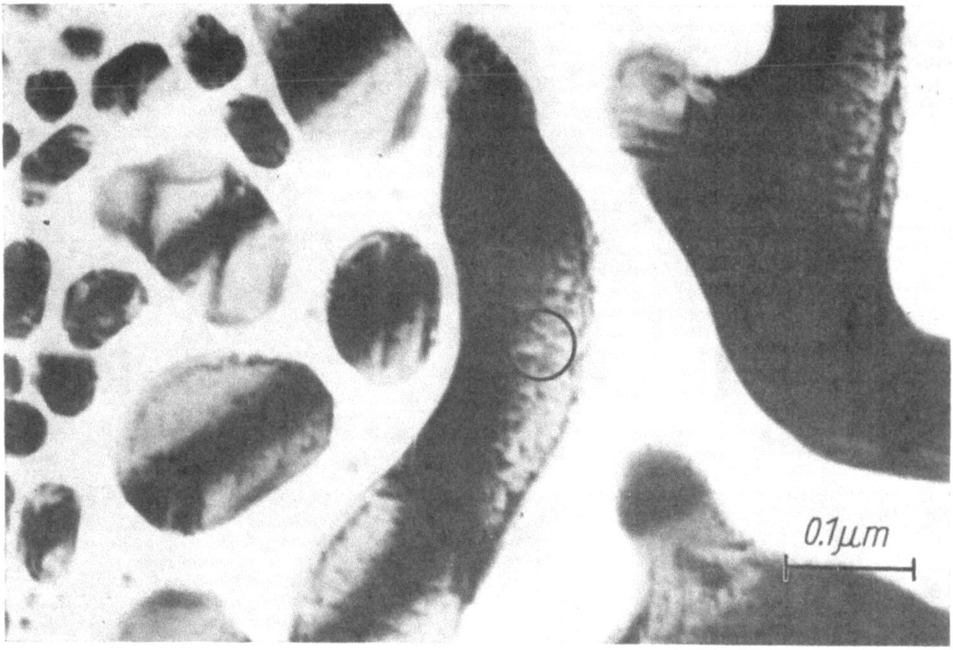

Fig. 2. Misfit dislocations (encircled) in Ag/Pd bicrystal islands ($h_{Pd} = 1.5$ nm; h_{Au} increases proportional to the island diameter)

film thickness and the corresponding island size decrease down to zero within the 1 to 2 µm shadowed region (see, e.g., Fig. 1). Pd layers of uniform thickness between 0.9 and 3 nm were deposited onto the Ag islands. Finally, the bicrystal islands imbedded in a carbon coating were dissolved from the rocksalt support to be observed in TEM.

The island thickness was measured in two different ways: On the one hand, it was presupposed that the island thickness increases linearly from zero to the deposited maximum thickness. On the other hand, an almost constant ratio of island thickness and diameter was assumed. Both methods lead to nearly the same island thickness.

The critical thickness of complete pseudomorphic growth characterized by the first appearance of misfit dislocations was measured by TEM images. Besides, the distance between bending contours [5] were estimated as a measure of the bicrystal bending.

2.2 Results

As shown in Fig. 2 edge dislocation half-loops are formed in Ag/Pd islands, in analogy to those observed in Au/Pd bicrystals (cf. [6, 7]).

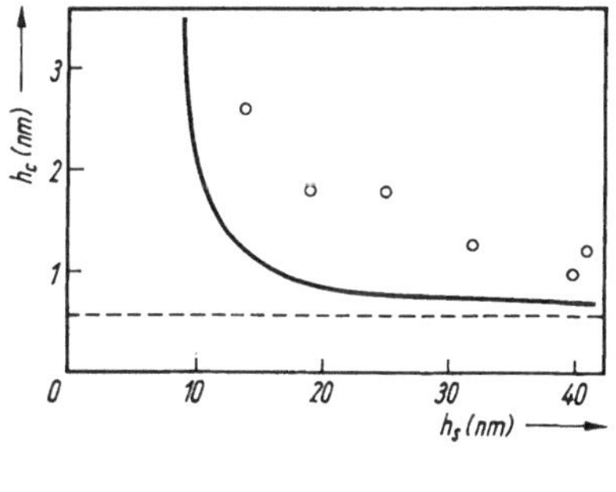

Fig. 3. Calculated dependence of critical deposit thickness ($h_c \triangleq h_d$) on substrate thickness (h_s) in Ag/Pd bicrystals (circles represent the experimental results)

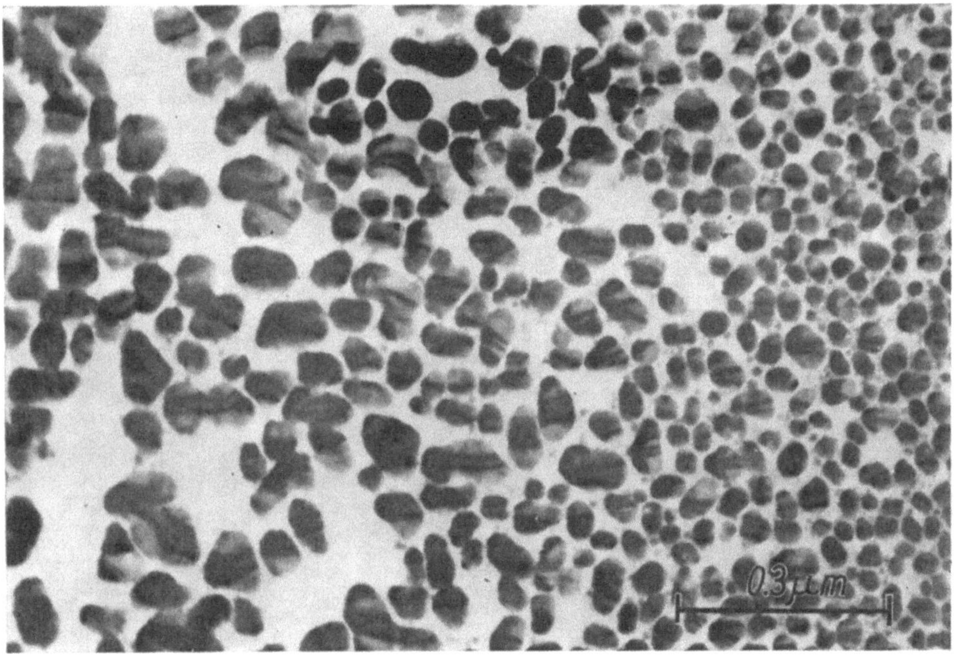

Fig. 4. Bending contours on Ag/Pd bicrystal islands

The critical thickness for complete pseudomorphism was demonstrated to increase with decreasing substrate thickness (island diameter). This can clearly be seen in Fig. 2: No misfit dislocations are generated below a certain island size, which is due to the increasing substrate relaxation, leading to a corresponding decrease of pseudomorphic strain in the deposit. Fig. 3 shows a graphic representation of the critical thickness (h_c) measured in dependence on the substrate thickness (h_s). Moreover, this figure shows the corresponding theoretical dependence which results from (22) and (23) of Part I. The ratio of Young's moduli is $E_{Ag}:E_{Pd} = 3:4$ [8]; the misfit resulting from the difference in the lattice parameters ($a_{Ag} = 0.4086$ nm, $a_{Pd} = 0.388$ nm [9]) amounts to $m = 4.9\%$. With regard to the restrictions made in Part I as well as in the Introduction, the experimental and theoretical thickness values in Fig. 3 correspond with each other in a satisfactory way. Above a thickness of 30 nm the influence of substrate relaxation is small enough for the substrate to be regarded as rigid.

Because the Pd lattice parameters are smaller than those of the substrate, the bicrystal must bend in the direction of the deposit to reduce misfit strain. As can be seen in Fig. 4 bending contours [5] are observed on the islands. Applying the relation

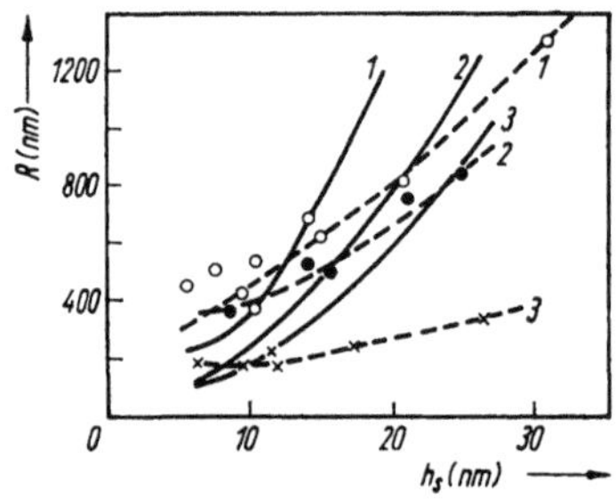

Fig. 5. Dependence of the radius of curvature (R) of Ag/Pd bicrystals on substrate thickness (h_s); ——— calculated by (17) of Part I; – – – experimental values: (1) $h_d = 0.9$, (2) 1.8, (3) 3 nm

$R = Wd_{hkl}\lambda^{-1}$ [10] (λ electron wavelength, d_{hkl} distance between the lattice planes hkl) the radius of curvature of the bicrystal can be estimated from the distance between the bending contours. Dependent on the substrate thickness the radii of curvature obtained in the experiments are shown in Fig. 5. In this figure the deposit thickness (h_d) is used as a parameter. Moreover, Fig. 5 includes the corresponding radii of curvature as calculated from the material parameters. In analogy with the theoretical results of Part I the experimental radii of curvature decrease with increasing deposit thicknesses, i.e. the islands are more strongly bent. The experiments also prove the relaxation to increase with decreasing substrate thickness.

3. Pseudomorphic Growth of Continuous Au/Pd Bicrystal Films Influenced by Substrate Relaxations

3.1 Experimental

Continuous substrate films were prepared by condensing Au on the (100) surface of monocrystalline Ag spheres (concerning their preparation, cf. [11]). The obtained monocrystalline Au films were resolved from the Ag support in a subsequent electrolytical process and mounted on 50 μm copper grids. It was possible to prepare such unrestrained Au films down to a thickness of 6 nm, where they, of course, contain a lot of holes.

Before the Pd deposition, the Au films were "cleaned" in vacuum (3×10^{-4} Pa) at a temperature of approximately 400 °C. The thickness of the epitaxially grown Pd deposit was in the range from 0.9 to 5 nm. The substrate temperature was adjusted to 160 °C, the deposition rate to $r_{Au} = 0.25$ nm/s, $r_{Pd} = 0.02$ nm/s.

In analogy to the Ag/Pd islands the Au/Pd bicrystal films were assumed to bend in the deposit direction $\left(a_{Au}\ (= 0.4074\ \text{nm}) > a_{Pd}\ (= 0.388\ \text{nm})\right)$; the misfit amounted to $m = 4.9\%$.

3.2 Results

First of all, edge dislocation half-loops are generated as misfit dislocations in unrestrained Au/Pd films similarly as in Ag/Pd bicrystal islands (cf. Fig. 2). As an example, Fig. 6 shows a TEM image of an Au/Pd bicrystal where $h_{Au} = 35$ nm and $h_{Pd} = 1.5$ nm (cf. [6, 7], where the bicrystal was prepared on compact substrates, however).

The critical substrate thickness for a 1.5 nm thick Pd deposit layer was estimated to be distributed in the region of $h_{Au} = 8$ to 10 nm for several experiments. Taking into account the critical thickness on rigid substrates in the present system being $h_c \approx 0.6$ to 0.7 nm [7], the substrate relaxation causes a doubling (!) of h_c. The experimental values ($h_{Pd} = 1.5$ nm; $h_{Au} = 8$ to 10 nm) correspond well to those ($h_d = 1.5$ nm; $h_s = 12$ nm) calculated with the help of (22) and (23) of Part I. Therefore, an estimation of the substrate influence using the model discussed in Part I also should be justified in the case of the preparation technique used here.

In the following a bicrystal will be discussed the deposit of which is thick compared with the substrate ($h_{Pd} = 3$ nm, $h_{Au} = 10$ nm). In such specimens the (220) reflection produces clear moiré fringes (cf. Section 4). Using the relation $d_M = a_1a_2[(a_1 - a_2) \times \sqrt{h^2 + k^2 + l^2}]^{-1}$ [5] the measured moiré distance $d_{M(220)} \approx 4$ nm corresponds to a difference of the lattice parameters of substrate and deposit being $(a_1 - a_2) \approx 0.014$ nm. As substrate and deposit are both distorted it is impossible to estimate the absolute lattice parameter only from the moiré distance. Besides, it is not possible to condense a standard without influencing the stress in the bicrystal. With regard to

Fig. 6. Typical arrangement of misfit dislocations in epitaxial Au/Pd bicrystals prepared on mostly unrestrained Au substrate films (h_{Au} = 35 nm, h_{Pd} = 1.5 nm)

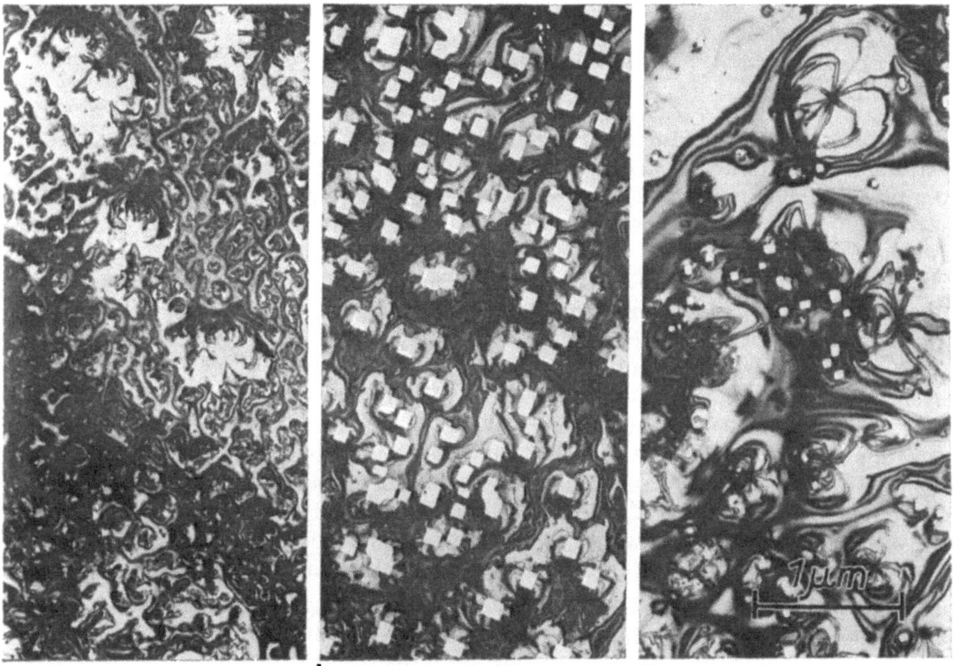

a b c

Fig. 7. Three different kinds of "microbending" in Au/Pd bicrystal films the natural lattice parameter difference (which amounts to $a_{Au} - a_{Pd} = 0.018$ nm) it follows from the measured moiré distance that the misfit is compensated by dislocations ($\delta = m \cdot 80\%$) and homogeneous strain ($\varepsilon = m \cdot 20\%$) in the present case. Thus, the remaining effective misfit is $m^* \equiv \varepsilon \approx 0.9\%$. Using this value of m^* (17) of Part I yields a radius of curvature of $R^* \approx 1.2$ µm. On the other hand, the distance of the experimentally estimated bending contours ($W = 60$ nm) leads to $R = 2.4$ µm. Both the radii, R and R^*, agree in their order of magnitude.

According to the model in Part I an unrestrained Au film should be bent regularly if deposited with Pd. However, the bending contours observed in TEM show three kinds of "microbending" in the order of 1 µm. This can be seen in Fig. 7: In Fig. 7a flat troughs are assumed to be caused by special growth effects of Au or Pd on the Ag support. Very thin substrate films (<10 nm) often contain rectangular holes. As is shown in Fig. 7b these holes especially influence the "microbending" of the film. Fig. 7c contains an example of a nearly spherical shaped bicrystal region. But its generation process and the nature of its border are unknown. It should be noted that the "microbending" in unrestrained bicrystals can change the radii of curvature by some hundred per cent in these small regions. This has to be taken into consideration if comparing theory and experiment.

4. Misfit Dislocations and Moiré Patterns

As already mentioned, in unrestrained Au/Pd bicrystals first of all complete edge dislocation half-loops occur to compensate the misfit (cf. Fig. 6 and Fig. 8a). Moreover, the contamination process in the electron microscope causes a further foil bending connected with the initiation of dislocation gliding in very thin bicrystals (for instance, $h_{Au} = 8$ nm, $h_{Pd} = 1.5$ nm). During this process 60°-mixed-type dislocations are

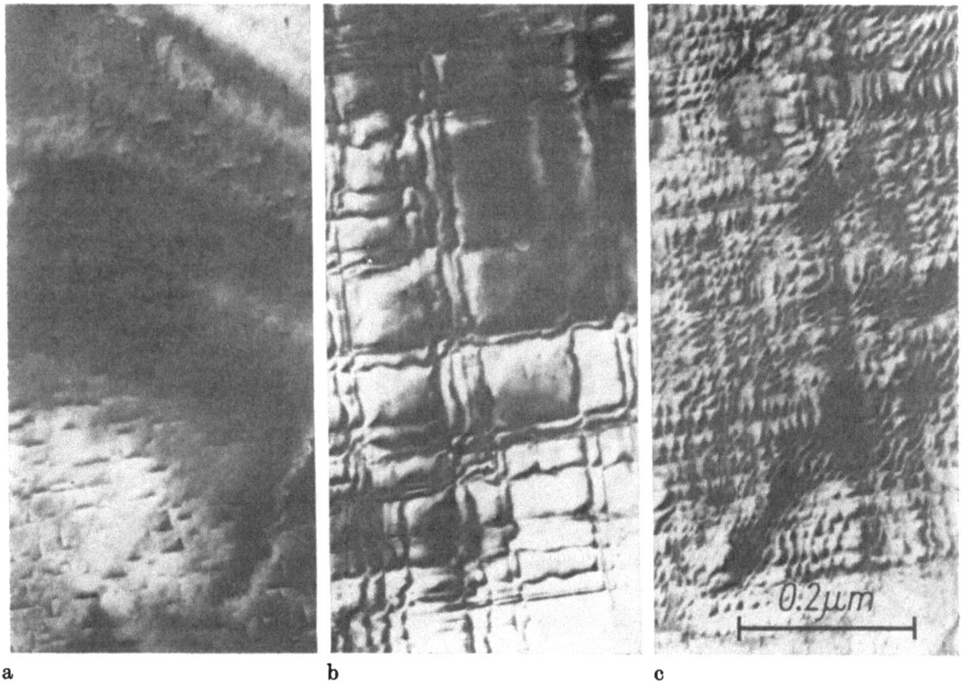

a b c

Fig. 8. Generation of misfit dislocations in the system Au/Pd during relaxation of foil deformation ($h_{Au} = 8$ nm, $h_{Pd} = 1.5$ nm)

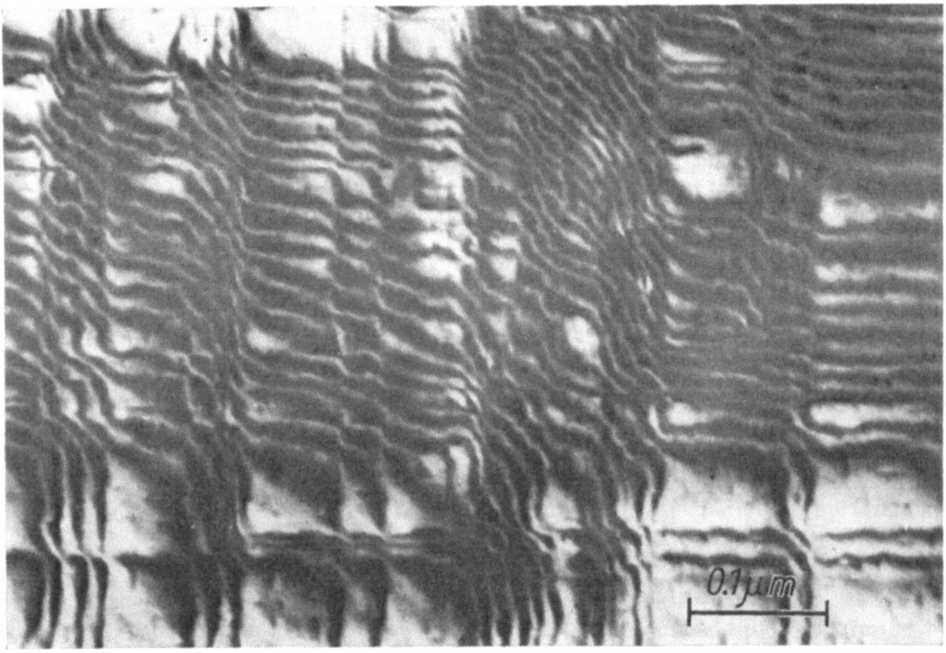

Fig. 9. Misfit dislocations and related effects in Au/Pd bicrystals (h_{Au} = 8 nm, h_{Pd} = 1.5 nm)

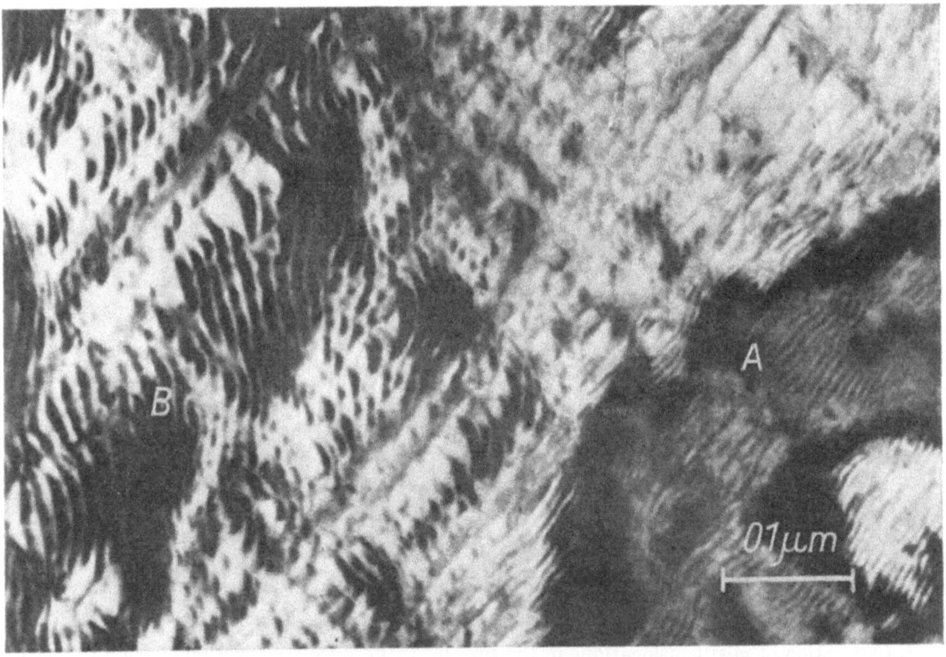

Fig. 10. Misfit dislocations and moiré fringes in Au/Pd bicrystals (h_{Au} = 10 nm, h_{Pd} = 3 nm)

generated in the interface (Fig. 8b) and the degree of pseudomorphism decreases [12]. As soon as the Pd layer obtains its own lattice parameter, line structures propagating at an angle of 45° to the dislocation lines are visible in the excited (200) reflection (Fig. 8c). These lines can be interpreted as moiré fringes. Fig. 9 shows that the angle between the dislocations and the moiré fringes strongly depends on the relative density of the crossing arrays of dislocations. Therefore, the fringes on Fig. 9 can be assumed to be a result of superposition of two dislocation arrays.

With increasing deposit thickness (for instance, $h_{Pd} = 3$ to 5 nm) most of the misfit is compensated by dislocations. In this case moiré fringes lying parallel to the dislocations also appear in the (220) reflection (cf. Fig. 10, region A).

With respect to their geometry moiré fringes and misfit dislocations have a characteristic ratio in epitaxial bicrystals. The following distance relations are valid in the present case of (100)-oriented f.c.c. bicrystals [5]: moiré fringes $d_{200} = a_1a_2/2(a_1 - a_2)$, $d_{220} = a_1a_2/2\sqrt{2}\,(a_1 - a_2)$; misfit dislocations $d_\alpha = a_1a_2/\sqrt{2}\,(a_1 - a_2)$, $d_\beta = a_1a_2/2\sqrt{2}\,(a_1 - a_2)$ (α edge dislocations, β 60°-mixed-type dislocations). It results from these relations that the knowledge of their distance alone is not sufficient for the decision on whether the lines represent moiré fringes or dislocations. Special contrast calculations made by Thölén [13] also describe a line pattern in the case of crossing dislocations which is difficult to distinguish from moiré fringes and corresponds to those seen in Fig. 10, region B.

A better distinction between misfit dislocations and moiré patterns by TEM needs a more detailed consideration of the contrast mechanism: Moiré fringes are the result of the interference of two overlapping undistorted crystals [14] where the best contrast appears at the exact Bragg position [5]. However, dislocations are connected with a local distortion of the regular crystal lattice. They show a strong TEM contrast if the distorted regions fulfil the Bragg condition (but not their surroundings). Some examples are shown in Fig. 10 (for the (220) reflection and the (200) reflection in the regions A and B, respectively).

Summarizing these considerations it seems to be unimportant to classify the line structure of region B in Fig. 10 exactly. Particularly, the line contrast will be generated by crossing dislocations in thin deposits ($\lesssim 1.5$ nm) where the dislocations strongly distort the whole layer. Otherwise, the moiré contrast of undistorted crystal regions will dominate in thicker crystals.

5. Summary and Discussion

Based on the theoretical considerations of Part I in the present paper the influence of substrate relaxations on pseudomorphic growth in epitaxial bicrystals was examinated by the help of TEM. To allow substrate relaxations Ag/Pd bicrystals were prepared on NaCl in form of small islands and Au/Pd bicrystals as unrestrained thin films.

Because the lattice parameter of the Pd deposit is smaller than that of the substrate in both systems the bicrystals bend in the direction of the deposit. The radii of curvature measured by the help of bending contours on Ag/Pd islands correspond well with the related theoretical values of Part I. Opposed to it the "microbending" in unrestrained Au/Pd films cannot be described by the model used here.

In both systems the critical thickness for pseudomorphic growth could be shown to increase by substrate relaxations. Particularly, a certain range exists where h_c is critical as well as a minimum substrate thickness below that h_c cannot be obtained at all. This is in agreement with the theoretical considerations of Part I. Further on, the substrate was stated to be compact if its thickness is larger than twenty times the deposit thickness in the present case of about 5% misfit.

However, the quantitative comparison of the experimental results (critical thickness, radius of curvature) with the theoretical values reveals errors caused by the structure of the bicrystals: a) influence of the island edges and b) "microbending" in unrestrained films. On the other hand, the model described in Part I cannot represent all aspects of epitaxial bicrystals: With increasing ratio $h_d:h_s$ a strain gradient is developed in the deposit. In this case there is no pseudomorphism in the common sense which excludes graded strain distributions [15]. Pseudomorphism now occurs to a certain extent only as an integral quantity resulting from the distance of moiré fringes or diffraction pattern. A strain gradient within the deposit excludes also the use of (22) in Part I for calculation of the critical thickness, as this equation presupposes likewise a regular deposit strain. In this connection the generation process of misfit dislocations should be reconsidered. Particularly, a smaller strain at the deposit surface could be insufficient to effect a spontaneous dislocation loop generation. Moreover, the deposition process is time-dependent in such cases: During the deposit growth the strain gradient will be developed step by step. With respect to these changing bicrystal properties it is important to point again to the fact that all present interface models only describe static conditions, and therefore they cannot be used without any restriction especially in the case of substrate relaxations.

Acknowledgement

The authors are grateful to Prof. H. Bethge who continuously supported the progress of this work.

References

[1] J. WOLTERSDORF and E. PIPPEL, phys. stat. sol. (a) 78, 475 (1983).
[2] E. PIPPEL, in: Strukturen kristalliner Phasengrenzen – Elektronenmikroskopischer Bildkontrast, Ed. H. G. SCHNEIDER and J. WOLTERSDORF, VEB Deutscher Verlag für Grundstoffindustrie, Leipzig 1977.
[3] J. WOLTERSDORF, A. S. NEPIJKO, and E. PIPPEL, Surface Sci. **106**, 64 (1981).
[4] G. HONJO and K. YAGI, in: Current Topics in Materials Science, Vol. 6, Ed. E. KALDIS, North-Holland Publ. Co., Amsterdam 1980.
[5] P. B. HIRSCH, A. HOWIE, R. B. NICHOLSON, D. W. PASHLEY, and M. J. WHELAN, Electron Microscopy of Thin Crystals, Butterworth, London 1965.
[6] D. CHERNS and M. J. STOWELL, Thin Solid Films **29**, 107 (1975).
[7] E. PIPPEL, phys. stat. sol. (a) **67**, 443 (1981).
[8] D. E. GRAY (Ed.), American Institute of Physics Handbook, McGraw-Hill Publ. Co., New York 1963.
[9] W. P. PEARSON, A Handbook of Lattice Spacings and Structures of Metals and Alloys, Pergamon Press, Oxford 1964.
[10] J. W. MATTHEWS and W. A. JESSER, J. Vacuum Sci. Technol. **6**, 641 (1969).
[11] H. BETHGE and M. KLAUA, Ann. Phys. (Leipzig) **17**, 177 (1966).
[12] J. WOLTERSDORF, Appl. Surface Sci. **11/12**, 495 (1982).
[13] A. R. THÖLÉN, phys. stat. sol. (a) **2**, 537 (1970).
[14] R. GEVERS, Phil. Mag. **7**, 181 (1962).
[15] J. W. MATTHEWS, Epitaxial Growth, Academic Press, London 1975.

(Received May 19, 1983)

phys. stat. sol. (a) 79, 199 (1983)

Subject classification: 1.6 and 10.2; 14.3.3; 22.1.2

Fundamental Research Laboratories, Nippon Electric Co., Ltd., Kawasaki[1])

Transient Behaviour of Intrinsic Gettering in CZ Silicon Wafers

By

H. TSUYA and F. SHIMURA[2])

The transient behaviour of intrinsic gettering (IG) is clarified for the first time on the silicon wafer with surface defects introduced by the pre-annealing in wet O_2 at 1140 °C. It is found that surface defects introduced by the pre-annealing in O_2 at higher temperatures, which are speculated to be heavy metal clusters in this work, migrate to an inner region and are gettered by inner defects during the post-annealing at lower temperatures such as 950 °C. These results suggest that surface defects introduced by pre-oxidation process can be gettered during device heat treatment at comparatively lower temperatures if microdefect nuclei are suitably given.

Das nichtstationäre Verhalten der Intrinsic-Getterung (IG) wird erstmalig geklärt an Siliziumscheiben mit Oberflächendefekten, die durch die Vortemperung in feuchtem O_2 bei 1140 °C eingeführt werden. Es wird gefunden, daß Oberflächendefekte, die durch eine Vortemperung in O_2 bei höheren Temperaturen eingeführt werden und von denen in dieser Arbeit angenommen wird, daß sie Schwermetallcluster sind, in einen inneren Bereich wandern und durch innere Defekte während der Nachtemperung bei tieferen Temperaturen, wie etwa 950 °C gegettert werden. Diese Ergebnisse zeigen, daß Oberflächendefekte, die durch einen Vor-Oxydationsprozeß erzeugt werden, während einer Wärmebehandlung bei vergleichsweise niedrigeren Temperaturen gegettert werden können, wenn geeignete Mikrodefektkeime gebildet werden.

1. Introduction

Recently, many experimental studies on the intrinsic gettering (IG) phenomenon in CZ silicon wafers and its application to IC devices have been published [1 to 8]. IG effectiveness is usually evaluated by the observation of ungettered defects in a surface region using a combination of heat treatment and etching procedure or related to the device characteristics.

In the former case, microdefects in the wafer surface [2, 3] or epitaxial layer [3, 4] introduced by process contamination have been examined. In order to evaluate these defects, a wafer must be subjected to proper heat treatments which increase the defect size to an observable one. Microdefects due to process contamination described above appear as saucer pits (S-pits) by a preferential etching after high temperature heat treatment [4, 9, 10]. In the latter case, device characteristics such as leakage current of a bipolar device [1], lifetime of MOS capacitor [5 to 7], holding time of dynamic RAM [8], and so on are studied in connection with the IG effect.

In order to apply the IG effect effectively to device processing, especially to low temperature process, it is important to clarify at which temperatures IG effect occurs and how microdefects are gettered. However, those methods described above are not suitable to examine the transient phenomenon of IG effect, because extra heat treatment processes are added or IG effect is not confirmed until device processing is completely finished.

[1]) 1-1, Miyazaki, Miyamae-ku, Kawasaki 213, Japan.

[2]) Present address: Monsanto Co., 800N Lindbergh Blvd. St. Louis, Mo. 63167.

This paper clarifies the nature of surface defects (S-spits) using an etching procedure and transmission electron microscopy (TEM) and also discusses the transient behaviour of IG effect observed by a combined heat treatment; pre-annealing to introduce surface defects, multistep and post-annealings.

2. Experimental

Wafers of 450 μm thickness were cut from the boron-doped ⟨111⟩ grown CZ silicon crystal of 76 mm diameter. Resistivity was 10 to 20 Ω cm. Interstitial oxygen content was 16 to $17.5 \times 10^{17}/cm^3$ by the ASTM procedure F-121 [11]. The wafers used in this study were without backside damage.

We examined the transient behaviour of IG effect using processes A, B, and C as shown in Fig. 1. First, all of wafers were pre-annealed in wet O_2 at 1140 °C for 2 h to introduce surface defects beforehand. Process A was performed to clarify at which stage IG effect occurs. Process B was carried out to examine the behaviour of surface defects at high temperatures in a wafer involving no inner defects. Process C was adopted to check the gettering ability of extrinsic gettering at elevated temperature.

In process A, some wafers were subjected to heat treatments in dry O_2 at temperatures of 520, 620, and 720 °C sequentially each for 16 h. This multistep annealing was performed to generate a high density of microdefect nuclei resulting in the enhancement of the IG effectiveness even in a low-oxygen wafer [12]. Subsequently, wafers were post-annealed at various temperatures of 950, 1050, and 1140 °C in wet O_2 or N_2 for 2 h to increase the size of inner microdefects induced by the multistep annealing. Process B does not involve a multistep annealing. The pre-annealed wafers were subjected to heat treatments in N_2 at 950 or 1140 °C for 2 h. In process C, after the wet O_2 annealing, a part of a wafer was scratched by a diamond point and subjected to the post-annealing at temperatures of 720, 950, 1050, and 1140 °C in N_2 for 2 h. After the post-annealing the distribution of S-pits was examined using the Sirtl etchant. At each stage testing wafers were extracted in order to examine the induced defects and the IG effectiveness by etching and TEM observation.

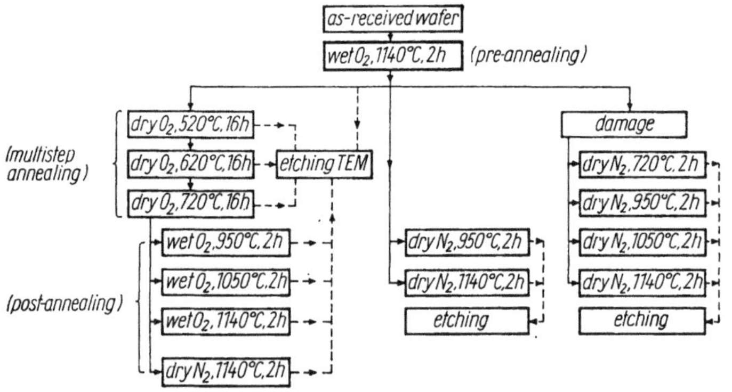

Fig. 1. Heat treatment process and evaluation procedures of IG

3. Results and Discussion

The dependences of S-pit density on annealing temperature and atmosphere were examined. Wafers were subjected to heat treatment at various temperatures in wet O_2 for 2 h and then etched using the Sirtl etchant for 30 s. Several wafers were heated in N_2 at 1140 °C for comparison. Observed result between the S-pit densities and the annealing temperatures in wet O_2 is shown in Fig. 2. The annealing time was 2 h. S-pit density increased drastically above 1140 °C. On the other hand, in the sample annealed at 1140 °C in N_2 instead of wet O_2, S-pit density was less than $10^4/cm^2$.

We have reported that surface defects are micro-stacking faults with heavy metal precipitates by TEM observation [13]. Recently, Stacy et al. [14] also confirmed that the intentional contamination with heavy metal followed by oxidation at 1100 °C produced extrinsic stacking faults decorated with heavy metal on the wafer surface using TEM combined with energy dispersive X-ray spectroscopy.

However, in the recently advanced LSI process with more cleaness, almost all of S-pits observed on the wafer surface or the epitaxial layer are found to be not due to lattice defects. For example, as shown in Fig. 3a, dense S-pits were usually delineated using the Sirtl etchant for 30 s in the sample annealed in wet O_2 at 1140 °C, but we could not observe lattice defects or precipitates corresponding to S-pits even by the TEM weak beam method.

So, we prepared TEM samples from a surface region of the sample slightly etched using the Sirtl etchant for 5 s. In this case, S-pits were not observed by optical microscope (x1000), but triangular-shaped saucer pits (TSP) were observed by TEM as shown in Fig. 3b. TSP's are recognizable to be surrounded by (110) planes of about 0.15 μm length. The density of TSP's is about $1.3 \times 10^7/cm^2$ and nearly equals to that of S-pits ($1 \times 10^7/cm^2$) as shown in Fig. 3a. So, it is considered that TSP's seen in Fig. 3b correspond to surface microdefects introduced by the annealing in wet O_2 at 1140 °C.

From the above TEM observation and the gettering phenomena reported previously [13 to 15], surface defects are reasonably considered to be heavy metal clusters. We believe that whether the surface defects become clusters or lattice defects depends on the degree of heavy metal contamination [13, 14, 16] and the heat treatment conditions. These results are reproducible.

We examined the transient behaviour of IG in the wafers in which surface defects (S-pits) were introduced beforehand as described above. Fig. 4 shows the etching patterns of surface and inner defects (about 30 μm from the front under the chemically removed portion) observed in the wafers which were pre-annealed in wet O_2 at 1140 °C for 2 h followed by a multistep annealing. After the pre-annealing dense S-pits of about $10^7/cm^2$ were observed, but no inner defects were generated. S-pits still remain at each stage of multistep annealing as shown in Fig. 4 b to d (left). On

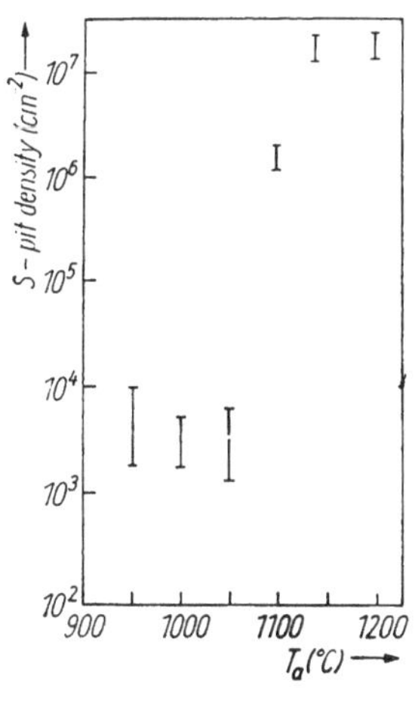

Fig. 2. S-pit density observed on the wafers pre-annealed in O_2 at different annealing temperatures

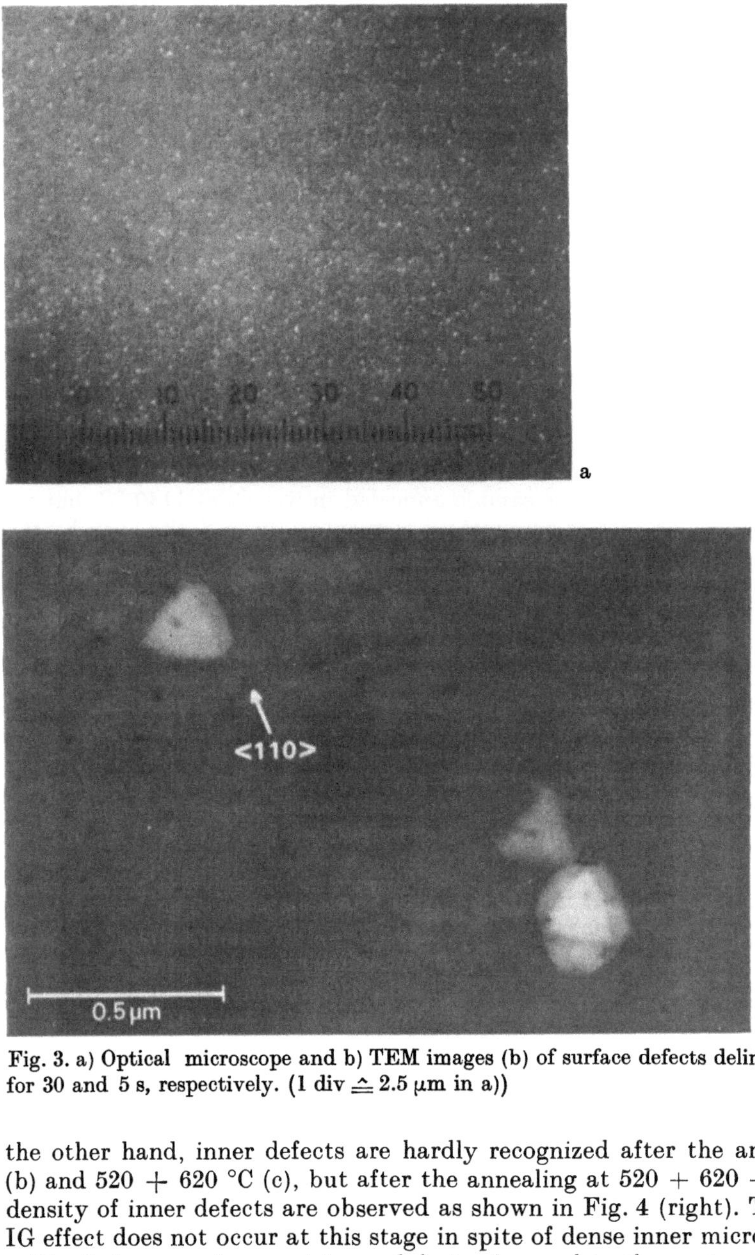

Fig. 3. a) Optical microscope and b) TEM images (b) of surface defects delineated by Sirtl etching for 30 and 5 s, respectively. (1 div ≙ 2.5 μm in a))

the other hand, inner defects are hardly recognized after the annealings at 520 °C (b) and 520 + 620 °C (c), but after the annealing at 520 + 620 + 720 °C (d) a high density of inner defects are observed as shown in Fig. 4 (right). This shows that the IG effect does not occur at this stage in spite of dense inner microdefects.

Fig. 5 shows surface and inner defects observed in the post-annealed wafer. After the post-annealing in wet O_2 at 950, 1050, or 1140 °C, a high density of inner defects were seen and the density of S-pits decreased drastically to less than $10^3/cm^2$. Also in the case of post-annealing in N_2 at 1140 °C, a similar result was obtained. This may show the IG effect. However, it is necessary to consider the possibility that surface defects are annihilated or shrink only during the post-annealing at high temperature, as the annihilation of oxidation-induced stacking faults (OSF) discussed by Sugita et al. [17].

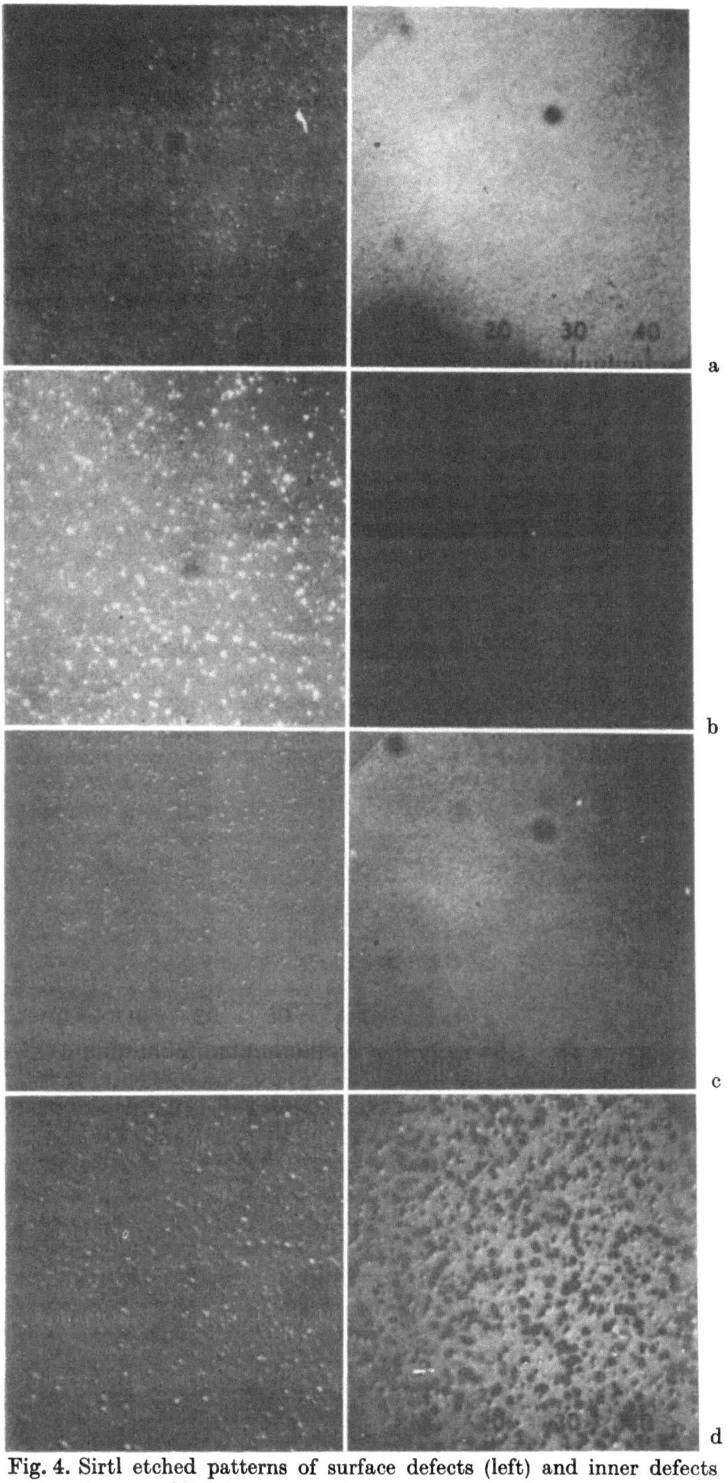

Fig. 4. Sirtl etched patterns of surface defects (left) and inner defects (right) for each process. a) wet O_2, 1140 °C (W); b) a) + dry O_2, 520 °C; c) b) +620 °C; d) c) +720 °C

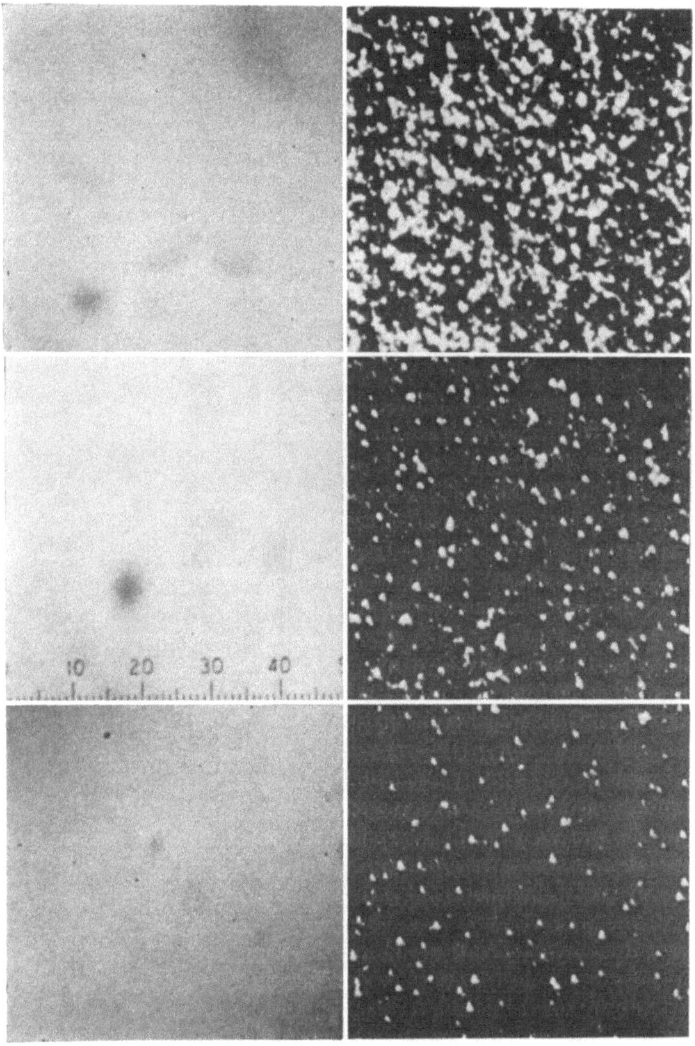

a

b

c

Fig. 5. Sirtl etched patterns of surface defects (left) and inner defects (right) for each process. a) W + MIG + wet O_2, 950 °C; b) W + MIG + wet O_2, 1050 °C; c) W + MIG + wet O_2, 1140 °C

To clarify the IG effectiveness, the behaviour of surface defects at high temperatures was examined using the process B without a multistep annealing. In this case, a high density of S-pits distinctly remained after the annealing in N_2 at 950 or 1140 °C. The density of inner defects was very low, less than $1 \times 10^3/cm^2$. The results of both processes A and B show that surface defects introduced at the first stage of pre-annealing in wet O_2 are gettered by inner defects introduced during the subsequent heat treatments of multistep and post-annealing, do not shrink, or are annihilated only at high temperatures.

Next, we shall discuss the gettering ability of inner defects generated by multistep annealing. In the samples annealed at 520 or 520 + 620 °C, inner defects were not observed by TEM because of the size smaller than detectable. On the other hand, dislocation dipoles and smaller dislocation loops mingled with a high density of micro-

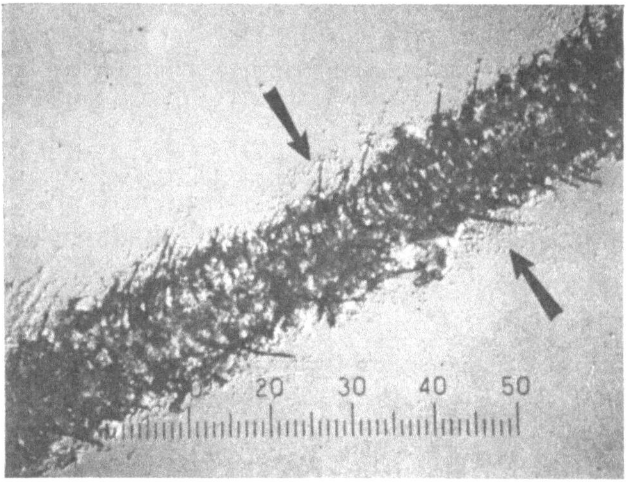

Fig. 6. Sirtl etched patterns of surface defects and scratches observed in process C. The magnification is the same as in Fig. 3a

precipitates were observed in the sample annealed at 520 + 620 + 720 °C. These dislocations are expected to be IG sinks [18]. But, at this stage the IG effect does not occur as shown in Fig. 4d. For the occurrence of the IG effect, not only the inner defects which are enough to getter surface defects, but also the temperature at which surface defects can migrate to an inner region are essentially necessary. However, it is uncertain, at this stage, whether the fact that no IG occurred during multistep annealing is due to the lack of IG ability or due to the less migration of surface defects at lower temperatures such as 720 °C.

Then, the migration of heavy metal clusters at elevated temperatures was examined using the process C. When no post-annealing was performed, S-pits were observed all over the wafer as shown in Fig. 3a, as well as in the region around the scratches. While, when the post-annealing was carried out at elevated temperatures, the migration of S-pits to scratches was observed. For example, after the post-annealing even at 720 °C for 2 h, an S-pit free region was observed in the environment of about 100 μm from the scratch. This result is interpreted with the diffusion length of copper calculated using the diffusion coefficient, $D = 2 \times 10^{-7}$ cm^2/s at 720 °C [19]. Elevating the post-annealing temperature expanded the area of the S-pit free region. Fig. 6 shows the etched pattern of S-pits observed in the wafer in the case of the post-annealing in N_2 at 1140 °C for 2 h. It is obviously observed that a defect free region was formed around scratches during the post-annealing. It is also noticed that S-pits gather collectively at scratches indicated by arrows shown in Fig. 6.

Therefore, from the results of processes A, B, and C, the following conclusion is derived. The fact that no IG occurred during the multistep annealing is not due to less migration of heavy metal clusters, but due to less IG ability of inner defects. That is, only the multistep annealing is insufficient to generate inner defects with effective IG sinks. The transient phenomenon described above reveals that the post-annealing at lower temperatures such as 950 °C provides the gettering ability to such inner defects, to say nothing of the thermal energy for the migration of surface defects.

4. Conclusions

From the investigation of the transient behaviour of IG effect in a wafer with surface defects introduced by pre-annealing in wet O_2 at 1140 °C, the following results are obtained.

(i) Surface defect density increases drastically by the annealing at temperatures higher than 1100 °C in wet O_2.

(ii) Almost all surface microdefects are heavy metal clusters, but they become micro-stacking faults with their precipitates under certain conditions described above.

(iii) Heavy metal clusters introduced by the pre-annealing migrate to an inner region during the post-annealing and are gettered by the sinks due to inner defects at lower temperature such as 950 °C.

(iv) These results suggest that IG effect is effective even for low temperature device processing.

Acknowledgement

The authors wish to acknowledge Dr. T. Kawamura for his encouragement in this work.

References

[1] T. Y. Tan, E. E. Gardner, and W. K. Tice, Appl. Phys. Letters **30**, 175 (1977).
[2] G. A. Rozgonyi and C. W. Pearce, Appl. Phys. Letters **32**, 747 (1978).
[3] L. E. Katz and D. W. Hill, J. Electrochem. Soc. **125**, 1151 (1978).
[4] H. Tsuya, K. Tanno, and F. Shimura, Appl. Phys. Letters **36**, 658 (1980).
[5] K. Yamamoto, S. Kishino, Y. Matsushita, and T. Iizuka, Appl. Phys. Letters **36**, 195 (1980).
[6] K. H. Yang, H. F. Kappert, and G. H. Schwuttke, phys. stat. sol. (a) **50**, 221 (1978).
[7] W. J. Patrick, S. M. Hu, and W. A. Westdrop, J. appl. Phys. **50**, 1399 (1979).
[8] H. Otsuka, K. Watanabe, H. Nishimura, H. Iwai, and H. Nihira, Extended Abstract No. 396, Electrochem. Soc. Meeting, Denver (Oct. 1981).
[9] D. Pomerantz, J. appl. Phys. **38**, 5020 (1967).
[10] G. A. Rozgonyi, R. P. Deysher, and C. W. Pearce, J. Electrochem. Soc. **123**, 1910 (1976).
[11] Annual Book of Standards (American Society for Testing Materials, Philadelphia, 1977). Document F-121, Part. 43.
[12] H. Tsuya, K. Ogawa, and F. Shimura, Japan. J. appl. Phys. **20**, L31 (1981).
[13] F. Shimura, H. Tsuya, and T. Kawamura, J. appl. Phys. **51**, 269 (1980).
[14] W. T. Stacy, D. F. Allison, and T.-C. Wu, Proc. 4th Internat. Symp. Silicon Mat. Sci. Tech. 1981 (p. 344).
[15] P. F. Schmidt and C. W. Pearce, J. Electrochem. Soc. **128**, 630 (1981).
[16] C. W. Pearce and R. G. McMahon, J. Vac. Sci. Technol. **14**, 40 (1977).
[17] Y. Sugita, H. Shimizu, A. Yoshinaka, and T. Aoshima, J. Vacuum Sci. Technol. **14**, 44 (1977).
[18] F. Shimura, H. Tsuya, and T. Kawamura, J. Electrochem. Soc. **128**, 1579 (1981).
[19] H. F. Wolf, Silicon Semiconductor Data, Pergamon Press, 1969 (p. 137).

(Received January 27, 1983)

phys. stat. sol. (a) 79, 207 (1983)

Subject classification: 14.2; 21

L.M.S.E.S., C.N.R.S., L.A. 306, Institut de Physique, Université Louis Pasteur, Strasbourg[1])
and Electrotechnical Institute, Warsaw

On the Flux Jumps in Hard Superconductors

By

J. SOSNOWSKI[2])

The results of the calculations of the higher stability limits appearing after the successive flux jumps are presented. The solutions of the recurrencive equations for two kinds of relations between the flux jump field and stability-limit field are given. In the first case, in which all instabilities develop into the flux jumps, the periodicity of the successive flux jumps is predicted, while in the second, in which the flux jump field is proportional to the stability-limit field, the distance between flux jumps is described by geometrical progress.

Dans cet article sont présentés les résultats de calculs de limites de stabilité, consécutives à des sauts de flux magnétique. Les solutions des équations récursives pour deux types de relations entre la valeur du champ ou saut de flux et la valeur du champ limite de stabilité sont proposées. Ainsi, dans le premier cas où toutes les instabilités sont accentuées par les sauts de flux, on établit la périodicité des sauts successifs; dans le second où la valeur du champ au saut de flux est proportionelle à la valeur du champ limite de stabilité, on montre que la distance entre les sauts de flux suit une loi de progression géométrique.

1. Introduction

Since the time of the construction of the first superconducting coils, the problem of the instabilities in the superconductors has both scientific as well as technical importance. The basic approach to this problem was given by Wipf [1] for the pinning forces described with Kim et al.'s [2] critical state model. In Wipf's paper the problem of the first stability limit field and flux jump have been accurately treated and the extension of the formalism to higher instabilities has been suggested. An example of a unique flux jump on the magnetization of Nb_3Al sample is presented in Section 2. In the numerous experimental papers (see, for example, [3 to 8]), however, beside the first flux jump, also many successive jumps have been observed at higher values of magnetic field. When the flux jump has appeared then the fluxoids will form a stable lattice. With the increase of the magnetic field a new stability limit might be reached, in which the next flux jump can occur.

In the present paper, we have applied Wipf's approach in order to describe the successive stability limits appearing after corresponding flux jump for two particular cases: when all instabilities are developing to flux jumps or if the flux jump field is proportional to the stability limit field. We show also as an illustration of the first flux jump the results of measurements on an A15 type sample, while the experimental verification of the present calculations on Laves phases is under progresss. In Section 2, we introduce the reader into the problems of higher instabilities and flux jumps, formulating the basic equation, the solution of which will be given in Section 3. Section 4 contains the concluding remarks.

[1]) 3, rue de l'Université, 67084 Strasbourg, France.

[2]) Permanent address: Department of Fundamental Research, Electrotechnical Institute, 04-703 Warsaw, Pożaryskiego 28, Poland.

2. Formulation of the Problem of Higher Instability Description

As an introduction to the problems concerned we present in Fig. 1 the measured hysteresis of magnetization curve of an A15-type sample, on which the existence of one flux jump has been observed at the magnetic field value 2.8 T. This experiment was performed at $T = 4.2$ K by the extraction method in a superconducting magnet which produced a magnetic field up to 6.5 T. The large irreversibility of the magnetization indicates that the investigated material is a hard superconductor with large value of critical currents. As is seen from Fig. 1, in agreement with previous experimental results [3 to 8], in the catastrophic flux jump process the magnetization falls to or almost to zero. It means that the magnetic induction inside the superconductor is then approximately equal to the external induction value [5]. The magnetic induction profile $B(x)$ for the external magnetic field B_e higher than the i-th flux jump field B_i^* can be described by the equation

$$\begin{aligned} B(x) &= -B^0 + \sqrt{(B_e + B^0)^2 - 8\pi\alpha x} \quad \text{for} \quad x \leqq x_i\,, \\ B(x) &= B_i^* \quad \text{for} \quad x > x_i\,, \end{aligned} \tag{1}$$

where the renormalized penetration depth x_i is defined by the condition

$$8\pi\alpha x_i = (B_e + B^0)^2 - (B_i^* + B^0)^2\,. \tag{2}$$

Index i numerates the subsequent flux jumps, while α and B^0 are the parameters describing the pinning force density F_p in the critical state model of Kim et al.

$$F_p = \frac{\alpha B(x)}{B(x) + B_0}\,. \tag{3}$$

The determination of the stability limit for a magnetic field higher than the $(i-1)$-th flux jump field needs, as it was suggested byWipf, the determination of the non-vanishing magnetic field disturbance $\beta_i(x)$, which fulfils the corresponding equilibrium equation

$$\frac{|B(x) + B^0|^2}{4\pi\alpha}\,\frac{d\beta_i}{dx} - \beta_i + \frac{1}{c}\,\frac{\partial\alpha}{\partial t}\int\limits_x^{x_{i-1}} \beta_i\,dx = 0\,, \tag{4}$$

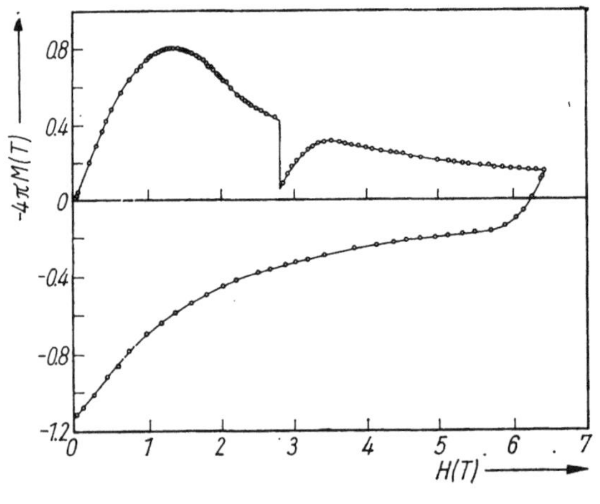

Fig. 1. The hysteresis of the magnetization curve of an Nb–Al sample measured at $T = 4.2$ K. The existence of flux jump is seen in the magnetic field equal to 2.8 T

where the penetration depth x_{i-1} is determined now by (2) and c is the specific heat per unit volume. The solution of (4) can be written in a form useful for higher stability limit descriptions,

$$\beta_i(x) = \frac{\sqrt{8\pi\alpha}}{B(x) + B^0} [C_{1i} \cos \{a[B(x) + B^0]\} + C_{2i} \sin \{a[B(x) + B^0]\}\,. \tag{5}$$

In (5) we have introduced the parameter a defined as

$$a = \sqrt{-\frac{1}{4\pi\alpha c}\frac{\partial\alpha}{\partial T}}\,. \tag{6}$$

We consider now the stability limit occurring after the $(i-1)$-th flux jump. The coefficients C_{1i} and C_{2i} as well as the disturbance function $\beta_i(x)$ not need be in principle the same for each stability limit, as it will be seen in Section 3. The determination of the unknown parameters in (4), (5) needs additional boundary conditions. Two conditions $\beta_i(0) = 0$, $\beta_i(x_{i-1}) = D_i > 0$ lead to the following equations, somewhat modified in comparison with Wipf's equations conducted for the first stability field:

$$C_{1i} \cos [a(B_i + B^0)] + C_{2i} \sin [a(B_i + B^0)] = 0\,, \tag{7}$$

$$C_{1i} \cos [a(B^*_{i-1} + B^0)] + C_{2i} \sin [a(B^*_{i-1} + B^0)] = \omega_i\,, \tag{8}$$

where

$$\omega_i = \frac{D_i(B^*_{i-1} + B^0)}{\sqrt{8\pi\alpha}}\,.$$

A third boundary condition is deduced from (5) and (4),

$$C_{1i} \sin [a(B^*_{i-1} + B^0)] - C_{2i} \cos [a(B^*_{i-1} + B^0)] = 0\,. \tag{9}$$

After simple transformations (7) to (9) may be presented in the following form:

$$A_i \cos (aB_i) + K_i \sin (aB_i) = 0\,, \tag{10}$$

$$A_i \cos (aB^*_{i-1}) + K_i \sin (aB^*_{i-1}) = \omega_i\,, \tag{11}$$

$$A_i \sin (aB^*_{i-1}) - K_i \cos (aB^*_{i-1}) = 0\,. \tag{12}$$

We have used in (10) to (12) the notations

$$A_i = C_{1i} \cos (aB^0) + C_{2i} \sin (aB^0)\,, \tag{13}$$

$$K_i = C_{2i} \cos (aB^0) - C_{1i} \sin (aB^0)\,. \tag{14}$$

3. Solution of the Problem of the Higher-Instability Description

Equations (10) to (12) form a system of three linear equations with four unknowns: B_i the field of the stability limit, occurring after previous flux jump field B^*_{i-1} and two coefficients A_i and K_i for each instability process. Therefore, for simultaneously solving these recurrence equations, one additional condition joining the stability limit field with the flux jump field is needed. In the general case such dependence is difficult to resolve analytically. However, in two simple physical cases, the solution of (10) to (12) can be easily obtained: (i) if all instabilities develop into the flux jump. It means that if the stability limit is reached, then the flux jump immediately appears. B^*_i is then equal to B_i; (ii) in the second case we assume that the flux jump field is proportional to the limited stability field with the coefficient of the proportionality, denoted as η, higher than unity: $B^*_i = \eta B_i$. The coefficient η can depend on various physical parameters (as temperature, magnetic field change, velocity), but should be independent of magnetic field values in the investigated region, in order to maintain the proportionality. The proportionality, in some cases, between the limited stability field and the flux jump field may be supposed among others on the basis of results of experiments in which the specific heat has been changed in Nb_3Sn porous material by filling it with liquid helium [9, 10].

In the first case the recurrence equations (10) to (12) have been solved by the method of induction. For the first stability limit field B^*_{i-1} is equal to zero and the solution has the form

$$i = 2n + 1\,, \quad n = 0\,, \quad K_i = 0\,, \quad A_i = \omega_i\,, \quad \cos(aB_i) = 0\,. \tag{15}$$

In this case, (15) reduces therefore to Wipf's result with the value of the first stability field given by the first solution of (15), i.e.

$$B_1 = \frac{\pi}{2a}\,. \tag{16}$$

For the determination of the second stability field we should find the solution of (10) to (12) for $i = 2$, while B^*_1 is given now by (16). This solution has the following form:

$$i = 2n\,, \quad n = 1\,, \quad K_i = \omega_i\,, \quad A_i = 0\,, \quad \sin(aB_i) = 0\,. \tag{17}$$

From (17) we receive the second stability limit field,

$$B_2 = \frac{\pi}{a} = 2B_1\,. \tag{18}$$

The third stability limit field is deduced after substitution the result (17) into (10) to (12). We again obtain (15) with the proper value of parameter $i = 3$, which in this case gives us the stability field equal to $3B_1$. So we can in a similar way, by the method of induction, prove that for odd index $i = 2n + 1$; $n = 0, 1, 2, 3, \ldots$ the solution of (10) to (12) is given by (15), while the solution of these equations describing the even stability limit ($i = 2n + 2$) will have the form of (17). Therefore, in the first case for which, when the stability limit is reached when the flux jump appears, the successive flux jump fields are described by the periodical dependence

$$B^*_i = iB_1\,. \tag{19}$$

Such periodicity of flux jump field is frequently observed experimentally [3 to 8].

It is interesting to notice that the solutions of (15) and (17) lead to two different families of disturbance function $\beta_i(x)$. In the case of odd stability limit described by (15) the disturbance function has the form

$$\beta_{2n+1} = C_{2,2n+1} \frac{\sqrt{8\pi\alpha}}{\sin(aB^0)\,|B(x) + B^0|} \cos\big(aB(x)\big)\,, \tag{20}$$

where

$$B(x) = -B^0 + \sqrt{|B^0 + (2n + 1)\,B_1|^2 - 8\pi\alpha x} \tag{21}$$

for $\sin(aB^0) \neq 0$. If $\sin(aB^0) = 0$, then as is easily seen from (14), the disturbance has the form

$$\beta_{2n+1}(x) = C_{1,2n+1} \frac{\sqrt{8\pi\alpha}\cos(aB^0)\cos(aB(x))}{B(x) + B^0}\,. \tag{22}$$

However, for the even stability limit described by (17) the disturbance may be written as

$$\beta_{2n}(x) = C_{2,2n} \frac{\sqrt{8\pi\alpha}\sin(aB(x))}{\cos(aB^0)\,[B(x) + B^0]}\,, \tag{23}$$

where $B(x)$ is now determined by the dependence

$$B(x) = -B^0 + \sqrt{(B^0 + 2nB_1)^2 - 8\pi\alpha x}\,. \tag{24}$$

For $\cos(aB_0) = 0$ the disturbance function will be equal to

$$\beta_{2n}(x) = -C_{1,2n} \frac{\sqrt{8\pi\alpha}\,\sin(aB^0)\,\sin(aB(x))}{B(x) + B^0}\,. \tag{25}$$

Both functions (20) and (23) fulfil of course the boundary conditions.

In the second approximated case ($B_i^* = \eta B_i$ and $\eta > 1$), if $\cos(aB_{i-1}^*) \neq 0$ we receive the solution of (10) to (12) describing the dependence of magnetic stability field B_i on previous flux jump field value B_{i-1}^* in the form

$$\operatorname{ctg}\{aB_i\} = -\operatorname{tg}(aB_{i-1}^*)\,. \tag{26}$$

The first solution of (26) gives us the recurrence result describing successive flux jumps fields,

$$B_i^* = \eta B_{i-1}^* + \eta B_1\,. \tag{27}$$

Equation (27) can be easily solved, because the following flux jump fields are in this case described by the sum of the geometrical progress:

$$B_i^* = B_1\eta \frac{\eta^i - 1}{\eta - 1}\,, \tag{28}$$

while the distance between the $(i - 1)$-th and i-th flux jumps is equal to

$$\Delta B_i^* = B_1\eta^i\,. \tag{29}$$

The disturbance function $\beta_i(x)$ describing the i-th stability limit field in this case will be expressed by the superposition of both terms from (5).

It is interesting to notice also that both above-described solutions (19) and (28) lead to different values of the full stability field, it means to the maximal number of flux jumps. This value can be estimated from the condition that the penetration depth x_i should be smaller than the half-width of the sample, $D/2$. In the first case described by (19) we receive the number of the flux jumps equal to

$$i_{\max} = E\left|\frac{4\pi\alpha D + B_1(B_1 - 2B^0)}{2B_1^2}\right|, \tag{30}$$

where $E(x)$ means the integer part of x. In the second case (28) $i_{\max}$ is limited by the condition

$$i_{\max} = E\left|\frac{\ln F - \ln(\eta + 1)\,B_1}{\ln \eta}\right|, \tag{31}$$

where F is equal to

$$F = \sqrt{|\eta(B^0 - B_1) - B^0|^2 + \pi\alpha D(\eta^2 - 1)} - \eta(B^0 - B^1) + B^0\,. \tag{32}$$

If $i_{\max} < 1$, then the sample is internally stable and flux jumps cannot appear. From the above equations it is seen that while the limited stability field values are dependent on the α-parameter value, the number of flux jumps is a function of both parameters describing the pinning force density, B^0 and α.

In many real physical situations both approximations applied here are, of course, too simple. Then the relation remains valid joining the i-th stability limit field with the previous $(i - 1)$-th flux jump field is

$$B_i = B_{i-1}^* + B_1\,. \tag{33}$$

4. Conclusions

In the present paper we have investigated the case of the existence of numerous flux jumps in the magnetization of hard superconductors. Such a phenomenon is frequently observed experimentally. For two simple relations between flux jump field and stability limit field, the analytical expressions describing the successive flux jump fields are given. Two families of the disturbance function, an expression for the full stability field, and a comparison with experimental papers are also presented.

5. Acknowledgement

The author would like to express his acknowledgements to Prof. R. Clad for his help in performing the measurements at Strasbourg University.

References

[1] S. L. WIPF, Phys. Rev. **161**, 404 (1967).
[2] Y. B. KIM, L. F. HEMPSTEAD, and A. R. STRNAD, Phys. Rev. Letters **9**, 306 (1962).
[3] Y. B. KIM and M. J. STEPHEN, in: Superconductivity, Vol. 2, Ed. R. D. PARKS, M. Dekker Inc., New York 1969 (p. 1107).
[4] J. H. P. WATSON, J. appl. Phys. **37**, 516 (1966).
[5] H. T. COFFEY, Cryogenics **7**, 73 (1967).
[6] J. SOSNOWSKI, phys. stat. sol. (a) **64**, 771 (1981).
[7] J. SOSNOWSKI and T. MYDLARZ, Internat. Conf. Physics of Transition Metals 80, Ed. P. RHODES, in: Inst. Phys. Conf. Ser. **55**, 527 (1980).
[8] Y. B. KIM, L. F. HEMPSTEAD, and A. R. STRNAD, Phys. Rev. **129**, 528 (1963).
[9] R. HANCOX, Phys. Letters (Netherlands) **16**, 208 (1965).
[10] R. HANCOX, Appl. Phys. Letters **7**, 138 (1965).

(Received May 24, 1983)

phys. stat. sol. (a) 79, 213 (1983)

Subject classification: 5 and 18.2; 22.8

Solid State Laboratory, Department of Chemistry (a) and Department of Physics (b), Indian Institute of Technology, Bombay[1])

Magnetic and Mössbauer Studies of Some Mixed Cobaltites of the Lanthanides and Alkali Metals

By

D. K. CHAKRABARTY[2]), A. BANDYOPADHYAY, S. B. PATIL (a), and S. N. SHRINGI (b)

Magnetic and Mössbauer studies are carried out on compounds $M_xLn_{1-x}CoO_3$ where M is Ba or Sr and Ln is La, Nd or Dy. All compositions with $x > 0.2$ with the exception of the dysprosium compounds are ferromagnetic at low temperatures. From hysteresis and initial susceptibility studies, it is concluded that they contain a number of single domain clusters in a paramagnetic matrix. Mössbauer spectra of $Ba_{0.3}Nd_{0.7}CoO_3$ and $Ba_{0.5}Dy_{0.5}CoO_3$ show two resonance lines and that of $Sr_{0.5}La_{0.5}CoO_3$ shows a single line spectrum at all temperatures, and no hyperfine spectra are shown even far below T_C. The results are interpreted in terms of short-range magnetic ordering in these systems.

An $M_xLn_{1-x}CoO_3$-Verbindungen, wobei M = Ba oder Sr und Ln = La, Nd oder Dy bedeuten, werden magnetische und Mößbaueruntersuchungen durchgeführt. Alle Zusammensetzungen mit $x > 0{,}2$ mit Ausnahme der Dysprosiumverbindungen sind ferromagnetisch bei tiefen Temperaturen. Aus den Untersuchungen der Hysteresis und der Anfangssuszeptibilität wird geschlossen, daß sie eine Anzahl von Ein-Domänenklustern in einer paramagnetischen Matrix enthalten. Die Mößbauerspektren von $Ba_{0,3}Nd_{0,7}CoO_3$ und $Ba_{0,5}Dy_{0,5}CoO_3$ zeigen zwei Resonanzlinien und die von $Sr_{0,5}La_{0,5}CoO_3$ bei allen Temperaturen ein Ein-Linienspektrum und bis weit unterhalb T_c werden keine Hyperfeinspektren beobachtet. Die Ergebnisse werden mit magnetischer Nahordnung in diesen Systemen interpretiert.

1. Introduction

The compound $LaCoO_3$ has generated much interest of chemists and physicists. The anomalous behaviour of the inverse magnetic susceptibility versus temperature (a plateau between 400 and 600 K) was attributed to the coexistence of diamagnetic low-spin ($S = 0$) $^1A_1$ state and high-spin ($S = 2$) $^5T_2$ state of trivalent cobalt [1 to 4]. Near 0 K, only the low-spin state that has slightly lower energy than the high-spin state (≈ 0.05 eV) exists. With increase in temperature, the low-spin state changes to the high-spin state and a temperature-dependent equilibrium exists between them. Following the notations of Raccah and Goodenough [5], they will be referred to as Co^{III} and Co^{3+}. The coexistence of low-spin (Co^{III}) and high-spin (Co^{3+}) cobalt was established by the Mössbauer studies of Bhide et al. [6]. They have also shown that above 200 K, an electron transfer from Co^{3+} to Co^{III} takes place and the Co^{3+} population decreases and completely disappears at 1210 K, the temperature of a first-order localized to collective electron transition observed by Raccah and Goodenough [5].

The electronic properties of the perovskites MTO_3, where M and T are non-transition and transition metals, respectively, is essentially controlled by the TO_3 frame (ReO_3 structure). The d-orbitals of the alternate T ions can form σ (between e_g of

[1]) Powai, 400076 Bombay, India.

[2]) To whom all correspondence should be addressed.

metal and p_σ of oxygen) and π (between t_{2g} of metal and p_π of oxygen) overlap with the intervening oxide ion. Raccah and Goodenough [5] considered two types of cation–anion–cation overlap integrals, Δ^σ_{cac} and Δ^π_{cac} such that $\Delta^\pi_{cac} < \Delta^\sigma_{cac}$. If these integrals are sufficiently large, they will form π^*- and σ^*-bands. There should be then a critical value Δ_c such that if $\Delta_c > \Delta_{cac}$, the d-electrons will be localized and if $\Delta_c < \Delta_{cac}$, they will form a band.

In the ideal perovskite structure, the T–O–T angle is 180° and Δ_{cac} will essentially depend on the T–O–T distance. In the $LnCoO_3$ series, this distance can be gradually varied by changing the lanthanide Ln, which makes this series attractive.

Ferromagnetism in $La_{1-x}Sr_xCoO_3$ ($x \geqq 0.15$) was first reported by Jonker and van Santen [7] and metallic conductivity by Menyuk et al. [8] was explained by the itinerant electron model [18]. The substitution of Sr^{2+} for La^{3+} creates charge imbalance leading to the formation of a d-hole at the cobalt for every Sr^{2+} ion. These d-holes are not localized at a single cobalt site, but are equally shared between the eight cobalt ions nearest to the Sr^{2+} ion. At $x > 0.125$, it results in an impurity band. The observation of a single line in the Mössbauer spectra above T_C for the ferromagnetic samples ($x \geqq 0.125$) as well as the magnetically split spectrum below T_C for the former was rationalized by Rao [9] by saying that the formation of an impurity band is connected with magnetic ordering.

The results, however, can also be artionalized by Zener double exchange [10] between configurations

$$Co^{3+} - O^{2-} - Co^{4+} \leftrightarrow Co^{4+} - O^{2-} - Co^{3+} .$$

The non-observation of two types of cobalt ions in the Mössbauer spectra may simply mean that the above transformation is very fast. The real difference between the itinerant electron model and fast double exchange is not clear.

Contrary to the observation of a single resonance line in the Mössbauer spectra of $La_{1-x}Sr_xCoO_3$ ($x \geqq 0.125$) above T_C by Bhide et al. [11], Patil et al. [12] observed a two-line spectrum for magnetic $La_{1-x}Ba_xCoO_3$ above T_C, suggesting that the d-electrons are localized in this system. This raises the question, which is the critical Co–O–Co distance below which the electrons are delocalized. To answer this question, we have studied various compositions $M_xLn_{1-x}CoO_3$.

Magnetic hysteresis and initial ac susceptibility of some of these compounds have been studied by us [13]. These showed the presence of single-domain clusters in these materials. The low-temperature Mössbauer spectra are expected to show the magnetically split spectrum or a relaxed spectrum below T_C, none of which is really observed. This leads us to reinvestigate the Mössbauer spectrum of $Sr_{0.5}La_{0.5}CoO_3$ for which such spectra were reported. Even in this case our work did not show a magnetically split spectrum or a relaxed spectrum. These results along with some other relevant information are presented and rationalized in this paper.

2. Experimental

The compounds were prepared from the respective oxalates by mixing them in required quantities, decomposing at 900 K, pelletizing and sintering by placing them on platinum grids. In the case of $Ba_xNd_{1-x}CoO_3$, firing at 1600 K for five days was necessary, whereas $Ba_xLn_{1-x}CoO_3$ (Ln: La, Nd, Sm) were formed after heating at 1450 K for 48 h. The sintered pellets were furnace cooled at 100 K/h, the surface was cleaned and polished. $DyCoO_3$ could not be prepared by this method and the cobalticyanide method of Gallagher was followed [14].

X-ray powder diffraction patterns were recorded on a Philips diffractometer (PW 1140) using CuK_α radiation at a scanning rate of 1°/min.

DC electrical resistivity was measured by the four-probe technique.

Magnetization measurements were carried out using a modified version of a hysteresis loop tracer described by Likhite et al. [15] and the initial susceptibility was measured by the double-coil method of Radhakrishnamurty et al. [16].

Mössbauer spectra were recorded using a multichannel analyzer (MCA 38A, ECIL, India). The movement was achieved by a transducer drive in constant acceleration mode. A 310 stainless foil was used as the absorber. For preparing the source, a well-polished pellet was chosen and a few drops of $^{57}CoCl_2$ in 0.1 M HCl (obtained from Amersham International, England) were added on the surface. After drying it under an infrared lamp, the activity was diffused into the sample by heating in UHP nitrogen at 1200 K for 2 h and finally at 1400 K for another 2 h. After cooling slowly, the surface was cleaned to remove any undiffused activity.

3. Results and Discussion

3.1 Structure

Structural parameters of the various compounds synthesized are given in Table 1. Barium neodymium cobaltites for $0 \leqq x \leqq 0.3$ could be indexed as pseudocubic to account for a superlattice line of low intensity. The observation is in agreement with

Table 1

Structural parameters of the various compounds

system	x	crystal type	cell parameter (nm)		
			a	b	c
$Ba_xNd_{1-x}CoO_3$	0	pseudocubic	0.7548		
	0.05	pseudocubic	0.7586		
	0.1	pseudocubic	0.7613		
	0.2	pseudocubic	0.7632		
	0.3	pseudocubic	0.7675		
	0.4	tetragonal	0.7768	0.7768	0.753
$Ba_xDy_{1-x}CoO_3$	0	orthorhombic	0.516	0.540	0.737
	0.5	orthorhombic	0.534	0.544	0.735
	0.6	orthorhombic	0.552	0.549	0.755
	0.7	orthorhombic	0.550	0.556	0.753
	0.8	orthorhombic	0.540	0.572	0.758
$Sr_xLa_{1-x}CoO_3$	0.5	pseudocubic	0.7662		
$Sr_xNd_{1-x}CoO_3$	0.5	pseudocubic	0.7607		
$Sr_xSm_{1-x}CoO_3$	0.5	orthorhombic	0.537	0.539	0.760

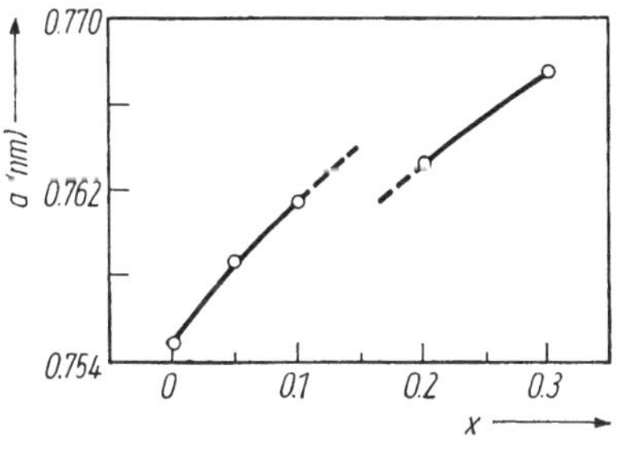

Fig. 1. Variation of lattice parameter with composition in the system $Ba_xNd_{1-x}CoO_3$

the results of Bertaut and Forrat [17] on $NdCoO_3$, which were attributed to an asymmetric displacement of the cobalt ion. The pseudocubic cell has an edge twice that of the smallest cubic cell representing $NdCoO_3$. At $x = 0.4$, the structure is tetragonal and biphasic beyond this. The cell dimension increases with x because of the larger Ba^{2+} ion. However, there is a discontinuity in the cell edge versus x plot (Fig. 1), an observation similar to that reported by Raccah and Goodenough for the system $Sr_xLa_{1-x}CoO_3$ [18]. Since ferromagnetic ordering is observed in these compounds with x near 0.2, it is possible that the appearance of magnetic ordering affects the lattice parameter.

Compounds $Ba_xDy_{1-x}CoO_3$ $(0.5 \leqq x \leqq 0.8)$ have orthorhombic unit cell. Even $DyCoO_3$ that could be obtained only at a lower temperature by the cobalticyanide method (this compound breaks down into several phases when the temperature is raised to sinter it) has orthorhombic cell. Although the Dy^{3+} is much smaller than Nd^{3+}, one would expect the substitution by a large percentage of the larger Ba^{2+} would reduce the structural distortion to give a cubic cell. This possibly does not occur because of the non-stoichiometry. By DTA and TG experiments it has been shown by Patil et al. [19] that the concentration of oxygen vacancies in the compounds $Ba_xLa_{1-x}CoO_3$ and $Ba_xNd_{1-x}CoO_3$ increases with x and the same has been found to be true for the systems reported here.

For the compounds $Sr_{0.5}Ln_{0.5}CoO_3$ the structure is pseudocubic for lanthanum and neodymium, but a smaller average A site cationic radius makes $Sr_{0.5}Sm_{0.5}CoO_3$ orthorhombic.

3.2 Ferromagnetic behaviour

Compounds $Ba_xNd_{1-x}CoO_3$ are ferromagnetic for $x \geqq 0.2$ and show hysteresis loops below T_C. Typical hysteresis curves for $Ba_{0.4}Nd_{0.6}CoO_3$ at various temperatures are shown in Fig. 2. Magnetization decreases with increasing temperature as expected, but a more interesting feature is the decrease of both coercive force (H_C) and relative remanence (σ_r/σ_s) with increasing temperature. These results as well as the magnetic parameters for the compounds $Sr_{0.5}Ln_{0.5}CoO_3$ (Ln: La, Nd or Sm) are presented in Table 2. The $Ba_xDy_{1-x}CoO_3$ system did not exhibit any ferromagnetism down to 77 K.

Initial susceptibility (χ_i) versus temperature plots for the barium neodymium cobaltites were reported by us earlier [19]. The cusp-like maxima in the plots were similar to those exhibited by spin glasses. Similar plots for $Sr_{0.5}Ln_{0.5}CoO_3$ compounds ear shown in Fig. 3. The Curie temperature cannot be determined precisely from these because of the tailing. The temperature at which χ_i is maximum is called the blocking temperature (T_b).

Ferromagnetic material may contain either multidomain (MD) or single-domain (SD) grains. If the grains are SD, i.e. they do not have domain walls, the direction of magnetization is fixed in space due to an effective magnetic anisotropy. The only

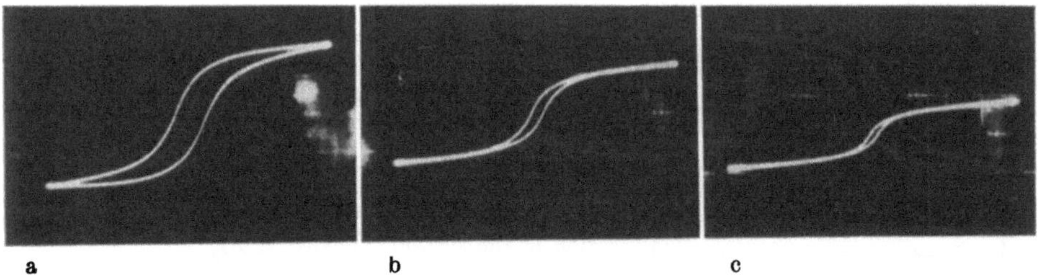

a b c

Fig. 2. Hysteresis loops of $Ba_{0.4}Nd_{0.6}CoO_3$ at various temperatures; a) 77, b) 104, c) 125 K

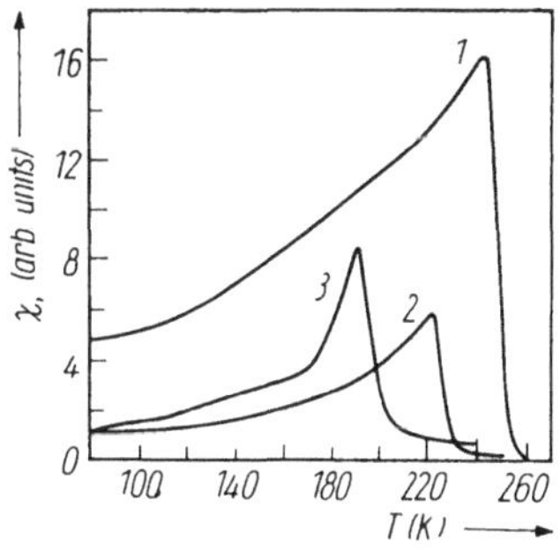

Fig. 3. Initial susceptibility as a function of temperature in the system $Sr_{0.5}Ln_{0.5}CoO_3$ (Ln = (1), La, (2) Nd, and (3) Sm)

way for the magnetization to align with an applied magnetic field is by domain rotation. Since this would need higher fields, SD materials show high coercivity H_c and high remanence (σ_r). On raising the temperature, when kT becomes comparable to the magnetic anisotropy energy, the magnetization direction can fluctuate between the easy axes. This state is known as superparamagnetism (SP). Since the direction of magnetization can fluctuate, the SP state will show zero H_c and zero σ_r. The temperature at which the SD state changes to SP is the blocking temperature.

The hysteresis loops clearly show that the above-cited materials are SD in nature. The relative remanence value suggests that K_1 dominates at low temperatures. Compositions $Ba_xDy_{1-x}CoO_3$ did not show any hysteresis loops. It may be recalled that the compound $Ba_{0.3}Sm_{0.7}CoO_3$ did not show hysteresis at 77 K, but showed a clear loop at 135 K [20]. This was attributed to high H_c at low temperature.

The sharp rise in χ_i with temperature followed by a rapid fall is similar to a behaviour normally exhibited by spin glasses. It is most likely that these crystallographically homogeneous phases have statistically fluctuating local compositions leading to barium-rich (or strontium) and lanthanide-rich regions. Such a suggestion, in fact, was made by Raccah and Goodenough [5]. The barium-rich region could be forming the SD clusters, which is in agreement with the observation by Kneller et al. [21] for $BaFe_4Ti_2O_{11}$ and Radhakrishnamurty et al. [22] for some titanomagnetites.

Table 2

a) Temperature dependence of the hysteresis loop parameters for $Ba_{0.4}Nd_{0.6}CoO_3$

temperature (K)	H_c (10^3 A/m)	σ_r/σ_s
77	31.84	0.50
104	15.92	0.25
125	7.96	0.19

b) Magnetization of the compounds $Sr_{0.5}Ln_{0.5}CoO_3$ (Ln: La, Nd, and Sm) at 77 K

compound	magnetization (e.m.u./g)
$Sr_{0.5}La_{0.5}CoO_3$	36.11
$Sr_{0.5}Nd_{0.5}CoO_3$	23.8
$Sr_{0.5}Sm_{0.5}CoO_3$	20.28

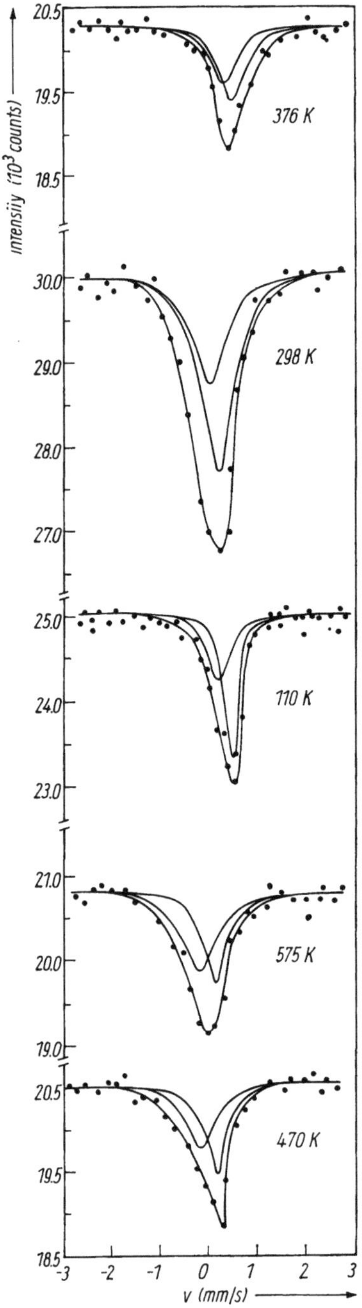

Fig, 4

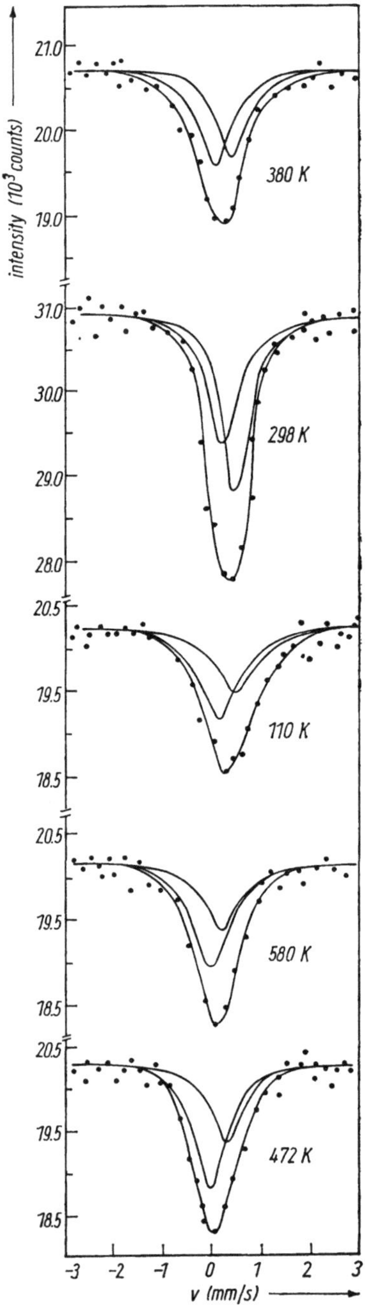

Fig. 5

Fig. 4. Mössbauer spectra of $Ba_{0.3}Nd_{0.7}CoO_3$: ^{57}Co matched against a 310 stainless steel absorber at various temperatures

Fig. 5. Mössbauer spectra of $Ba_{0.5}Dy_{0.5}CoO_3$: ^{57}Co matched against a 310 stainless steel absorber at various temperatures

3.3 Mössbauer spectra

The following compounds were chosen for Mössbauer study: $Ba_{0.3}Nd_{0.7}CoO_3$, $Ba_{0.5}$·$Dy_{0.5}CoO_3$, and $Sr_{0.5}La_{0.5}CoO_3$. The Mössbauer spectra of the first and the second compound are shown in Fig. 4 and 5, respectively. The spectra could be resolved into two Lorentzians. The relative areas under resonance, isomer shift (IS), and line-width data at various temperatures are given in Tables 3 and 4.

Based on the existing literature, it is reasonable to assume that the presence of high-spin Co^{3+} will show the resonance of high-spin Fe^{3+} (formed by electron capture from Co^{3+}) and that the presence of low-spin Co^{III} will show the resonance of low-spin Fe^{III}. Based on isomer shift systematics, the high-resonance Mössbauer line can be attributed to Co^{3+} and the low-resonance line to Co^{III}. It will be seen, however, that some more cobalt species contribute to the low-resonance line.

Table 3

Mössbauer parameters of $Ba_{0.3}Nd_{0.7}CoO_3$ at various temperature

temperature (K)	high-energy resonance line			low-energy resonance line		
	area (%)	line width (mm/s)	isomer shift (mm/s)	area (%)	line width (mm/s)	isomer shift (mm/s)
110	58	0.32	+0.55	42	0.53	+0.29
298	60	0.74	+0.46	40	0.80	+0.20
376	55	0.64	+0.46	45	0.64	+0.20
477	49	0.53	+0.29	51	0.80	−0.05
575	45	0.53	+0.20	55	0.80	−0.15

Since only Co^{III} is present near 0 K, with increase in temperature one expects some Co^{III} to change to Co^{3+}, which will increase the relative intensity of the high-resonance line. A marginal increase was noted in the case of $Ba_{0.3}Nd_{0.7}CoO_3$ in going from 110 to 298 K, but a fairly large change was observed for $Ba_{0.5}Dy_{0.5}CoO_3$. By increasing the temperature further, the intensity of the low-resonance line increased. Similar observations were made by Bhide et al. for $LaCoO_3$ [6]. This was explained as due to an electron transfer from Co^{3+} to Co^{III} as follows:

$$Co^{3+}(t_{2g}^4e_g^2) - O^{2-} - Co^{III}(t_{2g}^6e_g^0) \rightarrow Co^{iv}(t_{2g}^4e_g^1) - O^{2-} - Co^{II}(t_{2g}^6e_g^1) .$$

Both low-spin Co^{II} and intermediate-spin Co^{iv} will give resonance near zero velocity and this is responsible for the increase in the low-resonance line intensity with temperature. In $Ba_{0.3}Nd_{0.7}CoO_3$, this starts at lower temperature and that is why one

Table 4

Mössbauer parameters of $Ba_{0.5}Dy_{0.5}CoO_3$ at various temperature

temperature (K)	high-energy resonance line			low-energy resonance line		
	area (%)	line-width (mm/s)	isomer shift (mm/s)	area (%)	line-width (mm/s)	isomer shift (mm/s)
110	41	1.00	+0.46	59	1.00	+0.12
298	54	0.69	+0.38	46	0.80	+0.03
380	46	0.74	+0.29	54	0.80	−0.02
472	40	0.78	+0.20	60	0.80	−0.06
580	35	0.80	+0.20	65	0.90	−0.11

does not observe much increase in the intensity of the high-energy resonance at 298 K. In $Ba_{0.5}Dy_{0.5}CoO_3$, the electron transfer takes place at a relatively higher temperature and that is why in going from 110 to 298 K, a fair increase in the high-resonance intensity is observed. This observation is in agreement with that of Bhide et al. [6] who noted that by increasing the atomic number of the lanthanide no electron transfer occurs in $HoCoO_3$. Thus, there are two temperature-dependent processes occurring: in the low-temperature region Co^{III} changes to Co^{3+}, and above a certain temperature the electron transfer process from Co^{3+} to Co^{III} takes over.

Mössbauer spectra of all the compounds show a gradual centre shift with increase in temperature. Such observations were made by Bhide et al. [6] for $LaCoO_3$ and Rajoria et al. [23] for $NdCoO_3$. The decrease in isomer shift is due to in creased covalency from greater cation–anion–cation orbital overlap.

Ferromagnetism in $Sr_xLa_{1-x}CoO_3$ ($x \geqq 0.125$) was explained by Bhide et al. [11] by the itinerant electron model of Raccah and Goodenough [18]. A six-line spectrum was observed below T_C and a single-line one above this temperature, which is consistent with the itinerant electron model. Thus, $Sr_{0.5}La_{0.5}CoO_3$ showed a six-line spectrum and with increase in temperature, the central lines became more and more intense and finally at 202 K a relaxed spectrum was obtained. At room temperature and above, a single line was only observed. Since in all the barium substituted compounds we observed two distinct peaks, it is clear that the oxidation states of cobalt can be distinguished by Mössbauer spectroscopy and hence the d-electrons are not delocalized in these compounds. Further, even for the ferromagnetic compounds, we did not notice any hyperfine splitting at 110 K. This led us to study the spectrum of $Sr_{0.5}La_{0.5}CoO_3$ afresh.

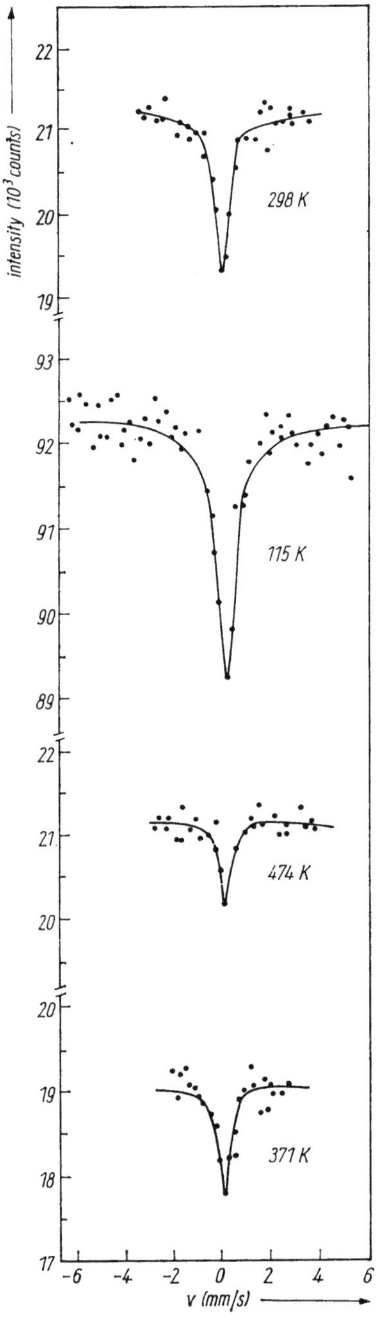

Fig. 6. Mössbauer spectra of $Sr_{0.5}La_{0.5}CoO_3$: ^{57}Co matched against a 310 stainless steel absorber at various temperatures

Mössbauer spectra of $Sr_{0.5}La_{0.5}CoO_3$ are shown in Fig. 6 and the Mössbauer parameters are given in Table 5.

Table 5
Mössbauer parameters of $Sr_{0.5}La_{0.5}CoO_3$ at various temperatures

temperature (K)	linewidth (mm/s)	isomer shift (mm/s)
115	0.86	+0.204
298	0.60	+0.300
371	0.60	+0.033
476	0.50	−0.140

It can be seen that from 115 to 474 K only a single resonance is shown. The rapid electron exchange between the various cobalt species or electron delocalization in $Sr_{0.5}La_{0.5}CoO_3$ is thus established. Although T_C is near 250 K, no hyperfine splitting was seen even at 115 K, contrary to the reports by Bhide et al. [11]. This experiment was done extremely carefully by collecting about 10^5 counts per channel. We conclude that the observation by Bhide et al. could be erroneous as they had employed a mechanical drive and single-channel analyzer. In fact, even in our experiment it was observed that if data were not collected for sufficient length of time, the random scattering of the points could make the identification of peaks difficult.

The question is then why the ferromagnetic spectra do not show hyperfine structure. We attribute this to the short-range magnetic order in these samples. A similar behaviour in magnetic spinels $Fe_{3-x}Ti_xO_4$ was observed by Nagarajan and Radhakrishnamurty [24]. The barium- or strontium-rich ferromagnetic clusters are distributed randomly in the paramagnetic matrix and each particle may contain a number of such clusters of varying size. SD ferromagnetic order prevails only within small-sized clusters not allowing any long-range ordering to be established.

From the observation of a single-line spectrum for $Sr_{0.5}La_{0.5}CoO_3$ and two-line spectra in this barium-substituted compounds, it may be concluded that electron behaviour in the former is itinerant, but in the latter electrons are localized. It is also interesting here to note that although the compound $Sr_{0.5}Dy_{0.5}CoO_3$ as well as the barium-substituted compounds are semiconducting, $Sr_{0.5}La_{0.5}CoO_3$ shows metallic conduction. Electron delocalization in the strontium compound is, to a certain extent, due to the smaller cation–anion–cation distance in the strontium-substituted compound. Ferromagnetism in the barium-substituted compounds is due to Zener double exchange between high-spin Co^{3+} and intermediate-spin Co^{iv}. This means a simultaneous electron transfer from the p-orbital of oxygen to the e_g-level of Co^{iv} and of another electron from the e_g-level of Co^{3+} to the oxygen p-orbital as follows:

$$Co^{3+}(t_{2g}^4e_g^2) - O^{2-} - Co^{iv}(t_{2g}^4e_g^1) \rightarrow Co^{iv}(t_{2g}^4e_g^1) - O^{2-} - Co^{3+}(t_{2g}^4e_g^2)\,.$$

Although $Ba_{0.5}Dy_{0.5}CoO_3$ has a unit cell smaller than that of $Sr_{0.5}La_{0.5}CoO_3$, it does not exhibit any ferromagnetism, a point that needs further clarification. This may be due to the deviation from 180° of the Co–O–Co angle in the orthorhombic structure of the barium compound.

References

[1] R. R. Heikes, R. C. Miller, and R. Mazelsky, Physica (Utrecht) **30**, 1600 (1964).
[2] G. H. Jonker, J. appl. Phys. **37**, 1424 (1966).
[3] C. S. Naiman, R. Gilmore, B. DiBartolo, A. Linz, and R. Santoro, J. appl. Phys. **36**, 879 (1965).

[4] N. Menyuk, K. Dwight, and P. M. Raccah, J. Phys. Chem. Solids **28**, 549 (1967).
[5] P. M. Raccah and J. B. Goodenough, Phys. Rev. **155**, 932 (1967).
[6] V. G. Bhide, D. S. Rajoria, G. Rama Rao, and C. N. R. Rao, Phys. Rev. B **6**, 1021 (1972).
[7] G. H. Jonker and J. H. van Santen, Physica (Utrecht) **19**, 120 (1953).
[8] N. Menyuk, P. M. Raccah, and W. Dwight, Phys. Rev. **166**, 510 (1968).
[9] C. N. R. Rao, J. Indian Chem. Soc. **41**, 979 (1974).
[10] C. L. Zener, Phys. Rev. **82**, 403 (1951).
[11] V. G. Bhide, D. S. Rajoria, C. N. R. Rao, G. Rama Rao, and V. G. Jadhao, Phys. Rev. B **12**, 2832 (1975).
[12] S. B. Patil, D. K. Chakrabarty, M. V. Babu, and S. N. Shringi, phys. stat. sol. (a) **65**, 65 (1981).
[13] A. Bandyopadhyay, S. B. Patil, D. K. Chakrabarty, and C. Radhakrishnamurty, phys. stat. sol. (a) **69**, 441 (1982).
[14] P. M. Gallagher, Mater. Res. Bull. **3**, 225 (1968).
[15] S. D. Likhite, C. Radhakrishnamurty, and P. W. Sahasrabudhe, Rev. sci. Instrum. **36**, 1558 (1965).
[16] C. Radhakrishnamurty and S. D. Likhite, Earth Planetary Sci. Letters **7**, 389 (1970).
[17] F. Bertaut and F. Forrat, J. Phys. Radium **17**, 129 (1956).
[18] P. M. Raccah and J. B. Goodenough, J. appl. Phys. **39**, 1209 (1968).
[19] S. B. Patil, A. Bandyopadhyay, D. K. Chakrabarty, and H. V. Keer, Thermochim. Acta **61**, 269 (1983).
[20] A. P. Murrani, J. Magnetism magnetic Mater. **5**, 95 (1977).
[21] E. Kneller, M. Velicescu, and F. Habery, J. Magnetism magnetic Mater. **7**, 49 (1978).
[22] C. Radhakrishnamurty, S. D. Likhite, and R. Nagarajan, J. Magnetism magnetic Mater. **15–18**, 195 (1980).
[23] D. S. Rajoria, V. G. Bhide, G. Rama Rao, and C. N. R. Rao, J. Chem. Soc. Faraday Trans. II **70**, 512 (1974).
[24] R. Nagarajan, and C. Radhakrishnamurty, Bull. Mater. Sci. **3**, 217 (1981).

(Received February 17, 1983)

phys. stat. sol. (a) 79, 223 (1983)

Subject classification: 14.3.3 and 14.3.4; 13.4; 22.6

Departament d'Electricitat i Electrònica, Universitat Autònoma de Barcelona[1])

Trap-Assisted Tunneling in MIS and Schottky Structures

By

F. CAMPABADAL, V. MILIÁN, and X. AYMERICH-HUMET

Trap-assisted tunneling current in metal–insulator–semiconductor and metal–semiconductor (Schottky) structures with traps in the insulator and in the semiconductor, respectively, is obtained, using the stationary resonance model. The two-band model is considered for the insulator gap of the MIS structure, introducing an effective potential barrier. A comparison between resonant and direct tunneling currents in MIS structures is performed, showing that the main conduction mechanisms is determined by the trap density, the oxide thickness, and the trap depth. For the Schottky diodes, the resonant current–voltage characteristic is analysed as function of the centre of resonance of the system. A discussion between resonant tunneling in MIS and Schottky structures is also included.

Der durch Haftstellen unterstützte Tunnelstrom in Metall–Isolator–Halbleiter- und Metall–Halbleiter(Schottky)-Strukturen mit Haftstellen im Isolator bzw. Halbleiter wird mit dem stationären Resonanzmodell erhalten. Für die Isolatorlücke wird das Zwei-Bandmodell benutzt und eine effektive Potentialbarriere eingeführt. Ein Vergleich zwischen den resonanten und direkten Tunnelströmen in den MIS-Strukturen wird durchgeführt und gezeigt, daß der Hauptmechanismus der Leitung durch die Haftstellendichte, die Oxiddicke und die Haftstellentiefe bestimmt wird. Für die Schottky-Dioden wird die resonante Strom–Spannungscharakteristik in Abhängigkeit vom Resonanzzentrum des Systems analysiert. Der Vergleich zwischen Resonanz-Tunnelung in MIS- und Schottky-Strukturen wird diskutiert.

1. Introduction

The current–voltage characteristics of metal–semiconductor structures has been extensively studied within the framework of thermionic-diffusion theory or by means of tunneling conduction [1]. In the last few years a new conduction mechanism related to the existence of defects in the interface of these structures has become very important in describing the characteristics experimentally observed in Schottky diodes as well as in MIS structures with thin insulator.

In this sense, Parker and Mead [2] show that the relative variation of the current with the bias voltage in CdTe Schottky diodes could be explained assuming the existence of trap states in the semiconductor depletion layer. Moreover, Helman and Sanchez-Sinencio [3] have established the current assisted by traps within a two-step simplified model, obtaining a relationship between energy and position that gives the location of the more efficient traps in the tunneling conduction. With these results one can compare both assisted and direct tunneling conduction mechanisms making evident the relevant role played by resonant tunneling in Schottky structures. An analysis within the two-step approximation using the Green's function formalism has been performed by Sarrabayrouse et al. [4] approximating the bottom of the conduction band in the depletion layer by a straight line, obtaining the resonant tunneling current in a reversely biased AsGa Schottky diode. Other works on resonant tunneling exist in the literature [5, 6]. It is important to remark the paper of Martínez et al. [7] in which direct evidence of the resonant tunneling in metal–silicon diodes (metal = Mo, Pt)

[1]) Bellaterra, Barcelona, Spain.

is shown. These authors discuss extensively the origin of the trap level responsible for the resonant tunneling, as well as a possible explanation of the excess current experimentally observed.

On the other hand, several authors have shown the existence of a thin insulator layer between the metal and the semiconductor which is intrinsically associated to the fabrication process. When the existence of this layer can be directly proved, we obtain a MIS structure. In these structures the conduction mechanism is similar to that in the ideal Schottky junction but in that case the tunneling barrier is performed by the insulator gap. Consequently, the insulator properties are decisive factors of the j–V characteristics. As in Schottky diodes, the existence of traps in the insulator gap contributes to the tunneling currents by means of trap-assisted tunneling. But in this case, the existence of traps is a fact widely established [8, 9] and their origin is intrinsically associated to the fabrication process. Direct evidence of the resonant tunneling has been observed in Al–SiO_2–Si structures [10 to 12]. The main difference between MIS and Schottky diodes, is that the former is an idealization of the metal–semiconductor junction whereas the MIS structure is a more realistic picture of this junction. Besides, another important difference is that the trap density is easier to control in the semiconductor than in a thin insulator layer.

In this work, an analysis of the trap-assisted tunneling in MIS and Schottky structures is performed. This study is worked out separately for each structure, since, as we will see later, the barrier model and the trap energy level must be differently considered in each case. Nevertheless, a discussion of the results obtained in MIS and in Schottky diodes is given, allowing to consider the role played by the insulator in these structures, provided that the trap existence is unavoidable.

The method used in the proposed analysis is the stationary resonance model (SRM) [13], which is applicable to a general potential barrier when the trap responsible for the resonant tunneling can be properly described by a short-range potential. This method has already been applied to analyse the zero bias anomaly in MIM structures [14, 15] and to describe the oscillatory behaviour of the MIM tunnel conductance at low temperatures [16].

In the present problem, the potential barrier in Schottky and MIS structures is slowly varying versus atomic potentials so that the W.K.B. approximation will be considered. Nevertheless, certain numerical limitations that appear in this approximation do not imply practical restriction in the analysed structures.

In the MIS structures, as the traps are located in the insulator, the potential barrier is basically given by the insulator gap whereas in the Schottky diodes it is given by the parabolic potential in the depletion layer. In both cases, the SRM method starts with the onedimensional Green's function of the barrier without traps. Once the Green's function is known the resonant tunneling current associated to one trap inside the barrier can be calculated in terms of the Green's function by the expression

$$I = \frac{q}{|\mathrm{P}g_\mathrm{B}(E_0, z_\mathrm{t}, z_\mathrm{t})|} \frac{j^+(E_0)\, j^-(E_0)}{j^+(E_0) + j^-(E_0)} \left[f^+(E_0) - f^-(E_0)\right], \tag{1}$$

where q is the absolute value of the electron charge, z_t the trap position, E_0 the energy of the resonant level, $f^\pm(E_0)$ the distribution functions at both sides of the barrier, $g_\mathrm{B}(E_0, z, z_\mathrm{t})$ the Green's function of the barrier, P stands for the Cauchy principal value, and

$$j^\pm(E_0) = \pm \frac{i\hbar}{2m} \int\limits_{-\infty}^{E_0} \mathrm{d}\varepsilon \left[g_\mathrm{B}^*(\varepsilon, z \gtrless z_\mathrm{t}, z_\mathrm{t}) \frac{\partial}{\partial z} g_\mathrm{B}(\varepsilon, z \gtrless z_\mathrm{t}, z_\mathrm{t}) - \mathrm{c.c.}\right], \tag{2}$$

m being the electron mass.

This method is used since its simplicity gives a powerful way to evaluate three-dimensionally a complex quantum mechanical conduction mechanism, without assuming unnecessary hypotheses which could be misleading [10]. Nevertheless, reasonable numerical approximations will be used in order to obtain analytical expressions which are more clarifying than numerical tables.

The aim of this work is to explain the excess current observed in metal–silicon junctions considering that a thin oxide layer exists at the interface, although in some structures the oxide thickness can be negligible. The calculation of the resonant tunneling current assisted by traps in the insulator or in the semiconductor, that will be performed, points out the importance of this conduction mechanism. So, the contribution of the traps must be taken into account in a realistic study of Schottky and MIS structures.

2. Resonant Tunneling Current in MIS Structures

In this section we will first establish the effective tunneling barrier of a MIS structure and then we will calculate the resonant tunneling current assisted by traps in the insulator.

2.1 Effective barrier model

In the study of tunneling currents through the insulator of a MIS structure, the basic point is to establish the potential barrier associated to this structure. A first attempt to estimate it can be made by assuming that the barrier is given by the insulator conduction band. However, in practical MIS structures (metal–SiO_2–Si) the electrons involved in the tunneling conduction mechanism are well below the conduction band of the insulator, hence the valence band of the insulator must also be taken into account. Considering a two-band model for the relationship between the energy and the dampng coefficient in the insulator gap, we can write [17]

$$k^{-2} = k_v^{-2} + k_c^{-2} = \left[\frac{\hbar^2}{2m_v^*(E - E_v)}\right] + \left[\frac{\hbar^2}{2m_c^*(E_c - E)}\right], \tag{3}$$

where m_c^*, m_v^* are the effective masses in the conduction and valence bands of the insulator and E is the energy referred to an arbitrary origin.

The above expression can be expressed as

$$k^2 = \frac{2\hbar^{-2}m^*(E)\,(E - E_v)\,(E_c - E)}{E_G^I}, \tag{4}$$

where E_G^I is the energy gap of the insulator and $m^*(E)$ an effective mass considered independent of energy, since we will deal with energies around the mid-gap.

Analysing equation (4) we see that the $E(k)$ relationship approaches the one-band model for values of E near the bottom of the conduction band, whereas the relationship is quadratic for values near the middle of the gap. Besides, for a metal–SiO_2–Si junction the conduction band of the semiconductor is 3.15 eV below the insulator conduction band [1, 19]. Therefore, as the SiO_2 gap is 8 eV [20], one should use a two-band model to calculate the tunneling current.

This fact has been used by many authors in the treatment of direct tunneling [18, 21], and it gives rise to an effective tunneling barrier that results in an enhancement of the tunneling probability.

In the resonant tunneling case we will proceed in the same way, so that the thermionic barrier will be replaced by an effective potential barrier $U_I(z)$ defined as

$$U_I(z) = E + \frac{\hbar^2 k^2}{2m^*}. \tag{5}$$

If a trap state with energy E exists in the insulator gap of a MIS structure, this trap will support resonant tunneling and the electrons with total energy E will contribute to the current. The potential barrier "seen" by these electrons is $U_{\mathrm{I}}(z)$, which, combining (4) and (5), can be put in the form

$$U_{\mathrm{I}}(z) = E_{\mathrm{c}}(z) - \frac{q^2 \varphi_{\mathrm{t}}^2(z)}{E_{\mathrm{G}}^{\mathrm{I}}}, \tag{6}$$

where $q\varphi_{\mathrm{t}}(z) = E_{\mathrm{c}}(z) - E$ is the energy depth of the trap.

We conclude from (6) that the effective potential barrier is flatter than the conduction band of the insulator and, besides, the barrier height is lower, i.e., a trap located at $\varphi_{\mathrm{t}}(z)$ acts as if it were located at $\varphi_{\mathrm{t}}'(z)$ which is related with $\varphi_{\mathrm{t}}(z)$ by

$$\varphi_{\mathrm{t}}'(t) = \varphi_{\mathrm{t}}(z) \left[1 - \frac{\varphi_{\mathrm{t}}(z)}{E_{\mathrm{G}}^{\mathrm{I}}}\right], \tag{7}$$

in other words, the effective barrier depends on the trap energy.

2.2 Resonant tunneling current assisted by traps in the insulator

In order to obtain the resonant tunneling current through the insulator of a MIS structure, with the effective potential barrier previously obtained, we need the Green's function of the system. This function is the solution of the differential equation

$$\left[E + \frac{\hbar^2}{2m} \frac{\mathrm{d}^2}{\mathrm{d}z^2} - U(z)\right] g(E, z, z') = \delta(z - z'), \tag{8}$$

where m is the insulator effective mass. Since we are interested in the resonant tunneling through traps in the insulator gap, and considering equations (1) and (2), $z' = z_{\mathrm{t}}$ where $-d < z_{\mathrm{t}} < 0$, d being the insulator thickness. Equation (8) must be solved in three different regions (metal, insulator, semiconductor), having each of them a potential energy $U_i(z)$ ($i = 1, 2, 3$) associated (cf. Fig. 1). Then, the discussion of the effective barrier only concerns $U_2(z) = U_{\mathrm{I}}(z)$.

It is useful to define

$$k_i(z) = \left\{\frac{2m}{\hbar^2} [U_i(z) - E]\right\}^{1/2}; \qquad i = 1, 2, 3, \tag{9}$$

$$\alpha(z) = \int_0^z \mathrm{d}\zeta\, k_2(\zeta), \tag{10}$$

and

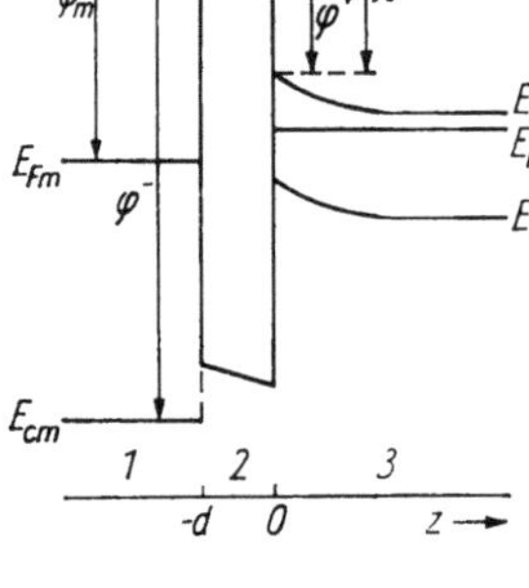

Fig. 1. Energy diagram for a MIS barrier. The dot-dashed line represents the effective barrier given by the two-band model

$$\beta(z) = \int_{-d}^{z} d\zeta\, k_2(\zeta)\,. \tag{11}$$

Note that $k_1(z)$ and $k_3(z)$ are imaginary whereas $k_2(z)$ is real.

For solving (8) the W.K.B. approximation is used. Assuming that $e^{\beta(0)}$, $e^{\beta(z_t)}$, $e^{-\alpha(z_t)} \gg 1$, three expressions for the retarded Green's function are obtained depending on the z-value.

For $z < -d$

$$g_1(E, z, z_t) = -\frac{e^{-\beta(z_t)}}{q\varphi^-}\left[\frac{k_2(-d)}{k_2(z_t)}\right]^{1/2} [k_2(-d) + k_1(-d)]\, e^{-k_1(-d)(z+d)}\,, \tag{12}$$

for $-d < z < 0$

$$g_2(E, z, z_t) =$$

$$= \frac{-m}{\hbar^2[k_2(z_t)\, k_2(z)]^{1/2}} \{[k_2(0) - k_3(0)]^{-1} [k_2(0) + k_3(0)] \exp [\alpha(z) + \alpha(z_t)] +$$

$$+ [k_2(-d) - k_1(-d)]^{-1} [k_2(-d) + k_1(-d)] \exp \left(-[\beta(0) + \beta(z_t) + \alpha(z)]\right) +$$

$$+ \exp\left(-|\alpha(z) - \alpha(z_t)|\right)\}\,, \tag{13}$$

and for $z > 0$

$$g_3(E, z, z_t) = -\frac{e^{\alpha(z_t)}}{q\varphi^+}\left[\frac{k_3(0)\, k_2(0)}{k_2(z_t)\, k_3(z)}\right]^{1/2} [k_2(0) + k_3(0)] \exp \int_0^z d\zeta\, k_3(\zeta)\,, \tag{14}$$

where $q\varphi^\pm$ are shown in Fig. 1.

Since the product $g^*(\partial/\partial z)\, g$ does not depend on the region we calculate it [14], in order to find $j^\pm$ defined in (2), we will use the simplest expressions of the Green's function, i.e. (12) and (14). Thus, we get

$$j^+(E_0) = \frac{2(2m)^{1/2}}{q\hbar^2\varphi^+} \int_{U_3(0)}^{E_0} d\varepsilon \left\{\frac{[\varepsilon - U_3(0)]\,[U_2(0) - \varepsilon]}{U_2(z_t) - \varepsilon}\right\}^{1/2} e^{2\alpha(z_t)} \tag{15}$$

and

$$(j^- E_0) = \frac{2(2m)^{1/2}}{q\hbar^2\varphi^-} \int_{U_3(0)}^{E_0} d\varepsilon \left\{\frac{[\varepsilon - U_1(-d)]\,[U_2(-d) - \varepsilon]}{U_2(z_t) - \varepsilon}\right\}^{1/2} e^{-2\beta(z_t)}\,, \tag{16}$$

where E_0 is the trap energy.

As only the values of ε near E_0 contribute to the integrals, we will only consider the linear terms in the Taylor's expansion of $\alpha(\varepsilon, z_t)$ and $\beta(\varepsilon, z_t)$ around E_0,

$$\alpha(\varepsilon, z_t) \approx \alpha(E_0, z_t) \left\{1 + \int_{z_t}^{0} d\zeta \frac{E_0 - \varepsilon}{2[U_2(\zeta) - E_0]}\right\}, \tag{17}$$

$$\beta(\varepsilon, z_t) \approx \beta(E_0, z_t) \left\{1 + \int_{-d}^{z_t} d\zeta \frac{E_0 - \varepsilon}{2[U_2(\zeta - E_0]}\right\}. \tag{18}$$

Taking into account that the nonexponential terms of (15) and (16) vary much more slowly than the exponentials, and that the traps which contribute more efficiently to the resonant tunneling process are located far from both electrodes, we obtain

$$j^+(E_0) = \frac{2}{(2m)^{1/2}\, q\varphi^+} \left\{ \frac{[E_0 - U_3(0)]\,[U_2(0) - E_0]}{U_2(z_t) - E_0} \right\}^{1/2} \left[\int\limits_{z_t}^{0} \frac{d\zeta}{k_2(\zeta)} \right]^{-1} e^{2\alpha(z_t)} \quad (19)$$

and

$$j^-(E_0) = \frac{2}{(2m)^{1/2}\, q\varphi^-} \left\{ \frac{[E_0 - U_1(0)]\,[U_2(0) - E_0]}{U_2(z_t) - E_0} \right\}^{1/2} \left[\int\limits_{-d}^{z_t} \frac{d\zeta}{k_2(\zeta)} \right]^{-1} e^{-2\beta(z_t)}. \quad (20)$$

Besides, we obtain for the principal value of the Green's function

$$Pg(E_0, z_t, z_t) \approx -\frac{m}{\hbar^2 k_2(z_t)}. \quad (21)$$

This expression is valid for any barrier shape if the traps are deep enough [14].

The expression of the current through the insulator due to a trap can be obtained, using (1), by means of the three last equations, yielding

$$I = \frac{4q^2\hbar}{m}\,[f^+(E_0) - f^-(E_0)] \left\{ \frac{\varphi^-}{\{[E_0 - U_1(-d)]\,[U_2(-d) - E_0]\}^{1/2}} \left(\int\limits_{-d}^{z_t} \frac{d\zeta}{k_2(\zeta)} \right) \times \right.$$

$$\left. \times\, e^{2\beta(z_t)} + \frac{\varphi^+}{\{[E_0 - U_3(0)]\,[U_2(0) - E_0]\}^{1/2}} \left(\int\limits_{z_t}^{0} \frac{d\zeta}{k_2(\zeta)} \right) e^{-2\alpha(z_t)} \right\}^{-1}, \quad (22)$$

where $f^{\pm}(E_0)$ are the distribution functions at the semiconductor and the metal, respectively.

When a bias voltage V is applied to the MIS structure, the energy band diagram shown in Fig. 1 is modified. The bias voltage has two contributions, i.e., V_I, the voltage change across the insulator and, V_S, the voltage change across the depletion layer of the semiconductor. However, we will assume that the voltage change across the insulator is negligible [22], and, consequently, all the voltage is applied to the semiconductor depletion layer. This fact causes the main dependence on bias voltage in (22) to appear in the difference of distribution functions.

Analysing (22) we observe that the resonant tunneling current due to one trap does not depend on the shape of the semiconductor conduction band at the depletion layer only its value at the insulator–semiconductor interface appears, i.e., the effective barrier height. Then, our results will be valid for any doping concentration of the semiconductor, provided the W.K.B. approximation is valid. The validity criterion of this approximation restricts the trap energy to be at least about 10 meV above the semiconductor conduction band at the interface. For lower energies the potential energy cannot be considered slowly varying.

We have obtained an analytical expression for the one-trap-assisted tunneling current; however, the fabrication process of thin insulating films leads to the existence of a large amount of impurities and defects. Calling $N_T(z)$ the trap density at a given depth, the current density will be the sum of the contributions of each trap,

$$j = \int\limits_{-d}^{0} N_T(z_t)\, I(z_t)\, dz_t\,. \quad (23)$$

The current due to a trap, which is given by (22), shows a sharp peak for the trap located at $z_t = \bar{z}$, solution of the equation

$$\frac{\varphi^-}{\{[E_0 - U_1(-d)]\,[U_2(-d) - E_0]\}^{1/2}} \left(\int_{-d}^{\bar{z}} \frac{d\zeta}{k_2(\zeta)}\right) e^{2\beta(\bar{z})} = \frac{\varphi^+}{\{[E_0 - U_3(0)]\,[U_2(0) - E_0]\}^{1/2}} \left(\int_{\bar{z}}^{0} \frac{d\zeta}{k_2(\zeta)}\right) e^{-2\alpha(\bar{z})} . \tag{24}$$

Neglecting the z_t-dependence on unimportant factors, we can write

$$j = N_T(\bar{z})\, I(\bar{z}) \int_{-d}^{0} dz_t \left\{\cosh 2 \int_{\bar{z}}^{z_t} d\zeta\, k_2(\zeta)\right\}^{-1} . \tag{25}$$

In order to obtain j we substitute the actual barrier by a mean rectangular one. This fact would not be justified in general, but on one hand, we are dealing with structures that are not too highly biased and on the other hand, the effective resonant tunneling barrier is much flatter than the actual one. So the approximation will be reasonable for a wide range of bias voltage in SiO_2–Si structures.

Then, $\bar{z}$ can be approximated by

$$\bar{z} = -\frac{d}{2}\left\{1 - \frac{1}{2}\left[1 + \frac{d(2mq\varphi_t')^{1/2}}{h}\right]^{-1} \ln \frac{\varphi^+}{\varphi^-}\left(\frac{\varphi^- - \varphi_t'}{\varphi^+ - \varphi_t'}\right)^{1/2}\right\}, \tag{26}$$

where $q\varphi_t'$ is the effective depth of the trap level. We conclude from this result that traps at positions far from both electrodes, i.e., near the middle of the insulator thickness give the main contribution to the resonant current. The same result has been obtained in MIM structures [23] and is in accordance with the two-step model [2].

Taking a Maxwell-Boltzmann distribution in both electrodes, and as $\bar{z} \approx -d/2$, we finally obtain the following expression for the resonant tunneling current density:

$$j = \frac{2hq\pi N_T}{md}\left(\frac{\varphi_t'}{\varphi^+\varphi^-}\right)^{1/2} [(\varphi^- - \varphi_t')\,(\varphi^+ - \varphi_t')]^{1/4} \exp\left[-\frac{1}{h}(2mq\varphi_t')^{1/2}\, d\right] \times \\ \times \exp\left[-\frac{q(\varphi_m' - \varphi_t)}{kT}\right]\left\{\exp\left(\frac{qV}{kT}\right) - 1\right\}, \tag{27}$$

where k is the Boltzmann constant and T the temperature, and other parameters are indicated in Fig. 1.

In this expression the typical diode current dependence on the applied bias can be recognized. However, it must be kept in mind that the saturation current varies slowly with V, because the barrier shape changes with voltage. Let us notice that the saturation current dependence on the oxide thickness (d), is mainly given by the exponential term which is characteristic of resonant tunneling. The direct tunneling current decreases with d faster than the resonant tunneling current. This fact causes that resonant tunneling can be the predominant contribution mechanism even when the oxide is so thick that direct tunneling is not observed. Besides, the energy trap level which contributes more efficiently to the resonant tunneling is that located at the silicon conduction band at the insulator–silicon interface, since the increase of distribution functions with trap depth is clearly uncompensated by the decrease of the transmission coefficient, because in the two-band model the effective trap depth is weakly dependent on trap depth.

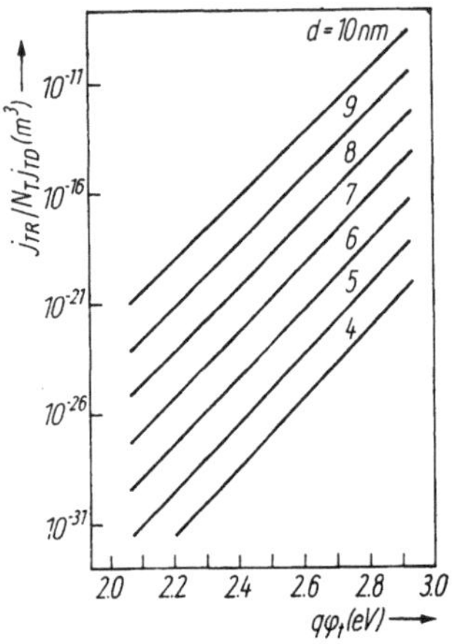

Fig. 2. $j_{\mathrm{TR}}/N_{\mathrm{T}}j_{\mathrm{TD}}$ vs. energy trap level ($q\varphi_t$) for different oxide thicknesses (d) in an Al–SiO_2–Si structure ($m^* = 0.42m_0$) at $T = 300$ K

One of the most interesting points is to compare the direct tunneling current through the insulator with the resonant current previously calculated. According to our model and taking into account the two-band model, the direct tunneling current in a MIS structure is given by [17]

$$j_{\mathrm{DT}} = A^* T^2 \exp\left[-\frac{2}{\hbar}(2mq\varphi^+)^{1/2}\,d\right]\exp\left[-\frac{q(\varphi_{\mathrm{m}}' - \chi)}{kT}\right]\left[\exp\left(\frac{qV}{kT}\right) - 1\right], \tag{28}$$

where A^* is the effective Richardson constant and χ the barrier height.

Note that the dependence on V is basically the same in both tunneling mechanisms, thus the analysis of the j–V curve for a given diode does not show clearly when the conduction is due to direct or to resonant tunneling. In this sense, a good fitting can be obtained with resonant tunneling considering an effective oxide thickness or trap density. If we cannot obtain accurate measurements of the oxide thickness, in order to distinguish the conduction mechanism, one must obtain the j–V characteristic for a wide range of temperatures, since, as can be seen from (27) and (28), the temperature dependence is slightly different in both mechanisms.

The relative importance of resonant and direct tunneling currents is indicated by the ratio $r = j_{\mathrm{TR}}/j_{\mathrm{TD}}$, which is straightforwardly obtained from (27) and (28),

$$r = M\frac{N_{\mathrm{T}}}{d}\exp\left[\frac{\varphi_{\mathrm{t}}}{kT} - P(\varphi_{\mathrm{t}}'^{1/2} - Q)\,d\right], \tag{29}$$

where M, P, and Q are almost independent of d and φ_{t}.

For a given MIS structure r depends on three parameters: N_{T}, d, and φ_{t}, since φ_{t}' is related to φ_{t} by (7). The numerical results for an Al–SiO_2–Si structure have been represented on Fig. 2, where one can see the importance of the resonant tunneling current with respect to the direct tunneling one; e.g. $N_{\mathrm{T}} = 10^{24}\ \mathrm{m^{-3}}$, $d = 5$ nm, $q\varphi_{\mathrm{t}} = 3$ eV, $j_{\mathrm{TR}} = 10^8 j_{\mathrm{TD}}$.

Besides, r increases exponentially as the trap depth or the oxide thickness increases.

From the above results we can conclude that a realistic study of MIS structures has to consider the trap-assisted current, since its contribution is several orders larger than the direct tunneling current.

3. Resonant Tunneling Current in a Schottky Structure

In this section the resonant tunneling current through the depletion layer of the semiconductor in an ideal metal–semiconductor junction will be studied. In order to do this, the same procedure as in Section 2.2 will be followed. However, in this case, the barrier has a smooth turning point, which causes some differences with respect to the MIS case.

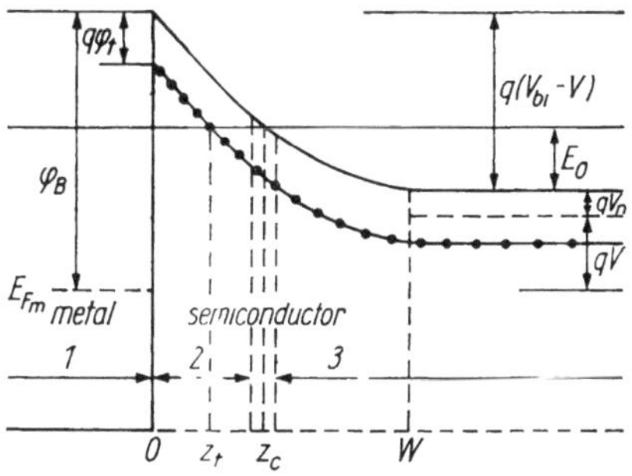

Fig. 3. Energy diagram for a Schottky structure with deep donor centres located at $q\varphi_t$

Henceforth, all energies will be referred to the bottom of the semiconductor conduction band at the bulk. The barrier potential energy $U(z)$ is given by (cf. Fig. 3)

$$U_1(z) = -q(V + V_n); \qquad z < 0,$$

$$U_2(z) = \frac{q^2 N_D}{2\varepsilon_S}(W - z)^2; \qquad 0 < z < z_t, \tag{30}$$

$$U_3(z) = \begin{cases} \dfrac{q^2 N_D}{2\varepsilon_S}(W - z)^2; & z_c < z < W, \\ 0; & z > W, \end{cases}$$

where N_D is the impurity doping concentration in the semiconductor, z the distance measured from the metal, ε_S the semiconductor permittivity, z_c the smooth turning point, and W the width of the semiconductor depletion layer,

$$W = \left[\frac{2\varepsilon_S}{qN_D}(V_{bi} - V)\right]^{1/2}, \tag{31}$$

where V_{bi} is the diffusion potential.

Since we are dealing with resonant tunneling due to one trap in the depletion layer, the one-dimensional Green's function must be obtained by solving (8) at $z' = z_t$, $0 < z_t < z_c$. Using the W.K.B. approximation and considering the trap far enough from both turning points, we obtain the following expressions for the retarded Green's function:

For $z < 0$

$$g_1(E, z, z_t) = A_1(E, z_t) \exp k_1(z) z, \tag{32}$$

for $0 < z < z_c$

$$g_2(E, z, z_t) = \frac{A_2(E, z_t)}{[\hbar k_2(z)]^{1/2}} \exp\left[\int_{z_c}^{z} d\zeta\, k_2(\zeta)\right] + \frac{B(E, z_t)}{[\hbar k_2(z)]^{1/2}} \times$$

$$\times \exp\left[-\int_{z_c}^{z} d\zeta\, k_2(\zeta)\right] - \frac{m}{\hbar^2}\frac{1}{[(k_2(z_t)\, k_2(z)]^{1/2}} \exp\left[-\left|\int_{z_t}^{z} k_2(\zeta)\, d\zeta\right|\right], \tag{33}$$

and for $z_c < z$

$$g_3(E, z, z_t) = \frac{A_3(E, z_t)}{[i\hbar k_3(z)]^{1/2}} \exp\left[-\int_{z_c}^{z} k_3(\zeta)\, d\zeta\right], \tag{34}$$

where

$$k_i(z) = \left[\frac{2m}{\hbar^2}\left(U_i(z) - E\right)\right]^{1/2}; \qquad i = 1, 2, 3\,, \tag{35}$$

and the coefficients A_i ($i = 1, 2, 3$) and B depend on the energy and the trap position. Besides, we must observe that, because of the W.K.B. approximation, $g_2(E, z, z_t)$ and $g_3(E, z, z_t)$ are only valid in points far from z_c.

In order to calculate the coefficients A_i and B, we have to connect the zones 2 and 3 at the left and right sides of the smooth turning point. We will remark that the connection formulas in the case of Green's functions are not the same as in the case of wave functions since these satisfy a homogeneous differential equation whereas Green's functions involve an inhomogeneous one due to a δ-Dirac term.

Taking into account this fact and following the same procedure as in the case of wave functions [24], we obtain the connection formula

$$A_3 = \frac{\sqrt{2}}{2}\left\{2A_2 - B + m[\hbar k_2(z_t)]^{-1/2} \exp\left[\int\limits_{z_c}^{z_t} d\zeta\ k_2(\zeta)\right]\right\}, \tag{36}$$

obtaining finally

$$g_1(E, z, z_t) \approx -\frac{2m}{\hbar^2}\frac{\left(k_2(0)/k_2(z_t)\right)^{1/2}}{k_1(z) + k_2(0)} \exp\left[-\int\limits_0^{z_t} d\zeta\ k_2(\zeta)\right] \exp\left(k_1(z)\, z\right), \tag{37}$$

$$g_3(E, z > W, z_t) \approx \sqrt{2}\frac{m}{\hbar^2} \times$$

$$\times \frac{[ik_2(z_t)\, k_3(W)]^{-1/2} \exp\left[\int_{z_c}^{z_t} d\zeta\ k_2(\zeta)\right] \exp\left(-k_3(W)\,(z - W)\right)}{(i + 1) \sin\left[i \int_{z_c}^{z_t} dz\, k_3(z)\right] + (i - 1) \cos\left[i \int_{z_c}^{z_t} dz\, k_3(z)\right]}, \tag{38}$$

$$g_2(E, z, z_t) \approx \frac{m}{\hbar^2}[k_2(z_t)\, k_2(z)]^{-1/2}\left\{\frac{-i}{2} \exp\left[\int_{z_c}^{z_t} k_2(\zeta)\, d\zeta\right] \times\right.$$

$$\times \exp\left[\int_{z_c}^{z} k_2(\zeta)\, d\zeta\right] + \frac{k_1(z) - k_2(0)}{k_1(z) + k_2(0)} \exp\left[\int_{z_t}^{0} k_2(\zeta)\, d\zeta + \int_{z}^{0} k_2(\zeta)\, d\zeta\right] -$$

$$\left. - \exp\left[-\left|\int_{z_t}^{z} k_2(\zeta)\, d\zeta\right|\right]\right\}. \tag{39}$$

With these results we can calculate the terms $j^{\pm}(E_0)$ defined in (2) which can be integrated in the same way as in the MIS case, obtaining

$$j^-(E_0) \approx \frac{2(2m)^{1/2}}{\hbar}\left[\frac{\left(E_0 - U_1\right)\left(U_2(0) - E_0\right)}{\left(U_2(0) - U_1\right)\left(U(z_t) - E_0\right)}\right]^{1/2} \times$$

$$\times \left[\int\limits_0^{z_t} m[2m\left(U(z) - E_0\right)]^{-1/2}\, dz\right]^{-1} \exp\left[-\frac{2}{\hbar}\int\limits_0^{z_t} [2m\left(U(z) - E_0\right)]^{1/2}\, dz\right]. \tag{40}$$

Analogously, we get $j^+(E_0)$ using (38)

$$j^+(E_0) \approx \frac{1}{2\hbar} [2m(U(z_t) - E_0)]^{-1/2} \left[\int_{z_t}^{z_c} [2m(U(z) - E_0)]^{-1/2} \, dz \right]^{-1} \times$$

$$\times \exp \left[\frac{-2}{\hbar} \int_{z_t}^{z_c} [2m(U(z) - E_0)]^{1/2} \, dz \right]. \tag{41}$$

On the other hand, the expression of $Pg(E_0, z_t, z_t)$ can be straightforwardly obtained from $g_2(E, z, z_t)$ and it is similar to the one obtained in (21) with the appropriate k_2.

Now we calculate the difference of the Boltzmann distribution functions at both electrodes for E_0. But in the case we are interested in, one assumes a deep level at $q\varphi_t$ below the conduction band edge with a trap concentration N_T (cf. Fig. 3), and the energy of a trap depends on its position z_t by the relationship

$$E_0(z_t) = U(z_t) - q\varphi_t \,. \tag{42}$$

Therefore, we can write this difference in terms of z_t as

$$F(z_t, V) \equiv f^+(E_0) - f^-(E_0)$$

$$= 2 \exp \left\{ - \frac{U(z_t) + qV_n}{kT} - \frac{qV}{2kT} \right\} \sinh \frac{qV}{2kT}. \tag{43}$$

Consequently, by substitution of (21), (40), (41), and (43), in (1), the current carried by a trap at certain energy level is

$$I(z_t) = I(\bar{z}_t) \, [\cosh 2 \int_{\bar{z}_t}^{z_t} k_2(z) \, dz]^{-1} \,, \tag{44}$$

where $\bar{z}_t$ is the position of the trap for which $I(z_t)$ is maximal, being its value given by

$$I(\bar{z}_t) = \frac{\hbar}{2} qF(z_t, V) \, [\int_{\bar{z}_t}^{z_c} m \, k_2^{-1}(z) \, dz]^{-1} \exp [-2 \int_{\bar{z}_t}^{z_c} k_2(z) \, dz] \,. \tag{45}$$

The optimal position of the trap at any energy can be obtained in the usual way,

$$\big(U_2(0) - U_1\big) \int_0^{\bar{z}_t} k_2^{-1}(z) \, dz \exp [-2 \int_{\bar{z}_t}^{z_c} k_2(z) \, dz] =$$

$$= 4[(U_2(0) - E) \, (E - U_1)]^{1/2} \int_{\bar{z}_t}^{z_c} k_2^{-1}(z) \, dz \exp [-2 \int_0^{\bar{z}_t} k_2(z) \, dz] \,. \tag{46}$$

This result is similar to the one obtained by Parker and Mead [2] and by Helman and Sánchez-Sinencio [3] considering a two-step process, but here the tunneling probability for each step is modulated by a coefficient.

Equation (46) is useful to obtain the trap position which provides a maximal contribution to the current. In fact, plotting (42) and (46) in a diagram, one obtains by intersection of them the resonant centre of the system (z_{tr}, E_{0r}), and the current carried by a trap at this point is given by (45) with $E_0 = E_{0r}$.

However, in order to calculate the current density through the junction we consider separately the contributions of all traps at the same depth $q\varphi_t$, with energies greater and less than E_{0r}, assuming that the integrals with limits z_t and z_c are constant. This approximation is analogous to the one carried out in the lineal barrier approximation performed in the literature [4].

For $E_0 > E_{0r}$ and from (45) we obtain

$$I_1(z_t) = \frac{\hbar}{m} q \left[\int_{z_t}^{z_c} k_2^{-1}(z)\, dz \right]^{-1} F(z_t, V) \exp\left[-2 \int_{z_t}^{z_c} k_2(z)\, dz \right] \quad (47)$$

which only depends on the term F.

The traps with $0 < E_0 < E_{0r}$ could contribute also to the current, but we have shown that its contribution is negligible in comparison with $I_1(z_t)$ when $N_D < 10^{24}\ m^{-3}$.

Considering that the trap density varies slowly enough, we can write for the total contribution to the current

$$j \approx N_T \int_0^{z_{tr}} I_1(z_t)\, dz_t = qN_T \left\{ \int_{z_{tr}}^{z_c} m \left[2m\left(U(z) - E_{0r}\right)\right]^{-1/2} dz \right\}^{-1} \times$$

$$\times \exp\left\{ \frac{-2}{\hbar} \int_{z_{tr}}^{z_c} \left[2m\left(U(z) - E_{0r}\right)\right]^{1/2} dz \right\} \int_{E_0(0)}^{E_{0r}} F(E_0, V) \frac{dz_t}{dE_0}\, dE_0 \,. \quad (48)$$

Solving the integral in energies by approximating the trap level by its tangent at $z_t = z_{tr}$, it follows:

$$\frac{j}{N_T} = \frac{qkT}{2} [2mq]^{-1/2} \left[\frac{\left(\frac{U_2(0)}{q} - \varphi_t\right)^{1/2}}{\left(\frac{U_2(0)^{1/2}}{q} - \varphi_t\right)^{1/2}} \right]^{\frac{2}{\hbar}\left(\frac{U_2(0)}{q} - \varphi_t\right)\left(\frac{m\varepsilon_S}{N_D}\right)^{1/2}} \times$$

$$\times \left(\frac{E_{0r}}{q} + \varphi_t\right)^{-1/2} \left[\ln \frac{\left(\frac{U_2(0)}{q} - \varphi_t\right)^{1/2}}{\left(\frac{U_2(0)}{q}\right)^{1/2} - \varphi_t^{1/2}} \right]^{-1} \exp\left(\frac{-qV_n}{kT}\right) \times$$

$$\times \left[\exp\left(-\frac{E_{0r}}{kT}\right) - \exp\left(-\frac{E_0(0)}{kT}\right) \right] \left[1 - \exp\left(\frac{-qV}{kT}\right) \right] \times$$

$$\times \exp\left[\frac{-2}{\hbar} \left(\frac{m\varepsilon_S}{N_D} \frac{U_2(0)}{q} \varphi_t \right)^{1/2} \right]. \quad (49)$$

This latter expression provides the value of the trap-assisted resonant tunneling current, and it will be used to study the j–V characteristics. But the dependence on V can be observed not only in $U_2(0)$ but also in E_{0r}. Therefore, one must consider an analytical expression of this dependence which can be obtained approximating the optimal position of the trap at any energy by a straight line, resulting in

$$E_{0r} = \frac{(b-1)^2 + 1}{2} U_2(0) - b \left[\frac{(b-2)^2}{4} U_2(0)^2 + q\varphi_t U_2(0) \right]^{1/2}, \quad (50)$$

where $b = 4.27$. This energy, when measured from the Fermi level of the metal remains practically constant with V if a reverse bias is applied.

In the forward-bias tunneling case, the resonant current expression (49) shows a maximum around the voltage for which E_{0r} vanishes, and as the voltage rises, the current decreases slowly as far as it goes quickly to zero in accordance with the fact that the number of traps contributing to the current decreases.

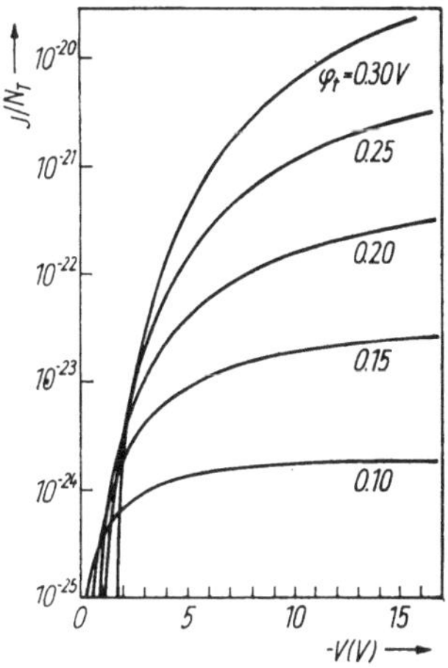

Fig. 4. Reverse j–V characteristics for different trap levels in an Al–Si Schottky diode ($q\varphi_B = 0.77$ eV, $T = 300$ K, $N_D = 10^{23}$ m^{-3}, $m^* = 0.16m_0$, and $\varepsilon_S = 11.8\varepsilon_0$)

For reverse bias, the results obtained are plotted in Fig. 4. The observed growth of the current with V for different trap levels can be explained since when V rises, the barrier narrows and the number of available states in the semiconductor increases.

Also we can observe that for lower bias voltages the resonant tunneling current decreases as $q\varphi_t$ increases. But current grows with the depth of traps as V is increased, and for higher voltages the opposite occurs.

4. Discussion and Results

We have analysed the resonant tunneling conduction through a MIS structure and through an ideal metal–semiconductor junction, within the framework of the stationary resonance model and using the W.K.B. approximation in the calculation of the Green's function.

In the case of the MIS structure, the influence of the two bands of the insulator in the tunnel transmission coefficient has been pointed out, showing that one must consider an effective trap depth in calculating the resonant tunneling current. A comparison between this current and the direct tunneling one has been performed. The analysis of the ratio between these currents shows that it depends linearly on the trap density and exponentially on the trap depth and the oxide thickness. Moreover, for the expected values of the trap density, our results show that, in usual MIS structures, resonant tunneling is the main conduction mechanism. This result is in accordance with the experimentally observed currents [11].

The analysis worked out in Schottky structures shows the existence of a current peak in a forward-biased junction. However, this peak only will be detected by the second derivative technique [25], since the thermionic current hides the erap-assisted contribution. When the Schottky barrier is reverse-biased the contribution of the resonant tunneling becomes determining in the j–V characteristics. These results are also in accordance with the excess current observed by some authors [2 to 7].

Analysing a metal–semiconductor junction one of the points of interest is to decide whether an oxide exists or not in the interface region. Usually the fabrication process implies the existence of an incontrollable, very thin layer which, in turn, has a large number of defects which create a trap density. The obtained analytical expressions (27) and (49) allow to analyse a given device as a Schottky or as a MIS structure. So, using our results for fitting experimental data one can decide about the properties of the interface region. Unfortunately, the resonant tunneling currents in metal–SiO_2–Si and metal–Si Schottky diodes, are comparable, so that this method becomes not decisive to characterize the insulator. In this case, the analysis of the interface parameters must be performed in a different way (C–V measurements, DLTS, optical response, etc.). However, as we have shown, the resonant tunneling contribution to the total current can prevail over other conduction mechanisms.

References

[1] S. M. SZE, Physics of Semiconductor Devices, John Wiley, New York 1981.
[2] G. H. PARKER and C. A. MEAD, Phys. Rev. **184**, 780 (1969).
[3] J. S. HELMAN and F. SANCHEZ-SINENCIO, Appl. Phys. Letters **28**, 34 (1976).
[4] G. SARRABAYROUSE, J. BUXÓ, and D. ESTEVE, J. Physique **38**, 1443 (1977).
[5] G. SARRABAYROUSE, J. BUXÓ, A. MYSZKOWSKI, and D. ESTEVE, Rev. Phys. appl. **12**, 433 (1977).
[6] A. N. KOROL, V. I. STRIKHA, and D. I. SHEKA, Soviet Phys. – Semicond. **14**, 698 (1980).
[7] J. MARTÍNEZ, E. CALLEJA, and J. PIQUERAS, phys. stat. sol. (a) **60**, 277 (1980).
[8] C. E. JONES and D. EMBREE, J. appl. Phys. **47**, 5365 (1976).
[9] D. J. DIMARIA, The Physics of the SiO_2 and Its Interfaces, Ed. S. T. PANTELIDES, Pergamon Press, New York 1968 (p. 160).
[10] F. CAMPABADAL, X. AYMERICH-HUMET, J. MILLÁN, and F. SERRA-MESTRES, 2nd Conf. Condensed Matter, EPS, ECA 6A, Manchester (G.B.) 1982, (p. 175).
[11] J. MILLÁN, F. SERRA-MESTRES, and X. AYMERICH-HUMET, Solid State Electronics **25**, 565 (1982).
[12] J. MILLÁN, F. SERRA-MESTRES, and X. AYMERICH-HUMET, 11th European Solid State Device Research Conf. ESSDERC 81, Toulouse (France) 1981.
[13] X. AYMERICH-HUMET and F. SERRA-MESTRES, phys. stat. sol. (a) **51**, 583 (1979).
[14] X. AYMERICH-HUMET and F. SERRA-MESTRES, phys. stat. sol. (a) **54**, 655 (1979).
[15] X. AYMERICH-HUMET, F. SERRA-MESTRES, and J. MILLÀN, phys. stat. sol. (a) **67**, 597 (1981).
[16] X. AYMERICH-HUMET and F. SERRA-MESTRES, Solid State Commun. **36**, 551 (1980).
[17] W. FRANZ, Hdb. Phys. , Vol. 17, Springer-Verlag 1956 (p. 155).
[18] H. C. CARD, Inst. Phys. Conf. Ser. **50**, 140 (1980).
[19] P. V. DRESSENDORFER and R. C. BARKER, Appl. Phys. Letters **36**, 933 (1980).
[20] R. WILLIAMS, J. Vacuum Sci. Technol. **14**, 5, 1106 (1977).
[21] S. E. D. HABIB and J. G. SIMMONS, Solid State Electronics **23**, 87 (1980).
[22] J. SHEWCHUN, A. WAXMAN, and G. WARFIELD, Solid State Electronics **10**, 1165 (1967).
[23] H. KNAUER, J. RICHTER, and P. SEIDEL, phys. stat. sol. (a) **44**, 303 (1977).
[24] P. PASCUAL and A. GALINDO, Mecanica Cuántica, Alhambra, Madrid 1978.
[25] E. CALLEJA and J. PIQUERAS, J. appl. Phys. **51**, 3980 (1980).

(Received April 8, 1983)

phys. stat. sol. (a) 79, 237 (1983)

Subject classification: 20.3; 13.4; 20.1; 22.5

Department of Physics, Post-Graduate College of Science, Osmania University, Saifabad[1] *(a)*
and Solid State and Materials Science Laboratories, Department of Physics, Osmania University, Hyderabad[2] *(b)*

Thermoluminescence of Thallium-Doped BaFCl

By

K. SOMAIAH (a) and V. HARI BABU (b)

The effect of doping of Tl upon thermoluminescence, its emission and optical absorption in flux grown BaFCl crystals is investigated. Besides an increase in luminescence output, X/γ-irradiations of BaFCl:Tl crystals produce three additional glow peaks at 450, 465, and 500 K and a TL emission spectrum at 460 nm. Optical absorption studies reveal two small bands at 340 and 650 nm. Room temperature annealing studies indicate that the three glow peaks at 450, 465, 500 K and the optical absorption band at 650 nm are more stable than others present in undoped crystals. Thus the additional glow peaks are attributed to the detrapping of electrons from the substantial number of metastable energy levels as populated by the thallium impurity and subsequent radiative recombination with holes. The trap depth and frequency factor are calculated employing various methods.

Der Einfluß einer Tl-Dotierung auf die Thermolumineszenz, die Emission und optische Absorption von aus dem Schmelzfluß gezogenen BaFCl-Kristallen wird untersucht. Neben einem Anwachsen des Lumineszenzsignals, ruft eine Röntgen/γ-Bestrahlung der BaFCl:Tl-Kristalle drei zusätzliche Glowmaxima bei 450, 465 und 500 K und ein Tl-Emissionsspektrum bei 460 nm hervor. Untersuchungen der optischen Absorption ergeben zwei kleine Banden bei 340 und 650 nm. Temperungsuntersuchungen bei Zimmertemperatur zeigen, daß die drei Glowmaxima bei 450, 465 und 500 K sowie die optische Absorptionsbande bei 650 nm stabiler sind als die anderen in undotierten Kristallen auftretenden Banden. Die zusätzlichen Glowmaxima werden deshalb der Befreiung von Elektronen aus einer beträchtlichen Zahl von metastabilen Energieniveaus, die mit Thalliumverunreinigungen besetzt sind und nachfolgender strahlender Rekombination mit Löchern zugeordnet. Die Traptiefe und der Frequenzfaktor werden mit verschiedenen Methoden berechnet.

1. Introduction

Recently there has been considerable interest in the defect properties of materials with fluorite structure [1 to 3]. Most phenomena observed in irradiated alkali halides have been properly related to the stabilization of radiation-induced anion Frenkel pairs. Traps, both electron and hole type, play a prominent role in the light emission phenomena. A study of the distribution of traps helps in understanding the luminescent properties of materials used in the fabrication of several optoelectronic devices like LED, electroluminescent panels. Experimental techniques like TSC (thermally stimulated current), photocapacitance decay, and thermoluminescence glow curve method are some of the conventional methods for studying the nature of traps. Thus thermoluminescence has gained enhanced importance within the last two decades because of its applications in diverse fields like archaelogy, geology, and dosimetry of nuclear radiations.

---

[1]) Saifabad, Hyderabad 500004, India.
[2]) Hyderabad 500007, India.

In our laboratory a comprehensive programme of study on the luminescence of PbFCl-type crystals has been undertaken with a view to understand the radiative recombination process associated with the thermoluminescence (TL) traps of these phosphors. In this paper the results of a study of thermoluminescence, its emission and optical absorption of BaFCl:Tl crystals irradiated at room temperature are presented, while these results on undoped BaFCl have been published earlier [4].

2. Experimental

Single crystals of BaFCl doped with thallium (2 mol%) were grown in our laboratory in air by flux technique. The details of the growth are given elsewhere [5]. Samples of the size 2 × 2 × 1 mm³ were irradiated either with X-rays from a tungsten target operated at 30 kV, 10 mA or γ-rays from a ^{60}Co source, the dose rate being 0.24 Mrd/h. The glow curves were recorded by heating the irradiated crystals at a constant rate of 25 K/min using an Indotherm TL-405 temperature programmer [4]. The optical absorption measurements were carried out by a Cary 14 double beam spectrophotometer. TL emission studies were made by using a 0.25 m Jarrel-Ash monochromator.

3. Results

3.1 Glow curves

In order to compare the thermoluminescence of undoped and doped crystals, crystals of equal dimensions were selected and glow curves were recorded under similar conditions. Fig. 1 shows thermoluminescence of undoped BaFCl and BaFCl doped with thallium impurity, the irradiation time being 10 min. From curves a and b it is observed that there are two thermoluminescence peaks situated at 345 and 365 K and the intensity in BaFCl:Tl crystals has enhanced by nearly 10 times that of undoped BaFCl crystal.

Glow curves recorded at higher amplifications showed the presence of additional shoulders. Hence to resolve these from the complex curve the thermal cleaning technique [6] has been employed at appropriate temperatures. Fig. 2 represents the glow

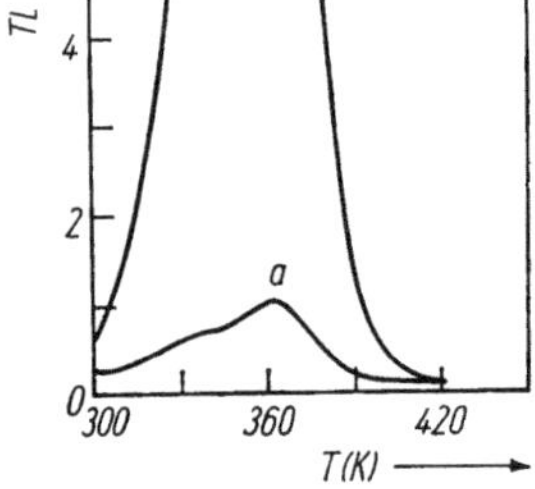

Fig. 1

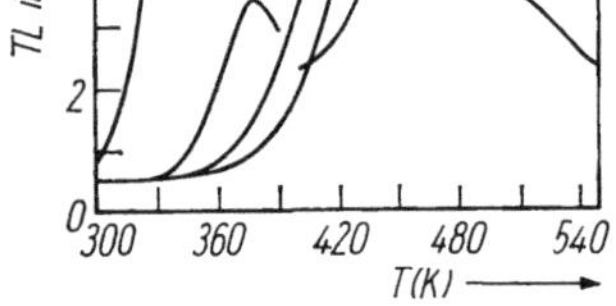

Fig. 2

Fig. 1. Glow curves of (a) BaFCl (pure) and (b) BaFCl:Tl (2%)

Fig. 2. Glow curves of BaFCl:Tl crystal resolved into component peaks by thermal cleaning technique

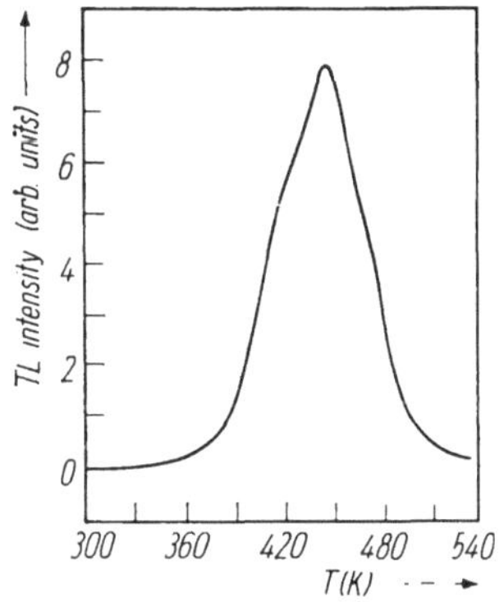

Fig. 3. Glow curves of γ-irradiated BaFCl:Tl annealed for 90 days

peaks of BaFCl:Tl obtained after thermal cleaning. From this figure, one can notice the presence of three additional peaks situated at 450, 465, and 500 K, respectively.

The stability of thermoluminescence peaks of BaFCl:Tl has been verified by room temperature annealing studies. Fig. 3 represents the glow curves of BaFCl which is γ-irradiated for 10 min and annealed for 90 days at room temperature. It consists of a peak at 445 K having shoulders at 410 and 465 K, on either side. Apart from these an underlying shoulder may also be noticed at about 500 K.

3.2 Thermoluminescence emission

On doping with thallium the intensity of the emitted light at RT has considerably enhanced. Fig. 4 represents the emission spectra of thallium-doped BaFCl at RT and higher temperatures. Curve a is recorded on a 10 min γ-irradiated sample immediately after irradiation at RT. It consists of a well-developed band centred at 410 nm (3.02 eV) and a shoulder at about 460 nm (2.7 eV). Emission curves recorded at 323 and 343 K are represented by curve b and c. They show that the intensity of the band 410 nm along with the shoulder rapidly increases with temperature. The intensity however decreases when the temperature is increased to 363 K (curve d). The existence of the shoulder at 460 nm can be seen in all the curves. On further increase of the temperature to 413 K (curve f) the 410 band decreases further, but the shoulder at 460 nm becomes prominent. Beyond 413 K the band becomes broader (curve f). A small shift towards the high-energy side may also be seen in the case of the 410 nm band.

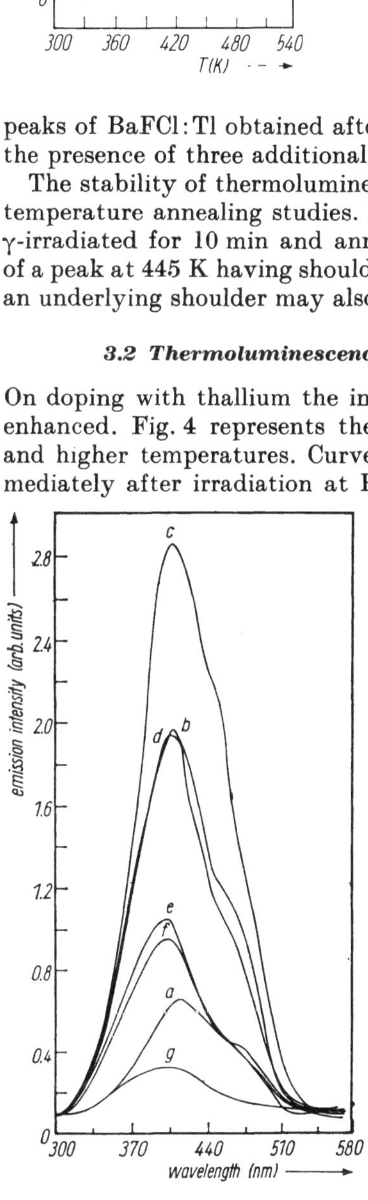

Fig. 4. Emission spectra of BaFCl:Tl (γ-irradiated) (a) at room temperature, (b) 323, (c) 343 (d) 363, (e) 393, (f) 413, (g) 523 K

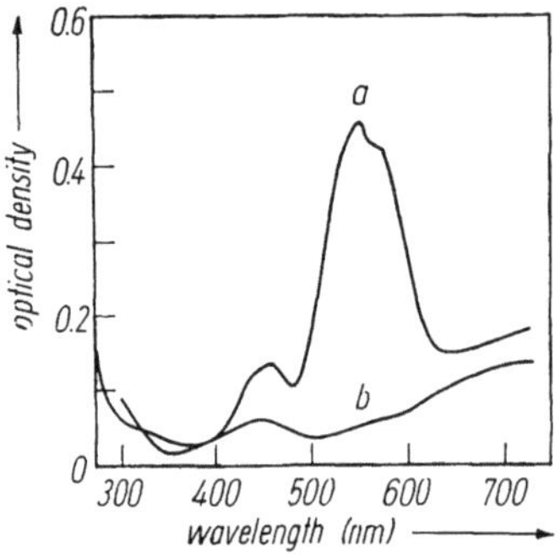

Fig. 5. Optical absorption spectra of γ-irradiated BaFCl:Tl (a) immediately after irradiation, (b) annealed for 90 days

3.3 Optical absorption

In Fig. 5 curve a shows the optical absorption spectra of 20 h γ-irradiated BaFCl:Tl immediately after irradiation, while curve b has been recorded after 90 days. Curve a consists of a small bulging at 340 nm, a small broad band at about 455 nm and a well-split double band at 550 and 565 nm. On annealing for 90 days at room temperature, the bands at 550 and 340 nm are completely eliminated, but the other band at 440 nm persists with decreased intensity. Diffused bulging beyond 650 nm can also be seen in both curves.

3.4 Analysis of the glow curves

In the present investigation the geometrical factor for the glow curves is found to be less than 0.52 and first-order kinetics is therefore employed. The trap depth and frequency factor are calculated using different equations and are listed in Table 1.

Table 1

Calculated trap parameters (E, p_0) using different methods for BaFCl:Tl

peak position (K)	E (eV), frequency factor p_0 (s^{-1})							
	a)		b)		c)		d)	
	E	$10^{-9}\,p_0$	E	$10^{-6}p_0$	E	$10^{-6}p_0$	E	$10^{-9}p_0$
345	0.75	2.5	0.5	0.4	—	—	0.65	2.4
370	0.81	2.3	0.62	5.4	—	—	1.27	2.1
410	0.87	2.2	0.74	4.4	—	—	—	
450	0.97	2.0	0.73	3.3	0.7	1.2	—	
465	1.00	2.0	0.78	7.2	0.69	0.5	—	
500	1.07	1.6	—		—		—	

a) Randall-Wilkin's method ($E = 25kT$).
b) Halperin and Braner's method.
c) Randall-Wilkins method (numerical curve fitting).
d) Different heating rates method.

4. Discussion

In our earlier paper [4] on TL of BaFCl we have shown that the glow peaks occur around 345, 365, 375, and 415 K. The last two peaks at 375 and 415 K which appeared at higher amplifications were attributed to the two types of F-centres, i.e. F($\overline{\text{Cl}}$) and

F($\bar{F}$), respectively. In the present investigation three additional glow peaks at 450, 465, and 500 K were observed on incorporation of thallium impurity. Optical absorption results revealed two additional bands at 340 and 650 nm while the other bands at 455 and 550 nm were also present in undoped BaFCl [4]. Thermoluminescence emission results have also shown an additional peak at 460 nm as the other one at 410 was present in undoped BaFCl [7]. The presence of additional glow curves, optical absorption band, and thermoluminescence emission peaks have been explained on the following lines.

BaFCl has the same layered tetragonal structure as PbFCl [8]. Incorporation of thallium impurity monovalent cations on barium ion sites requires for charge compensation either halide ion vacancies or interstitial barium ions. Charge compensation by interstitial barium ions most probably is not possible because of the small interstitial space available. Halide vacancies most probably are formed because of doping with monovalent cations. This was also confirmed [3] from ionic conductivity studies where the magnitude of the ionic conductivity was observed to increase with the increase of thallium impurity. Similarly it is expected to produce greater changes in the TL glow curve pattern of thallium-doped BaFCl as compared to undoped BaFCl.

A considerable amount of work has been reported on the optical properties of alkali halides containing thallium ion impurity [9, 10]. Four absorption bands called A(4.38 eV), B(5.06 eV), C(5.30 eV), and D(5.50 eV) are formed in the doped crystals on the longer-wavelength side of the intrinsic excitonic absorption. In the present investigation also small bulding at about 340 and 650 nm has been observed. The increase in thermoluminescence intensity output and the additional emission peak as seen from Fig. 1 and 4, confirm the presence and activation of thallium impurity in the lattice. Similar increase of TL output has been reported by earlier workers [11, 12] in NaCl due to thallium doping. This has been attributed to the presence of a substantial number of metastable energy levels as populated by the impurity thereby increasing the trap density. Thus the additional TL peaks which appeared at 450, 465, and 500 K may perhaps be due to the detrapping of electrons from the impurity centres and subsequent radiative recombination with holes.

Thermoluminescence emission has shown (Fig. 4) only one additional emission band at 460 nm at various temperatures. It suggests that the emission band is common for all the glow peaks as the recombination centres are the same for all glow peaks. However, this attribution needs further confirmation through low-temperature optical absorption which is in progress.

Acknowledgements

One of us (K.S) thanks the University Grants Commission, New Delhi, for the financial assistance through a minor research project to undertake the present investigation. The authors wish to thank Prof. Y. V. G. S. Murti, Department of Physics, I.I.T., Madras for helpful discussions and the Head of the Department of Physics, Osmania University for providing experimental facilities.

References

[1] A. F. Halff and J. Schoonman, phys. stat. sol. (a) **40**, 511 (1977).
[2] E. Nicklaus and F. Fischer, phys. stat. sol. (b) **52**, 453 (1972).
[3] K. Somaiah and V. Hari Babu, phys. stat. sol. (b) **117**, 75 (1983).
[4] K. Somaiah, P. Veeresham, K. L. N. Prasad, and V. Hari Babu, phys. stat. sol. (a) **56**, 737 (1979).
[5] K. Somaiah and V. Hari Babu, Indian J. pure appl. Phys. **14**, 702 (1976).

[6] N. M. GUPTA, J. M. LUTHRA, and J. SHANKAR, Indian J. pure appl. Phys. **11**, 684 (1973).
[7] K. SOMAIAH and V. HARI BABU, Radiat. Eff. Letters, to be published 1983.
[8] G. P. M. VAN DEN HEUVEL, G. BLASSE, and J. SCHOONMAN, Solid State Commun. **28**, 689 (1978).
[9] F. SEITZ, J. chem. Phys. **6**, 150 (1938).
[10] A. FUKUDA, Sci. Light (Japan) **13**, 64 (1964).
[11] R. V. JOSHI and T. R. JOSHI, Japan. J. appl. Phys. **10**, 1921 (1971).
[12] R. V. JOSHI and T. R. JOSHI, Z. Naturf. **27a**, 12 (1972).

(Received May 25, 1983)

phys. stat. sol. (a) 79, 243 (1983)

Subject classification: 9; 22.5.2

Department of Applied Physics, University of New South Wales, Kensington[1])

Diffusion of Water in NaCl Bicrystals

By

L. B. HARRIS and S. L. BEWLAY

Measurement of the variation in the light scattered from precipitates produced by moisture diffusing along a grain boundary is developed into a method for determining the grain boundary diffusion coefficient D of the water. Suitable treatment of data from nominally pure bicrystals enables an approximate value for D at room temperature to be obtained by two independent methods.

Die Messung der Änderung der Lichtstreuung von Präzipitaten, die durch Diffusion von Feuchtigkeit entlang von Korngrenzen erzeugt werden, wird zu einer Methode zur Bestimmung des Korngrenzen-Diffusionskoeffizienten D von Wasser weiterentwickelt. Die geeignete Auswertung der Daten von nominell reinen Bikristallen ermöglicht die näherungsweise Bestimmung des Wertes von D bei Zimmertemperatur nach zwei verschiedenen Methoden.

1. Introduction

Moisture in the atmosphere forms an adsorbed layer on the surface of a sodium chloride crystal, but as long as atmospheric humidity is less than the water vapour concentration above a saturated sodium chloride solution the adsorbed layer remains stable and produces no immediate change in bulk properties [1]. It is still possible, however, for deposited moisture to diffuse into the bulk of the crystal and produce eventual change [2, 3]. Diffusion of water into an alkali halide crystal, which has been directly measured for potassium halides [4 to 6] but not for sodium chloride, is materially affected by divalent cation impurities in the crystal. It has been found, for example, that the low-temperature solubility of water in KCl is increased by the presence of calcium [4], and that addition of calcium to NaCl crystals exposed to moisture produces recrystallisation on the surface and precipitate growth within the bulk [7]. Such changes merit further consideration.

Formation of second-phase particles by penetrating water has possible use as an indicator for estimating the diffussion coefficient of the water. Direct observation is not appropriate, however, since most of the precipitates remain below the limit of resolution of the optical microscope, and a more sensitive technique such as light scattering is required. The problem with light scattering is that it is difficult to eliminate extraneous background illumination. In particular, when observation is made of the region immediately below the exposed surface, where the effects of diffusion are greatest, excessive light is scattered directly from the surface, while regions free of surface scatter are usually beyond the range of bulk diffusion. Preferential transport along grain boundaries, however, can carry water much deeper into the crystal to where accurate measurement is possible. The present work reports on measurements that relate light scattering to grain boundary diffusion of water in NaCl at room temperature, at the same time providing a simple theoretical framework for a quantitative evaluation of results.

---

[1]) P. O. Box 1, Kensington, N.S.W. 2033, Australia.

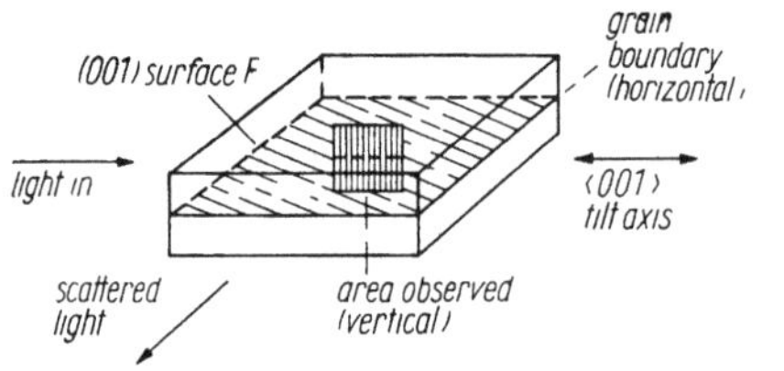

Fig. 1. Specimen configuration. Area in centre observed by repeated vertical traverses across specimen

2. Procedure

Bicrystal specimens were taken from ingots pulled from a melt of Merck analytical grade salt using a split seed under high purity nitrogen at one-third atmospheric pressure. Small samples (≈ 0.5 cm^3) with a 20° tilt boundary bisecting the sample were cleaved and sawn to size, given polished faces where necessary, and then mounted for observation. Light from a 5 mW He–Ne laser that was scattered through an angle of 90° was collected by a pinhole aperture at the surface of a microscope objective turret, amplified, counted by a single channel analyser, and displayed digitally. A raster scan of the laser beam, effected by screw threads driven by stepper motor under microprocessor control, enabled sample count rates to be collected over an area normal to the grain boundary and perpendicular to the direction of observation, as shown in Fig. 1. A two-dimensional scattered intensity profile could then be assembled from the data obtained. It had been established, during cutting of bicrystal ingots by wet-string saw [8], that water penetrated more rapidly along, than at right angles to, the tilt axis, and hence the specimen was oriented (Fig. 1) so that moisture penetrated from the (001) face along the [001] tilt axis direction.

Nominally pure specimens that contained approximately 10^{-5} mol fraction of calcium impurity were used, and these were taken from ingots in which impurity segregation was inhibited by relatively rapid cooling inside the switched-off growth furnace. Thus, grain boundary scattering due to effects other than moisture could be kept to a minimum, though a slow aggregation of residual impurity at room temperature over many months was found to increase background scatter. Measurements were made of the changes produced by storage in air containing water vapour at a partial pressure of approximately 300 Pa (humid air), a level that can be maintained without difficulty inside a desiccator.

3. Observations

Storage in humid air increased the scattering-centre concentration near the grain boundary and produced the hump shown in Fig. 2, while subsequent storage in dry air reduced this hump to the modest proportions shown in Fig. 3. Looking at the cattered light by optical microscope revealed ultramicroscopic scattering centres

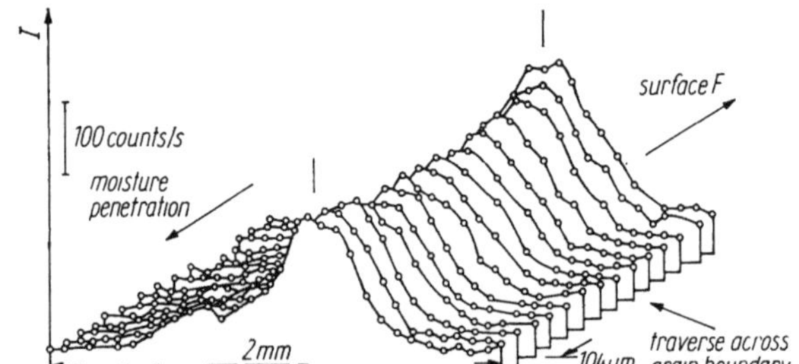

Fig. 2. Scattered intensity profile after exposure to humid conditions. Horizontal plane corresponds to area observed in Fig. 1

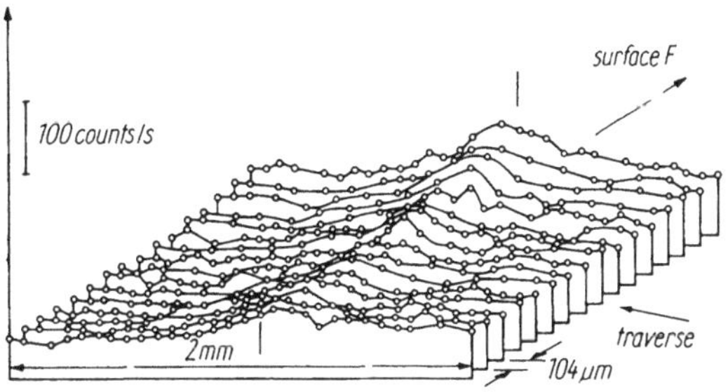

Fig. 3. Scattered intensity profile on specimen which produced Fig. 2, obtained three months later

in greater density near the grain boundary than elsewhere [9], as would be expected for scattering from precipitates. Storage in humid air of single crystals, by contrast, produced no observable change, as exemplified by the scattered intensity profile of Fig. 4, taken at a penetration depth comparable with that in Fig. 2 and 3, which remained unchanged throughout three months' exposure to moisture. Even just below the surface of single crystals, where there is spurious extra scatter from the surface, the measured scattered intensity was less than that observed at large penetrations in bicrystals. Therefore, under the conditions of the present experiments, diffusion in the bulk was negligible compared with that along grain boundaries.

Hump width was large (≈ 500 μm), due mainly to the coarseness of the light probe. Collimating apertures restricted the diameter of the laser beam to 0.5 mm, and though the receiving pinhole had a smaller diameter, it was able to collect light from a range of angles about the main scatter angle of 90°. Hence, measured hump width was larger than the spread of scattering sites about the boundary. Part of the width may also be attributed to segregation since, even before exposure to moisture, ultramicroscopic particles could be observed up to 50 μm from the boundary [9]. It was significant, however, that within experimental error, there was no variation in hump width with the observed changes in hump height, and hence the region of preferential diffusion can be considered, at least for that part of the crystal within which observations were made, as a parallel-sided slab from which lateral diffusion into the bulk was negligible. This is an approximation, but it is reasonable in view of the large penetration depth at which observations were made [10], and it accords with the observation, stated above, that diffusion in the bulk was slower than in the boundary.

An apparent anomaly was that scattered light intensity varied in the bulk of bicrystal specimens in the same way as light scattered from the boundary, being higher for vertical traverses closer to surface F, an observation which contradicts evidence from Fig. 3 and elsewhere that moisture does not penetrate the bulk. This variation was, in fact, found to be due to the spurious scatter emanating from a surface whenever an incident beam was directed parallel and close to that surface. The surface in this

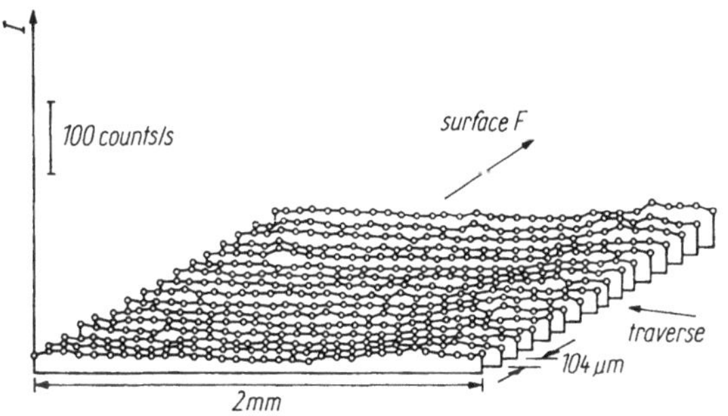

Fig. 4. Scattered intensity profile from single crystal exposed to same humid conditions as bicrystal of Fig. 2, observed at same penetration below surface F

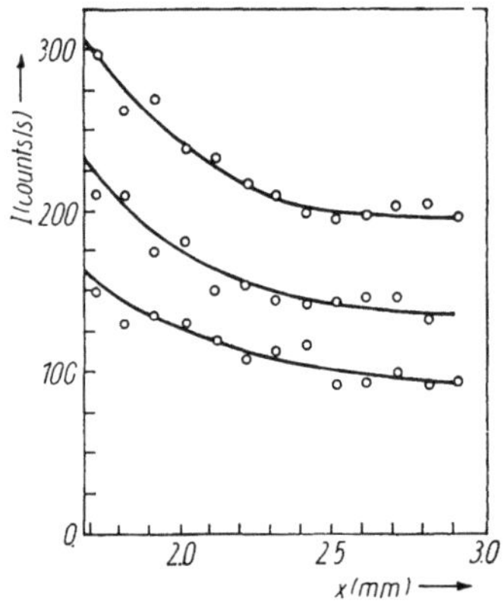

Fig. 5. Scattered intensity obtained at centre of grain of grain boundary hump at weekly intervals on same specimen after exposuret to humid air

case was the grain boundary containing precipitate, and since scatter is greater where precipitate concentration is higher, a variation with distance below the surface is to be expected. This is an example of how crystalline defects contribute to background scatter, another instance being the ragged background produced by the deteriorating crystal quality that resulted from continued use of a specimen (c.f. Fig. 2 and 3).

Quantitative assessment of the diffusion process was best obtained by study of the variation in scattered intensity along the tilt-axis direction at the centre of the hump, as shown in Fig. 5 by curves taken after successive exposures to humid air. Such curves can be related to diffusion along the tilt axis, assuming diffusing water is incorporated into the particles that scatter light, by consideration of the physical mechanisms involved.

4. Theory

The reaction between water and impurity that forms precipitate is similar to the reaction between water and indicator (impurity or F centre) that results in the migration of a plane of decolouration into the crystal [4 to 6]. This plane is sharply defined when the reaction between water and indicator takes place entirely at this plane, and in this case the diffusion coefficient of water can be obtained from the equation [6]

$$D = \frac{1}{2c_0} \frac{x_p^2}{t_p} \left(\frac{n_i}{l} + \frac{c_0}{2}\right), \tag{1}$$

where x_p is the penetration of the reaction plane after time t_p, n_i the concentration of indicator, and l the number of indicator centres reacting with one water molecule. c_0 is the concentration of water molecules in the surface at $x = 0$, on the assumption that a steady state has been reached. When $c_0 \ll n_i$ the second term in the brackets may be neglected [5], and when $c_0 \gg n_i$ equation (1) simplifies to

$$D = \frac{x_p^2}{4t_p} \tag{2}$$

which expresses the fact that the penetration depth x_p of the reaction plane equals the diffusion length $2(Dt_p)^{1/2}$.

There is no sharply defined reaction plane in Fig. 5, which means that precipitation occurs over a wide region of crystal, rather than being confined to a single plane, and hence the precipitation reaction needs different theory. The growth in scattered intensity with exposure to moisture comes about because diffusing water is trapped at scattering sites, this immobilised fraction of diffusant being the cause of the observed intensity distribution. Consider one-dimensional diffusion along the grain boundary slab, and let dJ' be the decrease in the flux of water molecules produced by traps over distance dx, where dJ' is proportional to total flux J and to dx. Then

$$\mathrm{d}J' = -n\sigma J \,\mathrm{d}x\,,$$

where n is trap concentration and σ is trap cross section. Hence

$$-\frac{\partial J'}{\partial x} = \frac{1}{k+1}\frac{\partial p}{\partial t} = n\sigma J = -n\sigma D\frac{\partial c}{\partial x}, \tag{3}$$

where $-\partial J'/\partial x$ is the rate of loss of water concentration due to traps, $\partial p/\partial t$ the rate of gain of precipitate concentration, i.e., the amount of precipitate formed in unit volume in unit time, c the concentration of unprecipitated water at any section x along the boundary slab, and D the boundary diffusion coefficient of water. The quantity $1/k$ is the ratio within the precipitate of the concentration of aqueous to non-aqueous constituents. Combining (3) with the continuity equation

$$\frac{\partial c}{\partial t} + \frac{1}{k+1}\frac{\partial p}{\partial t} = -\frac{\partial J}{\partial x} = D\frac{\partial^2 c}{\partial x^2}, \tag{4}$$

where $\partial J/\partial x$ is the divergence of flux due both to traps and to the concentration gradient, yields a differential equation for the concentration c of water,

$$D\frac{\partial^2 c}{\partial x^2} + n\sigma D\frac{\partial c}{\partial x} = \frac{\partial c}{\partial t}. \tag{5}$$

General solution of the above equations requires more information than is provided by the present experiments, but if steady state only is considered, a straightforward comparison may be made with the conditions represented by (1). Steady state occurs when $\partial c/\partial t = 0$ and hence, since c is unprecipitated water, all diffusing water is absorbed into second phase, equations (3) and (4) become identical and (5) is redundant. A sharply defined reaction edge is produced when absorption takes place in a narrow layer of crystal [4 to 6], but when each plane normal to diffusion absorbs only a fraction of the water passing through it, the layer of crystal needed for complete absorption is large. The sharply defined reaction plane is thus spread out into a "diffuse layer" with a width of the order of the mean free path, λ, of water molecules diffusing between traps, where $\lambda = 1/n\sigma$. This is illustrated in Fig. 6.

Water concentration c, which is zero at the reaction plane in Fig. 6a, has a finite but decreasing value across the diffuse layer in Fig. 6b. It is useful to assume that this decrease is exponential, in accordance with (3) when $n\sigma$ is constant, though this is only a rough approximation to what happens in practice. When the variation of precipitate concentration p across the diffuse layer is similar to that of c, as would be expected of (3) under steady state conditions, the diffusion coefficient D of water

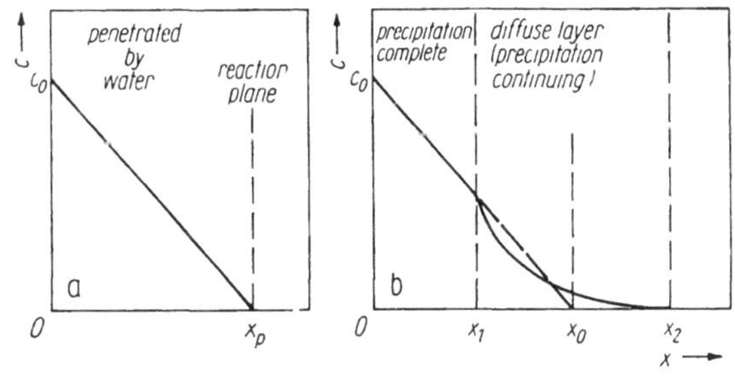

Fig. 6. Water concentration c as a function of penetration x when reaction between water and impurity takes place a) at plane $x = x_p$, b) over diffuse layer $x_1 \leq x \leq x_2$

equals the apparent diffusion D_p defined in terms of the slope and curvature of the precipitate distribution,

$$D_p = \frac{\partial p}{\partial t} \bigg/ \frac{\partial^2 p}{\partial x^2}. \tag{6}$$

In fact, insertion of (3) in (6) gives

$$D_p = \frac{\lambda^2}{T} = \bar{v}\lambda ,$$

where $\bar{v}$ is average speed of a particle diffusing with mean free path, λ, and T is diffusion time, which is the simple equation obtained for D of the particles actually diffusing [11], apart from a numerical factor $\frac{1}{3}$ resulting from use of three dimension in place of one.

Values of p are not directly available from light-scattering measurements, but application of Rayleigh-Gans theory [12] to the experimental conditions used in this work shows that scattered light intensity I is proportional to p^2, and hence that

$$D_p = \frac{\partial I}{\partial t} \bigg/ \left\{\frac{\partial^2 I}{\partial x^2} - \frac{1}{2I}\left(\frac{\partial I}{\partial x}\right)^2\right\}. \tag{7}$$

This equation provides an approximate determination of D, assuming $D_p = D$, for those experimental conditions in which penetration depth x_p in (1) is not precisely defined.

5. Analysis

Values of $\partial p/\partial t$ obtained from curves of scattered intensity I were fitted to (3) and (4) to check the conditions under which the experiments were carried out. This required a value for $n\sigma$. Assuming precipitation is caused by hydration of 10^{-5} mol fraction of calcium impurity [7], n may be taken as the average concentration of (unaggregated) calcium ions and σ as the cross section of a calcium chloride molecule to give $n\sigma = 10^4\ \text{m}^{-1}$. This value is too large because, in practice, light-scattering by a non-integrating detector, such as a photomultiplier, requires scattering centres that are aggregates of many impurity ions [13]. A more appropriate value of $n\sigma$, obtained from electron microscope observations on grain boundary aggregates approximately 100 nm in diameter [14], is $10^3\ \text{m}^{-1}$. A distribution of even larger particles (1 μm diameter) produced by moisture seeping into the grain boundary has been observed by optical microscope [15], giving $n\sigma = 10^2\ \text{m}^{-1}$, but this value is too small because it ignores small particles below the limit of optical visibility. The value of $10^3\ \text{m}^{-1}$ was acceptable because it gives a mean free path $\lambda = 1/n\sigma$ of 1 mm, which agrees with the magnitude of diffuse layer thickness inferred from Fig. 5. The value of $1/(1 + k)$ for hydration of calcium chloride was taken as approximately unity.

The above values could be fitted to (3) and (4) only when $\partial c/\partial t$ was very small, which indicates that precipitate was formed under quasi-steady state conditions and that Fig. 6b applies. The initial linear slope in Fig. 6b may be extended down to $c = 0$, giving the average penetration depth x_0 of the diffuse layer and x_0 may then be inserted in (1) to obtain an approximate value of D. In doing this, k may be used in place of l, but n_i must be taken as the impurity concentration and not as the number of precipitates per unit volume. For cation content of 10^{-5} mol fraction and external water partial pressure of 300 Pa it was found that c_0 and n_i had similar orders of magnitude (10^{-22} to $10^{-23}\ \text{m}^{-3}$), which means that (1) becomes

$$D = \frac{Ax_0^2}{T},$$

where T is diffusion time and A a numerical factor not greatly different from unity. x_0 cannot be determined precisely in Fig. 5, but may be taken as the depth at which scattered intensity increases significantly, say 2 mm. Taking $A = 1$ and T as one week gives a grain boundary diffusion coefficient D of 10^{-12} m^2 s^{-1}.

The above value may be compared with the magnitude of D, given by (7), derived from the shape of the intensity distribution across the diffuse layer. The value of D_p obtained from the two separate time intervals between the curves of Fig. 5 are $2 \times \times 10^{-12}$ and 7×10^{-13} m^2 s^{-1}, the lower value being obtained after the longer diffusion time, possibly as a consequence of the variation in σ and of the approach to saturation in the precipitation level. The main point of interest is that the same order of magnitude of D is obtained by the two independent methods used in this work. There is thus justification for asserting that the grain boundary diffusion coefficient of water at room temperature in the NaCl bicrystals used here is approximately 10^{-12} m^2 s^{-1}.

6. Discussion

The process that enhances light scattering can be seen directly on freshly cleaved bicrystal surfaces in the form of an increase in the size of deliquescent calcium chloride precipitates as they pick up water. Such precipitates are observed to be unevenly spread over the boundary, and therefore the irregularities seen in the profiles of Fig. 2, 3, and 5 represent genuine variation in scattered light output. This output is derived primarily from large particles (dimensions greater than 25 nm) since weak scatter from smaller particles is below detection limits. Average particle size, for the impurity content of the specimens used, is unlikely to exceed 100 nm [13], which means that the change in output corresponds to only a modest change in $n\sigma$, and that use of a fixed value of $n\sigma$ in calculations (Section 5) will not introduce gross error. Note that, since scattered intensity I is proportional to the square of the volume of a scattering particle [12], the relative change in light output is larger than relative change in particle size. A reduction in scattering level following exposure to dry air (Fig. 3) requires a reduction in precipitate size, which is possible when efflorescence occurs. Loss of water of crystallisation is to be expected since the range of water vapour pressure over which the experiments were operated, up to 300 Pa, extends well into the hydration–dehydration range of calcium chloride hydrates [17].

Absorption of water by precipitates increases aqueous solubility [4], but does not necessarily enhance diffusion. Boundary precipitates tend to form at compositional inhomogeneities, introduced during bicrystal growth, that develop into a network of defect paths that connect the interior of the boundary to the surface [16]. Diffusant finds its way into the bicrystal along these paths and, equally, water lost by dehydration uses the same pathways out. This means that only a fraction of overall grain boundary volume contributes to preferential diffusion, wheres diffusion coefficient D is defined in terms of total grain boundary volume, and hence high rates of diffusion in the network convert to only a modest enhancement of D over lattice diffusion coefficient, D_l. Comparing the present value of D with an earlier rough estimate of D_l as about 10^{-14} m^2 s^{-1} [7] suggests an enhancement of about 100,which is consistent with the modest enhancement obtained for self diffusion by tracer measurement [9]. Note that the low saturation vapour pressure of deliquescent precipitates [17] provides an explanation for the preferential dissolution of grain boundaries that is often observed [18].

Grain boundary volume is determined by the extent to which the impurity substructure spreads out from the lattice mismatch surface that defines the location of the boundary. As a result of deviations and irregularities this spread may be many µm [9, 16], and diffusion is thus related to a grain boundary width of similar extent [9].

A scattered light determination is a measurement that integrates over a substantial portion of the boundary, and for this reason the scattered light profile, though it emanates from a network of precipitates, appears to come from a slab of finite thickness.

Equation (7) enables an estimate to be made of D, which might not otherwise be possible, but the level of accuracy is low on account of the assumptions used to obtain the equation (Section 4) and of the inherent difficulty in obtaining reliable values of I. A further drawback is that the method works for only a limited range of specimen purity. When specimens are pure (i.e., $n\sigma$ small) precipitation takes place in the boundary under non-steady state conditions, and when specimens are doped ($n\sigma$ large) the diffuse layer in Fig. 6b is too narrow for (7) to be applied. These shortcomings suggest that a suitable procedure for obtaining better values of D is to use equation (1) and the more precise measurement of x_0 that is possible with doped specimens. There are, however, additional requirements in the form of reliable values for c_0 and k and, in particular, for a value of n_1 that takes account of segregation. Beyond this, it would also be desirable to investigate how well changes in the precipitation process that take place as the transition is made from pure to doped specimens can be fitted to the equations in Section 4. Experiments are, in fact, planned on calcium-doped specimens to check the variation of the saturation level of precipitation with the width of the diffuse layer. It is expected that the grain boundary diffusion coefficient will be structure sensitive, and that variations will be observed with change in bicrystal growth conditions and in specimen heat treatment.

The value for D of 10^{-12} m^2 s^{-1} obtained in this work, though large, is consistent with the high diffusion rates of water that have been measured in alkali halide single crystals [4 to 6] and with the rapid dissolution rates observed along grain boundaries in NaCl [18]. The observed value is still orders of magnitude below the upper limit at which diffusion processes cease to be described in terms of mechanisms involving the jumping of defect species between lattice sites [19].

References

[1] M. Hucher, A. Oberlin, and R. Hocart, Bull. Soc. Franc. Minér. Crist. **90**, 320 (1967).
[2] J. H. Beaumont and P. W. M. Jacobs, phys. stat. sol. **17**, K45 (1966).
[3] K. Kamiyoshi and Y. Nigara, phys. stat. sol. (a) **3**, 735 (1970).
[4] C. Rühenbeck, Z. Phys. **207**, 446 (1967).
[5] J. C. Penley and R. S. Witte, J. chem. Phys. **40**, 1550 (1964).
[6] T. J. Neubert and D. L. Roskelley, J. Phys. Chem. Solids **34**, 493 (1973).
[7] L. B. Harris and J. Fiasson, J. Mater. Sci **15**, 1456 (1980).
[8] L. B. Harris, J. Phys. E **2**, 432 (1969).
[9] L. B. Harris, J. Physique **37**, C7-365 (1976).
[10] A. E. le Claire, Brit. J. appl. Phys. **14**, 351 (1963).
[11] J. Jeans, Kinetic Theory of Gases, Cambridge University Press, 1952 (p. 202).
[12] H. C. van de Hulst, Light Scattering by Small Particles, John Wiley, 1957 (p. 85).
[13] M. Nottin, Acta cryst. **A26**, 636 (1970).
[14] L. B. Harris and N. G. Cutmore, Surface Sci. **65**, 1 (1977).
[15] L. B. Harris, V. R. Howes, and N. G. Cutmore, J. Amer. Ceram. Soc. **65**, 35 (1982).
[16] L. B. Harris, J. Crystal Growth **35**, 249 (1976).
[17] Gmelin's Handbuch der Anorganischen Chemie, 8th ed., No. 28, Ca Part BII, Ed. E. H. E. Pietsch, Verlag Chemie, GmbH., Weinheim 1957 (p. 461).
[18] N. Fuschillo, M. L. Gimpl, and A. D. McMaster, J. appl. Phys. **37**, 2044 (1966).
[19] H. Rickert in: Fast Ion Transport in Solids, Ed. W. van Gool, North-Holland Publ. Co., Amsterdam 1973 (p. 3).

(Received June 14, 1983)

phys. stat. sol. (a) 79, 251 (1983)

Subject classification: 13.4; 20.3

Department of Physics, Simon Fraser University, Burnaby[1])

Analysis of Thermoluminescence Data Dominated by Second-Order Kinetics

By

R. Chen[2]), D. J. Huntley, and G. W. Berger

In the analysis of the thermoluminescence glow curves, and in particular when applied to TL dosimetry and TL dating of archaeological and geological samples, a first order behaviour of the peaks involved is usually implicitly assumed. In this work the TL response for second order kinetics is described and shown that the usual analysis, in which a linear growth of TL with excitation dose is assumed, is correct to a good approximation provided that the curves are first shifted in temperature so that their peaks coincide. This is true for a single isolated peak and for a series of overlapping peaks.

Dans l'analyse des courbes de thermoluminescence (TL), particulièrement quand elle est appliquée à TL dosimetrie et la datation TL d'échantillons archéologiques et géologiques, on suppose régulièrement, quelquesfois implicitement, une conduite de premier ordre des pics TL. Nous considérons ici les autres possibilités de l'ordre de la cinétique et nous donnons, par example, une analyse détaillée des pics TL de deuxième ordre. Nous suggérons que pour établir la propriété de deuxième ordre d'un pic, il faut examiner quelques caractèristiques des resultats expérimentaux. Il faut considérer: a. la symétrie du pic, b. le changement de sa position quand on excite l'échantillon avec des doses différentes, et c. la dépendance non-linéaire de la dose des regions différentes du pic. Nous considérons aussi le „plateau-test" qui a été utilisé pour établir la stabilité thermique des pics de TL. Cet examen doit être modifié dans les cas où le deuxième ordre se trouvera dominant. La modification suggerée est l'étude du changement de position des pics lorsque les maxima coïncident, et l'examen du „plateau" dans les pics ainsi transformés. Cette methode est examinée pour un pic unique synthétique (calculée par ordinateur) de deuxième ordre, et pour une combinaison de quatre pics synthétiques de deuxième ordre.

1. Introduction

The thermoluminescence (TL) of minerals is widely used for the determination of past radiation doses. In archaeology and, to a limited extent, in geology, such doses when combined with dose rates (which are separately evaluated) are used to determine the age of past events in the technique of thermoluminescence dating. Details of the methods used are found in books by Aitken [1] and Fleming [2]. It is generally assumed implicitly that the mineral grains extracted from a specimen (pottery, brick, etc.) exhibit first order kinetics. The possibility of second order or other kinetics is seldom mentioned, and apparently seldom observed.

First order kinetics is characterized by a TL glow curve (light intensity versus temperature) which increases uniformly linearly with radiation dose. In such a case the determination of a past radiation dose is straightforward even if the TL results from a complex distribution of electron (or hole) traps. In the case of second order kinetics the glow curve resulting from a single trap shows a markedly different behaviour for which as yet no simple method of analysis exists. In this work we find that an ad hoc method in which the glow curves were shifted so that their peaks aligned, described in Berger and Huntley [3], has significant justification.

[1]) Burnaby, British Columbia VSA 1S6, Canada.

[2]) Permanent address: Department of Physics and Astronomy, Tel Aviv University.

The motivation for this work arose in the course of studies aimed at the development of a TL dating method for sediments (Berger and Huntley [3], Wintle and Huntley [4]) which has been hampered by second order kinetics behaviour shown by some samples. In this context, the work of Levy [10, 19] should be mentioned, in which second order kinetics TL peaks were reported in albite (NaAl Si_3O_8) [19, 26] as well as in feldspars and ocean sediments [10]. Putative second order and other non-first order peaks have also been reported in quartz and feldspar [23] and in quartz in some more recent works [24, 25].

Finally, we note that although a few authors studying well defined crystalline solids have suggested that their samples exhibited second order behaviour, the evidence presented is inadequate. We present here a set of criteria which we hope will allow more satisfactory identification.

2. Second Order Peaks — Basic Considerations

The relatively simple case of a single TL peak resulting from a single trap and a single recombination centre is mathematically represented by a set of three simultaneous differential equations in the three unknown functions: $n(t)$ — the concentration of electrons in traps at time t, $m(t)$ — the concentration of holes in centres and $n_c(t)$ — the concentration of free electrons in the conduction band (see Section 2.1 in [5]). Adding a simplifying assumption (e.g. Halperin and Braner [6]), that no substantial concentration of electrons builds up in the conduction band, the following expression emerges

$$I = -\frac{\mathrm{d}m}{\mathrm{d}t} = sn \exp\left(-\frac{E}{kT}\right) \frac{A_\mathrm{m} m}{[A_\mathrm{m} m + A_\mathrm{n}(N - n)]}, \tag{1}$$

where I is the TL intensity (in arbitrary units), s — the frequency factor (s^{-1}), E — the activation energy (eV), k — the Boltzmann factor (eVK^{-1}), T — the absolute temperature (K), N — the concentration of traps (m^{-3}) (out of which n have electrons at a given moment), and A_m and A_n are the recombination and retrapping probabilities respectively (m^3s^{-1}). Assuming dominating recombination, i.e. $A_\mathrm{m} m \gg A_\mathrm{n}(N - n)$ one obtains the first order case

$$I = -\frac{\mathrm{d}n}{\mathrm{d}t} = s \exp\left(-\frac{E}{kT}\right) n \tag{2}$$

first suggested by Randall and Wilkins [7] and extensively investigated later on. The solution of this equation for a linear heating function $T = T_0 + \beta t$ yields an asymmetric curve having a "symmetry factor" $\mu_\mathrm{g} \approx 0.42$ where $\mu_\mathrm{g} = \delta/\omega$, and where $\delta = T_2 - T_\mathrm{m}$, $\omega = T_2 - T_1$ and T_m is the temperature at the maximum; T_1 and T_2 are the low and high half-maximum-intensity temperatures [8].

Garlick and Gibson [9] were the first to investigate the case of non-negligible retrapping. Assuming equal recombination and retrapping probabilities $A_\mathrm{m} = A_\mathrm{n}$ as well as equal concentrations of electrons and holes in traps and centres respectively, $n = m$, they obtained

$$I = -\frac{\mathrm{d}n}{\mathrm{d}t} = \left(\frac{s}{N}\right) \exp\left(-\frac{E}{kT}\right) n^2 . \tag{3}$$

A similar equation which may be more likely to occur results from "dominating retrapping", namely $A_\mathrm{n}(N - n) \gg A_\mathrm{m} m$, along with the assumptions that the trap is far from saturation, $n \ll N$, and, again $n = m$. Considering (1) with these assumptions, one gets

$$I = -\frac{\mathrm{d}n}{\mathrm{d}t} = \left(\frac{sA_\mathrm{m}}{NA_\mathrm{n}}\right) \exp\left(-\frac{E}{kT}\right) n^2 . \tag{4}$$

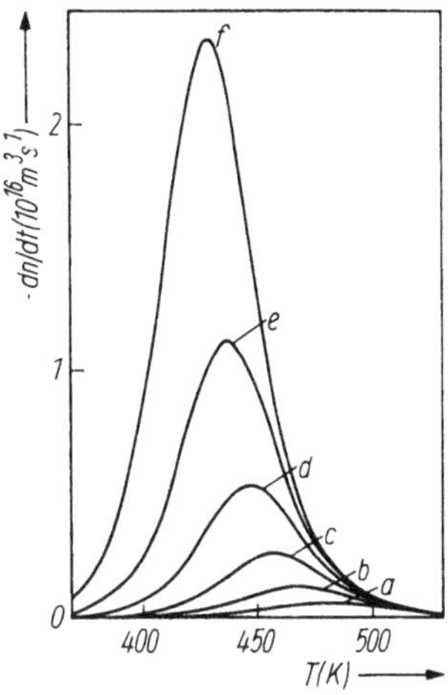

Fig. 1. A set of ideal second order glow curves with different initial trap fillings, (6). The parameters used were $E = 1.1$ eV, $s' = 10^{-6}$ m^{3}s^{-1}, $\beta = 1$ Ks^{-1}, (a) $n_0 = 2 \times 10^{16}$, (b) 4×10^{16}, (c) 8×10^{16}, (d) 1.6×10^{17}, (e) 3.2×10^{17}, (f) 6.4×10^{17} m^{-3}

Equations (3) and (4) can be written as

$$I = -\frac{dn}{dt} = s'n^2 \exp\left(-\frac{E}{kT}\right) \tag{5}$$

where s' is a constant having dimensions of m^{3}s^{-1}, which is s/N in the formre case and sA_m/NA_n in the latter. The solution of (5) for a constant heating rate is

$$I = n_0^2 s' \exp\left(-\frac{E}{kT}\right)\left[1 + \left(\frac{n_0 s'}{\beta}\right)\int_{T_0}^{T} \exp\left(-\frac{E}{kT'}\right) dT'\right]^{-2}, \tag{6}$$

where n_0 is the initial value of n at T_0. A set of curves of $I(T)$ for varying n_0 is shown in Fig. 1. Similar sets have been previously published [9, 10]. Such curves are nearly symmetric with $\mu_g \approx 0.52$.

We emphasize that by no means does one have to assume that a given TL peak is necessarily of first or second order. On the contrary, the sets of assumptions made above leading to the first or second orders were quite particular, and many kinds of intermediate cases are possible. The main reason for an in-depth study of second order kinetics is to point out features which may be common to non-first-order kinetics cases.

Characteristics of these second order curves which distinguish them from first order ones are:

(i) A decrease in the temperature of the peak, T_m, with increasing n_0.

(ii) A slightly superlinear increase of the peak intensity, I_m, with n_0.

(iii) A quadratic increase in intensity with n_0 at a given temperature well below the peak.

(iv) an asymptotic approach to a constant intensity with increasing n_0 at a given temperature well above the peak.

The increase in TL with n_0 at a fixed temperature is shown in Fig. 2; this shows both (iii) and (iv) as the peak moves from above to below this temperature.

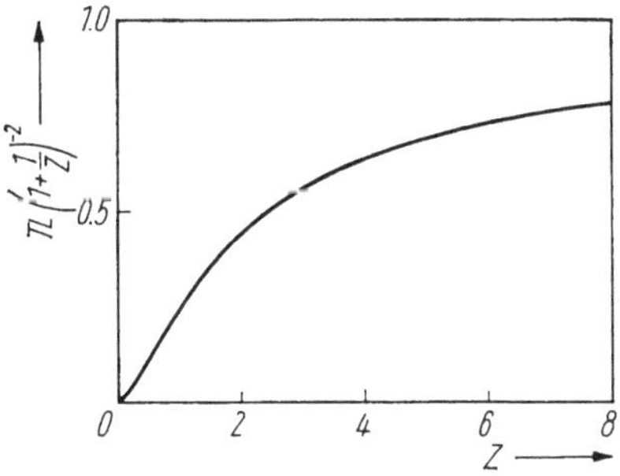

Fig. 2. Second order growth curve, (7) with $Z = n_0/C$. Note that approach to a constant TL at large n_0 is not due to saturation (traps all filled) and that this curve is very different from that in the first order case when non-linearity due to saturation may occur

3. Detailed Considerations

Let us consider the dependence of I on n_0 for a given value of T from (6). The integral on the right-hand side starts off from 0 at $T = T_0$ and grows rapidly so that at a high enough temperature (well above the peak) we have $(n_0 s'/\beta) \int_{T_0}^{T} \exp(-E/kT')\, dT' \gg 1$. Thus if T is well below the maximum, $I \sim n_0^2$ whereas if T is well above the maximum, I is independent of n_0. At a fixed T, (6) can be simplified to

$$I = A\left(1 + \frac{C}{n_0}\right)^{-2}, \tag{7}$$

where A and C are constants defined by (6) and (7). This is the curve shown in Fig. 2.

For dosimetry applications, the best measure is the area under the whole glow curve, because this is proportional to n_0. (Physically this arises because each trapped electron yields a photon in the model.) However, since in many cases TL peaks are accompanied by adjacent overlapping peaks, the behaviour of which may be different from the main peak, one quite often takes the intensity at the peak, $I_m = I(T_m)$, as a measure of the accumulated dose. Let us thus consider $I = f(n_0, T)$ with fixed s', E, and β. The expression for the differential at $T = T_m$ is

$$dI_m = \left(\frac{\partial f}{\partial n_0}\right) dn_0 \tag{8}$$

because $(\partial f/\partial T)_{T_m} = 0$ here. Using (6) and (8) we obtain

$$\frac{dI_m}{I_m} = \left[\frac{\beta E \exp(E/kT_m)}{(n_0 k T_m^2 s')}\right] \frac{dn_0}{n_0} \tag{9}$$

or which we can write a solution

$$I_m \sim n_0^{\gamma},$$

where

$$\gamma = \frac{\beta E \exp(E/kT_m)}{(n_0 k T_m^2 s')} \tag{10}$$

because γ is very nearly constant. One can test the near constancy of γ by choosing a set of parameters and solving numerically the condition for T_m (see e.g. [8]) and substituting in (10).

An alternative approach is to use the asymptotic series approximation (e.g. see [12]) for the integral in (6)

$$F(T, E) = \int_0^T \exp\left(-\frac{E}{kT'}\right) dT' = \left(\frac{kT^2}{E}\right) \sum_{i=0} \left(\frac{kT}{E}\right)^i (-1)^i\, i! = \left(\frac{kT^2}{E}\right) \eta\left(\frac{E}{kT}\right), \tag{11}$$

where $\eta(x)$ denotes the asymptotic series [8].

From this it is immediately seen that

$$\int_{T_0}^{T_m} \exp\left(-\frac{E}{kT}\right) dT = F(T_m, E) - F(T_0, E) \tag{12}$$

however, since $F(T, E)$ is a rapidly increasing function of T, the integral in (12) is very accurately represented by $F(T_m, E)$ provided that T_0 is not very close to T_m (which is in practice usually the case). Denoting kT_m/E by X_m, we obtain the peak condition

$$\frac{1}{n_0} = \left(\frac{kT_\mathrm{m}^2 s'}{\beta E}\right) \exp\left(-\frac{E}{kT_\mathrm{m}}\right) (1 + 2X_\mathrm{m} - 6X_\mathrm{m}^2 + 24X_\mathrm{m}^3 - \dots) \tag{13}$$

and therefore in (10)

$$\gamma \approx 1 + 2X_\mathrm{m} - 6X_\mathrm{m}^2 + 24X_\mathrm{m}^3 - \dots = 2 - \eta(X_\mathrm{m}) . \tag{14}$$

In most cases that are likely to be considered, $X_\mathrm{m} < 0.1$ (in the examples in Table 1, $0.03 < X_\mathrm{m} < 0.08$), thus we see immediately that for second order peaks the maximum intensity I_m varies slightly superlinearly with the initial filling of traps n_0 since γ in (14) is slightly larger than unity. The best accuracy in evaluating γ is by taking the terms down to the smallest one in absolute value and adding one half of the next, i.e.

$$\gamma = 1 + \sum_{i=1}^{K} (i + 1)!\,(-1)^{i+1} X_\mathrm{m}^i + \tfrac{1}{2}(K + 2)!\,(-1)^{K+2} X_\mathrm{m}^{K+1} , \tag{15}$$

where K is the integer part of $1/X_\mathrm{m}$. The procedure for evaluating and checking its approximate constancy is to choose a set of parameters, solve (13) numerically to get T_m, find $X_\mathrm{m} = kT_\mathrm{m}/E$ and insert into (15).

Three points are to be considered here. First, looking at (6) and (10) one can immediately see that the relevant parameter to be varied is $s'n_0/\beta$ rather than s', n_0 or β separately. Changing s' and n_0 (for given E and β) in such a way that the product $s'n_0$ remains the same would leave T_m unchanged. In this sense, $s'n_0$ is a "frequency factor" having units of s^{-1} similarly to the s parameter in the first order case (equation (2)), but unlike the first order case, this parameter is dependent on the initial filling n_0.

The second point has to do with the range of values of $s'n_0$ that are physically relevant. As discussed above, the value of s' is given either by s/N or by $(sA_\mathrm{m}/NA_\mathrm{n})$ when retrapping dominates, i.e. A_n is large as compared to A_m. Since $n_0 \leqq N$ in all cases ($n_0 = N$ when the traps are filled to saturation), $n_0 s'$ should be lower than the frequency factor s. Since s is known (e.g. [8]) to be up to $10^{13}\ \mathrm{s}^{-1}$, $s'n_0$ should be smaller. In Table 1, some results of T_m, X_m and γ are given for $s'n_0 = 10^4$, 10^8, $10^{12}\ \mathrm{s}^{-1}$ and

Table 1

Table values of peak temperature, T_m, $X_\mathrm{m} = kT_\mathrm{m}/E$, and peak growth exponent, γ, for several values of $s'n_0$ and E in second order kinetics. A heating rate $\beta = 1\ \mathrm{K\,s^{-1}}$ was assumed. For other heating rates the table can be used if the first column is relabelled $s'n_0/\beta$ in units of K^{-1}

$s'n_0$ (s^{-1})	E (eV)	T_m (K)	X_m	γ
10^4	0.5	450.4	0.0776	1.1273
10^4	1.0	860.7	0.0742	1.1226
10^4	1.5	1258.3	0.0723	1.1200
10^4	2.0	1647.9	0.0710	1.1182
10^8	0.5	275.4	0.0475	1.0836
10^8	1.0	534.7	0.0461	1.0814
10^8	1.5	788.6	0.0453	1.0801
10^8	2.0	1039.1	0.0448	1.0793
10^{12}	0.5	196.1	0.0338	1.0615
10^{12}	1.0	383.9	0.0331	1.0604
10^{12}	1.5	568.7	0.0327	1.0597
10^{12}	2.0	751.6	0.0324	1.0592

$E = 0.5, 1.0, 1.5, 2.0$ eV. It is readily seen that the value of γ is hardly dependent on the activation energy and does depend more appreciably on the parameter $s'n_0$.

The third point to be considered is that on one hand the value of γ is the power of dependence of I_m on n_0, namely it expresses the variation of I_m with n_0. On the other hand, for a given value of the constant s', γ depends on n_0 as seen in Table 1. Note however, that γ can be considered to be practically a constant when n_0 is varied by say, an order of magnitude which is in many cases the range of variation of the excitation dose in experiments, whereas the appreciable variation of γ with $s'n_0$ shown in Table 1 is related to changes of factors of 10^4 in n_0.

4. Application of the Plateau Test to Single Second Order Peaks

Because of the nonlinear growth curve, and the shift in temperature (Fig. 2), it is evident that application of the plateau test [1] as used in first-order cases is quite inappropriate. Berger and Huntley [3] found that shifting a set of experimental curves along the temperature axis in such a way that the peaks were aligned and then applying the plateau test, gave encouraging results. This procedure lacked proper theoretical justification and we explore this now. The problem is not amenable to algebraic analysis so instead we use theoretical curves generated from (6) and using parameters which give curves similar to those we usually observe in sediments.

The parameters we use are $E = 1.1$ eV, $s' = 10^{-6}$ m^{3}s^{-1}, $\beta = 1$ Ks^{-1} and n_0 between 10^{16} and 10^{17} m^{-3}. Fig. 3 shows curves for three values of n_0, shifted to have their maxima coincide. Curve d shows the ratio of the values for b and a. The variation of this ratio with temperature is rather small, about 10% in a range where the intensities of a and b change by a factor of ≈ 6. The approximately constant ratio, curve d, is to be compared to the exactly constant ratio for first order TL peaks.

An alternative way of discussing the matter is to examine the growth curves at a fixed temperature (after shifting). Fig. 4 shows four of these; that at the peak is curve b in which the slight superlinearity ($\gamma \approx 1.07$) is seen. The dependence of the intensity on n_0, 30 K below and above T_m are shown in curves a and c respectively. The dependence of n_0 is practically linear in both. The behaviour at 60 K above T_m is shown in curve d. This is repeated in curve d′ on a (X10) scale where it is seen to be slightly sublinear. It is to be noted that both the superlinearity and sublinearity depicted mainly in curves b and d′ respectively are rather minor because once the peaks are shifted to have maximum temperatures coincide, they have rather similar shapes as seen in Fig. 3.

In cases intermediate between first and second order it seems likely that linear growth curves will result if the TL curves are first shifted to align their maxima. We make this suggestion because it is true in both extremes, there being linearity in first order (no shift required) and near linearity in second order.

In this context, the work by Balarin [12] should be mentioned. Balarin solved the extension of (5) in which n^2 is replaced by n^b where b is a general kinetic order, not necessarily 1 or 2 (see also (16) below). He further normalized the resulting curves with

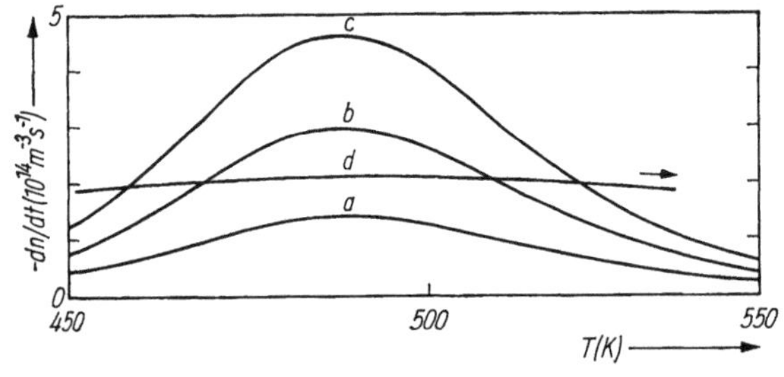

Fig. 3. Three calculated second order curves with $E = 1.1$ eV, $s' = 10^{-6}$ m^{-3}s^{-1} and $\beta = 1$ Ks^{-1}. Curve (a) is for $n_0 = 10^{16}$ m^{-3}, (b) for $n_0 = 2 \times 10^{16}$ m^{-3}, and (c) for $n_0 = 3 \times 10^{16}$ m^{-3}. Curve b is shifted by 12 K upwards and c by 18 K so that the maxima coincide. Curved represents the approximate plateau, namely, the ratio of values of (shifted) b to a

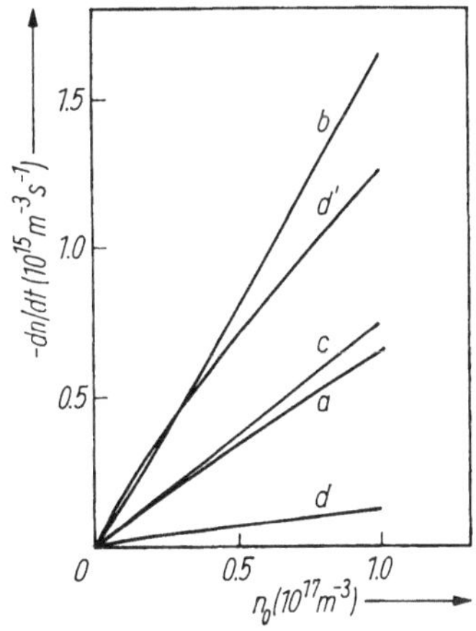

Fig. 4. The dependence of different parts of shifted second order calculated peaks on the initial filling. The peaks are all shifted to have their maxima at 453.2 K. The dependence 30 K below the maxima (a), at the maximum (b), 30 K above the maximum (c) and 60 K above the maximum (d) are shown. (d') gives the same data as (d) on a scale (× 10). The parameters chosen were $E = 1.1$ eV, $s' = 10^{-6}\ m^3s^{-1}$ and $\beta = 1\ Ks^{-1}$. n_0 varied between 1.47×10^{16} and $10 \times 10^{16}\ m^{-3}$

respect to their maxima. It is possible that results similar to that depicted by curve d in Fig. 3 can be more analytically deduced from his results for $b = 2$. This is not entirely straighforward, however, since the element of shifting the peak is not so simple. We therefore feel that at present, the empirical-numerical conclusion suffices.

5. Application to a Distribution of Peaks

In real cases which appear to be of second order kinetics ([2, 3, 10]) the observed curve may and usually does consist of combinations of peaks rather than a single peak. In order to simulate this situation, we have taken a combination of four second order peaks, the sum of which agrees reasonably well with an experimental curve. It is shown as curve a of Fig. 5. The calculation was repeated with all the initial concentrations multiplied by 6.8 and the curve shifted 30 K to make the peaks coincide. In order to compare the shapes of the curves, we have normalized the results by dividing the intensity values by the same factor of 6.8; this is shown as curve b. Although the two curves do not coincide, the difference is within 10% from 40 K below to 100 K above the peak. The growth curves in this case looked very much like those of Fig. 4.

We thus conclude that the plateau test using shifted curves should be useful when second order kinetics is present, and probably also in cases intermediate between first and second order.

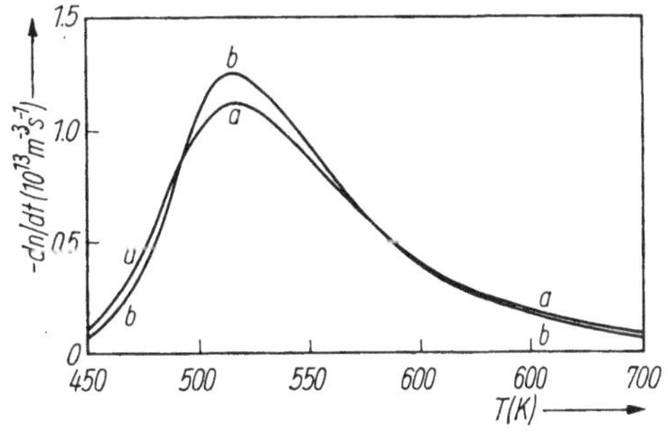

Fig. 5. Calculated combinations of four independent second order curves. The parameters chosen for curve (a) were: $E_1 = 1.1$ eV, $s_1' = 5 \times 10^{-6}\ m^3s^{-1}$, $n_0 = 7 \times 10^{14}\ m^{-3}$; $E_2 = 1.175$ eV, $s_2' = 5 \times 10^{-6}\ m^3s^{-1}$, $n_{02} = 4 \times 10^{14}\ m^{-3}$; $E_3 = 1.25$ eV, $s_3' = 5 \times 10^{-6}\ m^{-3}s^{-1}$ $n_{03} = 1.5 \times 10^{14}\ m^{-3}$; $E_4 = 1.35$ eV, $s_4' = 5 \times 10^{-6}\ m^3s^{-1}$, $n_{04} = 0.8 \times 10^{14}\ m^{-3}$. In curve (b) all, the values or n_{0i} were increased by a factor of 6.8. In order to see clearly the variation in the shape, b is normalized by dividing the final results by the same factor of 6.8. Also curve b is shifted by 30 K upwards in temperature

6. Criteria Indicating Second Order Kinetics

Some of the features of second order kinetics curves have been discussed above. If one has a clean single second order peak, each of these features can serve to classify the peak as second order. In practice, however, the situation may be complicated by two aspects briefly mentioned above.

1. Even if the peak is single, it is not necessarily represented by either (2) or (5). Two kinds of intermediate cases have been discussed in the literature, the "general order" (e.g. [12, 13])

$$I = -\frac{dn}{dt} = s'n^b \exp\left(-\frac{E}{kT}\right) \tag{16}$$

with $1 \leqq b \leqq 2$ the (effective) order of the kinetics, and the "mixed order" [14]

$$I = -\frac{dn}{dt} = s'n^2 \exp\left(-\frac{E}{kT}\right) + s'Cn \exp\left(-\frac{E}{kT}\right). \tag{17}$$

Both approaches are, however, empirical in nature and have physical justification only in specific cases, the general behaviour may be expected to be even more complicated. Some curves for different A_m/A_n ratios and different initial m/n ratios are published by Levy [10].

2. Some of the features of a curve consisting of, say, a main second order peak and accompanying satellites may not be those of a pure second order peak due to the effect of the additional ones. In addition, it is possible for a second-order characterstic to arise from a more complicated trap and recombination centre situation than that assumed. One should, therefore, demonstrate several properties of second order kinetics in order to establish that a peak is of second order.

For these reasons we present now a set of criteria which should be checked before claiming a particular peak follows second order kinetics.

I. Peak shape: In the (rare) case that one has a single peak, the shape of the peak provides sufficient information on the kinetics. $\mu_g \approx 0.52$ is characteristic of second order and $\mu_g \approx 0.42$ of first order; values between these are expected for intermediate orders [5, 11].

II. Peak shift: In second order peaks, or in fact, in most non-first order peaks, the maximum shifts to lower temperatures at higher initial trap fillings n_0 or experimentally, at higher doses. This shift to lower temperature is however only an indication of second order kinetics. Since the shift may be due to a build-up of accompanying peaks, a quantitative test of the amount of shift as related to the parameters involved is desirable. One method of doing this is explained in IV below.

III. Superlinearity: As explained above, a second order peak is characterized by a slight superlinear dependence of the maximum intensity I_m on the initial filling n_0. If the latter is linear with the dose, which is usually expected when one is relatively far from saturation, a behaviour of the form $I_m \sim D^\gamma$ is expected with γ being typically around 1.08.

IV. I_m–T_m method: Chen and Winer [15] suggested that when one has a TL curve of general order (e.g. (16)) with a known kinetics order b, one can plot

$$\ln\left[I_m^{b-1}\left(\frac{T_m^2}{\beta}\right)^b\right]$$

as a function of $1/T_m$ and get a straight line, with a slope of E/k. For a second order peak, $b = 2$, and a plot of ln $(I_m T_m^4)$ versus $1/T_m$ should give a straight line (assuming a constant heating rate).

V. Isothermal decay methods: If a sample is held at a constant temperature, the number of trapped electrons $n(t)$ decreases with time. This can be studied experimentally by measuring the light emitted during this time, the phosphorescence. Alternatively, if the heating is stopped at time t, the sample cooled, and a glow curve measured, the area of the last is proportional to $n(t)$. A series of such measurements for various values of t gives the shape of $n(t)$ and is called the isothermal decay method. These time dependencies are different in first and second order and may be readily found by solving (2) and (5).

In the case of first order kinetics, a plot of $\ln I$ versus t will yield a straight line with slope $-s\exp(-E/kT)$; here I is the TL intensity of the phosphorescence curve of the TL peak area (or intensity) in the isothermal decay measurements. For second order phosphorescence, one must plot $1/\sqrt{I}$ versus t to obtain a straight line with slope $[s'\exp(-E/kT)]^{1/2}$. In the second order isothermal decay case, a plot of $1/n$ versus t will yield a straight line of slope $s'\exp(-E/kT)$; here n is the trapped electron density at the end of the isothermal treatment and can be represented by the area of subsequent glow curve or, to a good approximation, by its maximum intensity. In each case, the variation of slope with holding temperature can be used to calculate the trap depth E. Two more isothermal phosphorescence methods which can provide information on the kinetics order will be briefly mentioned. May and Partridge [16] suggested a method in which one plots $\lg(-\mathrm{d}I/\mathrm{d}t)$ as a function of $\lg I$. The expected results should form a straight line with a slope of $(2 - 1/b)$ where b is the kinetics order (see (16)). For the special case of second order, the slope should be 1.5. Recently, Sinha and Mukherjee [17] utilized this method for glow in CdF_2: Pr crystals and found a straight line which yielded $b = 1.6$. Another method discussed by Visocekas [18] (see also [5] paragraph 6.6) is to plot $y = tI$ versus $x = \ln t$. The result should be a TL-like curve which resembles a first order TL peak ($\mu_g \approx 0.42$) in the first order case and a second order TL peak ($\mu_g \approx 0.52$) in the second order case.

7. Discussion

The properties of second order thermoluminescence curves have been further investigated in this paper. In particular the behaviours of the peak shape and maximum intensity while changing the initial filling of the traps have been studied. The main purpose of this research has been to expand the information one can extract from TL results in geological samples in cases where the second order property of the results can be established.

The method as applied to the "synthetic" curve depicted in Fig. 5 and to experimental ones of similar nature involves the shift of the whole glow curve by an amount determined basically by the shift of the main TL peak involved. This may be slightly altered if one has a series of separate second order peaks. If the shift is different for the various peaks, one can deal with each of the peaks separately and shift each of them by the correct amount so that the temperatures of the maxima coincide.

A word should be said about the feasibility of having a series of second order peaks from a given sample. Suppose that one has a single material with, say, a series of four traps and one centre which yield a series of four TL peaks which may be quite separate or may form a single composite glow peak. According to this picture, one should have

$$\sum_{i=0}^{4} n_{0i} = m_0\,, \tag{18}$$

where m_0 is the initial concentration of holes in centres and n_{0i} is the initial concentration of electrons in the i-th trap, where $i = 1 \dots 4$. As mentioned above, however, a necessary condition for a peak to be of second order is that $m = n$ which cannot be exactly true for each peak separately if (18) holds. In fact, according to this line of thought, only for the last peak in such a series may one have $n_0 = m_0$ (where m_0 here is the remaining concentration at the end of the first three peaks), and thus a second order behaviour.

There still are, however, a number of possibilities of having a series of second order peaks. First of all, the sample in hand may not be a single material, in fact, in the archaeological and geological samples, the usual case is that many minerals are involved each of which may emit a non-first order TL peak. Another possibility is that several traps as well as several centres are involved and that traps and centres are exhausted more or less in pairs, so that within each pair one has $n_0 \approx m_0$. Also, it is possible that some of the peaks result from electron traps and hole centres and others from hole traps and electron centres, which make the two sets of peaks independent of each other, which in turn increases the likelihood of the occurrence of second order kinetics. Finally, even within the framework of a series of electron traps and one hole centre discussed above, one can think of a situation in which $n_{01} \gg n_{02} \gg n_{03} \gg n_{04}$ which is, more or less, the case discussed with respect to Fig. 3 above. In this case n_{01} is related to the shallowest trap and the others are, in turn, deeper. If this is the case, we have $m_{01} \approx n_{01}$ which may make the first peak approximately of second order. At the end of the first peak, one should have the remaining concentration of holes in centres m_{02} and this may be rather close to n_{02} and so on. Of course, this argument is not very strong when the peaks overlap, in which case second order kinetics would be only an approximation.

Due to the factors mentioned above, but mainly the interference of additional peaks, we would like to stress the need for utilizing as many criteria as possible for establishing the second order nature (or more generally — the non-first order one) of TL peaks. Some experimental results previously published (e.g. [17], [19 to 21]) utilized only one or two of these criteria for determining the second order property. We note that some of the considerations made here may be applied rather easily to other thermally stimulated processes (e.g. see [5], [11]) provided that the second order kinetics can be established. Finally, we would like to reiterate that first and second order kinetics are only limiting cases and many kinds of intermediate situations are possible. In this respect, we would like to mention a recent paper by Moharil [22], who defined general order kinetics by utilizing the ratio between recombination and retrapping probabilities A_n/A_m. His definition of "general order" is not the most general possible, however, since he assumed $n = m$ which is not necessarily always the case [5, 6]. The method suggested here may possibly be refined in the future to take into account the exact process that takes place in various cases of non-first order kinetics.

References

[1] M. J. Aitken, Physics and Archaeology, 2nd ed., Oxford University Press, London 1974.
[2] S. J. Fleming, Thermoluminescence Techniques in Archaeology, Clarendon Press, Oxford (1979).
[3] G. W. Berger and D. J. Huntley, Council of Europe PACT Journal 6, 495 (1982).
[4] A. G. Wintle and D. J. Huntley, Canad. J. Earth Science 17, 348 (1981).
[5] R. Chen and Y. Kirsh, Analysis of Thermally Stimulated Processes, Pergamon Press, Oxford 1981.
[6] A. Halperin and A. A. Braner, Phys. Rev. 117, 408 (1960).
[7] J. T. Randall and M. H. F. Wilkins, Proc. Roy. Soc. A184, 347, 366, 390 (1945).

[8] R. Chen, J. appl. Phys. **40**, 570 (1969).
[9] G. F. J. Garlick and A. F. Gibson, Proc. Phys. Soc. **60**, **574** (1948).
[10] P. W. Levy, Council of Europe PACT Journal **6**, 224 (1982).
[11] R. Chen, J. Mater. Sci. **11**, 1521 (1976).
[12] M. Balarin, J. Therm. Anal. **17**, 319 (1979).
[13] R. Chen, J. Electrochem. Soc. **116**, 1254 (1969).
[14] R. Chen, N. Kristianpoller, Z. Davidson, and R. Visocekas, J. Lum. **23**, 293 (1981).
[15] R. Chen and S. A. A. Winer, J. appl. Phys. **41**, 5227 (1970).
[16] C. E. May and J. A. Partridge, J. Chem. Phys. **40**, 1401 (1964).
[17] R. K. Sinha and M. L. Mukherjee, phys. stat. sol. (b) **105**, 69 (1981).
[18] R. Visocekas, Thèse, Université P. et M. Curie, Paris 1978.
[19] P. W. Levy, Council of Europe PACT Journal **3**, 466 (1979).
[20] M. Prokić, J. Phys. Chem. Solids **38**, 617 (1977).
[21] V. S. Sivasankar and P. W. Whippey, Canad. J. Phys. **57**, 128 (1979).
[22] S. V. Moharil, phys. stat. sol. (a) **73**, 509 (1982).
[23] G. Hutt and A. Smirnov, Proc. III Specialist Seminar on TL and ESR Dating, Helsignor (Denmark), Council of Europe PACT Journal, in the press.
[24] A. G. Wintle, Soil Science **134**, 164 (1982).
[25] R. G. Schwartzman, J. A. Kierstead, and P. W. Levy, III Specialist Seminar on TL and ESR Dating, Helsignor (Denmark), Council of Europe PACT Journal, in the press.
[26] E. S. Pasternack and P. W. Levy, Geol. Soc. Amer. Abstracts with Programs **10**, 486 (1978).

(Received June 29, 1983)

phys. stat. sol. (a) 79, 263 (1983)

Subject classification: 1.2; 6; 22.8

Department of Physics, Kyushu University 33, Fukuoka[1]) *(a)*
and Faculty of Engineering (b) and *Faculty of Education (c), Oita University*[2])

Structural Phase Transition of $CsPbCl_3$ below Room Temperature

By

M. HIDAKA (a), Y. OKAMOTO (b), and Y. ZIKUMARU (c)

Studies of the structural phase transition of $CsPbCl_3$ are carried out, using X-ray and neutron diffraction, NQR and Raman scattering below room temperature. It is confirmed that there is a second-order transition at about 200 K (T_{c_4}). From the structural analysis, it is determined that, at room temperature, the structure has an orthorhombic symmetry D_{2h}^{16}–Pbnm, and, below T_{c_4}, it has a monoclinic symmetry C_{2h}^{5}–$P2_1/c$. The transition at T_{c_4} is not associated with the condensation of rotational modes of $PbCl_6$ octahedra around any principal axes, but associated with the elastic soft mode.

Mittels Röntgenstrahlen- und Neutronenbeugung, NQR und Ramanstreuung werden unterhalb Zimmertemperatur Untersuchungen des strukturellen Phasenübergangs von $CsPbCl_3$ durchgeführt. Es wird ein Übergang zweiter Ordnung bei etwa 200 K (T_{c_4}) festgestellt. Aus der Strukturanalyse wird gefunden, daß bei Zimmertemperatur die Struktur die rhombische Symmetrie D_{2h}^{16}–Pbnm, und unterhalb T_{c_4} monokline Symmetrie C_{2h}^{5}–$P2_1/c$ besitzt. Der Übergang bei T_{c_4} ist nicht mit der Kondensation von Rotationsmoden der $PbCl_6$-Oktaeder um eine der Hauptachsen, sondern mit der elastischen Soft-Mode verknüpft.

1. Introduction

There have been many investigations of the structural phase transition (SPT) of perovskite compound ABC_3; notable examples are $SrTiO_3$, $KMnF_3$, $RbCaF_3$, $CsPbCl_3$, et al. These compounds are found to undergo an antiferrodistorted structural phase transition, which can be related to the condensation of some kinds of soft phonons. The lower-symmetry phase is characterized by a slight displacement of the cations. The displaced direction is represented by a tilting scheme of octahedra BC_6 about the principal axes.

As for $CsPbCl_3$, Fujii et al. [1] using neutron diffractions, reported that the phase transitions at 47, 42, and 37 °C were related to the condensation of the M_3 phonon, the Z_5^x phonon, and the Z_5^y phonon, respectively. Below room temperature, Cape et al. [2] and Cohen et al. [3] utilizing EPR reported that there was a structural phase transition at nearly 190 K. However, there has been no confirmation of this transition using other experimental methods. Furthermore, the structural analysis of the lower-symmetry phases has not been carried out because a poly-domain structure appeared in these phases.

In the present paper, we will describe the experimental results of X-ray and neutron diffraction, NQR and Raman scattering of $CsPbCl_3$ in order to study its phase transition below room temperature.

---

[1]) Fukuoka 812, Japan.
[2]) Oita, Japan.

2. Experimental Results of X-Ray and Neutron Diffraction Studies

In order to get structural information about the lower-symmetry phases, X-ray and neutron diffraction studies were carried out. At room temperature, $CsPbCl_3$ is generally a poly-domain structure caused by the antiferrodistorted transitions. The distribution of the domains is strongly dependent on the size of the sample, and on the temperature hysteresis when the transitional temperature is surpassed.

Fig. 1 a, b show zeroth-layer Weissenberg photographs taken about $[010]_p$ at 300 and 20 K, respectively, where the suffix p means a pseudo-perovskite cell. Since a large splitting of the reflections is observed below about 200 K, it may be expected that there is a structural transition at around 200 K. As regarding the poly-domain structure of the sample [11], the structure was analyzed by using a Weissenberg photograph. At room temperature, the systematic absence rule of the reflections was as follows:

$$h0l;\ h + l \quad \text{even}, \qquad 0k0;\ k \quad \text{even},$$
$$h00;\ h \quad \text{even}, \qquad 00l;\ l \quad \text{even},$$

where a primitive unit cell volume was equal to $\sqrt{2}\, a_p \sqrt{2}\, b_p\, 2c_p$, and the structure probably had a D_{2h}^{16}-Pbnm orthorhombic symmetry. On the other hand, below about 200 K, the absence rule of the reflections was as follows:

$$hk0;\ k \quad \text{even}, \qquad 00l;\ l \quad \text{even},$$

where a primitive unit cell volume was equal to $2a_p 2b_p 2c_p$, $\beta \neq 90°$, and the structure probably had a C_{2h}^5-$P2_1/c$ monoclinic symmetry.

In order to study the transitional temperature, the temperature dependence profiles of the $(\frac{5}{2}, \frac{3}{2}, \frac{1}{2})_p$ reflection were analyzed by using a neutron diffraction method, since $CsPbCl_3$ has a large X-ray absorption. The result is shown in Fig. 2. The double peak corresponds to two kinds of domains, whose populations are dominant in the sample used. From Fig. 2, it is reasonable to assume that there is a structural transition at about 200 K. The peak separation was measured accurately in order to confirm the transition. The result is shown in Fig. 3, where $\Delta\theta_{sp}$ denotes the peak separation and $\Delta\theta_{FWHM}$ denotes the full width at half maximum height. Since there is no double peak close to T_{c_4}, only $\Delta\theta_{FWHM}$ was measured. From Fig. 3, it is clear that a second-order structural transition occurs at about 200 K, as suggested by Cape et al. [2] and Cohen et al. [3].

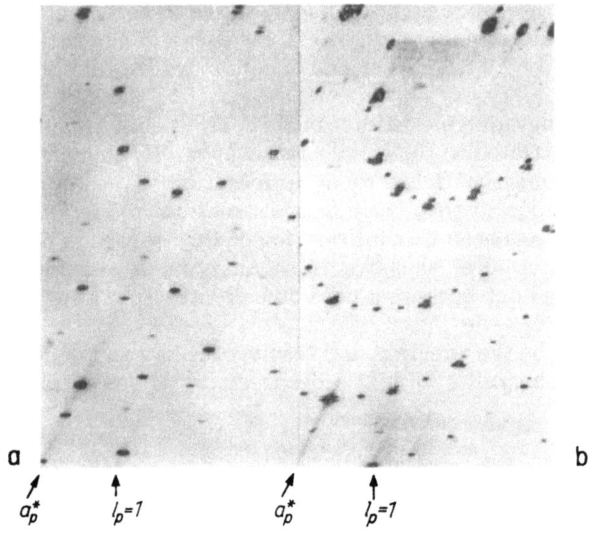

Fig. 1. Zeroth-layer Weissenberg photograph taken around $[010]_p$ at a) 300 b) 20 K

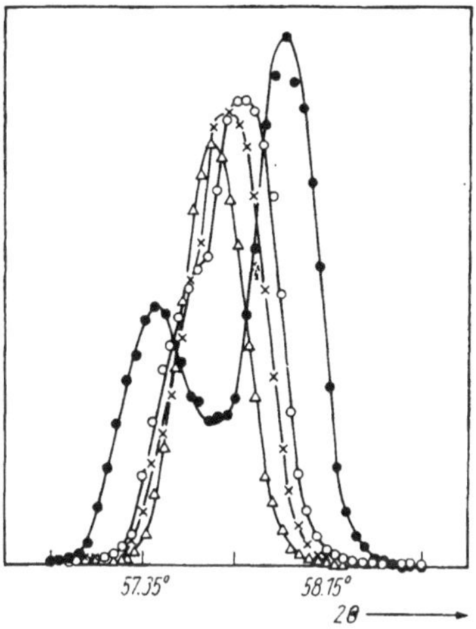

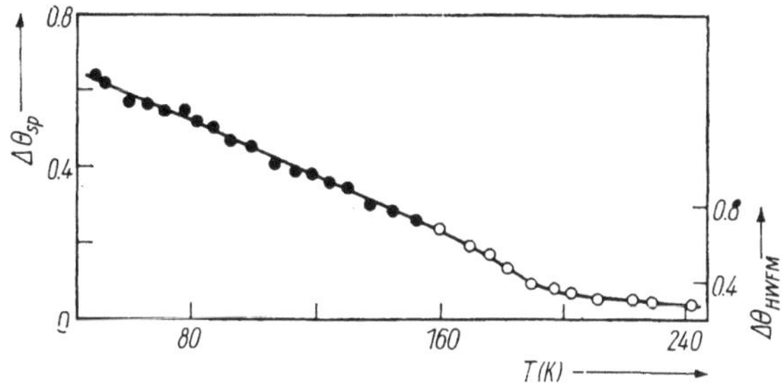

Fig. 2

Fig. 3

Fig. 2. Profile of the $(\frac{5}{2}\,\frac{3}{2}\,\frac{1}{2})_p$ reflection. ● 75, ○ 160, × 190, △ 240 K

Fig. 3. Temperatur dependence of $\Delta\theta_{sp}$ (●) and $\Delta\theta_{FWHM}$ (○) of the $(\frac{5}{2}\,\frac{3}{2}\,\frac{1}{2})_p$ reflection

3. Experimental Results of Raman Scattering

The Raman spectrum of $CsPbCl_3$ was investigated in a temperature range from 113 to 300 K. The data presented here were obtained from a crystal mounted in the optical cell of a conventional cold nitrogen gas flow cryostat for low- and high-temperature studies. The sample used in the Raman scattering experiment was a crystal of approximately $3 \times 4 \times 5$ mm³ with faces perpendicular to the cubic crystallographic axes. Temperatures were measured with a chromel–constantan thermocouple[3]) and were controlled to within ± 0.1 K. The thermocouple was mounted on the sample. The sample was excited by an argon laser light set at 50 mW at 514.5 nm, and the 90° scattered light was analyzed by a double monochromator. The photomultiplier detector pulses were processed by a digital data collection system linked to the monochromator. The system controlled the accumulation period per spectral position to measure the low-level light signals [4]. At the successive phase transitions in $CsPbCl_3$, a crystal in general has a poly-domain structure on cooling below T_{c_2}, so that the scattering geometry of a Raman experiment below room temperature cannot be uniquely defined. In this experiment the incident light was introduced along what was originally one of the cubic axes, and scattered light was collected along another cubic axis. These axes do not correspond to orthorhombic *a*-, *b*-axes, so the corresponding Raman tensors for the orthorhombic crystal class must be rotated 45° about the *c*-axis. In a lower-symmetry phase of the crystal, birefringence can cause depolarization of divergent or convergent light travelling along the *c*-axis.

The above effects lead to a number of complications in the execution and analysis of a Raman experiment. Raman spectra were obtained in the region 10 to 150 cm⁻¹ over a wide temperature range on several scattering geometries. The $X(YX)\,Z$ and $X(ZZ)\,Y$ spectra are shown in Fig. 4 a, b, which were measured at 171 and 113 K, respectively. The labels X, Y, Z refer to the cubic $[100]_p$, $[010]_p$, and $[001]_p$ directions, respectively.

[3]) This thermocouple was offered by the Chemical Thermodynamics Laboratory of the Faculty of Science at Osaka University.

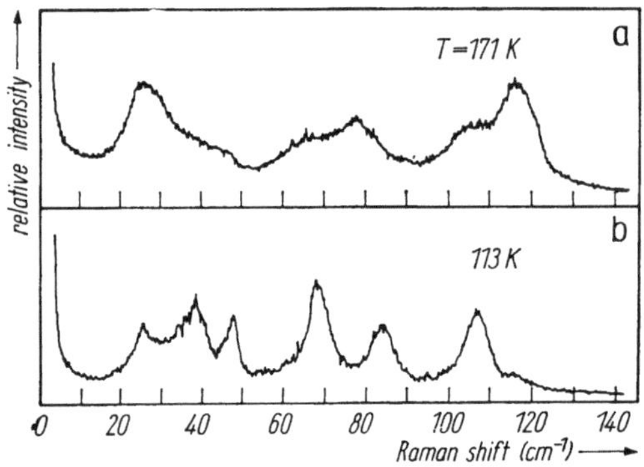

Fig. 4. The Raman spectra of $CsPbCl_3$ obtained by scattering geometries of a) $X(YX)Z$ and b) $X(ZZ)Y$

Raman spectra obtained by several scattering geometries are almost always classified into the two types shown in Fig. 4. The result presented in Fig. 4 is consistent with the Raman spectrum reported by Hirotsu [5] except for our frequency values for corresponding peaks. In the present work the three low-frequency temperature-dependent modes with energies 26, 38, and 48 cm^{-1} at 113 K were clearly observed by $X(ZZ)\ Y$ geometry. The temperature dependence of the low-frequency portion of the $X(ZZ)\ Y$ Raman spectrum is shown in Fig. 5. It is apparent from this figure that the three modes shift to the lower-frequency side and the broad wing of the central line appear above about 230 K with increasing temperature. In Fig. 6 the phonon frequencies of the three low-energy modes are plotted against temperature. It is apparent from these data that, with increasing temperature, the lowest-frequency mode decreases only to 17 cm^{-1} at T_{c_4} and maintains a constant frequency above T_{c_4}. The linewidth is almost independent of temperature, whereas the intensity decreases as temperature increases. The lowest-frequency phonon mode shows anomalous temperature dependence around 200 K, which suggests that there exists the phase transition and the soft mode is not an optical phonon.

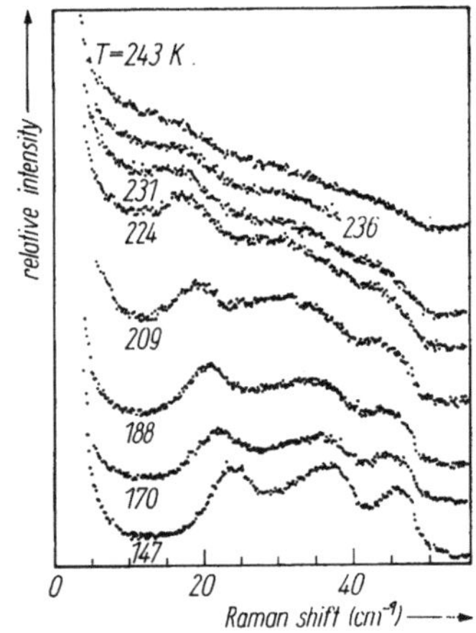

Fig. 5

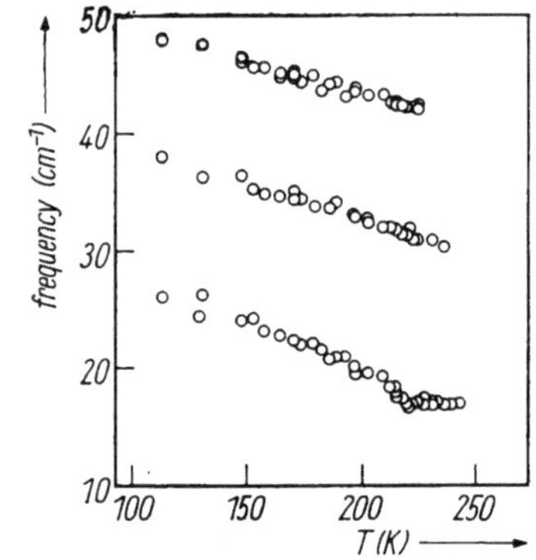

Fig. 6

Fig. 5. Temperature dependence of the three low-frequency modes of the $X(ZZ)Y$ Raman spectrum

Fig. 6. Phonon frequencies vs. temperature for the three low-frequency modes of $CsPbCl_3$

As described in Section 2, X-ray and neutron diffraction experiments on $CsPbCl_3$ indicate the occurrence of a structural phase transition from a D_{2h}^{16} orthorhombic symmetry to a C_{2h}^{5} monoclinic symmetry, which is a second-order transition. The phase transition is produced by a phonon mode at the centre of the Brillouin zone. On the basis of symmetry arguments it is expected that a phase transition from D_{2h} to C_{2h} point group symmetry is caused by the B_{2g} symmetry phonon in the D_{2h} [6]. The B_{2g} mode in the D_{2h} phase has the same Raman tensor $B_{2g}(zx)$ with the A_g mode in the C_{2h} phase. The lowest-frequency mode shown in Fig. 6 is the soft mode with the B_{2g} symmetry in the D_{2h}^{16} phase and the A_g symmetry in the C_{2h}^{5} phase.

4. Experimental Results of NQR

The ^{35}Cl and ^{37}Cl nuclear quadrupole resonance (NQR) in $CsPbCl_3$ was measured over a temperature range from 77 to 360 K. The NQR absorption spectrum was recorded as a second-derivative curve with a frequency-modulated spectrometer of Robinson type. The temperature dependence of NQR frequency is shown in Fig. 7; a great part of the data for ^{35}Cl is the same as already known [7, 8]. In the following description, the Cl species on the Pb–Pb line in the tetragonal phase (320 to 325 K) is designated as Cl_I and the other two Cl species are designated as Cl_{II} and Cl_{III}. In Fig. 7, the branch of high frequency corresponds to Cl_I and that of low frequency to Cl_{II} and Cl_{III}. The temperature-dependent curves of frequency are continuous below room temperature and the signal intensities are nearly constant for both NQR lines; its ratio is about one to two, which equals the ratio of the numbers of resonant atoms. Such an aspect seems to mean no existence of a discrete structure change below room temperature.

However, t he temperature dependence of NQR linewidth, as well as frequencies and intensities, reflects the change in Cl sites in the crystal structure. Fig. 8 shows that the width $\delta\nu$ of low-frequency lines rapidly increases by about 20% around 200 K with decreasing temperature, while the width of high-frequency lines keeps a constant value of 2.3 kHz; where the width $\delta\nu$, to be exact, means the interval between the two lobes of an observed curve shown in Fig. 9. On the other hand, since the spin–lattice relaxation time T_1 has been known [9] to be more than 100 ms at temperature below 250 K it can be imagined that the width observed around 200 K is almost independent of the lattice vibrations.

Thus it is to be considered that both Cl_{II} and Cl_{III} nuclei lose the equivalency of positions in the crystal structure and that their quadrupole interactions with surroundings differ from each other at temperatures below about 200 K. As a natural result a splitting should appear in the spectrum. However, if the difference in interactions is small compared with the linewidth, the inequivalency is observed not as a splitting but as an increase in width. Actually the increase of 20% in the linewidth is to be obtained instead of a splitting of 800 Hz, assuming that the shape of NQR lines is of Gaussian type.

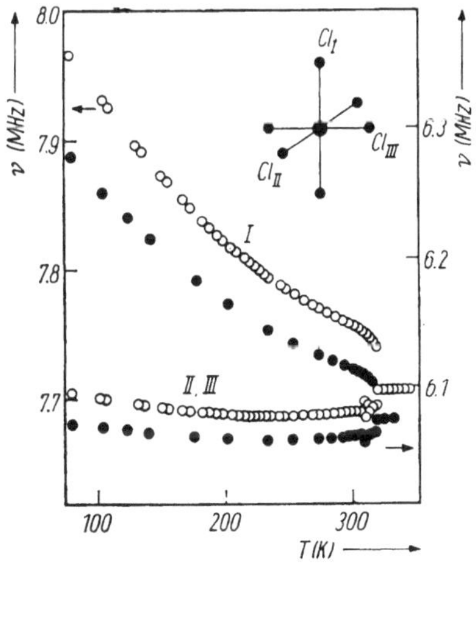

Fig. 7. Temperature dependence of ^{35}Cl and ^{37}Cl NQR frequencies in $CsPbCl_3$. The notation ○ and ● is represented for ^{35}Cl and ^{37}Cl, respectively

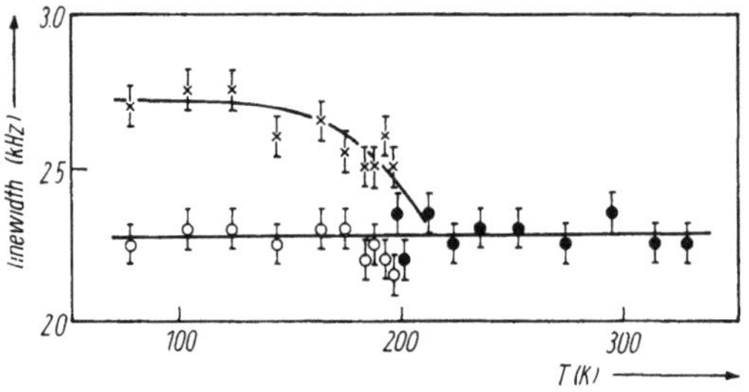

Fig. 8

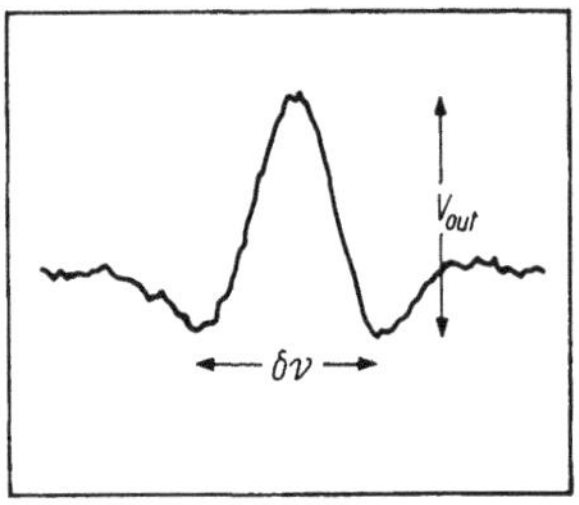

Fig. 9

Fig. 8. Temperature dependence of the linewidth

Fig. 9. A second-derivative curve observed for ^{35}Cl NQR absorption

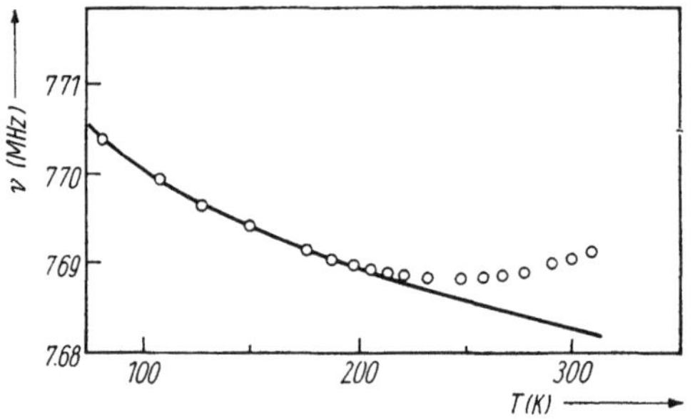

Fig. 10. Temperature dependence of the lower frequency fitted with the $\nu(T)$ curve

The temperature dependence of the low-frequency line is examined upon the fitting to the following simple expression on the basis of the Bayer's theory:

$$\nu(T) = \nu_0(1 + aT + b/T) .$$

When ν_0, a, and b are determined as exactly as possible to fit the data in the lower-temperature range, an anomalous tendency above 200 K is revealed as shown in Fig. 10. At the experiment, ν_0, a, and b was 7.6996 MHz, -9.11×10^{-6} K^{-1}, and $9.18 \times \times 10^{-2}$ K, respectively. The experimental points deviate from the above expression, and the temperature coefficient of NQR frequency turns positive and goes on increasing. Furthermore, the temperature dependence of T_1 above 200 K has been reported [9] to deviate rapidly from the T^{-2} law which means the stable lattice vibrations.

These facts suggests the following picture of the sites for resonant Cl nuclei. Below 200 K, the Cl_{II} and Cl_{III} nuclei are considered to vibrate about the inequivalent sites of minimum potential, in the very near neighbourhood of which there are the empty sites of minimum potential. With rising temperature, the lattice vibrations make the resonant nuclei go across the two close sites of minimum potential. Then, if the average positions of the two close sites for both Cl_{II} and Cl_{III} are very similar in the crystal structure, the $PbCl_6$ octahedra, which are distorted below 200 K, come to be apparently symmetric so that both Cl_{II} and Cl_{III} have the same quadrupole interactions with the surroundings. Therefore, the excess extent of the linewidth disappears above 200 K.

5. Discussion

From the present experiments, it is clear that there is a structural transition at about 200 K. Although $CsPbCl_3$ shows successive transitions at 310, 315, and 320 K, these transitions are associated with the condensation of soft phonon modes (M_3^z, R_{25}^x, and R_{25}^y) existing in an ideal structural phase. The perovskite compounds, having a space group D_{2h}^{16} as the lowest symmetry, generally show transitions related to the condensation of some kinds of soft phonon modes. As for $CsPbCl_3$, the lowest symmetry is C_{2h}^5. Therefore, the transition (T_{c_4}) might be considered to be caused by the condensation of other soft phonon modes. The result of Raman scattering experiment as shown in Fig. 6 suggests that any kind of optical phonon modes is not condensed at T_{c_4}.

Usually, the condensation of the M_3 and R_{25} soft phonon modes in the perovskite-type compounds ABC_3 makes a rotation of BC_6 octahedra around the principal axes $\langle 100\rangle_p$ in the lower symmetry phase. The structure can be, therefore, described with the tilting scheme of the octyhedra around the axis, below the transitional temperature [10]. But, such a representation of the structure is based on the assumption that the octahedron has a regular shape or a slightly distorted one in each phase even if the structure of crystal is not a cube. From the results of the X-ray diffraction, it is found that the structure is D_{2h}^{16} at room temperature and C_{2h}^5 below T_{c_4}. D_{2h}^{16} suggests that Cl_{II} and Cl_{III} as shown in Fig. 7 have equivalent atomic positions in the primitive cell unit, and it is possible to say that the octahedron has only a slight distortion from the regular shape [11]. On the other hand, C_{2h}^5 suggests that Cl_{II} and Cl_{III} occupy non-equivalent atomic positions, and the octahedron is largely distorted.

As mentioned in the preceding section, such a fact gives a reasonable explanation to the data of NQR, i.e., the temperature dependence of resonance frequency, linewidth, and spin–lattice relaxation time. From the results of the observation of optical phonons in Raman spectra of both phases above and below T_{c_4}, as shown in Fig. 6, it is possible that the phase transition at T_{c_4} is the elastic transition as pointed out by Aubry and Pick [12]. For the lowest-frequency mode decreases its energy near the transition point but has a still finite frequency at T_{c_4}. This mode is observed in both phases. In the rare-earth pentaphosphates which have the same D_{2h}–C_{2h} symmetry phase transition as $CsPbCl_3$, soft optical modes (B_{2g}-A_g) were observed in both the paraelastic and ferroelastic phases by Raman scattering [13]. The optical mode with B_{2g} symmetry in the D_{2h} phase couples to the strain of the soft acoustic mode. Some anomaly of the optical mode caused by such a coupling was observed by Raman scattering on $BiVO_4$ [14] and $LaNbO_4$ [15], too. This suggests that in $CsPbCl_3$ an anomaly of the optical phonon around T_{c_4} exhibits the phase transition and the phase transition is caused by the shear strain (x_5) associated with a soft acoustic phonon which belongs to the same B_{2g} symmetry.

References

[1] Y. Fujii, S. Hoshino, Y. Yamada, and G. Shirane, Phys. Rev. 9, 4549 (1974).
[2] J. A. Cape, R. L. White, and R. S. Feigelson, J. appl. Phys. 40, 5001 (1969).
[3] M. I. Cohen, K. F. Young, Te-Tse Chang, and W. S. Brower, Jr., J. appl. Phys. 42, 5267 (1971).
[4] Y. Okamoto, Rep. Fac. Engin. Oita Univ. 4, 107 (1978) (in Japanese).
[5] S. Hirotsu, J. Phys. Soc. Japan 31, 552 (1971).
[6] G. F. Koster, I. O. Dimmok, R. G. Wheeler, and H. Statz, Properties of the Thirty-Two Point Groups, M.I.T. Press, 1963 (p. 39).
[7] N. Tovborg-Jensen, J. chem. Phys. 50, 559 (1969).
[8] R. L. Armstrong, J. magnetic Resonance 20, 214 (1975).
[9] H. M. van Driel and R. L. Armstrong, Phys. Rev. B 12, 839 (1975).
[10] A. M. Glazer, Acta Cryst. B28, 3384 (1972); A31, 756 (1975).
[11] M. Hidaka and S. Hosogi, J. Physique 43, 1227 (1982).
[12] S. Aubry and R. Pick, J. Physique 32, 657 (1971).
[13] D. L. Fox, J. F. Scott, and P. M. Bridenbaugh, Solid State Commun. 18, 111 (1976).
[14] A. Pinczuk, G. Burns, and F. H. Dacol, Solid State Commun. 24, 163 (1977).
[15] M. Wada, Y. Nakayama, A. Sawada, S. Tsunekawa, and Y. Ishibashi, J. Phys. Soc. Japan 47, 1575 (1979).

(Received May 24, 1983)

phys. stat. sol. (a) 79, 271 (1983)

Subject classification: 1.2; 10.2; 22.4.4

Department of Physics, Banaras Hindu University, Varanasi[1])

Mechanism of Solid State Transformations in Single Crystals of $Zn_xCd_{1-x}S$

By

M. T. SEBASTIAN and P. KRISHNA

The mechanism of solid state transformations in single crystals of $Zn_xCd_{1-x}S$ is investigated by annealing metastable 2H crystals at different temperatures, arresting the solid state transformation by quenching in cold water, and examining the structure of the intermediate state by X-ray diffraction. In the range $x = 0.93$ to 0.95, the 2H structure is found to transform to a disordered 6H structure at temperatures around 600 °C. This 6H structure does not transform further to the 3C structure on annealing at higher temperatures. It is observed that the mechanism of the 2H to 6H transformation in these crystals is very different from that proposed by Pandey, Lele, and Krishna for a similar transformation in SiC. A model of transformation involving the non-random insertion of stacking faults is developed using a fault probability α for random nucleation of the 6H phase and a fault probability β for the growth of the 6H nuclei. The diffraction effects calculated from such a model for both, deformation faults and layer-displacement faults are compared with those recorded on a single crystal diffractometer using partially transformed crystals. It is concluded that the model of transformation by a non-random insertion of deformation faults with $\beta \gg \alpha$, closely approximates the actual mode of transformation in these crystals.

Der Mechanismus von Festkörpertransformationen in $Zn_xCd_{1-x}S$-Einkristallen wird durch Temperung von metastabilen 2H-Kristallen bei verschiedenen Temperaturen untersucht, wobei die Festkörperübergänge durch Abschrecken in kaltem Wasser fixiert und die Struktur des Zwischenzustandes durch Röntgenbeugung analysiert wird. Im Bereich von $x = 0,93$ bis 0,95 wird gefunden, daß die 2H-Struktur sich in die fehlgeordnete 6H-Struktur bei Temperaturen um etwa 600 °C umwandelt. Diese 6H-Struktur ändert sich nicht weiter zur 3C-Struktur nach Temperung bei höheren Temperaturen. Es wird beobachtet, daß der Mechanismus des 2H → 6H-Übergangs in diesen Kristallen sich sehr stark von dem von Pandey, Lele und Krishna für einen ähnlichen Übergang in SiC vorgeschlagenen Modell unterscheidet. Ein Modell des Übergangs, das eine nichtstatistische Einführung von Stapelfehlern einschließt, wird entwickelt, wobei eine Fehlerwahrscheinlichkeit α für statistische Keimbildung der 6H-Phase und eine Fehlerwahrscheinlichkeit β für das Wachsen der 6H-Keime benutzt wird. Beugungseffekte, die aus diesem Modell sowohl für Deformationsfehler als auch Fehler verlagerter Ebenen berechnet werden, werden mit denen mittels eines Einkristall-Diffraktometers an teilweise transformierten Kristallen aufgenommenen Fehlern verglichen. Es wird angenommen, daß das Modell der Transformation durch eine nichtstatistische Einführung von Deformationsfehlern mit $\beta \gg \alpha$ sehr genau das tatsächliche Transformationsverhalten in diesen Kristallen approximiert.

1. Introduction

It is known [1, 2] that the structural disorder observed in single crystals of ZnS grown from the vapour at temperatures above 1020 °C, occurs in the course of a solid-state transformation from the high-temperature h.c.p. (2H) structure to the low-temperature c.c.p. (3C) structure. The transformation is believed to occur by a process of periodic slip caused by the expansion of basal plane stacking faults around a screw dislocation [3, 4]. There is evidence to suggest that in crystals which do not grow around a screw dislocation the transformation takes place [2, 5] by the non-random nucleation of

[1]) Varanasi 221005, India.

deformation faults. When ZnS is doped with CdS to form solid solutions of the type $Zn_xCd_{1-x}S$, the disorder effects cease [6, 7] to occur for $x < 0.94$ and the 2H phase gets stabilized. In the present work we have studied the transformation behaviour of single crystals of $Zn_xCd_{1-x}S$ by annealing crystals with a 2H structure at different temperatures, quenching them in cold water, and examining the structure of the intermediate state by X-ray diffraction.

2. Experimental Results

It is observed that for $x \geqq 0.98$ the crystals undergo the 2H–3C transformation in a manner analogous to that for pure ZnS, but crystals with a larger Cd content ($0.95 < x < 0.98$) display an intermediate state with a disordered 6H structure around 600 °C. In the range $x = 0.93$ to 0.95 this 6H state is stable and does not transform to a 3C structure on annealing at higher temperatures. The 2H to 6H transformation is evident from Fig. 1 which shows the $10.L$ reciprocal lattice row of a typical 2H $Zn_{0.94}Cd_{0.06}S$ crystal recorded before and after successive annealing runs at 800 and 900 °C. Fig. 2 shows a plot of the point intensity distribution along the $10.L$ row of the partially transformed, disordered 2H + 6H structure obtained after annealing at 800 °C. The intensity was recorded on a single-crystal X-ray diffractometer by the method described earlier [8]. It is observed that the 6H peaks appear at an early stage and co-exist with the 2H peaks, building up as the transformation proceeds. The crystals studied by us showed no macroscopic kinks after the transformation indicating that they had not transformed by the periodic-slip (screw dislocation) mechanism but by the non-random nucleation of stacking faults.

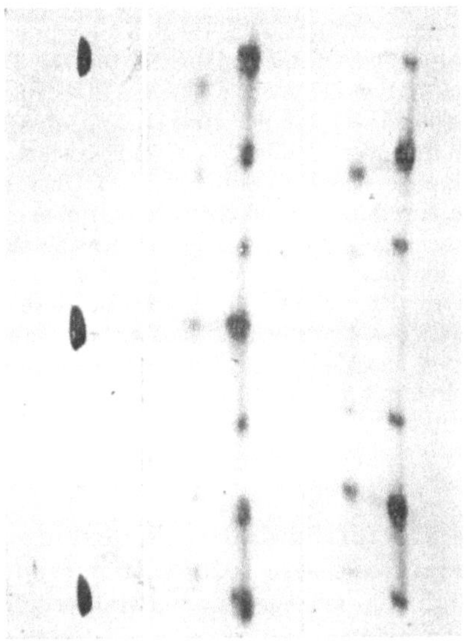

Fig. 1. The $10.L$ reciprocal lattice row of a 2H $Zn_{0.94}Cd_{0.06}S$ crystal recorded on 15° c-axis oscillation photographs after annealing at successively higher temperatures for 1 h and quenching; a) room temperature, b) 800, c) 900 °C. The 2H structure in a) has transformed partially to a faulted 6H structure in b) and completely in c); camera radius 3 cm, CuK radiation (4 ×)

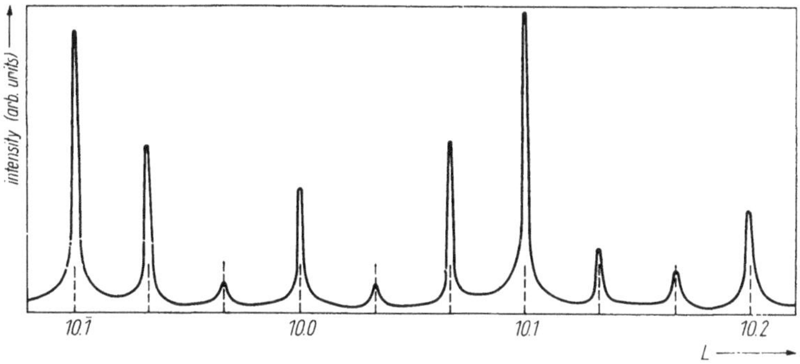

Fig. 2. Single crystal diffractometer plot of intensity versus L along the $10.L$ reciprocall attice row recorded after annealing a perfect 2H crystal of $Zn_{0.94}Cd_{0.06}S$ at 800 °C for 1 h. Vertical dashed lines indicate the 6H positions. The corresponding photographic record of the $10.L$ row is shown in Fig. 1b

3. Mechanism of the 2H to 6H Transformation

The 2H to 6H transformation in a close packed structure has been discussed in detail by Pandey et al. [8]. It can occur either by the non-random insertion of deformation faults or by the non-random insertion of layer displacement faults, the faults nucleating preferentially at three-layer separations in each case as depicted below.

(i) The deformation mechanism

initial structure (2H): AB|ABABABABAB...
CAC|ACA CACA...
BAB|ABAB...
CAC|A...
B...
Final structure (6H): ABCACBABCACB...

The positions of the deformation faults are shown by vertical lines.

ii) The layer displacement mechanism
initial structure (2H): ABABABABABABAB...
final structure (6H): ABCBACABCBACAB...

The positions of the displaced layers are underlined. Since the nucleation of faults is a statistical process, they will not always occur at three layer separations and therefore (he resulting structure would invariably be a faulted 6H.

Pandey et al. [8] have theoretically calculated the X-ray diffraction effects to be expected from crystals undergoing transformation by each mechanism and concluded that the 2H–6H transformation in SiC occurs by the non-random insertion of layer displacement faults. In their calculation they employed a single fault probability α which was assumed to be zero at smaller than three-layer separations and constant for fault separations equal or greater than three layers. Such a model qualitatively explained the diffraction effects of peak broadening and peak shift recorded experimentally from a SiC crystal undergoing the 2H to 6H transformation.

The main differences in the diffraction effects computed for the two mechanisms were as follows:

(i) the 2H reflections with $H - K \neq 3n$ and $L = 0, \pm 1 \bmod 2$ are broadened in the case of deformation mechanism, while these are not broadened in the layer displacement mechanism;

(ii) for transformation by the deformation mechanism the new 6H reflections $H - K \neq 3n$ and $L = \pm\frac{1}{3}, \pm\frac{2}{3} \bmod 2$ are shifted away from each other at an intermediate stage, whereas these are shifted towards each other for the layer displacement mechanism.

However, a comparison of the observed intensity distribution as recorded from 2H crystals of $Zn_xCd_{1-x}S$ undergoing transformation to the 6H structure with those calculated by Pandey et al. [8] shows that the above models are not applicable to the transformation in this system. While the direction of the shift of the 6H peaks observed in the initial stages of the transformation agrees with that expected for transformation by deformation faulting, the broadening of the 2H peaks is much less than that predicted by the above theory and the 6H peaks develop at a much earlier stage of transformation.

We have therefore computed the diffraction effects on the following model of the transformation, involving two fault probabilities α and β, as postulated by Lele and Pandey [9] to explain the 3C–2H transformation in cobalt. Let α denote the probability of a stacking fault nucleating randomly in the structure and β the probability of the faults nucleating at three-layer separations. Then α represents the probability of the fault creating a fresh 6H nucleus in the 2H structure, whereas β represents the probability of the fault causing the growth of an existing 6H nucleus. Following Pandey et al. [8] we assume that the probability of faulting at smaller than three-layer separations is zero, although it is clear that this is an artificial restriction imposed for mathematical simplicity. The consequences of this assumption will be discussed later. Other assumptions made in the development of the theory of X-ray diffraction are (i) that the crystal is infinite and free of distortions, (ii) that the scattering power of all the layers is the same, (iii) that the faults are spread over the entire basal plane and cause no change in the spacing of the layers, and (iv) that effects of absorption, thermal diffuse scattering, and scattering from point defects are negligible.

According to this model, the transformation commences with the insertion of random faults, which may be either deformation faults or layer displacement faults. Subsequent faulting tends to take place preferentially at three-layer spacings, causing the growth of several 6H nuclei. The transformation is completed when the different 6H regions have grown till they meet each other. In general, the interfaces between the 6H regions would be faulted, and therefore the end product would be a disordered 6H structure containing a random distribution of faults. The theory of X-ray diffraction from randomly faulted close-packed structures can therefore be applied at the very initial stages of transformation, when the crystal corresponds to a randomly faulted 2H structure, and again at the end of the transformation when the crystal represents a randomly faulted 6H structure. During the intermediate stages, the theory of random faulting cannot be applied to compute the diffraction effects. We therefore develop a theory of non-random faulting based on the model described above, both for deformation faults and for layer displacement faults, and predict the diffraction effects in each case.

3.1 X-ray diffraction from crystals undergoing 2H to 6H transformation for the deformation mechanism

Employing the notations used by Pandey et al. [8], the probability trees for a crystal undergoing 2H to 6H transformation by the deformation mechanism can be written as

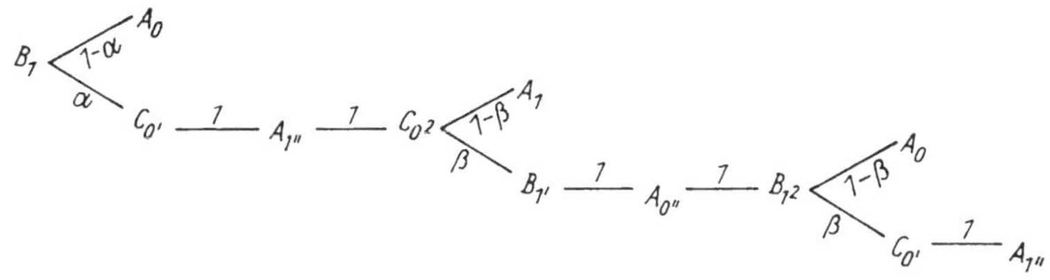

(m − 1)-th layer m-th layer (m − 1)-th layer m-th layer

$$B_1 \;\Big\{\; \xrightarrow{1-\alpha} A_0, \quad \xrightarrow{\alpha} C_{0'} \qquad\qquad C_{0^2} \;\Big\{\; \xrightarrow{1-\beta} A_1, \quad \xrightarrow{\beta} B_{1'}$$

$$A_0 \;\Big\{\; \xrightarrow{1-\alpha} B_1, \quad \xrightarrow{\alpha} C_{1'} \qquad\qquad C_{1^2} \;\Big\{\; \xrightarrow{1-\beta} B_0, \quad \xrightarrow{\beta} A_{0'}$$

$$C_{0'} \xrightarrow{\;1\;} A_{1''} \qquad\qquad A_{1''} \xrightarrow{\;1\;} C_{0^2}$$

$$C_{1'} \xrightarrow{\;1\;} B_{0''} \qquad\qquad B_{0''} \xrightarrow{\;1\;} C_{1^2}$$

From this we obtain the following difference equations:

$$J_{(m,0)} = (1-\alpha)\,J_{(m-1,1)}\omega_2 + (1-\beta)\,J_{(m-1,1^2)}\omega_2\,,$$

$$J_{(m,1)} = (1-\alpha)\,J_{(m-1,0)}\omega_1 + (1-\beta)\,J_{(m-1,0^2)}\omega_1\,,$$

$$J_{(m,0')} = \alpha J_{(m-1,1)}\omega_1 + \beta J_{(m-1,1^2)}\omega_1\,,$$

$$J_{(m,1')} = \alpha J_{(m-1,0)}\omega_2 + \beta J_{(m-1,0^2)}\omega_2\,,$$

$$J_{(m,0'')} = J_{(m-1,1')}\omega_2\,,$$

$$J_{(m,1'')} = J_{(m-1,0')}\omega_1\,,$$

$$J_{(m,0^2)} = J_{(m-1,1'')}\omega_2\,,$$

$$J_{(m,1^2)} = J_{(m-1,0'')}\omega_1\,.$$

On trying a solution of the form $J_{(m,j)} = C_j\varrho^m$; $m \geqq 0$ and eliminating the various C's we get the following characteristic equation:

$$\varrho^8 - (1-\alpha)^2\,\varrho^6 + \alpha(1-\beta)\,\varrho^4 - \beta^2\varrho^2 + (\alpha-\beta)^2 = 0\,.$$

Following Holloway [10], the expression for the diffracted intensity from crystals undergoing 2H–6H transformation can be written as

$$I = \{\tfrac{1}{2} + [T_1\,e^{7i\pi L} + T_2\,e^{6i\pi L} + (T_3 + a_6T_1)\,e^{5i\pi L} + (T_4 + a_6T_2)\,e^{4i\pi L} + $$
$$+ (T_5 + a_6T_3 + a_4T_1)\,e^{3i\pi L} + (T_6 + a_6T_4 + a_4T_2)\,e^{2i\pi L} +$$
$$+ (T_7 + a_6T_5 + a_4T_3 + a_2T_1)\,e^{i\pi L} - a_0]/(a_0 + a_2\,e^{2i\pi L} + a_4\,e^{4i\pi L} +$$
$$+ a_6\,e^{6i\pi L} + e^{8i\pi L})\} + \text{complex conjugate}\,,$$

where

$$a_6 = -(1-\alpha)^2\,, \qquad a_4 = \alpha(1-\beta)\,, \qquad a_2 = -\beta^2\,, \qquad a_0 = (\alpha-\beta)^2\,,$$

$$T_1 = -\tfrac{1}{2}\,,$$

$$T_2 = \frac{2-2\beta}{2(1-\beta+3\alpha)}\,,$$

$$T_3 = -\frac{1-\beta}{2(1-\beta+3\alpha)}\,,$$

$$T_4 = \frac{2-6\alpha-2\beta+6\alpha\beta}{2(1-\beta+3\alpha)}\,,$$

$$T_5 = -\frac{1-3\alpha-\beta-3\alpha^2+6\alpha\beta+3\alpha^3\beta}{2(1-\beta+3\alpha)}\,,$$

$$T_6 = \frac{2-12\alpha-2\beta+12\alpha^2+18\alpha\beta-6\alpha^3-12\alpha^2\beta+6\alpha^3\beta}{2(1-\beta+3\alpha)}\,,$$

$$T_7 = -\frac{1-6\alpha-\beta+3\alpha^2+12\alpha\beta-3\alpha^2\beta-3\alpha\beta^2-3\alpha^4+3\alpha^4\beta}{2(1-\beta+3\alpha)}\,.$$

Using the above expressions the diffracted intensity along the 10. L reciprocal lattice row was computed for values of L varying in steps of 0.01 using various values of α (ranging from 0.02 to 0.4) and different values of β (ranging from 0.2 to 0.9). Fig. 3 shows the calculated intensity distribution along the 10.L reciprocal lattice row as obtained for $\alpha = 0.04$ and β varying from 0.2 to 0.9.

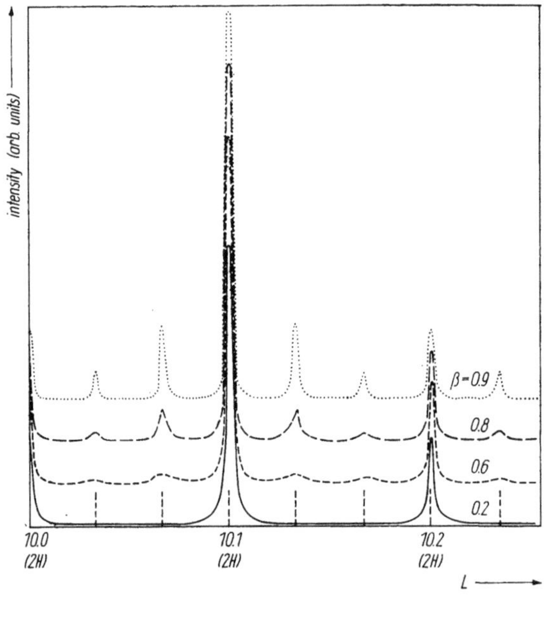

Fig. 3. Calculated variation of the diffracted intensity along the 10. L reciprocal lattice row for a crystal undergoing transformation by the deformation mechanism (two-parameter model with $\alpha = 0.04$, $\beta = 0.2$, 0.6, 0.8, and 0.9). The dashed vertical lines indicate the 6H positions. The calculated curves have been shifted vertically for the different values of β for clarity. The observed intensity distribution shown in Fig. 2 corresponds closely with the calculated curves for $\alpha = 0.04$ and β between 0.8 and 0.9

3.2 X-ray diffraction from crystals undergoing 2H to 6H transformation by the layer displacement mechanism

The probability trees for the 2H to 6H transformation by the layer displacement mechanism can be written as

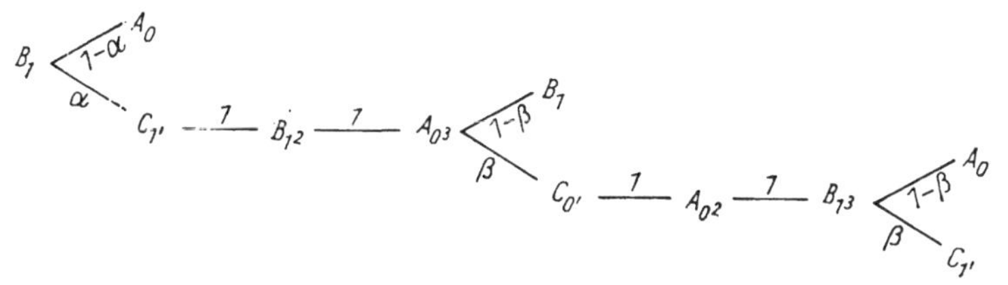

$(m-1)$-th layer		m-th layer	$(m-1)$-th layer		m-th layer
B_1	$1-\alpha$	A_0	B_{1^3}	$1-\beta$	A_0
	α	$C_{1'}$		β	$C_{1'}$
A_0	$1-\alpha$	B_1	A_{0^3}	$1-\beta$	B_1
	α	$C_{0'}$		β	$C_{0'}$
$C_{1'}$	1	B_{1^2}	B_{1^2}	1	A_{0^3}
$C_{0'}$	1	A_{0^2}	A_{0^2}	1	B_{1^3}

We obtain the folllowing difference equations from the probability trees:

$$J_{(m,0)} = (1-\alpha)\, J_{(m-1,1)}\omega_2 + (1-\beta)\, J_{(m-1,1^3)}\omega_2 ,$$

$$J_{(m,1)} = (1-\alpha)\, J_{(m-1,0)}\omega_1 + (1-\beta)\, J_{(m-1,0^3)}\omega_1 ,$$

$$J_{(m,0')} = \alpha J_{(m-1,0)}\omega_2 + \beta J_{(m-1,0^3)}\omega_2 ,$$

$$J_{(m,1')} = \alpha J_{(m-1,1)}\omega_1 + \beta J_{(m-1,1^3)} J\omega_1 ,$$

$$J_{(m,0^2)} = J_{(m-1,0')}\omega_1 ,$$

$$J_{(m,1^2)} = J_{(m-1,1')}\omega_2 ,$$

$$J_{(m,0^3)} = J_{(m-1,1^2)}\omega_2 ,$$

$$J_{(m,1^3)} = J_{(m-1,0^2)}\omega_1 .$$

On trying a solution of the form $J_{(m,j)} = C_j\varrho^m$; $m \geqq 0$, and eliminating the various C's we finally get the following characteristic equation

$$\varrho^8 - (1-\alpha)^2 \varrho^6 - 2\alpha(1-\beta)\,\varrho^4 - \beta^2\varrho^2 + (\alpha-\beta)^2 = 0$$

for which two of the roots have unit modulus.

The applicability of Holloway's method is restricted by the condition that none of the roots of the characteristic equation have unit modulus. Recently Lele [11] has obtained an analytical solution for the diffuse diffracted intensity when the characteristic equation found from the difference equation has roots with unit modulus. Following Lele [11] and Holloway [10] we get the intensity expression for crystals undergoing 2H–6H transformation by layer displacement mechanism as

$$I = \{\tfrac{1}{2} + [K_1\, e^{5i\pi L} + K_2\, e^{4i\pi L} + (K_3 + a_4K_1)\, e^{3i\pi L} + (K_4 + a_4K_2)\, e^{2i\pi L} + $$

$$+ (K_5 + a_4K_3 + a_2K_1)\, e^{i\pi L} - a_0]/(a_0 + a_2\, e^{2i\pi L} + a_4\, e^{4i\pi L} + e^{6i\pi L})\} +$$

complex conjugate ,

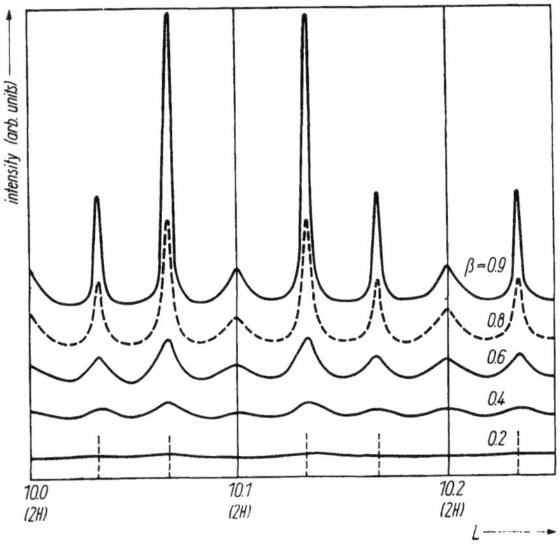

Fig. 4. Calculated variation of the diffracted intensity along the 10. L reciprocal lattice row of a crystal undergoing 2H to 6H transformation by the layer displacement mechanism (two-parameter model $\alpha = 0.1$, $\beta = 0.2$, 0.4, 0.6, 0.8, and 0.9). The dashed vertical lines indicate the 6H positions. The thick vertical lines represent positions of 2H reflections at which the calculated intensity has a singularity and becomes infinite. The calculated curves have been shifted vertically for different values of β for clarity

where

$$a_4 = 2\alpha - \alpha^2\,, \qquad a_2 = -\alpha^2 + 2\alpha\beta\,, \qquad a_0 = -(\alpha - \beta)^2\,,$$

$$K_1 = T_1 - (C_6 - C_7)\,,$$

$$K_2 = T_2 - (C_6 + C_7)\,,$$

$$K_3 = T_3 - (C_6 - C_7)\,,$$

$$K_4 = T_4 - (C_6 + C_7)\,,$$

$$K_5 = T_5 - (C_6 - C_7)\,,$$

$$C_6 = \frac{1 - \alpha - \beta + \beta^3 - \alpha\beta^2 - \beta^2 + 2\alpha\beta}{4(1 - \beta + 3\alpha)\,(1 + 2\alpha - 3\alpha^2 + 4\alpha\beta - \beta^2)}\,,$$

$$C_7 = \frac{3 - 3\beta + 9\alpha - 12\alpha^3 + 3\beta^3 + 24\alpha^2\beta - 3\beta^2 + 6\alpha\beta - 15\alpha\beta^2}{4(1 - \beta + 3\alpha)\,(1 + 2\alpha - 3\alpha^2 - \beta^2 + 4\alpha\beta)}\,,$$

$$T_1 = -\tfrac{1}{2}\,,$$

$$T_2 = \frac{2 - 2\beta}{2 - 2\beta + 6\alpha}\,,$$

$$T_3 = -\frac{1 - \beta + 3\alpha - 3\alpha\beta}{2 - 2\beta + 6\alpha}\,,$$

$$T_4 = \frac{2 - 2\beta + 6\alpha^2 - 6\alpha^2\beta}{2 - 2\beta + 6\alpha}\,,$$

$$T_5 = -\frac{1 - \beta + 3\alpha - 3\alpha^2 + 3\alpha^3 + 3\alpha^2\beta - 3\alpha^3\beta}{2 - 2\beta + 6\alpha}\,.$$

Using the above expressions the intensity distribution along the 10.L reciprocal lattice row was calculated for different values of α and β in steps of $\Delta L = 0.01$. Fig. 4 plots the intensity distribution for $\alpha = 0.1$ and β varying between 0.2 and 0.9.

4. Discussion of Results

A comparison of the actually observed intensity distribution along the 10. L reciprocal lattice row as recorded from a partially transformed crystal (see Fig. 2) with that theoretically computed for the two different mechanisms of transformation (see Fig. 3 and 4) shows that the observed intensity distribution in this crystal corresponds closely to that calculated for $\alpha = 0.04$ and $\beta = 0.8$ to 0.9 for crystals undergoing the 2H–6H transformation by the deformation mechanism. Similar fits were obtained for other crystals with different α and β but always for $\beta \gg \alpha$. For crystals with values of > 0.1 there are observable peak shifts in the position of the 6H reflections in the early stages of transformation. The direction of these shifts agrees with that predicted for the deformation mechanism. The proposed model explains the observed co-existence of the 2H and 6H reflections throughout the transformation. The peak broadening and peak shifts depend on the relative magnitudes of α and β. For small values of α ($\alpha < 0.1$) there is no appreciable peak shift or peak broadening. The case $\alpha = \beta$ is equivalent to the one-parameter model used by Pandey et al. [8] for the 2H to 6H transformation in SiC. The 2H to 6H transformation by the two-parameter model discussed above implies that a few randomly distributed 6H nuclei (probability α) are able to grow into thick 6H regions (probability β). During transformation both α and β vary independently and the quenched intermediate state corresponds to a particular value of α and β.

A comparison of the theoretically calculated intensity profiles with those observed experimentally shows that the model of deformation faulting with $\beta \gg \alpha$ represents a good approximation to the actual situation in these crystals though it does not give an exact fit. The deviations may be ascribed to the artificial assumption that the probability of faults occurring at smaller than three-layer separations is zero. In fact it is observed that there is a considerable enhancement of the intensity near positions of 3C peaks (see Fig. 1c) which suggests that there may be considerable probability of faulting at two-layer separations. A more complex fault probability distribution must therefore be assumed in order to obtain a still closer approximation to the actual mode of transformation in these crystals.

5. Conclusions

(i) Single crystals of $Zn_xCd_{1-x}S$ with 2H structure when annealed undergo a solid-state transformation to a twinned 3C structure for $x \geqq 0.98$ in a manner analogous to that [2] for pure ZnS.

(ii) 2H crystals with larger cadmium content transform to a disordered 6H state which is observed around 600 °C in the range $x = 0.93$ to 0.95. This 6H structure is stable and does not transform to the 3C structure on annealing at higher temperatures.

(iii) In crystals which do not contain an axial screw dislocation, solid-state transformations occur not by the periodic slip mechanism [3, 4], but by the non-random insertion of deformation faults [2].

(iv) A two-parameter model of deformation faulting with a probability α for the random nucleation of the 6H phase and a probability β for the growth of the 6H nuclei with $\beta \gg \alpha$ represents a good approximation to the actual mechanism of solid-state transformation in these crystals.

Acknowledgements

The authors are grateful to the Department of Science and Technology, New Delhi, for financial assistance and to Dr. S. Lele for his help in the theoretical deduction of the diffraction effects. One of us (P.K.) is grateful to Prof. H. Schulz for providing the use of the single-crystal diffractometer facilities at the Max-Planck-Institute (FKF), Stuttgart.

References

[1] W. L. Roth, Faulting in ZnS, General Electric Research Laboratory, Report No. 60-RI-2563 m, Schenectady (N.Y.) 1960.
[2] M. T. Sebastian, D. Pandey, and P. Krishna, phys. stat. sol. (a) **71**, 633 (1982).
[3] B. K. Daniels, Phil. Mag. **14**, 487 (1966).
[4] E. Alexander, Z. H. Kalman, S. Mardix, and I. T. Steinberger, Phil. Mag. **21**, 1237 (1970).
[5] F. Secco d'Aragona, P. Delavignette, and S. Amelinckx, phys. stat. sol. **14**, K115 (1966).
[6] M. J. Kozielski, J. Crystal Growth **30**, 86 (1975).
[7] M. T. Sebastian and P. Krishna, Solid State Commun. (1983) to be published.
[8] D. Pandey, S. Lele, and P. Krishna, Proc. Roy. Soc. **A369**, 435, 451, 463 (1980).
[9] S. Lele and D. Pandey, Proc. Internat. Conf. Solid to Solid Phase Transformation, Carnegie-Mellon University, Pittsburgh 1982 (p. 1487).
[10] H. Holloway, J. appl. Phys. **40**, 4313 (1969).
[11] S. Lele, Acta cryst. **A36**, 584 (1980).

(Received June 1, 1983)

phys. stat. sol. (a) 79, 281 (1983)

Subject classification: 12.1; 1.4; 21.1

Department of Metallurgy, Mechanics and Materials Science, Michigan State University, East Lansing[1])

Deformation of α–β Brass with Oriented and Equiaxed Duplex Boundaries

By

C. F. NILSEN[2]) and K. N. SUBRAMANIAN

As an extension of investigations dealing with two-phase bicrystals of α–β brass, tensile testing is carried out at various strain rates on α–β brass specimens with oriented duplex and equiaxed duplex boundary geometries. The equiaxed boundary is more effective in blocking the propagation of deformation from α to β phase. A strain-rate sensitivity is found in all specimens regardless of boundary geometry. At low-strain-rates, both boundaries are found to be ineffective barriers to the propagation of deformation. At high strain-rates both boundaries fully resisted the passage of deformation. In the α phase at low strain-rates coarse slip is observed, while at high strain-rates fine slip occurs. Optical and scanning electron microscopic techniques are used to study the microstructural features of deformation.

Als Erweiterung der Untersuchungen über Zweiphasen-Bikristalle von α–β-Messing, werden durch Zugversuche bei verschiedenen Dehnungsgeschwindigkeiten die Verfestigungskurven für Proben mit orientierter bzw. gleichachsiger Duplex-Korngrenzengeometrie bestimmt. Für die Blockierung der Verformungsfortpflanzung aus der α-Phase in die β-Phase ist die gleichachsige Grenze wirksamer als die orientierte Grenze. Unabhängig von ihrer Grenzengeometrie weisen alle Proben eine Empfindlichkeit bezüglich der Dehnungsgeschwindigkeit auf. Bei niedrigen Geschwindigkeiten stellt keine der Grenzen ein wirksames Hindernis gegen die Verformungsfortpflanzung dar. Bei hohen Geschwindigkeiten jedoch, widerstehen beide Grenzen dem Fortschreiten der Verformung. Während bei niedrigen Geschwindigkeiten grobe Gleitung beobachtet wird, findet bei hohen Verformungsgeschwindigkeiten feine Gleitung in der α-Phase statt. Die Untersuchung der mikrostrukturellen Merkmale wird unter Anwendung der Licht- und Rasterelektronenmikroskopie durchgeführt.

1. Introduction

Deformation studies of two-phase materials have dealt with bulk poly-crystalline or two-phase bicrystalline specimens [1 to 31]. Macroscopic compatibility conditions require that strains across the phase boundary in each phase be equal. This condition requires the operation of a certain number of slip systems in both crystals [26]. On the other hand, microscopic compatibility conditions must be satisfied at the phase boundary, and have been explained by employing the concepts of dislocation pile-ups, or dislocation density [32].

One of the first systematic investigations on the deformation of bulk polycrystalline two-phase materials was made by Unckel [1], using alloys of copper. Unckel reported that, as the amount of deformation increased, the deformation of the matrix and

---

[1]) East Lansing, Michigan 48824, USA.

[2]) Present address: Department of Mechanical Engineering, Rochester Institute of Technology, Rochester, New York 14623, USA.

particle tended to become more homogeneous. Chao and van Vlack [2, 3], confirmed Unckel's conclusions that the extent of the deformation of the second phase depends on the relative hardness of the particles compared to the matrix.

The first detailed metallographic study was undertaken by Honeycombe and Boas [4] in which they deformed bulk specimens of α–β brass containing 40% zinc. Clarebrough [5] subsequently studied silver–magnesium alloys and found that these alloys deformed in a manner similar to two-phase α–β brass. Clarebrough and Perger [6] observed similar behavior in alloys of copper and zinc. The mechanical properties of a ductile material containing either hard or soft particles have been determined by Gurland and Norton [7]. A systematic study of the role of the second phase particles in two-phase materials has also been made by Edelson and Baldwin [8]. They attempted to identify the parameters that control the mechanical properties of copper-based composites by varying indepenently the volume fraction and size of the second phase particles. Most recently Margolin and co-investigators [9 to 15] have studied the deormation of α–β titanium alloys [9 to 15].

In order to understand the basic mechanisms involved in the deformation of two-phase alloys, the number of variables must be minimized. This may be accomplished by studying the deformation behavior of two-phase bicrystals and multicrystals [16 to 31]. The purpose of the present investigation is to study the effects of interfaces with specific crystallographic relationships between α and β phases on the deformation of α–β brass in order to extend the understanding of deformation of two-phase bicrystals to polycrystalline two-phase materials.

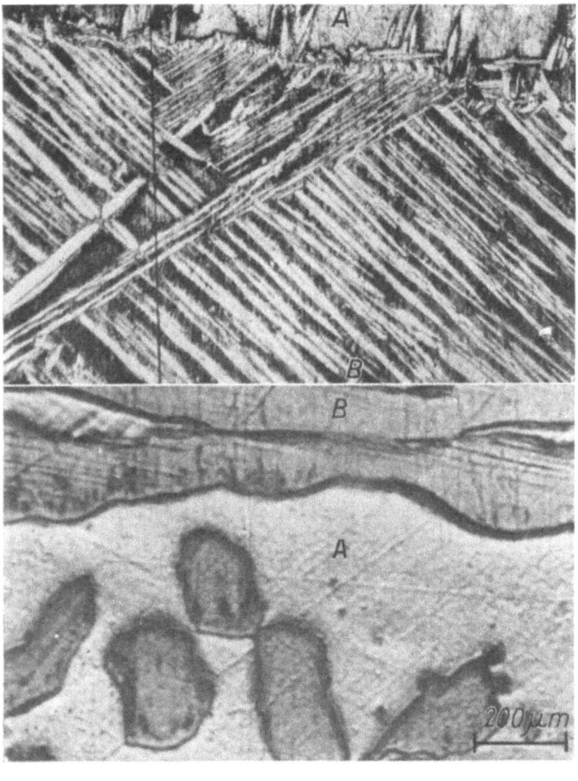

a

b

Fig. 1. Phase boundary geometry. a) Oriented duplex interface; b) equiaxed duplex interface

2. Experimental Procedure

Hingwe and Subramanian [16] have developed experimental techniques for producing fundamental bicrystal and duplex crystal units of α–β brass. Some of these interfaces may result as oriented duplex, equiaxed duplex, corrugated or flat configuration. For the present investigation, specimens with the first two types of boundaries were prepared as tensile specimens by techniques described elsewhere [17]. In these tests interface will be normal to the tensile axis so as to impose same stress due to applied load in all regions.

Tensile tests were carried out using an Instron testing machine with cross-head speed ranging from 0.1 to 3.0 cm/min. The objectives of the tests were to determine the strain-rate sensitivity, and to observe the deformation behavior. The oriented duplex boundary geometry, as shown in Fig. 1 a, has a definite crystallographic orientation relationship between the α and β phases. The relationship of the Widmanstatten plates of α phase in the β phase matrix is $\{111\}_\alpha \parallel \{110\}_\beta$ and $\langle 110\rangle_\alpha \parallel \langle 111\rangle_\beta$. In the unoriented duplex or equiaxed geometry there is no definite orientation relationship. An example of this boundary geometry is presented in Fig. 1 b.

Invariably slip in these specimens starts in the α phase in a region away from the interface and approaches the boundary. Such a fundamental unit is ideal for study of slip interactions with the phase boundary structure. In this study strain-rate is presented in terms of cross-head speed since gage length, length of α and as well as β in the gage length, were kept constant in all the specimens. The specimens deformed first in the softer α phase. Only when the plastic deformation of α was very heavy did the β phase deform plastically. Under such conditions the strain-rate can be assumed to be equal to the cross-head speed divided by the gage length of α phase.

3. Results

The tabulated results of the tensile tests are given in Tables 1 and 2. In all of the micrographs TA, A, and B represent tensile axis, α and β phases, respectively.

Table 1

Tensile test results for specimens with an oriented duplex interface

cross-head speed cm/min)	stress (σ) and strain (ε) at which slip progressed through the interface		resolved shear stress in α to propagate deformation across the interface (MN/m^2)	maximum stress (σ_{max}) and maximum strain (ε_{max}) experienced by the specimen during testing		specimen condition at maximum stress
	σ (MN/m^2)	ε*) (%)		σ_{max} (MN/m^2)	ε_{max}*) (%)	
0.2	42.2	2.8	19.2	121	132.3	failed in α
0.3	94.6	37.2	47.2	112.1	95.4	failed in α
2.0	119.0	64.1	57.0	139.2	82	no fracture

*) Strain values are based on the assumption that all the deformation was accommodated in the α phase.

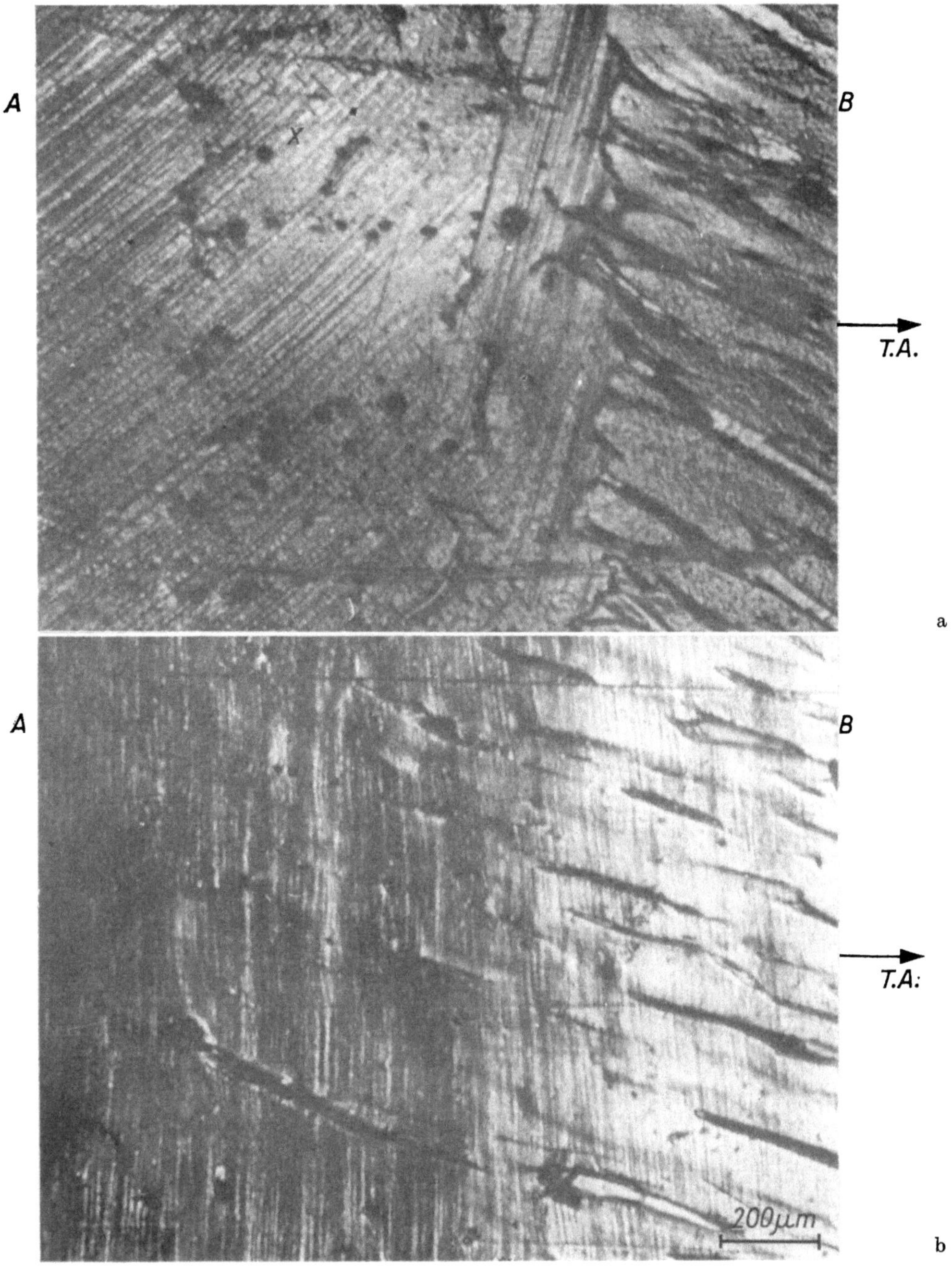

Fig. 2

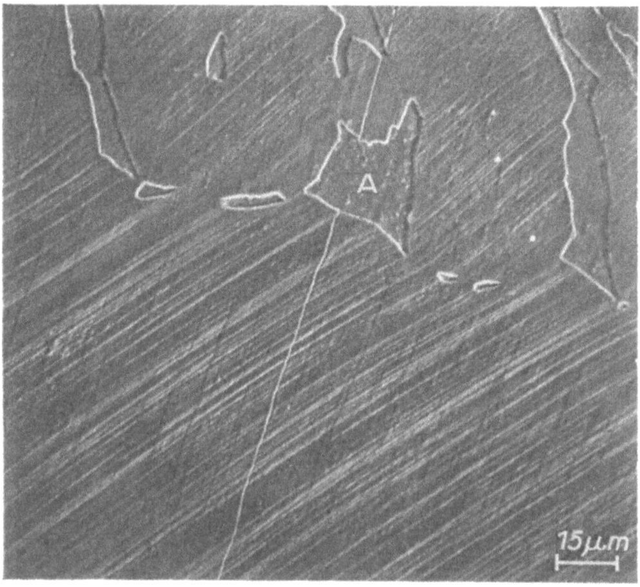

c

Fig. 2. Slip interactions with oriented duplex interface. a) Cross-head speed 0.2 cm/min. Condition for progress of slip through interface: stress 42.2 MN/m^2, strain 28%. Micrograph was taken after subsecting the specimen to stress 121 MN/m^2, strain 132.3%. b) Cross-head speed 0.3 cm/min. Condition for progress of slip through interface: stress 94.6 MN/m^2, strain 37.2%. Micrograph was taken after subjecting the specimen to stress 112.1 MN/m^2, strain 95.4%. c) Cross-head speed 2.0 cm/min. Scanning electron micrograph was taken after subjecting the specimen to stress 139.2 MN/m^2, strain 82%. Condition for progress of slip through the interface: stress 119 MN/m^2, strain 64.1%

3.1 Oriented duplex boundary

In oriented duplex specimens, slip progressed through the interface region at the lowest stress levels. The multiple slip system build-up at the interface of an oriented duplex specimen tested at 0.2 cm/min cross-head speed is presented in Fig. 2a. Fine cross-slip in the phase alpha is evident in a region away from the interface marked X.

Table 2

Tensile test results for specimens with an equiaxed (unoriented) duplex interface

cross-head speed (cm/min)	stress (σ) and strain (ε) at which slip progressed through the interface		resolved shear stress in α to propagate deformation across the interface	maximum stress (σ_{max}) and maximum strain (ε_{max}) experienced by the specimen during testing		specimen condition at maximum stress
	σ (MN/m^2)	ε*) (%)	(MN/m^2)	σ_{max} (MN/m^2)	ε^*_{max} (%)	
0.1	76.8	44.1	36.5	127	48.9	failed in α
0.5	slip did not go through the interface			72.2	41.46	failed in β near the grip
0.5	140.8	40.7	65.3	178.6	77.9	failed in α
1.0	slip did not go through the interface			135	98.7	no fracture
2.0	slip did not go through the interface			142.7	90.2	failed in β near the grip
3.0	slip did not go through the interface			77.9	29.1	no fracture

*) Strain values are based on the assumption that all the deformation was accomodated in the α phase.

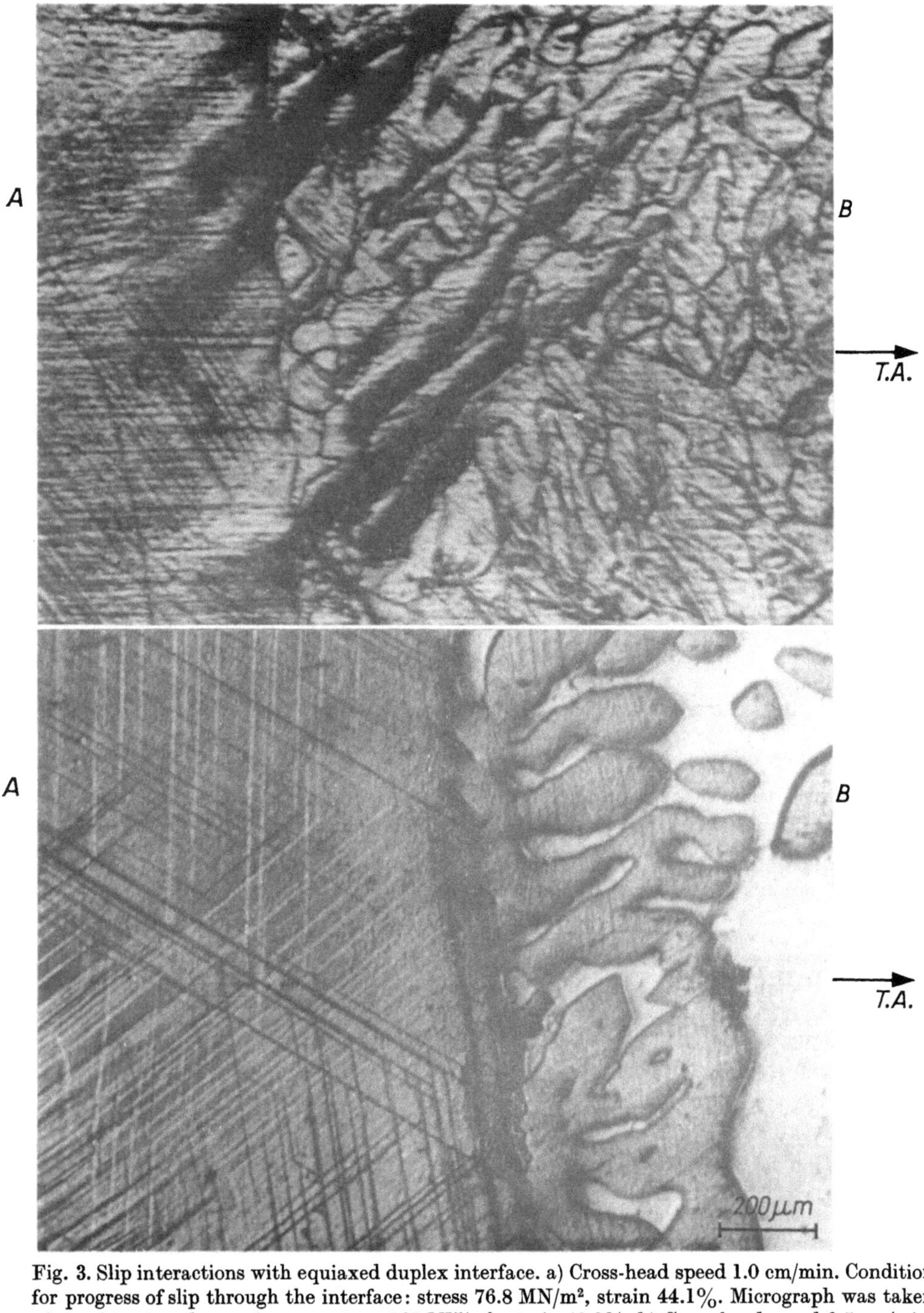

Fig. 3. Slip interactions with equiaxed duplex interface. a) Cross-head speed 1.0 cm/min. Condition for progress of slip through the interface: stress 76.8 MN/m^2, strain 44.1%. Micrograph was taken after subjecting the specimen to stress 127 MN/m^2, strain 48.9%. b) Cross-head speed 0.5 cm/min. Condition for progress of slip through the interface: stress 140.8 MN/m^2, strain 40.7%. Micrograph was taken after subjecting the specimen to stress 178.6 MN/m^2, strain 77.9%. c) Cross-head speed

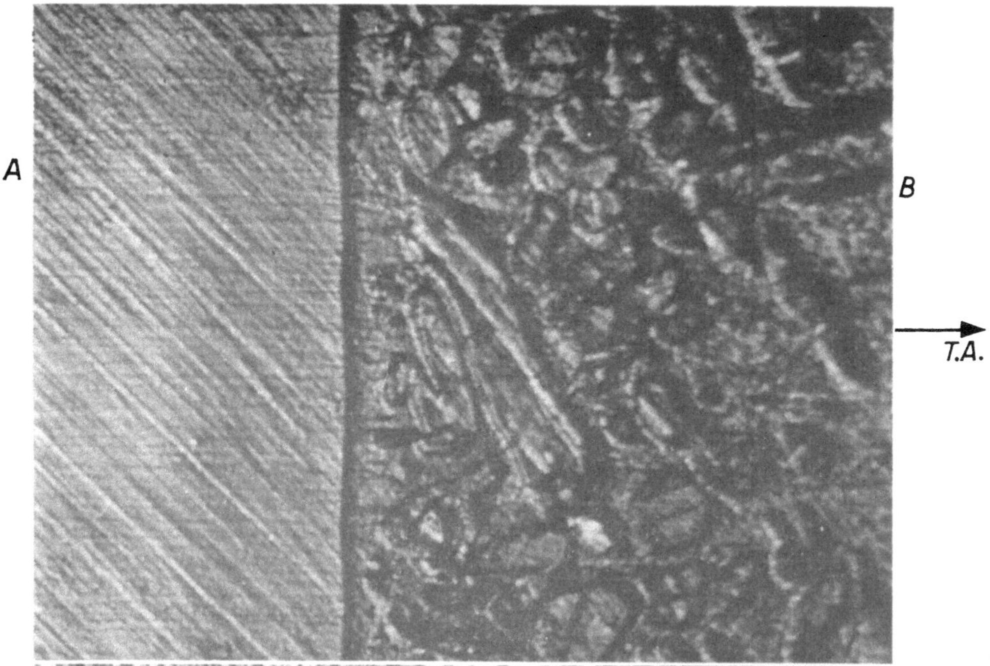

c

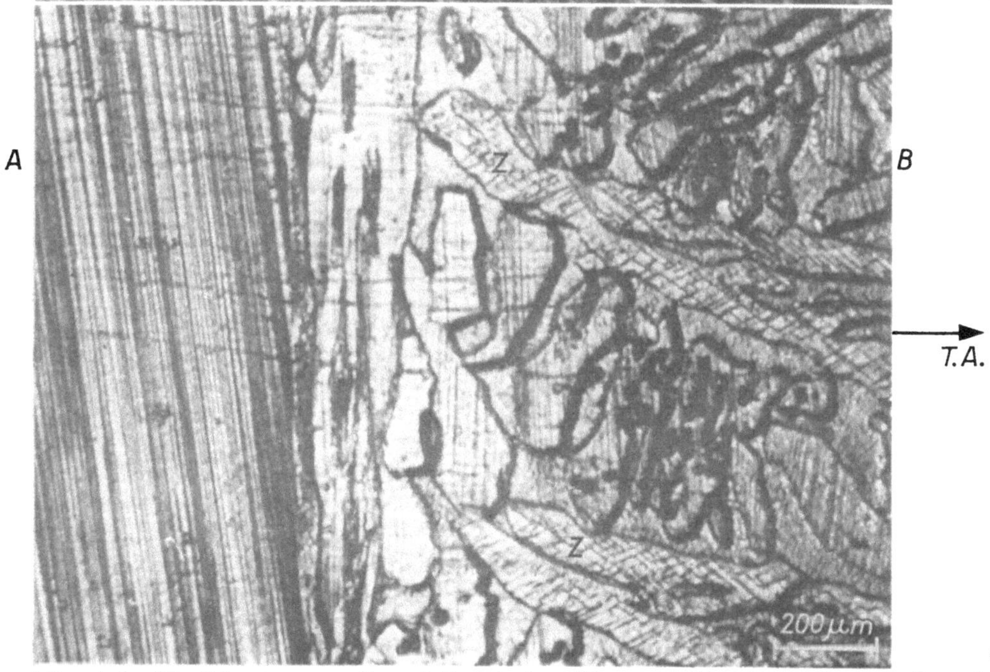

d

1.0 cm/min. Slip did not progress through the interface. Slip occurred in β in a region away from the interface at stress 82.5 MN/m², strain 45%. Micrograph was taken after subjecting the specimen to stress 135 MN/m², strain 98.7%. d) Cross-head speed 2.0 cm/min. Slip did not progress through the interface. Slip occurred in β in a region away from the interface at stress 69.8 MN/m², strain 33.7%. Micrograph was taken after subjecting the specimen to stress 142.7 MN/m², strain 96.2% (note extensive cross-slip in regions marked Z)

Slip progressed through the boundary at a moderate stress level. The slip interaction with the interface for another specimen tested at 0.3 cm/min is given in Fig. 2b. Due to the higher strain-rate and the particular orientation of the slip lines in this specimen, slip did not progress through the interface region until the stress level was approximately twice that for the specimen tested at 0.2 cm/min. A specimen tested at a relatively high cross-head speed of 2.0 cm/min is presented in Fig. 2c and it does not show evidence of progress of slip through the interface region until the stress level had increased an additional one-third beyond that required for a *specimen* deformed at a cross-head speed of 0.3 cm/min. This trend indicates that the stress level necessary to propagate slip through the interface region increases with increasing strain-rate. Generally in the oriented duplex interface samples tested, no appreciable amounts of cross-slip build-up was observed at the interface.

3.2 Equiaxed duplex boundary

The micrograph shown in Fig. 3a is that of an equiaxed (unoriented) duplex boundary sample tested using a cross-head speed of 0.1 cm/min. Slip progressed through the interface at a moderate stress level without appreciable cross-slip build-up at the boundary. The specimen shown in Fig. 3b however, tested at a cross-head speed of 0.5 cm/min, shows evidence of cross-slip and multiple slip systems operating and building up at the interface. The stress level necessary to initiate slip through the interface region was nearly twice that of the stress level for the specimen deformed at a cross-head speed of 0.1 cm/min. Micrograph of a specimen tested at a cross-head speed of 1.0 cm/min is shown in Fig. 3c. A fine slip pattern was observed near the boundary with no evidence of multiple or cross-slip build-up. Slip did not go through the interface in this specimen, even after undergoing nearly 100% strain in the α phase. Rather, slip was initiated in the β phase in a region away from the interface. A specimen tested at yet a higher rate of 2.0 cm/min, shown in Fig. 3d, also failed to show any evidence of slip progressing through the interface. Slip was also initiated in the β phase away from the interface in this specimen, at nearly 100% strain in the α phase. Extensive cross-slip in the equiaxed α regions designated by Z, and very heavy slip in the α region may be observed, although there is no evidence of slip in the β phase at the interface.

The reluctance of the slip lines in the α phase regions in this type of interface structure to progress through the β phase regions is further pointed out in Fig. 4. Here the α phase cross-slips, as shown along A–A, whenever it approaches too close to β phase regions marked B.

The series of strain-rate tests utilizing cross-head speeds from 0.1 to 3.0 cm/min with specimens having equiaxed duplex type of interface indicates that with increasing strain-rate the stress level to propagate slip across the interface also increases. The tests also indicate that at high strain-rates, slip does not progress across the boundary, rather, slip is initiated in the beta phase in a region away from the interface. A comparison of the values tabulated in Tables 1 and 2 show that the equiaxed duplex type of interface structure was capable of sustaining higher resolved shear stresses before slip propagated across the boundary. In general, specimens with equiaxed duplex structured interface proved to be stronger than the ones with oriented duplex structured interface.

In these strain-rate tests, none of the samples tested showed any evidence anywhere of impending failure in the region of the boundary. Slip and deformation was initiated somewhere in the mid-section or near the steel nut in the α single crystal. Slip progressed towards the boundary, and definite Luders type bands were observed to move toward the interface. Most often the slip bands initially contacted the bicrystal phase

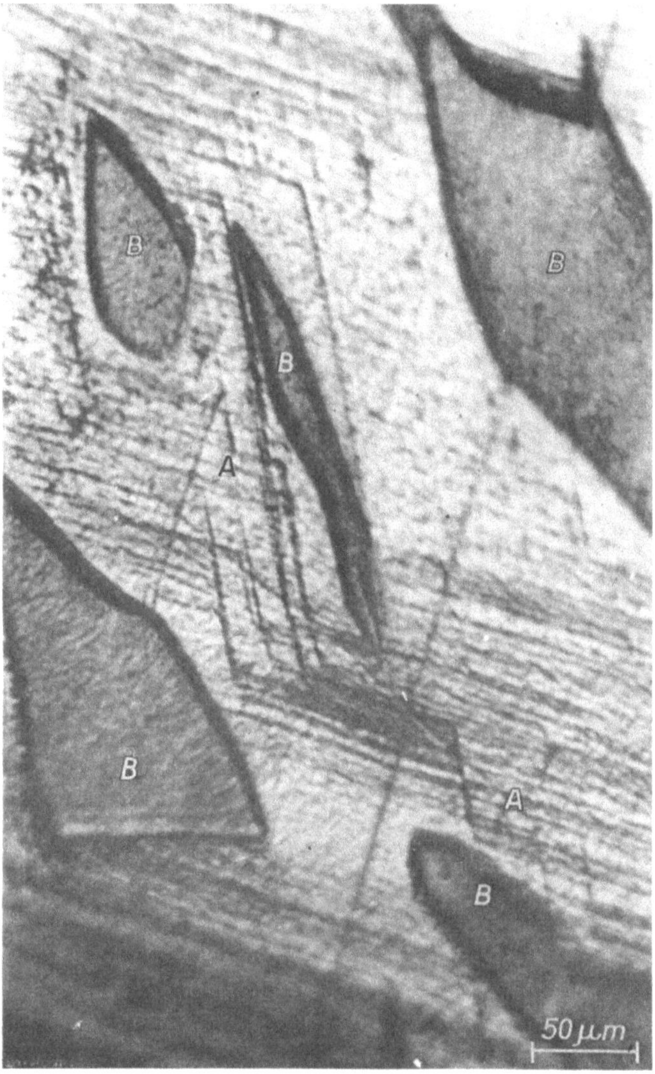

Fig. 4. Extensive cross-slip (points marked A) in α regions caused by island (marked B) in equiaxed interface. Cross-head speed 0.5 cm/min. Condition for progress of slip through the interface: stress 140.8 MN/m², strain 40.7%. Micrograph was taken after subjecting the specimen to stress 178.6 MN/m², strain 77.9%

boundary at an angle and first reached a point where the boundary met the specimen edge. With continued loading the slip lines gradually moved across the boundary face, and the previously formed slip-free triangular zone at the boundary became filled with slip lines. Continued stress application caused further elongation and produced heavy corrugated slip at low strain-rates, while creating very fine slip in the α phase at high strain-rates. As the original slip system became inactive owing to work-hardening, another slip system became active somewhere in the α single crystal, and this system propagated to the boundary as the original system had. Upon reaching the boundary, further work-hardening of the material occurred and cross-slip became an active means of allowing further deformation to take place. As the cross-slip became more extensive, a stress concentration was created at the interface, and at low-strain-rates, slip eventually progressed through the phase boundary into the beta phase where it quickly dissipated.

4. Discussion

4.1 Role of the type of boundary

The unoriented or equiaxed type of boundary geometry has been found to be more effective in preventing slip propagation from the α phase to the β phase than the oriented duplex boundary.

4.1.1 Oriented duplex boundary

In the oriented duplex type of boundary there is an orientation relationship between the Widmanstatten α platelets and the β matrix. Such crystallographic relationships have been observed by Greninger [33]. This type of interface structure results from heat-treating an extensive Widmanstatten type of duplex zone that occurs as a result of the joining process [16]. In the model of this type of structure shown in Fig. 5, the ⟨110⟩ slip direction in the α phase is nearly parallel to the ⟨111⟩ slip direction in the α phase. Further, the close packed planes in the α and β phases are parallel. This immediately provides the condition for the slip plane in α, $\{111\}_\alpha$, to be parallel to the slip plane in β, $\{110\}_\beta$. This most favorable orientation facilitates the progress of slip through the phase boundary with a minimal resistance. Honeycombe and Boas [4] have observed this behavior in α–β brass. Propagation of slip in the β phase is relatively difficult because it requires moving super-lattice dislocations on the slip planes of the ordered β phase. As a result slip does not propagate freely to great distances once it has overcome the phase boundary barrier. Observations of this behavior in the β phase have been made by Brown [34].

As shown in Fig. 5, however, for the case of oriented duplex interfaces, slip need not propagate over large distances in the β phase to progress through to the adjacent β phase regions. As a result of slip propagation through these β platelets, a considerable relaxation of the stress and deformation build up at the interface can occur. As the deformation proceeds from the β platelet, through an α region, and encounters another β platelet, progressively more of the slip is blocked. This behavior further aids the relaxation of stresses at the interface. The oriented duplex structure appears to be a less effective barrier to slip propagation however, at both low and high strain-rates, than the equiaxed duplex type of boundary. The oriented duplex specimens tested have shown a strain-rate sensitivity. The ineffectiveness of the oriented duplex interface as a barrier to slip propagation at all strain-rates (relative to the equiaxed boundary) is undoubtedly due to the existence of the orientation relationship between the α and β phases.

4.1.2 Equiaxed boundary

The unoriented or equiaxed duplex interface structure seems to be the most effective geometry to inhibit the progress of deformation across a boundary. Whenever a slip

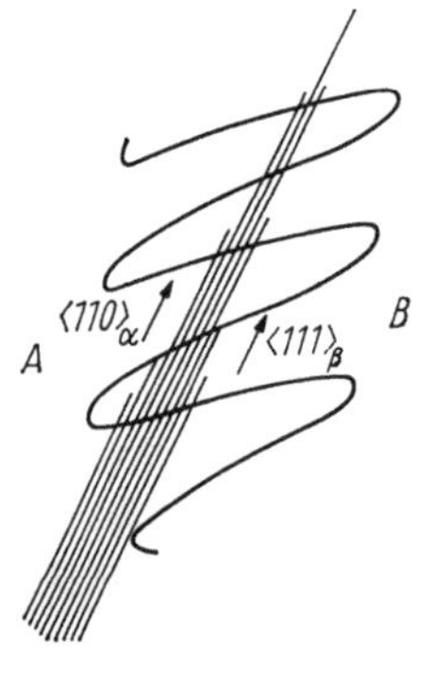

Fig. 5. Model for interaction of slip in α with oriented-duplex boundary

line interacts with an α–β equiaxed boundary, it is either turned away in another direction to continue in the α phase by multiple slip, or it cross-slips to a new plane because of the stress concentration created at the boundary. This type of behavior is clearly shown in Fig. 4. As the deformation of the α phase near the boundary continues the existence of the large boundary surface area causes extensive multiple slip and cross-slip in adjacent α regions. In this way a large amount of deformation can be accomodated in the α regions without the necessity of building large stress concentrations in some areas. Further, the extent of the pile-ups at the interfaces will be less. This will make them relatively ineffective in causing slip to propagate across the boundary.

Once the continous α phase regions in the boundary struture have been heavily deformed and reached a stage at which they cannot accomodate any further deformation, slip appears to progress across the boundary, especially at low strain-rates. The stess level to reach this condition is higher than that needed for the oriented dublex boundary specimens. The equiaxed boundary specimens also showed rather strong strain-rate sensitivity. At low strain-rates, slip did eventually progress across the equiaxed boundary, but at a stress level higher than that required for the oriented boundary samples tested at the same strain-rate. With increasing strain-rate, this stress level also increased until, at a moderate rate, slip was totally blocked at the interface. Rather, slip was initiated in the β phase in regions well away from the boundary. This means that the stress level imposed on the boundary was higher than the critical resolved shear stress in the β phase when high strain-rates were used. This can be explained by the plastic constraints existing at the boundary region between the phases, and the multi-axial stress state created by these constraints.

4.2 Strain-rate effects

α brass deforms by coarse slip at low strain-rates, and by fine and multiple slip at high strain-rates. Repeated cross-slip observed in low strain-rate deformation causes wide slip bands. In a recent article the strain-rate dependence of the slip propagation across phase boundary has been explained on the basis that the wide slip bands produce a more effective stress concentration so as to activate slip in beta regions across the boundary [35]. Fine slip and multiple slip with associated jogs in dislocations are not believed to produce effective stress concentration to cause slip across the phase boundary in β at high strain rates. Further, slip propagation in β is controlled by the motion of super-lattice dislocations. Strain-rate dependence of critical resolved shear stress of beta is not considered to be of any consequence for strain-rate dependence of slip propagation across the interphase boundary since slip has been observed to take place in β regions away from the interface in specimens deformed at high strain-rates. All these hypotheses hold good for the equiaxed and oriented duplex interfaces.

5. Summary

1. For the two types of boundaries considered in this study, the equiaxed (unoriented) type of duplex interface structure is the most effective type of barrier to the progress of slip across the phase boundary. All of the bicrystals, regardless of the type of boundary geometry, were found to be strain-rate sensitive.

2. Lower strain-rate tests exhibit extensive cross-slip near the boundary in α. Cross-slip appears to help in creating wider slip bands in the α phase regions near the interface, and cause higher stress concentration at the boundary.

3. Higher strain-rates tend to form very fine slip by activating many dislocation sources on numerous planes. Interaction of such fine slip with the boundary causes multiple slip which immobilizes a large fraction of the dislocations and consequently makes the initiation of deformation across the boundary very difficult.

4. In specimens having an equiaxed boundary, slip usually is initiated in the β phase in a region away from the boundary, indicating that the interface regions are very effective barriers to slip in the absence of definite crystallographic orientation between the phases.

5. Irrespective of the nature of the boundary, no void formation was observed at the interface because of slip interaction in any of the boundaries during this study.

6. The results of bicrystal tests suggest that the dislocation sources in the β phase become operative owing to dislocation pile-ups formed at the boundary in the α phase region, since the slip lines observed in the β regions lie at various crystallographic orientations and were not parallel to the trace of the primary slip planes operative in the α crystal.

7. Phenomena such as initiation of slip at regions far away from the boundary, especially in equiaxed specimens, indicate that the activation of sources in β region is more likely than the mechanism which will utilize grain boundary sources for initiation of slip.

References

[1] H. Unckel, J. Inst. Metals **61**, 171 (1937).
[2] H. C. Chao and L. H. van Vleck, Trans. MS AIME **233**, 1277 (1965).
[3] H. C. Chao and L. H. van Vleck, Mater. Res. Standards **5**, 61 (1965).
[4] R. W. K. Honeycombe and W. Boas, Austral. J. sci. Res. **A1**, 70 (1948).
[5] J. M. Clarebrough, Austral J. sci. Res. **A3**, 72 (1950).
[6] L. M. Clarebrough and G. R. Perger, Austral. J. sci. Res. **A5**, 114 (1952).
[7] J. Gurland and J. T. Norton, Trans. AIME **194**, 1051, (1952).
[8] B. I. Edelson and W. M. Baldwin, Trans. ASM **55**, 230 (1962).
[9] H. Margolin, P. A. Farrar, and M. A. Greenfield, The Science, Technology, and Applications of Titanium, Ed. R. I. Jaffe and N. E. Promisel, Pergamon Press, New York 1970 (p. 795).
[10] S. Ankem and H. Margolin, Metallurg. Trans. **8A**, 1320 (1977).
[11] J. Jinoch, S. Ankem, and H. Margolin, Mater. Sci. Engng. **34**, 203 (1978).
[12] J. L. Swedlow and R. E. Smelser, Mater. Sci. Engng. **40**, 139 (1979).
[13] J. Jinoch, S. Ankem, and H. Margolin, Mater. Sci. Engng. **40**, 141 (1979).
[14] S. Ankem and H. Margolin, Metallurg. Trans. **11A**, 963 (1980).
[15] S. Ankem and H. Margolin, Metallurg. Trans. **13A**, 595 (1982).
[16] A. K. Hingwe and K. N. Subrambnian, J. Crystal Growth **21**, 287 (1974).
[17] A. K. Hingwe and K. N. Subramanian, J. Mater. Sci. **10**, 1983 (1975).
[18] T. Takasugi, O. Izumi, and N. Fat-Halla, J. Mater. Sci. **13**, 2013 (1978).
[19] T. Takasugi, N. Fat-Halla, and O. Izumi, Acta metall. **26**, 1453 (1978).
[20] N. Fat-Halla, T. Takasugi, and O. Izumi, J. Mater. Sci. **13**, 2462 (1978).
[21] N. Fat-Halla, T. Takasugi, and O. Izumi, J. Mater. Sci. **14**, 1651 (1979).
[22] N. Fat-Halla, T. Takasugi, and O. Izumi, Metallurg. Trans. **10A**, 1341 (1979).
[23] T. Takasugi, N. Fat-Halla, and O. Izumi, Strength of Metals and Alloys, Vol. 1, 5th Internat. Conf., Aachen (FRG) 1979 (p. 199).
[24] T. Takusagi, O. Izumi, and N. Fat-Halla, J. Mater. Sci. **15**, 945 (1980).
[25] N. Fat-Halla, T. Takasugi, and O. Izumi, Trans. Japan. Inst. Metals **20**, 493 (1979).
[26] N. Fat-Halla, T. Takasugi, and O. Izumi, J. Mater. Sci. **15**, 3071 (1980).
[27] B. Baudelet and A. Eberhardt, Phil. Mag. **A41**, 843 (1980).
[28] T. Takasugi and O. Izumi, Acta metall. **28**, 465 (1980).
[29] H. Suzuki, T. Takasugi, and O. Izumi, Acta metall. **30** 1647 (1982).
[30] H. Kawazoe, T. Takasugi, and O. Izumi, Acta metall. **28**, 1253 (1980).
[31] T. Takasugi, N. Fat-Halla, and O. Izumi, Acta metall. **27**, 111 (1979).
[32] J. P. Meakin and H. G. F. Wilsdorf, Trans AIME **218**, 737 (1960).
[33] A. Greninger, Trans. AIME **124**, 379 (1937).
[34] N. Brown, Phil. Mag. **4**, 693 (1959).
[35] C. F. Nilsen and K. N. Subramanian, to appear in J. Mater. Sci.

(Received June 17, 1983)

phys. stat. sol. (a) 79, 293 (1983)

Subject classification: 1.2; 14.4; 22.5; 22.5.1

Faculty of Applied Science, Coventry (Lanchester) Polytechnic[1])

The Silver Iodide–Thallium Iodide Pseudo-Binary System

By

J. W. BRIGHTWELL, L. S. MILLER, A. MUNDAY, and B. RAY

A form of the T–x section of the equilibrium phase diagram for the system AgI–TlI is proposed on the basis of thermal and X-ray diffraction data. Two room temperature stable intermediate phases $AgTlI_2$ and $AgTl_2I_3$ are confirmed, and evidence is found for two other phases, stable at higher temperatures. The electrical conductivity is determined principally by the fraction of AgI present; fast ion conduction occurred only when α-AgI was present.

Une forme du section du diagramme de phases pour le système AgI–TlI est proposé sur la base d'accidents thermiques et de diffraction de rayons X. Deux phases intermedianes $AgTlI_2$ et $AgTl_2I_3$ de temperature ambiente etaient verifiés, et evidence était trouvré pour deux autres phases stables à temperatures plus hautes. Le conductivité electrique était determiné au primaire pour le fraction de AgI présent aux matériaux conduction vite ionique n'occurait que quand α-AgI était présent.

1. Introduction

AgI and TlI have, superficially, features in common; both are polymorphic with yellow low temperature and red high temperature forms, having transition temperatures of 147 and 175 °C respectively. The nature of the two transitions is, however, different in the two cases. For AgI, it is a rapid event from a β-(wurtzite) form to an α-form based upon a body centred cubic iodine sub-lattice and silver in various sites of low co-ordination number; this transition is accompanied by a large increase in electrical conductivity of some five orders of magnitude. Conductivity in both forms is due to the mobility of Ag^+ ions.

For TlI, the transition is from an orthorhombic low temperature form, with Tl having the unusual co-ordination number of 7 [1], to a CsCl lattice [2]; this transition is slow and, very unusually, to a higher density form [3]. This transition is accompanied also by increase in electrical conductivity, but of only about two orders of magnitude [4, 5]. The electrical conductivity in both forms of TlI is less than that in β-AgI.

There are reports on the phase reactions of mixtures of AgI and TlI. There is some conflict [6, 7] about the form of the liquidus for the system, but it is generally accepted that there are two intermediate compounds formed, corresponding to $AgTlI_2$ (δ) and $AgTl_2I_3$ (ϰ); it is not clear whether the latter compound forms congruently from the melt, although it is certain that the former does not. Both of these intermediate phases are stable at room temperature and can be formed by co-precipitation from aqueous solution [8, 9]. The structure of $AgTlI_2$ has been shown to be tetragonal [8, 10], whereas no satisfactory indexing of the diffraction lines of $AgTl_2I_3$ has been reported. The electrical conductivity of these intermediate phases has been studied [11] and they have been shown to be Ag^+ ion conductors, although of low conductivity.

---

[1]) Priory Street, Coventry CV1 5FB, Great Britain.

The work described here is directed at the study of the T–x section of the AgI–TlI equilibrium phase diagram using differential scanning calorimetry (DSC), differential thermal analysis (DTA) and X-ray diffraction (XRD), reinforced on occasions by hot stage microscopy. Electrical conductivity versus temperature data is essentially consistent with the proposed phase diagram.

2. Preparation and Experimental Techniques

2.1 Preparation

Samples were prepared by co-precipitation from aqueous solution of silver nitrate and thallium (I) acetate or thallium (I) nitrate in appropriate mole ratios with a slight stoichiometric excess of aqueous potassium iodide. The solution and precipitate were then heated together for about 10 min to allow equilibriation and particle size growth, the precipitate was filtered off, dried at 130 °C, intimately ground and then stored in an air oven at 80 °C for several days until required. The samples were yellow except for that corresponding to $AgTl_2I_3$ which was nearly white with a slight yellow cast, which microscopic examination showed to be due to presence of a trace ($\ll 1\%$) of an amber phase.

2.2 Thermal and chemical analysis

The composition of the samples was monitored by X-ray fluorescence analysis, using as control materials discs pressed at 8 kbar from mixtures of AgI and TlI seperately prepared and intimately mixed. It was found that the sample preparation technique led to slightly low TlI contents, although the differences were only of substantial significance at low TlI levels. Thermal analysis investigations were conducted with a Perkin-Elmer DSC1B and a Stanton Redcroft DTA 673-4 using α-Al_2O_3 reference material with heating/cooling rates of 8 and 10 K min^{-1}, respectively, for the two

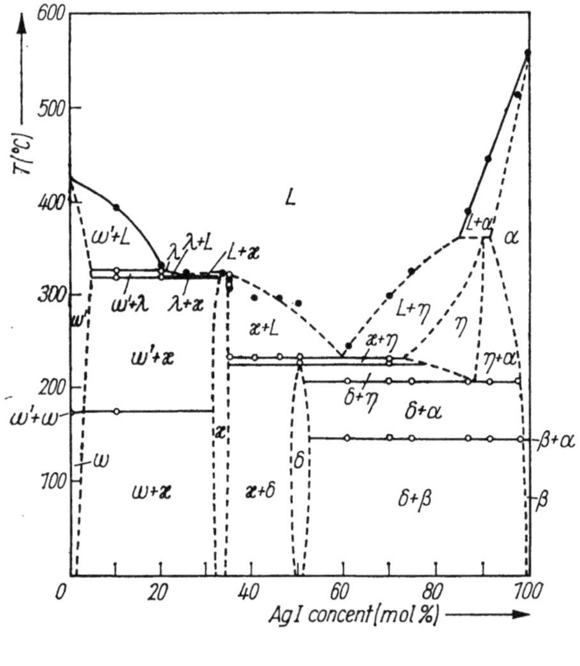

Fig. 1. AgI–TlI equilibrium phase diagram. ● DTA cooling events, ○ DSC heating events, α α-AgI, β β-AgI, η high temperature AgI rich phase, δ $AgTlI_2$, ϰ $AgTl_2I_3$, λ high temperature TlI rich phase, ω low temperature TlI, ω' high temperature TlI, L liquid phase

techniques. There were no significant differences between the two sets of results. The events shown on the equilibrium diagram (Fig. 1) are all DSC events up to ≈ 300 °C, since the resolution of this instrument is slightly better than the DTA, and all events, except for those on the liquidus, are derived from the heating mode. Cross reference to cooling data was, however, a necessary aid to interpretation of the results.

2.3 Structural studies

X-ray diffractometer studies (using a Philips diffractometer PW1730-1710 with CuKα radiation) showed all the samples to be single or two phased. The diffraction patterns observed for $AgTlI_2$ and $AgTl_2I_3$ are given in Tables 1 and 2, respectively.

Attempts to quench in the high temperature from of TlI by rapid cooling to −196 °C failed, but it was found that after repeated runs in the DSC, TlI showed a sharp transition at 175 °C; after thermal soaking at 200 °C, in order to eliminate completely the low temperature form, the reverse transition did not occur during cooling, at a rate of 8 K min^{-1}, to room temperature. The high temperature red form then took several weeks to undergo a complete transition to the yellow form at room temperature. Table 3 shows the diffraction pattern recorded at 20 °C for this high temperature form.

The samples mass used in these structural investigations was small (≈ 20 mg), so that the I/I_1 values obtained may differ from the ideal values.

Table 1

X-ray diffraction data for $AgTlI_2$

d (Å) (observed)	I/I_1	*hkl*	*d* (Å) (calculated)
5.91	1	110	5.904
4.18	12	200	4.175
3.36	54	211	3.357
3.215	10	112	3.215
2.953	53	220	2.952
2.824	100	202	2.823
2.641	16	310	2.640
2.339	9	222	2.339
2.109	12	213	2.109
2.088	6	400	2.087
1.959	17	411	1.958
1.916	10	004	1.918
1.868	12	420	1.867
1.834	10	402	1.833
1.752	9	332	1.751
1.680	11	422	1.678
1.608	10	224	1.607
1.588	7	413	1.587
1.552	4	314	1.552
1.507	3	423	1.507
1.433	4	530	1.432
1.419	3	indexing uncertain	
1.393	4		
1.339	6		
1.335	2		
tetragonal	$a_0 = 8.35$ Å, $c_0 = 7.665$ Å		

Table 2

X-ray diffraction data for $AgTl_2I_3$

d (Å)	I/I_1	d (Å)	I/I_1
cluster 9 to 10			
4.87	3	2.82	60
4.54	6	2.805	90
4.405	10	2.716	8
4.106	20	2.597	3
3.677	11	2.490	6
3.580	40	2.432	20
3.335	42	2.233	8
3.235	100	2.035	20
3.120	35	2.034	12
3.015	80	1.976	12

Many others at lower *d* values.

Table 3

X-ray diffraction data for the high temperature cubic form of TlI from measurements at room temperature

d (Å) (observed)	I/I_1	hkl	d (Å) (calculated)
4.197	6	100	4.211
2.970	100	110	2.978
2.427	2	111	2.431
2.104	29	200	2.105
1.882	1	210	1.883
1.719	45	211	1.719
1.489	15	220	1.489
1.404	1	300	1.404
1.333	22	301	1.332
1.216	4	222	1.216

Calculated density 7.37 g cm^{-3}.

2.4 Electrical conductivity versus temperature

Electrical conductivity studies were made on pressed blocks approximately $15 \times 4 \times \times 4$ mm^3 with contact to the 4×4 mm^2 face being made through fine wire contacts secured with silver dag.

A Portland Electronics conductivity meter operating at 1 kHz was used for the electrical conductivity versus temperature measurements; the method has been described elsewhere [12]. All measurements were taken with the temperature being increased, and time being allowed for the sample to attain thermal equilibrium before each measurement. Before any measurements were taken the pressed blocks were stored at 80 °C for several days in order to eliminate any transient effects due to pressing, and then heated in the measurement apparatus to a temperature of about 130 °C in order to reduce any trace water content. The samples were then cooled to room temperature in a closed environment. No attempt to distinguish between the ionic and electronic components of the electrical conductivity was made.

3. The Equilibrium Phase Diagram

The thermal analysis events suggest a phase diagram of the form shown in Fig. 1; clearly there are substantial differences from previous published information [6, 7]. At AgI levels of below 20%, the transition between the two forms of TlI (ω and ω') is observed, although with some difficulty. It does not appear clearly using the dynamic DSC and DTA techniques, whereas a change in electrical conductivity at temperatures of about 180 °C under pseudo-equilibrium conditions indicates the ω-to-ω' transition to still occur in the 10% AgI sample. The nature of the liquidus indicates that a phase (λ) of composition $\leqq$ 20 mol% AgI forms peritectically at 325 °C. The events at 315 to 325 °C are clearly resolved into doublets both on the heating and cooling cycles; the lower temperature of these events is taken as the eutectoid decomposition temperature of this λ-phase to TlI (ω') and $AgTl_2I_3$ ($\varkappa$).

The liquidus region between 20 and 35 mol% AgI is difficult to resolve, thermal events observed being multiple in nature. At a composition close to $AgTl_2I_3$, there is evidence of congruent melting, indicating that a eutectic reaction occurs at between 20 and 33 mol% AgI. Detailed study of this region is difficult and probably beyond the resolution of the instruments, but a feasible form of the reaction is shown in Fig. 1 consistent with experimental observations. In the region between 35 and 60 mol% AgI the liquidus may have greater detail than shown on the figure. The tetragonal phase $AgTlI_2$ (δ) is not congruently melting and dissociates to $AgTl_2I_3$ (λ) and a silver rich phase at 225 °C, close to the eutectic reaction at 231 °C; these two events were, therefore, not always resolved across the whole range shown.

There is evidence, from the form of the liquidus at high AgI levels, for the peritectic formation of a phase containing $\geqq$ 85 mol% AgI, at 360 °C; this phase (η) reacts eutectoidally at 205 °C to give α-AgI and $AgTlI_2$ (δ). Additional evidence for the formation of this phase is found in the DTA cooling data on samples with $<$ 70 mol% AgI, where the peritectic reaction at 360 °C is clearly observed. Samples with AgI content $>$ 50 mol% show the expected β- to α-AgI transition at 147 °C.

Between 60 and 85 mol% AgI the size of the thermal events on the liquidus were small. Hot stage microscopic examination of a sample of composition 70 mol% AgI, showed, however, growth of crystallites of triangular cross section at 280 °C, although the temperature of the onset of nucleation events at the liquidus was difficult to pinpoint.

The solubility of TlI in AgI is less than 2 mol%, and in that solubility region there was no observable effect on the structure of the β-AgI.

There may be scope for some alternative interpretations of these data from that shown in Fig. 1; however, the system is considerably more complex than revealed previously as a result of the mode detailed examination. The reactions identified in this study are summarised as follows:

(i)	$L + \alpha \rightleftharpoons \eta$	(peritectic at 360 °C),
(ii)	$\omega' + L \rightleftharpoons \lambda$	(peritectic at 325 °C),
(iii)	$L \rightleftharpoons \lambda + \varkappa$	(eutectic at 318 °C),
(iv)	$\lambda \rightleftharpoons \omega' + \varkappa$	(eutectoid at 315 °C),
(v)	$L \rightleftharpoons \varkappa + \eta$	(eutectic at 231 °C),
(vi)	$\varkappa + \eta \rightleftharpoons \delta$	(peritectoid at 225 °C),
(vii)	$\eta \rightleftharpoons \delta + \alpha$	(eutectoid at 205 °C),
(viii)	$\omega' + \varkappa \rightleftharpoons \omega$	(peritectoid at 180 °C),
(ix)	$\delta + \alpha \rightleftharpoons \beta$	(peritectoid at 147 °C).

4. Electrical Conductivity versus Temperature Behaviour

In the temperature range investigated only four samples, 1, 2, 5 and 6, showed sudden transitions of conductivity characteristic of phase changes in this class of material, see Fig. 2; these electrical transition temperatures were observed at typically 10 °C above the thermal transition temperatures. On the basis of the proposed equilibrium phase diagram, the other samples would not be expected to show transitions in this temperature range.

Table 4
Electrical conductivity data

sample	composition (mol% AgI)	low temperature activation energy (eV)	conductivity at 150 °C (S m^{-1})
1	0	0.51	3.2×10^{-5}
2	10	0.66	2.3×10^{-4}
3	33	0.68	2.0×10^{-3}
4	50	0.82	2.7×10^{-3}
5	67	0.46	4.4×10^{-2}
6	95	0.40	4.2×10^{-2}
7	100	0.44	3.4×10^{-2}

For each sample there is a (linear) region of activated conductivity, in what may be considered a low-temperature region. The activation energy (E_a) does not vary markedly as the composition is changed (Table 4). It is greatest (0.82 eV) for sample 4 (50 mol% AgI) and is lowest (≈ 0.4 eV) for the samples with high AgI content. The data for similarly prepared pure AgI gave a (σ versus $1/T$) curve very similar to that for sample 6. For pure AgI, Hoshino and Shimoji [13] obtain $E_a > 0.7$ eV for β-AgI and 0.26 eV for γ-AgI; in this investigation values obtained are consistently an intermediate between these two extremes. A possible interpretation is that both phases are present in our non-equilibriated samples. The X-ray data indicates that the addition of a few mol% of TlI promotes the growth of the β-phase, but this appears to have little effect on the conductivity.

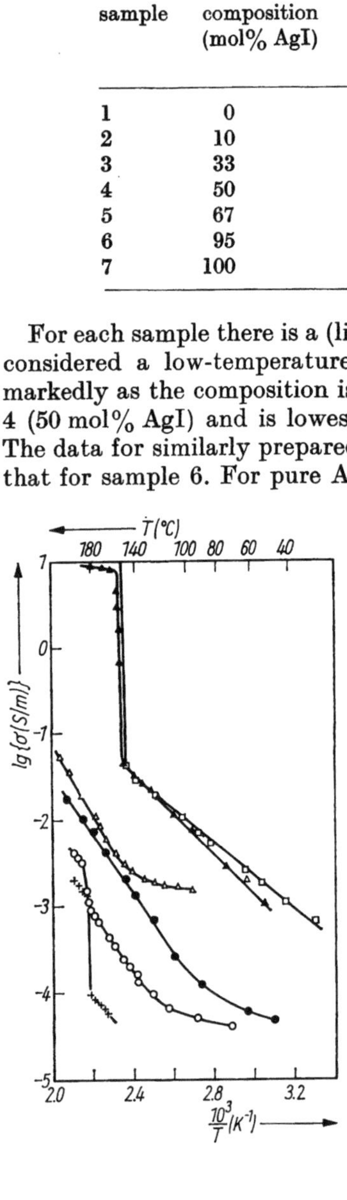

Fig. 2. Electrical conductivity σ vs. reciprocal temperature for AgI:TlI in various ratios. AgI content (mol%): + 0, ○ 10, ● 33, △ 50, ▲ 67, □ 95

It is interesting to consider the variation with composition of σ in the low-temperature region. By choosing a temperature of 150 °C, little extrapolation is necessary, and so these values have been listed in Table 4. The most striking feature is the rapid reduction of σ that occurs as the mol% of AgI is reduced to 50% and below. This is consistent with the general trend of lower mobility of Tl ions, as reported by Ammlung et al. in which 50 mol% TlI composition had the highest TlI content that did not contain free AgI [14].

This trend extends to the high temperature forms, where samples 5 and 6 have transitions to very high conductivity phases of low activation energy. The transition temperature in these cases is essentially the β–α AgI transition at 147 °C, implying the presence of α-AgI above this temperature. The thallium-rich samples on the other hand, have an activation energy of ≈ 0.5 eV above their transition temperature, which is around 175 to 180 °C. The conductivity change across the phase boundary is only one order for pure TlI and even less with 10 mol% AgI added (sample 2).

For pure TlI, our results may be compared with those of Samara [5], Morlin [4], and Shiraldi et al. [11]. Our value of activation energy below the transition temperature is close to Morlin's value of 0.53 eV but above the transition our value of ≈ 0.5 eV is closer to that of Samara (0.68 eV). Also, our actual conductivity values agree with those of Samara to within about 25%. The agreement with the TlI data of Schiraldi et al. is also good: further, present data supports their value at 33% AgI (but differs substantially for that at 50% AgI). It remains to discuss the low-temperature non-linear region of Fig. 2 for samples 2, 3, and 4. It is our opinion at this time that these do not represent different phases but are regions where a different mechanism dominates the conduction process. If the conductivity is activated, the activation energy is low; the conductivity in these regions is thus too low to be associated with an intrinsic process and an extrinsic mechanism is therefore suggested. Samara [5] has proposed that all the conduction in the low-temperature (orthorhombic) phase is electronic (whereas Schiraldi et al. [11] suggested it was partially so); we are not yet in a position to verify this but would suggest electronic conduction as a possible mechanism in these non-linear regions. Schiraldi et al. (for 33 and 50 mol% AgI) did not observe these regions for temperatures down to 90 °C.

5. Discussion and Conclusions

It is apparent from X-ray data that the addition of small quantities of TlI does not result in a structure different from that of β-AgI. However, in our experience, AgI samples pressed at 8 kbar form a pure γ-(zinc blende) phase, probably via conversion to the high pressure (NaCl) form which then relaxes back via a tetragonal to a cubic phase [15]. Since all of the strong diffraction lines of the γ-phase coincide with the strong lines of the β-phase there may be some restricted possibility of a mixture of β- and γ-AgI occurring. It is unlikely though that such transitions were induced in the samples used for electrical conductivity measurements since the pressure used was limited to approximately 2 kbar, which for pure AgI is below the pressure at room temperature for the β-to-γ transition.

The diffraction trace for the δ-phase ($AgTlI_2$) is given in Table 2 and is included primarily to illustrate the phase purity obtained from the preparative technique. The ϰ-phase ($AgTl_2I_3$) exhibits a much more complex diffractogram (Table 3) which has not been indexed; the large number of lines indicates a form of low symmetry or a large unit cell. It has been reported [10] that elimination of lines due to AgI and $AgTlI_2$ from such a pattern leaves a set which can be indexed to hexagonal with $a_0 = 10.37$ Å and $c_0 = 20.12$ Å, alternatively rhombohedral with $a_0 = 8.97$ Å and $\beta = 70°36'$.

The apparent purity of this obviously white phase, whilst AgI and $AgTlI_2$ are both yellow, and the generally poor fit to this cell make such a possibility remote. The structure of Tl_2CuI_3 has been reported from single crystal studies [10, 16] to be monoclinic with $a_0 = 13.427$ Å, $b_0 = 9.316$ Å, $c_0 = 9.587$ Å, and $\beta = 136°33'$; in view of the close structural relationship between CuI and AgI, which both exhibit zinc blende and wurtzite type phases, it may be that this material provides an analogue.

The determination of the structure of $AgTl_2I_3$ will probably have to await single crystal studies.

The data for the red, high temperature, form of TlI presented in Table 3 is clearly that of a CsCl type structure, for which the calculated density at ≈ 20 °C is 7.17 gcm^{-3}, with $a_0 = 4.211$ Å, a value which is in good agreement with that of (4.2099 ± 0.0003) Å extrapolated from measurements at higher temperature [3].

The principal feature of the electrical conductivity results is that the conductivity of the low temperature phases is dominated by the mole fraction of AgI present. In the temperature range investigated there is no indication of fast ion conduction, except in the region where α-AgI is present.

Acknowledgement

The authors wish to acknowledge the contribution made by Mr. R. Hollyoak through analysis of the samples and provision of facilties.

References

[1] R. W. WYCKOFF, Crystal Structures, Vol. 1, Interscience Publishers, 1963.
[2] L. HELMHOLTZ, Z. Krist. **96**, 129 (1936).
[3] G. A. SAMARA, L. C. WALTERS, and D. A. NORTHROP, J. Phys. Chem. Solids **28**, 1875 (1967).
[4] Z. MORLIN, phys. stat. sol. (a) **8**, 565 (1971).
[5] G. A. SAMARA, Phys. Rev. B **23**, 575 (1981).
[6] L. G. BERG and I. N. LEPESHKOV, Izv. Fiz.-Chim. Anal. Inst. Obshchei Neorg. Chim. **15**, 144 (1947).
[7] PLATONOV, Tr. Moscov, Selsko Khoz. Akad. Im Ka-Timiryazeva **36**, 13 (1946).
[8] H. HIRSCH, J. Chem. Soc. **246**, 1318 (1963).
[9] H. HIRSCH, J. Photo. Soc. **11**, 321 (1963).
[10] P. MESSIEN, Bull. Soc. Roy. Sci. Liege **38**, 490 (1969).
[11] A. SCHIRALDI, A. MAGISTRIS, and E. PEZZATI, Z. Naturf. **29**, 782 (1974).
[12] J. W. BRIGHTWELL, C. N. BUCKLEY, G. FOXTON, L. S. MILLER, and B. RAY, J. Mater. Sci., Letters **1**, 429 (1982).
[13] H. HOSHINO and M. SHIMOJI, J. Phys. Chem. Solids **33**, 2303 (1973).
[14] R. AMMLUNG, R. P. SCARINGE, J. A. IBERS, D. SHRIVER, and D. WHITMORE, J. Solid State Chem. **29**, 401 (1979).
[15] W. A. BASSETT and T. TAKAHASHI, Amer. Mineralogist **50**, 1576 (1965).
[16] L. GUEN, R. MARCHAND, and M. TOURNOUX, C.R. Acad. Sci. (France) **284**, 447 (1977).

(Received June 20, 1983)

phys. stat. sol. (a) 79, 301 (1983)

Subject classification: 1.6 and 8

Institute of Physical Chemistry, Bulgarian Academy of Sciences, Sofia[1])

Heat Capacity of Diatomic Molecules Adsorbed on a Periodic Substrate and the Localized-to-Mobile Transition

By

A. Milchev

Heat capacities and related thermodynamic functions of a system of isolated diatomic molecules adsorbed on a periodic substrate are calculated by means of classical statistical mechanics. It is shown that the thermal transition from localized-to-mobile behaviour of the system of dimers is clearly reflected by the characteristic shape of the specific heat vs. temperature curve. According to the present theoretical investigation the maximum in the specific heat curve marking the localized-to-mobile transition generally depends both in position and in height on the values of the energy barriers W_x and W_y along the X- and Y-axes. For substrates with $W_x/W_y < 1/4$ the single maximum splits into two maxima with different heights. Further, for a given substrate, one obtains qualitatively different specific heat curves (with a single or with two separate maxima) if adsorption of dimers with different length is considered. The physical backgrounds of the problem are elucidated in more detail in the case of 1D system.

Spezifische Wärmen und damit verbundene thermodynamische Funktionen eines Systems von isolierten zweiatomaren Molekülen, die auf einer periodischen Unterlage adsorbiert sind, werden im Rahmen der klassischen statistischen Mechanik berechnet. Es zeigt sich, daß der thermisch bedingte Übergang vom lokalisierten zum mobilen Zustand der Dimere aus der Temperaturabhängigkeit der spezifischen Wärme deutlich bestimmt werden kann. Die Untersuchung zeigt, daß das charakteristische Maximum des Übergangs von lokalisierter zu mobiler Adsorption sowohl seine Lage als auch Größe mit der Änderung der Energiebarriere der Oberflächendiffusion W_x und W_y in Richtung der X- und Y-Achse beträchtlich ändert. Für Unterlagen mit $W_x : W_y < 1/4$ spaltet dieses Maximum in zwei Maxima mit verschiedener Höhe auf. Qualitativ unterschiedliche Kurven der spezifischen Wärme werden vorausgesagt, wenn Gase aus diatomaren Molekülen mit verschiedener Länge auf ein und derselben Unterlage adsorbiert werden (zwei getrennte Maxima oder nur ein einziges). Eine ausführliche Diskussion der physikalischen Zusammenhänge wird am Beispiel des eindimensionalen Modells durchgeführt.

1. Introduction

The changes in the thermodynamic behaviour of a system of isolated atoms adsorbed on a crystalline surface, which reflect the transition from localized-to-mobile behaviour have been studied originally in the classic paper of Hill [1]. Using a simple model for the gas–solid interaction, Hill concluded that the transition from vibrational to free translational movement of the adatoms along the substrate leads to a considerable decrease in the heat capacity of the system (due to the loss of degrees of freedom). The heat capacity curve itself goes through a minimum at temperature $T = W/5k$, which, if detected, provides a rather direct method for measuring the height W of the energy barrier for surface diffusion.

The analytical expressions, derived by Hill, were later examined and precisely specified by Doll and Steele [2] by means of numerical analysis. Despite some important corrections their work confirms to a great extent the original conclusions of Hill.

---

[1]) 1040 Sofia, Bulgaria.

A natural extension of the investigations quoted above appears to study the far more complex problem of adsorption of dimers on a periodic substrate which is the subject of the present paper. The motivation for such a study could be viewed as twofold:

1. It provides deeper insight into the specifics of gas-solid interactions at crystalline surfaces as a function of temperature, and
2. it is intimately connected with the dynamics of the migration of small clusters on a periodic substrate whereby due to the nonlinearity of the problem and due to the existence of parametric resonances in the system a rather strange and unexpected behaviour of the molecules may result [3].

The present paper is organized as follows: In Section 2 for the sake of clarity the more transparent case of one-dimensional (1D) motion of the dimer is more thoroughly treated and some important physical relations are elucidated. Then in Section 3 for the case of a 2D substrate the derivation of the formulae is briefly traced and the theoretical results obtained are presented.

2. Theoretical Consideration

2.1 The 1D partition function

The potential energy of a diatomic molecule, adsorbed on a 1D periodic substrate is given by the Hamiltonian

$$H = \mu(x_1 - x_2 - b)^2 + \frac{W}{2}\left(1 - \cos\frac{2\pi x_1}{a} + 1 - \cos\frac{2\pi x_2}{a}\right). \tag{1}$$

In (1) μ is the elastic constant and b the distance between the atoms of the diatomic molecule, a denotes the lattice constant of the substrate, and W is the energy barrier for surface diffusion.

In order to evaluate the partition function of the system, q_{class}, one has to carry out the integration over the coordinates x_1 and x_2, specifying the position of the molecule, in

$$q_{\text{class}} = \frac{1}{\Lambda^2}\int_{-L}^{L}\int_{-L}^{L} \exp\left(\frac{-H}{kT}\right) dx_1\, dx_2 . \tag{2}$$

Here Λ is the de Broglie thermal "wavelength", $\Lambda = h/\sqrt{2\pi mkT}$, m is the mass of a single atom in the dimer and L denotes the half-length of the substrate (L is a large multiple of a).

Instead of x_1 and x_2, it is convenient to substitute in (2) new relative coordinates, $x_- = x_1 - x_2$ and $x_+ = x_1 + x_2$, so that after a simple transformation of the cosine potential one obtains

$$q_{\text{class}} = \frac{1}{2\Lambda^2}\int_{0}^{2L}\int_{-2L}^{2L} \exp\left[-\frac{\mu}{kT}(x_- - b)^2 - \frac{W}{kT} - \frac{W}{kT}\cos\frac{x_-}{a}\cos\frac{x_+}{a}\right] dx_-\, dx_+ . \tag{3}$$

After performing the integration over the centre of mass variable, x_+, one obtains from (3)

$$q_{\text{class}} = \frac{e^{-W/kT}}{2\Lambda^2}\,\frac{L}{\pi}\,2\pi\int_{0}^{2L} I_0\left(\frac{W}{kT}\cos\frac{\pi}{a}\,x_-\right) e^{-(\mu/kT)(x_- - b)^2}\, dx_- , \tag{4}$$

where $I_0(x)$ is the modified Bessel function [4]. Taking into account the properties of the sharply peaked Gauss function, the integral in (4) may be represented as

$$\approx I_0\left(\frac{W}{kT}\cos\frac{\pi b}{a}\right)\int\limits_0^{2L}\exp\left[-\frac{\mu}{kT}(x_- - b)^2\right]dx_-$$

$$= I_0\left(\frac{W}{kT}\cos\frac{\pi b}{a}\right)\sqrt{\frac{kT}{\mu}}\frac{\sqrt{\pi}}{2}\left[\operatorname{erf}\sqrt{\frac{\mu(2L-b)^2}{kT}} + \operatorname{erf}\sqrt{\frac{\mu b^2}{kT}}\right], \tag{5}$$

where $\operatorname{erf}(x) = (2/\sqrt{\pi})\int\limits_0^x \exp(-t^2)\,dt$.

Thus one obtains for the partition function of a diatomic molecule adsorbed on a periodic substrate at classical (not too low) temperatures,

$$q_{\text{class}} = \frac{L\,e^{-W/kT}}{\Lambda^2} I_0\left(\frac{W}{kT}\cos\frac{\pi b}{a}\right)\sqrt{\frac{\pi kT}{\mu}}\,[\operatorname{erf}(M_L) + \operatorname{erf}(M_b)] \tag{6}$$

with $M_L = \sqrt{\mu(2L-b)^2/kT}$ and $M_b = \sqrt{\mu b^2/kT}$.

The physical content of (6) becomes clearly visible if one takes into account the asymptotic properties of the Bessel function: $I_0(x) \to 1$ for $x \to 0$ and $I_0(x) \to e^x/\sqrt{2\pi x}$ for $x \to \infty$, and those of the error function: $\operatorname{erf}(\infty) = 1$, $\operatorname{erf}(x) \to 2x/\sqrt{\pi}$ for $x \to 0$. It follows from (6):

$$q_{\text{class}}(T) \to \begin{cases} \dfrac{(2L)^2}{\Lambda^2} \quad \text{at} \quad T \to \infty\,, & \text{(7a)} \\[2ex] \dfrac{2L\,e^{-(W/kT)(1-\cos(\pi b/a))}}{\Lambda^2\sqrt{\dfrac{2\pi W}{kT}}\left|\cos\dfrac{\pi b}{a}\right|}\sqrt{\dfrac{\pi kT}{\mu}} \quad \text{at} \quad T \to 0\,. & \text{(7b)} \end{cases}$$

According to (7a), at high temperatures the expected behaviour of the diatomic molecule will be like that of particles which perform a free translational movement along the substrate. In the more involved case of low temperatures, (cf. (7b)), the partition function $q_{\text{class}}(T)$ reflects the existence of two independent oscillatory motions which contribute to the complex behaviour of the diatomic molecule:

1. Oscillations of the length of the dimer with basic frequency $\nu_{\text{osc}} = (1/2\pi)\sqrt{2\mu/m}$ which corresponds to the characteristic Einstein temperature θ_E of a diatomic molecule in the gas phase.

2. Oscillations of a completely stiff diatomic molecule (a dumbbell) with length b in the potential wells of the substrate with ground state frequency ν_{sub}.

The latter may be seen from the following consideration: The equation of motion of a stiff dimer on a corrugated substrate reads as

$$2m\ddot{x}_1 = -\frac{\pi W}{a}\sin\frac{2\pi x_1}{a} - \frac{\pi W}{a}\sin\frac{2\pi}{a}(x_1 + b)\,, \tag{8a}$$

where the cosine potential of (1) has been used.

A transformation of the right-hand side of (8a) gives

$$2m\ddot{x}_1 = -\frac{2\pi W}{a}\cos\frac{\pi b}{a}\sin\frac{\pi}{a}(x_1 + b)\,. \tag{8b}$$

Renaming the variable $C = (\pi/a)(2x_1 + b)$, (8b) transforms then into the familiar sine-Gordon equation

$$\ddot{C} = A \sin C; \qquad A = -\frac{2\pi^2 W}{ma^2} \cos \frac{\pi b}{a} \tag{9}$$

with a trivial solution $C = n\pi$ $(n = 0, \pm 1, \pm 2, \ldots)$, and with a "soliton" solution $C(t) = 4 \tan^{-1} (\exp - \sqrt{A}\, t)$. Small oscillations about the ground state positions will be performed then with frequency $\nu_{\text{sub}} = (1/2a) \sqrt{(W/m) |\cos (\pi b/a)|}$. Thus it is seen that the ratio b/a may influence significantly the vibrational frequency of the rigid dimer, and, e.g. in the case of $b/a = n/2$ it even happens that $\nu_{\text{sub}} = 0$, i.e. the diatomic molecule will move translationally as on a completely smooth surface.

The present analysis shows that (7b) may be clearly interpreted if it is represented as

$$q_{\text{class}}(T) \sim e^{-(W/kT)((1-\cos(\pi b/a))} \frac{(kT)^2}{h\nu_{\text{osc}}\, h\nu_{\text{sub}}}, \tag{10}$$

where the power of the exp function contains the ground state energy of the rigid dimer in the cosine potential.

2.2 Quantum corrections

The partition function (6) derived in the previous section will not be valid at low temperatures where quantum corrections become important. A simple method, suggested by Pitzer and Gwinn [5] for the problem of hindered rotation and used with success subsequently [6, 7], may be employed in order to extend the validity of the present approach also for low temperatures.

The general partition function, $q(T)$, may be constructed as

$$q(T) = q_{\text{class}} \frac{q_{\text{harm osc-quant}}}{q_{\text{harm osc-class}}}, \tag{11}$$

where

$$q_{\text{harm osc-quant}} = \frac{\exp(-h\nu_{\text{osc}}/2kT)}{1 - \exp(-h\nu_{\text{osc}}/kT)} \frac{\exp(-h\nu_{\text{sub}}/2kT)}{1 - \exp(-h\nu_{\text{sub}}/kT)}$$

is the familiar Einstein partition function, and

$$q_{\text{harm osc-class}} = \frac{kT}{h\nu_{\text{osc}}} \frac{kT}{h\nu_{\text{sub}}}.$$

At low temperature, T, $q_{\text{class}} \to q_{\text{harm osc-class}}$ (cf. (7b)) and $q(T) \to q_{\text{harm osc-quant}}$. At high T the ratio in (11) tends to unity and one has $q(T) \to q_{\text{class}}$ from (6).

2.3 Heat capacity and related thermodynamic properties of the system

From the partition function (6), improved for quantum behaviour of the adsorbed molecules at low temperature all related thermodynamic functions of the system may be derived. Several of them which refer to a system of N isolated diatomic molecules are given below:

internal energy

$$\frac{E}{N} = kT^2 \frac{\partial \ln q(T)}{\partial T} = \frac{3kT}{2} + W - W \frac{\mathrm{I}_1\left(\frac{W}{kT} \cos \frac{\pi b}{a}\right)}{\mathrm{I}_0\left(\frac{W}{kT} \cos \frac{\pi b}{a}\right)} \cos \frac{\pi b}{a} - \\ - \frac{kT}{\sqrt{\pi}} \frac{M_L \, \mathrm{e}^{-M_L^2} + M_b \, \mathrm{e}^{-M_b^2}}{\operatorname{erf}(M_L) + \operatorname{erf}(M_b)} + \text{quant. corr.}\,, \tag{12}$$

entropy

$$\frac{S}{Nk} = \frac{\partial [T \ln q(T)]}{\partial T} = \ln \left\{ \frac{2\pi^{3/2} m L (kT)^{3/2} \, \mathrm{e}^{3/2}}{h^2 \sqrt{\mu}} \, \mathrm{I}_0[\operatorname{erf}(M_L) + \operatorname{erf}(M_b)] \right\} - \\ - \frac{W}{kT} \frac{\mathrm{I}_1}{\mathrm{I}_0} \cos \frac{\pi b}{a} - \frac{M_L \, \mathrm{e}^{-M_L^2} + M_b \, \mathrm{e}^{-M_b^2}}{2\big(\operatorname{erf}(M_L) + \operatorname{erf}(M_b)\big)} + \text{quant. corr.}\,, \tag{13}$$

specific heat c_v

$$\frac{c_v}{Nk} = \frac{1}{Nk} \frac{\partial E}{\partial T} = \frac{3}{2} + \left(\frac{W}{kT} \cos \frac{\pi b}{a}\right)^2 \left[1 - \frac{1}{\frac{W}{kT} \cos \frac{\pi b}{a}} - \left(\frac{\mathrm{I}_1}{\mathrm{I}_0}\right)^2 \right] - \\ - \frac{1}{\pi} \frac{(M_L/2 + M_L^3) \, \mathrm{e}^{-M_L^2} + (M_b/2 + M_b^3) \, \mathrm{e}^{-M_b^2}}{\operatorname{erf}(M_L) + \operatorname{erf}(M_b)} - \frac{1}{2\sqrt{\pi}} \times \\ \times \frac{\left[\frac{\mu}{kT} (2L - b) \exp(-M_L^2) + \frac{\mu b}{kT} \exp(-M_b^2) \right] [(2L - b) \, \mathrm{e}^{-M_L^2} + b \, \mathrm{e}^{-M_b^2}]}{[\operatorname{erf}(M_L) + \operatorname{erf}(M_b)]^2} + \\ + \text{quant. corr.} \tag{14}$$

Throughout in (12) to (14), I_1 denotes the first derivative of the modified Bessel function I_0. The respective quantum corrections to the energy E, the entropy S, and the specific heat c_v may be easily obtained from (11) and are omitted here for brevity.

Since our main concern is the motion of the particles along the substrate, (12) to (14) neglect contributions to the thermodynamic functions which originate from oscillations vertical to the surface. These oscillations could be, however, additionally incorporated into the present model.

2.4 Results and conclusions (1D case)

The contributions to the specific heat of the system reflecting the state of one-dimensional motion of diatomic molecules on a periodic surface are plotted in Fig. 1. Varying the "misfit" ratio b/a, we consider two different cases — of a stiff and of a loosely bonded dimer. In the former case we take the frequency $h\nu_{osc} = 4 \times 10^{-14}$ erg (i.e. as that of iodine I_2) and in the latter $h\nu_{osc} = 4 \times 10^{-13}$ erg. The energy barrier between adjacent wells, $W = 1 \times 10^{-14}$ erg, and in both cases $h\nu_{sub} = 4 \times 10^{-16}$ erg.

The 1D analysis leads to the following conclusions:

1. If quantum corrections are allowed for, the contribution to the specific heat of the system, c_v, depends strongly on the competition between the two main frequencies,

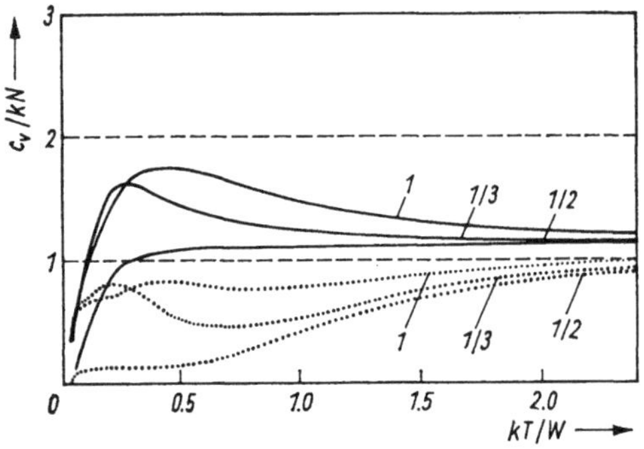

Fig. 1. Dependence of the (1D) specific heat on temperature (in dimensionless units). Full line: loosely bonded diatomic molecules, dotted line: dimers with "stiff" bonds. The "misfit" ratio b/a is given as a parameter

ν_{soc} and ν_{sub}. Generally, stiff dimers contribute much less to the specific heat of the system than loose ones.

2. If quantum corrections are not included, it may be shown that the position of the maximum of the specific heat curve (but not the height of the maximum) depends markedly on the ratio b/a. The temperature of the maximum is very nearly $T_{max} = (2W/5k)\,|\cos(\pi b/a)|$.

3. At very high temperatures all specific heat curves approach the value of $1R$ ($R = 1.987$ cal/mol) which corresponds to free translational motion of the dimer along the surface.

3. Results for the Case of a 2D Periodic Substrate

The case of a 2D periodic substrate with isolated diatomic molecules adsorbed on it, although more complex, may be dealt with within the same approach as in Section 2.1. The potential energy of a diatomic molecule adsorbed on a 2D periodic substrate with arbitrary energy barriers and lattice constants in the X- and Y-directions is given by the Hamiltonian

$$H = W_0 + \frac{W_x}{2}\left(1 - \cos\frac{2\pi x_1}{a_x}\right) + \frac{W_x}{2}\left(1 - \cos\frac{2\pi x_2}{a_x}\right) + \frac{W_y}{2}\left(1 - \cos\frac{2\pi y_1}{a_y}\right) + \frac{W_y}{2}\left(1 - \cos\frac{2\pi y_2}{a_y}\right) + \mu\left[\sqrt{(x_1 - x_2)^2 + (y_1 - y_2)^2} - b\right]^2 . \tag{15}$$

In (15) W_0 is the desorption energy, W_x and W_y are the energy barriers for surface diffusion, a_x and a_y the respective lattice constants along the X- and Y-axes, and μ the elastic constant of the diatomic molecule. A procedure similar to that described in Section 2.1 leads to the following expression for the classical partition function, q_{class}:

$$q_{class} = \frac{4L^2\, e^{-(W_0+W_x+W_y)/kT}}{\Lambda^4}\, J\left(\frac{W_x}{kT}, \frac{W_y}{kT}\right) F\left(\frac{\mu b^2}{kT}\right), \tag{16}$$

where $4L^2$ is the total area of the substrate. In (16)

$$J\left(\frac{W_x}{kT}, \frac{W_y}{kT}\right) = \int_0^{2\pi} I_0\left[\frac{W_x}{kT}\cos\left(\frac{\pi r_m}{a_x}\cos\psi\right)\right] I_0\left[\frac{W_y}{kT}\cos\left(\frac{\pi r_m}{a_y}\sin\psi\right)\right] d\psi \tag{17}$$

is the term which reflects the transition from a complex oscillatory behaviour of the dimer in the state of localization to that of translational motion

$$F\left(\frac{\mu b^2}{kT}\right) = \frac{b^2}{2M_b^2}\left(e^{-M_b^2} - e^{-M_L^2}\right) + \frac{b^2\sqrt{\pi}}{2M_b}\left(\operatorname{erf}(M_L) + \operatorname{erf}(M_b)\right). \tag{18}$$

The function $F(\mu b^2/kT)$ is, roughly speaking, responsible for the internal vibrations of the length of the dimer. Like everywhere in the calculations we consider only the thermodynamic limit of a microscopically large substrate ($4L^2 \gg 1$). In (16), (17) one has to operate in principle with an effective length r_m of the diatomic molecule,

$$r_m = \frac{b}{2}\left(1 + \sqrt{\frac{2kT}{\mu b^2}}\right). \tag{19}$$

Although r_m appears from (19) as a temperature dependent quantity, due to $M_b^2 = \mu b^2/kT \gg 1$, which is normally the case (e.g. for N_2 $\mu b^2/k \approx 10^6$ K), it coincides practically with the equilibrium length, b, of the dimer. Using (16), one may derive expressions for the thermodynamic functions in the same way as this has been done in Section 3 (cf. (12) to (14)).

As far as our main concern is the transition from localized-to-free translational motion of the dimer with increasing temperature, we consider here only the contributions to the heat capacity of the system which are described by (17).

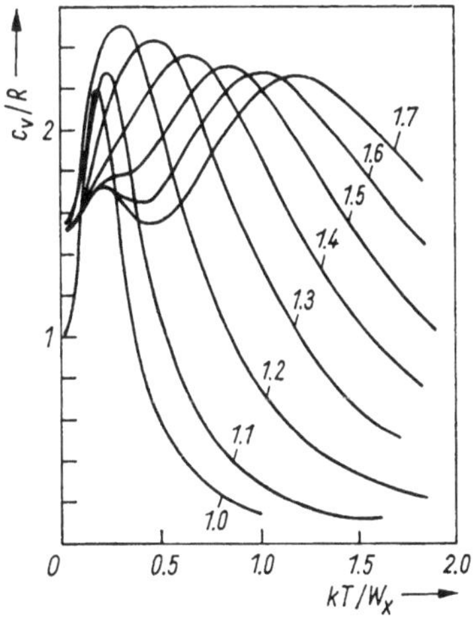

Fig. 2

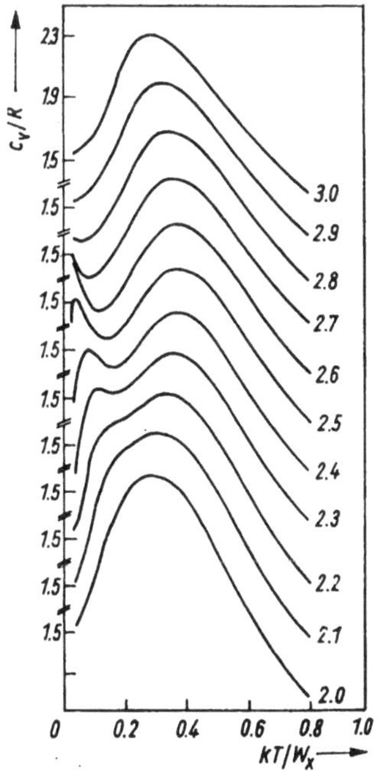

Fig. 3

Fig. 2. Variation of the heat capacity with temperature in the localized-to-mobile region. The lattice constants are $a_x = a_y = 3$ Å, and the length of the dimer $b = 2$ Å. The energy barriers for surface diffusion along the X- and Y-axes are given as parameters (in arb. units) for each curve

Fig. 3. Heat capacity vs. temperature dependence for diatomic molecules with different lengths b (given as a parameter). The periodic substrate has $a_x = a_y = 3$ Å and $W_x : W_y = 1:2$

In Fig. 2 the heat capacity versus temperature dependence is shown for a square lattice with $a_x = a_y = 3Å$, and length of the dimer, $b = 2$ Å. The heights W_x and W_y of the energy barriers for surface diffusion along the X- and Y-axes are given as parameters. The main conclusions following from the results given in Fig. 2 are:

1. The transition from localized-to-mobile adsorption is marked by a characteristic maximum (or, even by two maxima) at temperature T_{max}. Thereby both height and position of the maximum depend on the energy barriers W_x and W_y. Approximately, one has $T_{max} \approx (\lambda/6) \sqrt{W_x^2 + W_y^2}$, where the constant λ depends on a_x, a_y, and b.

2. If the heights W_x and W_y differ considerably, i.e. for $W_x : W_y < 1:4$ two separate maxima are observed while for $W_x : W_y \geqq 1:4$ these maxima merge into a single one whose height appears strongly dependent on the degree of interference between the various oscillatory motions, which the diatomic molecule performs in the state of localization.

Besides oscillations of the length of the dimer with frequency $\nu_{osc} = (1/2\pi) \sqrt{2\mu/m}$, one should consider three basic frequencies connected with vibrations of the molecule in the XY plane. For example on a grooved substrate, say $W(211)$, these would be vibrations along and perpendicular to the groove as well as hindered rotational motion with an immobile centre of mass of the dimer. At low temperature, $T \ll W_x/x$, they give a contribution of $\frac{3}{2} R$ to c_v, as is seen in Fig. 2.

Fig. 3 shows several dependences of specific heat on temperature (again in dimensionless units) for $a_x = a_y = 3$ Å, $W_x = 1$ and $W_y = 2$ (in arbitrary units) and varying length b of the dimer given at each curve as a parameter. Evidently, not only height and position of the heat capacity maximum depend on b, but at some ratios of b/a_x two separate maxima are clearly distinguished. The separation between these two maxima as well as the height of each are visibly changing with varying b.

Thus, on the ground of the results shown in Fig. 3 one could predict qualitatively different specific heat curves for diatomic molecules which differ only in length when they are adsorbed on a certain crystalline face. Moreover, the present study also shows that a given shape of the heat capacity curve may result either from the competition between the energy barriers W_x and W_y (cf. Fig. 2, $W_x = 1$, $W_y = 6$), or it may be due to the specific ratio of b/a_x and b/a_y (cf. Fig. 3, with $b = 2.4$ Å). This is certainly a manifestation of the much more complex nature of the system studied in this work as compared to that examined by Hill [1], and Doll and Steele [2]. It is clear that further development of the present approach should be expected to cover the more complicated (and difficult) case of larger clusters or macromolecules adsorbed on periodic substrates.

References

[1] T. L. HILL, J. chem. Phys. **14**, 441 (1946).
[2] J. J. DOLL and W. A. STEELE, Surface Sci. **44**, 449 (1974).
[3] A. MILCHEV, to be published.
[4] H. BATEMAN, in: Higher Transcendental Functions, Ed. A. ERDÉLYI, Vol. 2, § 7.2.2., Mc Graw-Hill Publ. Co., New York 1953.
[5] K. S. PITZER and W. D. GWINN, J. chem. Phys. **10**, 428 (1942).
[6] A. MILCHEV and M. PAUNOV, Surface Sci. **108**, 38 (1981).
[7] A. MILCHEV and I. GUTZOV, Phys. Chem. Liquids **11**, 25 (1981).

(Received May 17, 1983)

phys. stat. sol. (a) 79, 309 (1983)

Subject classification: 14.4.1; 2; 22.9

Institute of Molecular Physics, Polish Academy of Sciences, Poznań[1])

Determination of Pitch in Chiral Smectic C DOBAMBC

By

S. A. RÓŻAŃSKI

Selective reflection, Lagarde, Cano, and light diffraction methods are used for the determination of the pitch in chiral smectic C DOBAMBC {i-amyl-4-[4-decyloxybenzylideneamino]-cinnamate}. All four methods give consistent results. In the selective reflection, the rotation of sample enables the observation of the second peak previously predicted by Berreman, situated at about twice the wavelength of the first peak.

Anhand der Methoden der selektiven Reflexion, der Methoden von Lagarde und Cano und der Lichtdiffraktionsmethode wird die Helixganghöhe im smektischen C-Flüssigkristall DOBAMBC bestimmt. Alle vier Methoden ergeben gleiche Resultate. Zwei Linien der selektiven Reflexion werden beobachtet, was mit Vorhersagen der Theorie von Berreman übereinstimmt.

1. Introduction

The appearance of a spontaneous polarization in the plane of smectic layers, for liquid crystals in chiral smectic C phase, and the ferroelectric properties of these crystals were firstly described by Meyer et al. [17]. Since then, the so-called ferroelectric liquid crystals were extensively studied. Among them the most well known and widely studied is the i-amyl-4[4-decyloxybenzylideneamino]-cinnamate (DOBAMBC). From many parameters characterizing the chiral smectic C such as pitch, tilt, and spontaneous polarization, the pitch has a particular significance for the elucidation of ferroelectric properties, and applications of this type of liquid crystals. Therefore, the dependence of pitch on temperature and external fields as well as its actual numerical value are very important for theoretical and practical reasons. Recently, many results of pitch measurements have been reported [1 to 4, 6], however, some of them are still controversial [3, 6, 18]. Some divergencies in values of pitch are probably caused by different interpretation of experimental data, especially those obtained with a polarizing microscope [6, 18]. Present measurements are designed to explain these divergences.

2. Experiment

The well-known ferroelectric liquid crystal DOBAMBC was used. Within the temperature range 336 to 368 K it exhibits a SmC phase with helicoidal structure. To obtain a suitable orientation of DOBAMBC, an external magnetic field and glass plates with specially prepared surfaces were used.

We used the following experimental methods:

2.1 Lagarde method [18]

A sample about 120 μm thick was sandwiched between two glass plates, mounted in a brass heater, kept in a strong magnetic field (about 1.6 T), and then cooled from isotropic to smectic A phase with the rate of 0.1 K/min. The temperature was stabilized with the accuracy of 0.01 K. The well-known method of cooling in a magnetic

[1]) Smoluchowskiego 17/19, 60-179 Poznań, Poland.

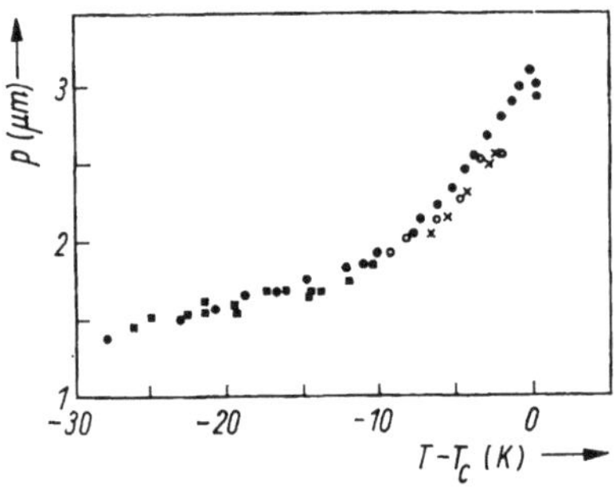

Fig. 1. Temperature dependence of pitch for DOBAMBC determined by different methods: ■ selective reflection method, × Cano method, ○ Lagarde method, ● diffraction method

field [5, 6] gave a homogeneous sample of good quality. Then, the temperature dependence of distance between stripes in SmC phase was measured using a polarizing microscope. The results are shown in Fig. 1.

2.2 Cano method [7]

The liquid crystal was placed between a lens and a glass plate. Their surfaces were rubbed in one direction with diamond powder and covered with chloroform solution of CTAB. The temperature dependence of the pitch was determined by observing a number of very delicate concentric circles with the polarizing microscope (Fig. 1).

2.3 Diffraction method

Many authors used a diffraction method to determine the pitch [8 to 10]. A sample with planar orientation was prepared in the same way as in Lagarde method. The sample was located in the focused He–Ne laser beam. The diffraction pattern of SmC structure was visible on the screen. The picture consisted of zero-, first-, and second-order strips. From the temperature dependence of the distances between the strips, the pitch and its temperature characteristics were determined (Fig. 1).

2.4 Selective reflection method

In the selective reflection method the homotropically oriented liquid crystal was used. The sample was contained between Spectrosil glass plates covered with CTAB with 120 μm copper spacer. By slow cooling from isotropic to SmA and SmC phase we obtained a homogeneous sample of good quality, which was confirmed by a distinct conoscopic figure. The sample was located in the IR spectrophotometer and the temperature dependence of selective reflection peak within the range from 2.2 to 5 μm was measured. Using Bragg's condition [16] for selective reflection the pitch and its temperature dependence were determined. The result is shown in Fig. 1. The observation of selective reflection lines in DOBAMBC is hindered by its own absorption bands. The second line, situated at doubled wavelength of the first one which, according to Berreman [11, 15], should be visible when the sample is rotated, is difficult to observe. However, we managed to see it. A typical picture of selective reflection peaks is shown in Fig. 2. The background within the region of 2.5 to 4 μm is caused by the difference in transmission for SmC and SmA phases, and the absorption bands of DOBAMBC. The theoretical conclusions [15] interpret the first line as connected with half the pitch and the second line with the full pitch. A similar effect was observed by Hori [13] in the visible range.

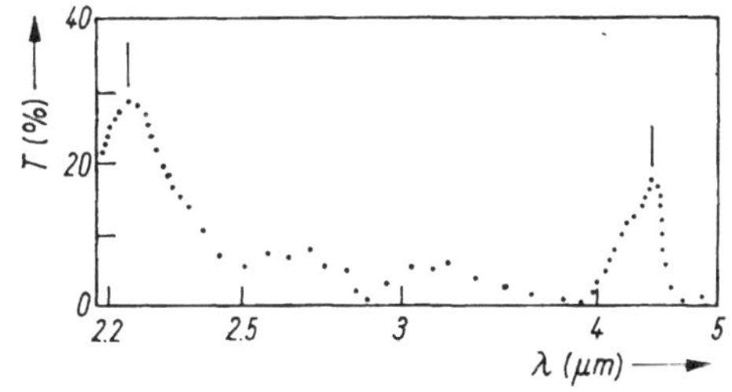

Fig. 2. Selective reflection in SmC phase of DOBAMBC for oblique light propagation ($\theta = 45°$) in temperature $T = 352$ K. Marks indicate the first peak at 2.23 μm and the second peak at 4.40 μm

3. Results and Discussion

In the Lagarde and diffraction methods the liquid crystal has a planar orientation. In this case, the influence of boundary conditions on the observed macroscopic picture may be essential, since molecules lying almost parallel to the glass plates strongly interact with their surfaces. This may lead to the formation of dislocations and defects of the type described by Glogarova et al. [14], as well as to the observation of different kinds of lines depending on the location of the focus plane of the objective in the sample [12]. In the methods of Sections 2.1 and 2.3 the surface effect can strongly deform the bulk effect, and thereby may influence the measured value of pitch. Thus, these methods cannot give a decisive answer concerning the value of pitch for the whole volume of the sample.

The lines observed in the Cano method have dislocative nature. Distances between these lines are connected with the radius of the lens used. The influence of boundary conditions is similar to that in the selective reflection method (homotropic orientation) what enables the direct comparison of results. However, in all three methods (Lagarde, Cano, and diffraction method) the question whether the lines (or spots in diffraction pattern) are connected with the full pitch or its half, remains to be answered. The answer to this question can be obtained from the selective reflection method. In this method, when light falls perpendicularly to smectic layers only one peak is observed. In all other configurations the second peak, situated at twice the wavelength of the first one, is also clearly visible (Fig. 2). The only possible explanation of this result is to assume that the first peak is related to half the pitch and the second to the full pitch. Then the real value of the pitch for chiral smectic C structure of DOBAMBC can be determined. If one assumes that in the Lagarde, Cano, and diffraction method the distances between lines (spots) are connected with the full pitch, a very good agreement of results from all four methods (Fig. 1) will be achieved. Our results also agree very well with those obtained by Takezoe et al. [2].

Acknowledgements

The author would like to thank Prof. J. Małecki, Dr. W. Kuczyński, and Dr. K. Wojciechowski for their helpful comments.

References

[1] G. Durand and Ph. Martinot-Lagarde, Ferroelectrics **24**, 89 (1980).
[2] H. Takezoe, K. Kondo, A. Fukuda, and E. Kuze, Japan. J. appl. Phys. **21**, L627 (1982).
[3] Ph. Martinot-Lagarde, R. Duke, and G. Durand, Mol. Cryst. liquid Cryst. **75**, 249 (1981).
[4] K. Kondo, F. Kobayashi, A. Fukuda, H. Takezoe, and E. Kuze, Japan. J. appl. Phys. **19**, 2293 (1980).
[5] K. Kondo, F. Kobayashi, A. Fukuda, and E. Kuze, Japan. J. appl. Phys. **20**, 1773 (1981).

[6] B. I. Ostrovskii, A. Z. Rabinovich, A. S. Sonin, and B. A. Strukov, Zh. eksper. teor. Fiz 74, 1748 (1978).
[7] R. Cano, Bull. Soc. Franc. Minér. Crist. **90**, 333 (1967).
[8] K. Kondo, H. Takezoe, A. Fukuda, and E. Kuze, Japan. J. appl. Phys. **21**, 224 (1982).
[9] E. Sackmann, S. Meiboom, L. C. Snyder, A. E. Meixner, and E. Dietz, J. Amer. Chem. Soc. **90**, 3567 (1968).
[10] J. W. Shelton and Y. R. Shen, Phys. Rev. A **5**, 1867 (1972).
[11] D. W. Berreman, Mol. Cryst. liquid Cryst. **22**, 175 (1973).
[12] K. Kondo, A. Fukuda, and E. Kuze, Japan. J. appl. Phys. **20**, 1779 (1981).
[13] K. Hori, Mol. Cryst. liquid Cryst., Letters **82**, 13 (1982).
[14] M. Glogarova, L. Lejcek, J. Pavel, V. Janovec, and J. Fousek, Czech. J. Phys. **B32**, 943 (1982).
[15] D. W. Berreman and T. J. Scheffer, Mol. Cryst. liquid Cryst. **11**, 395 (1970).
[16] P. G. de Gennes, The Physics of Liquid Crystals, Clarendon Press, Oxford 1974.
[17] R. B. Meyer, L. Liebert, L. Strzelecki, and P. Keller. J. Physique, Lettres **36**, 69 (1975).
[18] Ph. Martinot-Lagarde, J. Physique **37**, C3-129 (1976).

(Received May 17, 1983)

phys. stat. sol. (a) 79, 313 (1983)

Subject classification: 1.6; 10.2; 22.1.2; 22.6

Institute of Electron Technology, Warsaw[1])

A Mechanism for the Effect of Doping on the Silicon Native Oxide Thickness

By

A. WOLKENBERG

A theoretical basis of surface oxidation kinetics of silicon of semiconductor purity are presented for normal ambient conditions and a relation is proposed between the native oxide layer thickness and the concentration of defects in the silicon crystallographic structure. This provides the possibility for interpreting the experimental data relation between the oxide layer thickness and the impurity concentration. The experimental data are verified and composed using various measuring techniques. The discussed hypothesis of doping-introduced defects is employed to explain the kinetics of anodic polarization on silicon electrodes.

Die theoretische Grundlage der Oberflächenoxydationskinetik von Silizium mit Halbleiterreinheit wird für normale Umgebungsatmosphären aufgestellt und eine Beziehung zwischen der Eigenoxidschichtdicke und der Konzentration der Defekte in der kristallographischen Struktur des Silizium angegeben. Dies ermöglicht, die experimentellen Werte bezüglich Oxid-Schichtdicke und Störstellenkonzentration zu interpretieren. Die experimentellen Daten werden mit verschiedenen Meßmethoden verifiziert. Die diskutierte Hypothese von durch Dotierung eingebrachten Defekten wird benutzt, um die Kinetik der anodischen Polarisierung auf Silizium-Elektroden zu erklären.

1. Some Aspects of Physico-Chemical Model of Silicon Oxidation

A pure silicon surface, when exposed to the oxygen containing ambient, becomes instantly covered with a thin oxide film. The kinetics of this initial oxidation process is not evidently simple. In particular, one needs to calculate the magnitude of two interactions:

initial heat of adsorption of gaseous oxygen molecules on the bare silicon surface,

and the Si–O interaction energy, that is, the strength of this single bond when, e.g., a monolayer of "oxide" has been formed on the silicon surface.

The silicon–oxygen interaction energy of an isolated single bond may be estimated by the empirical equation

$$D(\text{Si–O}) = \tfrac{1}{2}\,\{D(\text{Si–Si}) + D(\text{O–O})\} + (\chi_{\text{Si}} - \chi_0)^2\,, \tag{1}$$

where $D(\text{Si–O})$ is the required silicon–oxygen interaction energy, that is, the bond strength of this single bond; $D(\text{O–O})$ is the bond dissociation energy of the interatomic bond in the oxygen molecule, $= 1.513$ eV/particle; χ_{Si} and χ_0 are the electronegativities, on the Pauling scale, of silicon and oxygen, respectively; the term $(\chi_{\text{Si}} - \chi_0)^2$ takes into account the extra strength due to the partial ionic character of the essentially covalent Si–O bond; $= 2.89$ eV/particle.

It may be added that the Pauling equation is valid only for calculating the strength of an isolated bond in the gaseous phase; it is not applicable to solids. However, it is frequently used, with some success, for estimating roughly the relative interaction energies of a given gaseous atom on a series of elements, for the purpose of a comparative analysis.

---

[1]) al. Lotników, 02-668 Warsaw, Poland.

For the purpose of estimation of the initial heat of adsorption of a gas on a bare semiconductor, the Eley equation based on the Pauling procedure for computing single bond energies is frequently used. According to this Pauling-Eley equation (ΔH_{ads}), the initial (exothermic) heat of adsorption of oxygen gas on silicon is given by

$$-(\Delta H_{ads})_{O_2} = 2D(\mathrm{Si{-}O}) - D(\mathrm{O{-}O})\,, \tag{2}$$

where as mentioned above $(\Delta H_{ads})_{O_2}$ is the initial heat of adsorption of O_2 on Si D(Si–O) the Si–O bond strength, D(O–O) the bond strength of the bond joining two oxygen molecule.

The calculated value for the silicon–oxygen interaction energy D(Si–O) after (1) is equal to 4.568 eV/particle and, consequently the calculated initial heat of adsorption of O_2 on Si is equal to 7.623 eV/particle.

Silicon surfaces with ⟨100⟩ and ⟨111⟩ orientations contain 6.78×10^{14} and 7.85×10^{14} Si atoms cm^{-2}, respectively. The number of Si–Si accessible to an oxidant at a ⟨111⟩ oriented surface is $11.76 \times 10^{14}\ cm^{-2}$; on a ⟨100⟩ oriented surface the corresponding number of bonds is $6.78 \times 10^{14}\ cm^{-2}$ [1]. The number of nonoxidized Si–Si bonds in thermally grown, amorphous SiO_2 was found to be less than $10^{13}\ cm^{-2}$ [2].

During the oxidation process, crystalline silicon substrate surfaces are converted into amorphous SiO_2 films. The total Si atom concentration per unit volume decreases by a factor 2.2, when Si is oxidized to form SiO_2. This conversion involves an interfacial chemical reaction in which the oxidant is absorbed in the transition region. The nature of the oxygen adsorption and dissociation processes within the transition region is not known. When an oxidation reaction occurs, considerable strain should arise at the interface region as a result of direct insertion of oxygen atoms between Si–Si bonds. Bond distances change from 2.3 Å for a Si–Si bond in the substrate to 3.0 Å for a Si–O–Si group in an oxide film.

Incorporation of oxygen within the silicon substrate lattice to form an oxide film, while still retaining a predominantly three-dimensional structure, resembles displacement of silicon atomes from their original sites. The changes at the substrate surface are, of course, facilitated if the oxidation reaction occurs via a multistep process.

The multistep oxidation processes are considered by Raider and Berman [3]. A two-step oxidation reaction is proposed that involves Si–Si bond breaking and siloxane (Si–O–Si) group formation. The Si–Si bonds within the transition region under tension and oxygen molecules or atoms, both of which contain unpaired electrons, interact with accessible nonoxidized Si–Si bonds.

This initial oxidation reaction, in which Si–Si bonds are broken, is only weakly thermally activated with an activation energy, E_a, near zero. For a given substrate orientation, the extent of this interaction is determined by C_{ox} (concentration of excess oxygen). Oxygen solubility is presumably enhanced by formation of a complex with nonoxidized Si. Raider and Berman [3] suppose that a reaction occurs with formation of two different electrically active intermediates such as

$$\mathrm{Si{-}Si} + (\mathrm{O}) \rightarrow \underset{\downarrow}{\overset{\mathrm{O}\atop\downarrow}{\mathrm{Si{:}Si}}} \rightarrow {>}\mathrm{Si.} + (\mathrm{SiO})^+ + e\,. \tag{3}$$

Ion scattering data indicate that there is 20% excess silicon in a layer 10 to 15 Å thick in front of the interface [4]. Johannessen et al. show that the change from Si–O bonding to Si–Si bonding takes place within an interface layer of 50 Å [5]. A principle sketch of their interface model is shown in Fig. 1. The total width of the interface region is approximately 35 Å, of which 20 Å are due to fluctuations of the SiO_2–Si boundary.

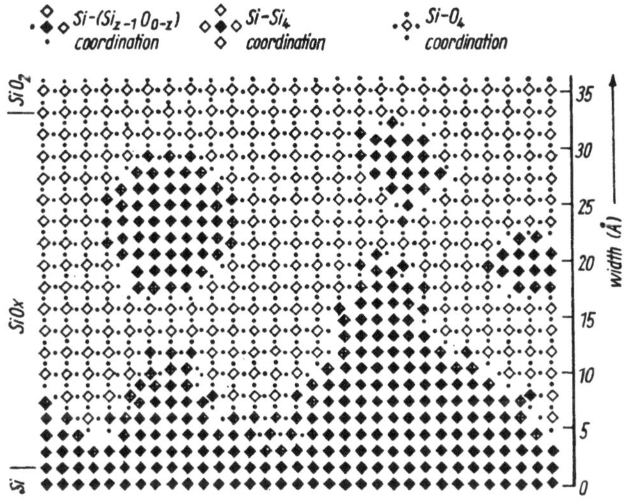

Fig. 1. Schematic SiO_2–Si interface morphology model showing how protrusions of SiO_2 grow into the silicon substrate, leaving inclusions of Si in the SiO_2 matrix at the interface [5]; ◇ silicon atoms, ● oxygen atoms

Sigmon et al. estimate the excess amount of silicon in a silicon-rich transition layer at the interface to be 10^{16} atoms/cm^2 [6].

The results presented by Wager and Wilmsen [7] indicate that there is at the interface either a graded transition layer of width starting from close-to-zero and growing up to 20 Å with the oxide becomes thicker or, that the SiO_2–Si transition region is abrupt but the oxide is nonuniformly thick [8]. He-backscattering experiments by Offermann suggest a transition region width below 25 Å.

Raider and Flitsch, using X-ray photoelectron spectroscopy, have estimated the measured thickness of the transition region present on ⟨111⟩ oriented substrates to be 10 Å and to possess Si atoms in lower oxidation states than the 7.5 Å thick transition region present on ⟨100⟩ oriented substrates [9]. The compositions and widths of the transition regions are unaffected by changes in the oxidation conditions or the oxidant.

The formation of intermediates in a transition region during the Si oxidation reaction presumably occurs with other oxidants, such as H_2O, HCl, and Cl_2. These reactants form nonbridging, or thermally bonded, intermediates during the Si–Si bond breaking reaction as that shown in [3],

$$\text{Si–Si} + H_2 \rightarrow \text{Si–H} + \text{Si–OH}\,. \tag{4}$$

The second oxidation reaction step, involving oxygen as an oxidant, is the thermally activated reaction in which siloxane groups are formed [10],

$$\rangle\!\!-\text{Si}. + (\text{Si–O})^+ + e \rightarrow \text{Si–O–Si}\,. \tag{5}$$

In the oxidation mechanism introduced here, the experimentally observed effect of doping on the oxidation rate of silicon is omitted. A new glance at the oxidation problems has been reported by Ho et al. [11 to 13]. They have proposed a physical model based on the statistics of silicon point defects to explain the commonly observed enhanced oxidation rates of doped silicon. The physical effect of high doping levels on the oxidation kinetics of the Si/SiO_2 interface is postulated to be primarily electrical in nature. The high doping levels shift the position of the Fermi level toward the conduction band (n-type) or toward the valence band (p-type), even at the oxidation temperatures causing an increase in the equilibrium concentration of point defects (vacancies) in the silicon substrate. These point defects may provide the reaction sites for the chemical reaction converting Si into SiO_2 and, thereby, increase the rate at which this reaction occurs. The authors describe the theoretical basis for this model and predict quantitatively the expected oxidation rates for n- and p-doped silicon over a wide range of oxidation conditions.

2. Influence of Crystal Defects on Oxidation Rate Factors

It was suggested [11] that the interface reaction rate constant, and therefore, the linear rate constant B/A commonly used to describe silicon oxidation kinetics [14], could include additive components. These components result from different contributions of the silicon substrate to the Si/SiO_2 interface oxidation reaction. Ho and Plummer [13] have postulated a silicon vacancy contribution to the increase of B/A. The effective linear rate constant may be expressed as the product of intrinsic $(B/A)^{i}$ for lightly doped silicon and a vacancy dependent multiplication factor [11],

$$\frac{B}{A}=\left(\frac{B}{A}\right)^{i}\left[1+\left(\frac{K}{C_1}\right)C^{i}_{V_T}\,e^{2/kT}\left(\frac{C_{V_T}}{C^{i}_{V_T}}-1\right)\right], \tag{6}$$

where the normalized quantities may be defined as $(B/A)/(B/A)^{i}=(B/A)'$, $K/C_1 = K'$, $C_{V_T}/C^{i}_{V_T}=C'_{V_T}$ charged vacancy states in silicon, $K'C^{i}_{V_T}=2.62\times10^{3}\,e^{-3.1/kT}$ experimental equation [11].

Rewriting equation (6) and substituting $K'C^{i}_{V_T}=2.62\times10^{3}\,e^{-3.1/kT}$ produces

$$\frac{\left(\frac{B}{A}\right)}{\left(\frac{B}{A}\right)^{i}}=1+2.62\times10^{3}\,e^{-1.1/kT}\,(C'_{V_T}-1)\,. \tag{7}$$

The analysis in [11] suggests that the vacancy contribution is negligible for low substrate doping, i.e., less than the intrinsic carrier concentration at oxidation temperatures.

Silicon vacancies are believed to exist in neutral V^{x} and charged V^{+}, V^{-}, and V^{2-} states. Since the neutral vacancy is a nonionized flaw, its concentration C_{V^x} in thermal equilibrium is a function of temperature only, independent of the position of the Fermi level in the silicon band gap [15]. The equilibrium concentration of neutral vacancies is after van Vechten [16]

$$C_{V^x}=C_x\exp\left(-\frac{E_x}{kT}\right), \tag{8}$$

where C_x is the number of silicon atoms in 1 cm^3 equal 5×10^{22}, $E_x=(2.6\pm0.2)$ eV.

Therefore, the intrinsic vacancy concentration in silicon at 300 K is equal to $\approx 1\times10^{-21}$ cm^{-3}, which is practically inessential. At 1000 K this value increase to $\approx 4\times10^{9}$ cm^{-3} and at 1300 K to $\approx 4\times10^{12}$ cm^{-3}.

The other contribution to concentration C_{V^x} comes from the intentionally introduced impurity content in silicon [11 to 13, 17, 18]. These flaw concentration, first defined by Shockley and Moll [15] and more precisely by van Vechten and Thurmond [17, 18] as charged vacancy states, will have thermal equilibrium concentrations that are dependent not only on temperature but also on C_{V^x}, and the relation of the Fermi level position, E_F, to the ionized vacancy energy levels in the band gap is

$$C_{V^+} = C_{V^x} \exp\left(\frac{E_{V^+} - E_F}{kT}\right), \tag{9}$$

$$C_{V^-} = C_{V^x} \exp\left(\frac{E_F - E_{V^-}}{kT}\right), \tag{10}$$

$$C_{V^{2-}} = C_{V^x} \exp\left(\frac{2E_F - E_{V^-} - E_{V^{2-}}}{kT}\right). \tag{11}$$

The ionized vacancy is formed in the reaction

$$V^x \rightarrow V^+ + e^- , \tag{12}$$

$$V^x \rightarrow V^- + e^+ , \tag{13}$$

$$V^- \rightarrow V^{2-} + e^+ . \tag{14}$$

The vacancy energy levels in silicon versus E_c are given by van Vechten and Thurmond [17, 18]. They are at 300 K: $E_{V^+} = 0.755$ eV, $E_{V^-} = 0.56$ eV, and $E_{V^{2-}} = 1$ eV. Fig. 2 plots the resulting concentrations (equations (9) to (11)) of the vacancy states as functions of the Fermi level at 300 K. Thus, if the oxidizing interface consumes vacancies to provide reaction sites for incoming oxidant species, a significant effect of substrate doping on the interface reaction could be electrical in nature, i.e., shift of the Fermi level to alter charged vacancy concentrations.

Since the effects of vacancy charge states on this contribution are not immediately apparent, various charged vacancy species are considered together and the total vacancy concentration, C_{V_T}, becomes one variable (to fulfil (6) and (7)).

According to [12, 19, 20] for extrinsic nondegenerate silicon, the concentration of vacancies in the charge state is related to the free carrier concentration as

$$C_{V_T} = C_{V^x} + C_{V^+} + C_{V^-} + C_{V^{2-}} , \tag{15}$$

where the vacancy concentration values are given by (9) to (11).

In Fig. 2 to 4 the calculated values (equation (7)) of linear rate constant for 300 K are illustrated, based on the vacancy contribution model fitted with two parameters $C_{V^x} = n$ or p and the experimental equation

$$K'C'_{V_T} = 2.62 \times 10^3 \exp(-3.1/kT) .$$

3. Experimental

It has been assumed for monocrystalline silicon that the energy barrier height for the oxidation reaction depends on the deviation from crystal perfection. The present

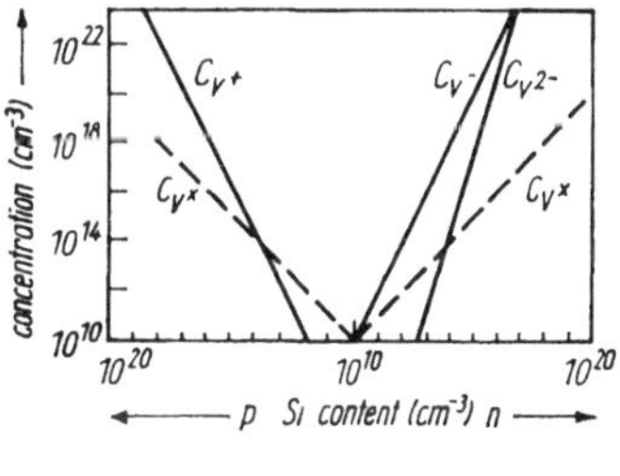

Fig. 2. Concentration of vacancy states as a function of the Fermi level position at 300 K ($C_{V^x} \approx p \approx n$) after (9) to (11)

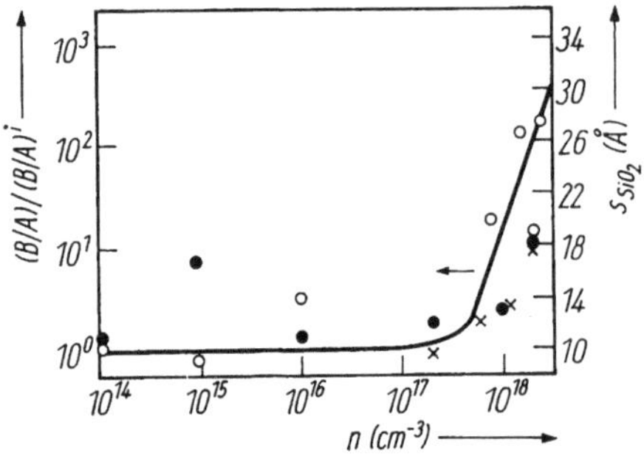

Fig. 3

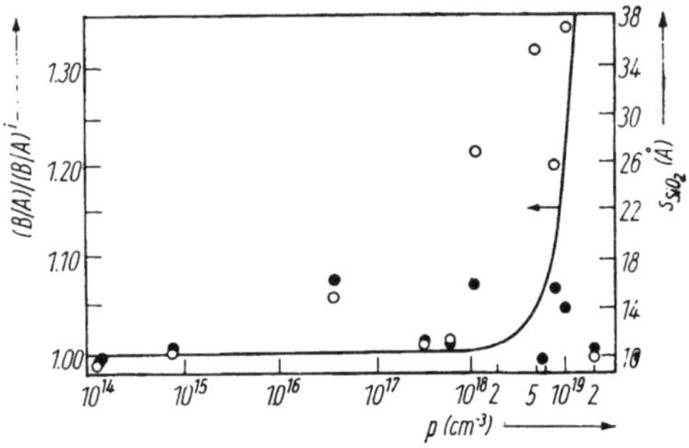

Fig. 4

Fig. 3. Oxide thickness vs. n-type doping in silicon, curve evaluated after (7) including $(B/A)/(B/A)^i$ vs. doping at 300 K; × experimental oxide thicknesses evaluated after (16) for Si–Au barriers, ○ experimental oxide thicknesses evaluated after (16) for Si–Hg barriers, ● experimental oxide thicknesses measured by an ellipsometer

Fig. 4. Oxide thickness vs. p-type doping in silicon, curve evaluated after (7) including $(B/A)/(B/A)^i$ vs. doping at 300 K; ○ experimental oxide thicknesses evaluated after (16) for Si–Hg, ● experimental oxide thicknesses measured by an ellipsometer

paper refers to the crystal imperfection in terms of doping and crystallographic periodicity. It has been assumed that the thickness of the so-called natural oxide layer on the silicon surface is a measure of this imperfection.

The above-mentioned native oxide layer thickness can be evaluated in three independent ways:

(i) electrical method: current–voltage and capacitance–voltage characteristics of silicon–0.1 M KCl, silicon–gold, and silicon–mercury barrier structures (the silicon–gold barrier height has also been evaluated by photoelectric method), (ii) optical method using an ellipsometer.

For the Si–0.1 M KCl structure the barrier height has been determined [21] together with the trapping state density at the silicon surface [22, 23] and the oxide layer thickness.

For the Si–Au structure the barrier height has been determined by current, capacitance, and photoelectric methods [24] and the oxide layer thickness from the analysis of the barrier ideality coefficient, n [25 to 27].

For the Si–Hg structure the barrier height has been determined from the current–voltage characteristic [24] and the oxide layer thickness from the analysis of the barrier ideality coefficient, n [25 to 27].

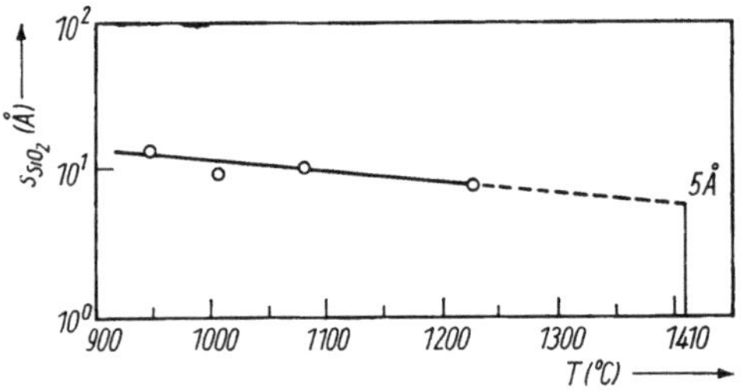

Fig. 5

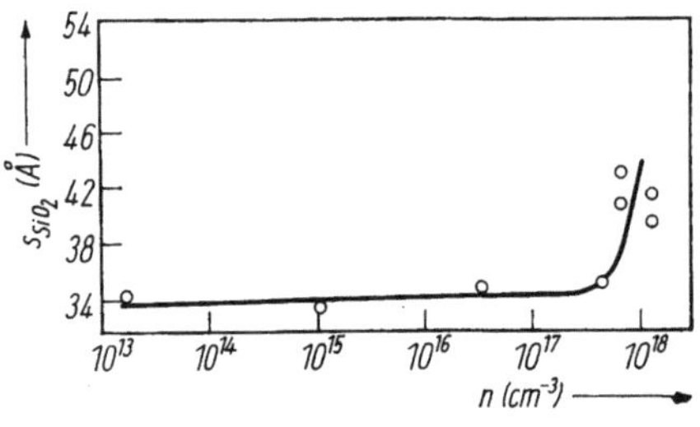

Fig. 6

Fig. 5. Oxide thickness evaluated after (16) vs. epitaxy temperature

Fig. 6. Oxide thickness measured electrochemically in 0.1 M KCl vs. silicon electrode doping

The measurements were performed on dislocation-free n-type antimony-doped silicon samples as well as samples with epitaxial layers being crystallized at different temperatures.

Fig. 5 illustrates the oxide thickness, evaluated after the equation derived in terms of the Card-Rhoderick theory [25 to 27],

$$\delta = \sqrt{\frac{8.854 \times 10^{-14}}{1.602 \times 10^{-19} N_t^2}} \left(\sqrt{\frac{11.8 N_D}{8} + (n-1)\, 2.15 N_t} - \sqrt{\frac{11.8 N_D}{8}} \right), \tag{16}$$

where N_t is the electrochemical trap density $= 2 \times 10^{13}\ \text{cm}^{-2}$ [23], N_D the donor impurity density in a semiconductor, n the Au–Si or Hg–Si barrier ideality coefficient, versus the silicon layer epitaxy temperature (layer resistivity 15 to 18 Ω cm; 0.007 to 0.008 Ω cm in the substrate).

In Fig. 3 and 4 the relation between the oxide layer thickness on the silicon surface and the carrier concentration is presented; the curves are calculated from (16) including the n coefficient values from Au–Si, In–Si, and Hg–Si barrier measurements as well as from ellipsometric measurements.

Fig. 6 shows the natural oxide thickness, measured electrochemically in 0.1 M KCl, versus silicon doping level.

The barrier height (for electrons) versus the carrier concentration, obtained from electrochemical (0.1 M KCl) and Au–Si, Hg–Si structure measurements, is plotted in Fig. 7.

The theoretical curve was calculated from an integrated equation (below) in terms of Crowell's hypothesis discussed by Archer and Yep [28],

$$\frac{d\Phi}{dE} = -q_0 \varepsilon_s \left(\frac{\varepsilon_{SiO_2}}{s_{SiO_2}} + 4\pi q_0^2 n_{ss} \right)^{-1}, \tag{17}$$

where

$$E = \sqrt{\frac{8\pi q_0 N_D \Phi_D}{\varepsilon_s}},$$

Φ_D is the diffusion potential, ε_s the dielectric permittivity of the semiconductor, ε_{SiO_2} the dielectric permittivity of SiO_2 film equal 2.15, s_{SiO_2} the oxide layer thickness $= \delta$, q_0 the electronic charge.

After integration with the proper boundary conditions we obtain

$$\Phi_B = \Phi_{BO} - \varepsilon_s \left[q_0 n_{ss} + \left(\frac{\varepsilon_{SiO_2}}{s_{SiO_2}} \right) \right]^{-1} \sqrt{\frac{2 q_0 N_D}{\varepsilon_s} \left(\Phi_{BO} - \frac{kT}{q_0} - E_F \right)}, \tag{18}$$

where Φ_{BO} is the barrier height for intrinsic material, $n_{ss} = N_t$ from (16).

The measured Φ_B value was introduced into (18) and the s_{SiO_2} value was fitted so as it was needed to obtain the proper value of Φ_{BO}. Then, using these calculated oxide

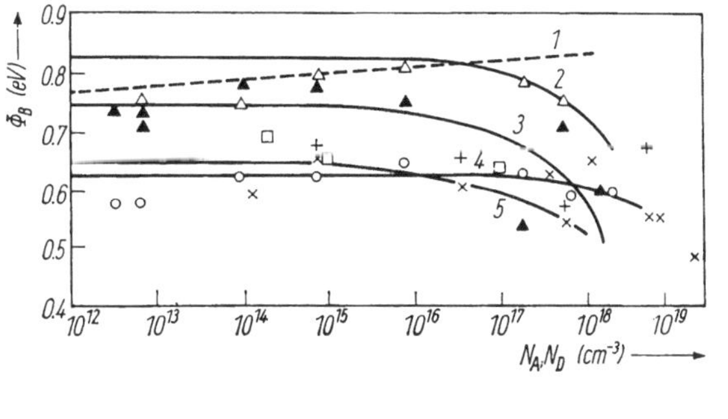

Fig. 7. Barrier height, Φ_B, vs. silicon doping (1) theoretical curve; (2) △ n-type Au–Si, barrier height from photovoltage measurements; (3) ▲ n-type Si, barrier height from I–U measurements; (4) + p-type Si, barrier height from C–U measurements; (5) × p-type Si, barrier height from I–U measurements; ▲, +, × Hg–Si interface; ○, n-type Si, barrier height from I-U measurements in 0.1m KCl; □n-type Si-In barierr height measurements.

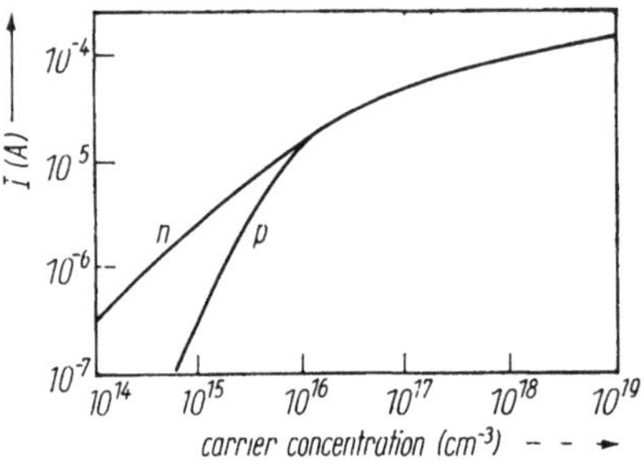

Fig. 8

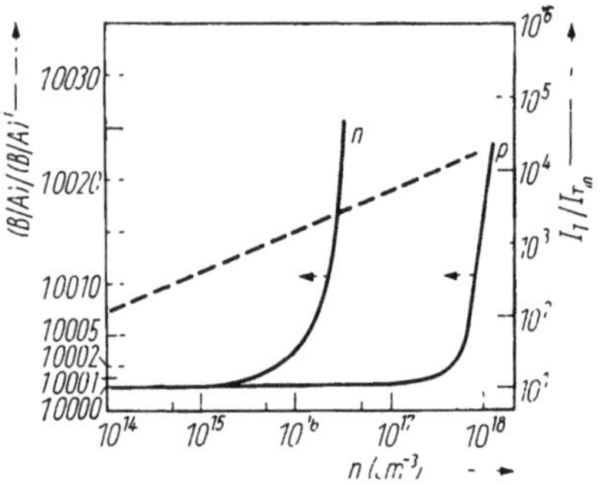

Fig. 9

Fig. 8. Anodic current at silicon electrodes in 1 M LiI vs. doping

Fig. 9. Comparison between evaluated and experimental curves for anode reaction velocity. ——— Si_n-electrode, experimental curve

thickness values (Fig. 3 and 4) theoretical curves Φ_B versus N_D were plotted (curves 2 and 3 in Fig. 7). Thickness values which could not agree with the so calculated Φ_B — — N_D curves were obtained from (16).

In connection with these results we can say that after (16) it is possible to calculate thickness values having only n (ideality factor) and from (18) it is possible to calculate thickness values which agree with all assumptions.

Fig. 8 presents the data collected during the measurements of anodic current density at n-type and p-type silicon electrodes in 1 M LiI (electrochemical oxidation) versus carrier concentration. Evaluated and experimental curves for the electrode reaction velocity are compared in Fig. 9. Experiments described in Fig. 8 and 9 were performed to compare the oxidation processes in electrolytes with those of the native oxide increase in the gaseous phase.

4. Discussion

The results presented prove that at 300 K the native oxide thickness on silicon surface depends upon the silicon doping level. The rate of oxide layer growth is quite small and remains stable during about 2 h after the oxide layer has been etched in HF [29]. That is why the measurements have been performed after that period.

If we consider the curves illustrating the oxidation rate for both doped and intrinsic silicon with the experimental data collected during electrochemical and ellipsometric measurements and those calculated from electrical measurements, we come to the conclusion that a reasonably precise theoretical description of the process of formation of very thin oxide films on silicon surfaces (Fig. 3 and 4) is possible in terms of the oxidation mechanism involving defects.

It has been assumed that each impurity atom stimulates the creation of a silicon vacancy in the crystal lattice that is active in a manner discussed earlier, while the activation energy is of the order of several kT.

The thickest layer of native oxide has been found on the silicon electrode in 0.1 M KCl (Fig. 6), slightly thinner layers covered the surfaces measured with the mercury-barrier probe, and thinner layers were measured with gold and indium electrodes (Fig. 3 to 5). The ellipsometric measurement correlate with that employing the mercury probe (the same time period passed from the moment of oxide etching and the same measuring conditions were preserved).

The contribution of defects manifested in the variations of a constant linear rate of oxide growth is further proved by a distinct effect of the temperature of silicon layer epitaxy on the native oxide thickness (Fig. 5).

The silicon oxidation rates in solutions of aqueous electrolytes (1 M LiI, 0.1 M $NaNO_3$, 0.1 M KCl) are proportional to C_{V_T} (Fig. 6), while the total amount of current carriers flowing during the anodic polarization of the silicon electrode in the above-mentioned solutions is proportional to the impurity concentration in silicon (Fig. 9). Therefore the impurity concentation affecting the current density, appears to be the concentration of ionized donors or acceptors.

The anodic current flowing through the silicon electrode dipped in an electrolyte solution can be divided into two parts:

current consumed for the formation of an oxide layer,

current consumed during the electrode reactions, i.e., neutralization of anions set apart on the anode and direct solving of the electrode material in the solution.

The amount of current consumed during the formation of a thin oxide layer (excluding intentional polarization to obtain thick oxide layers), when the layer thickness is its measure (Fig. 6), is consistent with the mechanism presented in Fig. 3 and 4. The participation of the current consumed during the other electrode reactions is much more pronounced (Fig. 9) and is proportional to the impurity concentration,

$$I_T = I_{ox} + I_n \,, \tag{19}$$

where I_{ox} is the current consumed according to the function C_{V_T} (15), I_n the current consumed during the other electrode reactions directly proportional to the impurity concentration.

The above hypothesis can be illustrated with the reaction given below [30],

$$A^- + V^+ \rightarrow A^-V^+ \,, \tag{20}$$

where A^- is a single impurity atom of acceptor type and

$$D^+ + V^- = D^+V^- \,, \tag{21}$$

where D^+ is a single impurity atom of donor type.

The A^-V^+ and D^+V^- pairs form complexes which easily dissociate during oxidation and thus act as the factors affecting the kinetics of the electrode process.

The plots and experimental results presented in Fig. 7 indirectly justify the hypothesis on the relation between the oxide layer thickness and the impurity concentration. The shape of curve 1, representative for silicon cleaved in vacuum [31], is reasonably well reproduced by experimental data collected for silicon samples of carrier concentrations less than 10^{16} cm^{-3}, i.e., those of a relatively thin oxide layer.

In the case of samples with impurity concentration exceeding 10^{16} cm^{-3} the experimental results are consistent rather with (18) derived under the assumption of an oxide layer and surface states present on the silicon surface [24] and their contribution in the measured barrier heights. As is evident from curves 2, 3 of Fig. 7 the barrier height for impurity concentrations above 10^{17} cm^{-3} should be drastically lower. However, the experimental data for samples with concentration above 10^{17} cm^{-3} do not seem to confirm the above. In view of the latter, the assumptions made to derive (18) should probably be revised, which actually is not the aim of this paper.

5. Conclusions

To summarize one can conclude that the proposed methods of analysis of electrophysical characteristics of the structures tested provide the possibility of detecting the presence of a native oxide layer on n-type and p-type silicon and reveal its thickness. They confirm also the dependence of this thickness upon the silicon doping level. Our results correlate with those reported by Ho and Plummer [11 to 13] on the silicon oxidation rate in dependence upon a very high doping level (above 10^{19} cm^{-3}).

Electrically active dopants in silicon can as well be involved in the electronic exchange between silicon and the oxidizing medium. In the case of n-type silicon, for example, the substitutional phosphorus atom (n-type dopant) acts as a momentary electron transporting agent, while in p-type silicon the boron atom (p-type dopant) acts as a momentary source of electrons. Such charge transporting mechanism, involving impurity atoms, may considerably reduce the energetic barrier for the electrode reaction and favour the formation of native oxide films.

The hypothesis presented does not explain completely which factors affect the anodic reaction rate at the silicon electrode employed for measuring the carrier concentration in silicon [32], although it offers an insight into the understanding of the process itself.

Acknowledgements

The author hereby expresses his deep appreciation to Mr. W. Jung, MSc., for performing electrical measurements of the structures with metal–semiconductor barriers, to Mrs. A. Kaminska, MSc., for ellipsometric measurements, to Mr. K. Klima, MSc., for photoelectric measurements of barrier heights, and to Mrs. B. Wasilewska, MSc., for some electrochemical measurements.

References

[1] J. R. LIGENZA, J. phys. Chem. **65**, 2011 (1961).
[2] R. FLITSCH and S. J. RAIDER, IBM J. Res. Developm. **22**, 294 (1978).
[3] S. J. RAIDER and A. BERMAN, J. Electrochem. Soc. **125**, 629 (1978).
[4] W. L. HARRINGTON and A. M. GOODMAN, Appl. Phys. Letters **25**, 531 (1974).
[5] J. S. JOHANNESSEN, W. E. SPICER, and Y. E. STRAUSER, J. appl. Phys. **47**, 3028 (1976).
[6] T. W. SIGMON, W. K. GHU, E. LUGUJJO, and J. W. MAYER, Appl. Phys. Letters **24**, 105 (1974).
[7] J. F. WAGER and C. W. WILMSEN, J. appl. Phys. **50**, 874 (1979).
[8] P. OFFERMANN, J. appl. Phys. **48**, 1890 (1977).
[9] S. J. RAIDER and R. FLITSCH, IBM J. Res. Developm **22**, 294 (1978); AIP Proc. Internat Topical Conf. Phys. Silicon Dioxide and Its Interfaces, Yorktown Heights (New York), 21 to 23 March (1978).
[10] S. J. RAIDER and R. FLITSCH, J. Vacuum Sci. Technol. **13**, 58 (1976).
[11] C. P. HO, J. D. PLUMMER, J. D. MEINDL, and B. E. DEAL, J. Electrochem. Soc. **125**, 665 (1978).
[12] C. P. HO and J. D. PLUMMER, J. Electrochem. Soc. **126**, 1516 (1979).
[13] C. P. HO and J. D. PLUMMER, J. Electrochem. Soc. **126**, 1523 (1979).
[14] B. E. DEAL and A. GROVE, J. appl. Phys. **36**, 3770 (1965).
[15] W. SHOCKLEY and J. L. MOLL, Phys. Rev. **119**, 1480 (1960).
[16] J. A. VAN VECHTEN, Phys. Rev. B **10**, 1482 (1974).
[17] J. A. VAN VECHTEN and C. D. THURMOND, Phys. Rev. B **14**, 3539 (1976).
[18] J. A. VAN VECHTEN and C. D. THURMOND, Phys. Rev. B **14**, 3551 (1976).
[19] D. SHAW, phys. stat. sol. (b) **72**, 11 (1975).
[20] R. B. FAIR and J. C. C. TSAI, J. Electrochem. Soc. **124**, 1107 (1977).
[21] A. WOLKENBERG, phys. stat. sol. (a) **48**, 203 (1978).
[22] A. WOLKENBERG, phys. stat. sol. (a) **50**, 249 (1978).
[23] A. WOLKENBERG, Surface Sci. **2**, 502 (1979).
[24] S. M. SZE, Physics of Semiconductor Devices, Wiley, New York 1968.
[25] H. C. CARD and E. H. RHODERICK, J. Phys. D **4**, 1589 (1971).
[26] H. C. CARD and E. H. RHODERICK, J. Phys. D **4**, 1602 (1971).
[27] H. C. CARD and E. H. RHODERICK, Solid State Electronics **16**, 365 (1973).
[28] R. J. ARCHER and T. O. YEP, J. appl. Phys. **41**, 303 (1970).
[29] S. J. RAIDER, R. FLITSCH, and M. J. PALMER, J. Electrochem. Soc. **122**, 413 (1975).
[30] A. S. JORDAN, A. VAN NEIDE, R. CARUSO, and C. K. KIM, J. Electrochem. Soc. **121**, 153 (1974).
[31] R. J. ARCHER and M. N. ATALLA, Ann. New York Acad. Sci. **101**, 697 (1963).
[32] A. WOLKENBERG, J. Electroanal. Chem. **85**, 311 (1977).

(Received April 4, 1983)

phys. stat. sol. (a) 79, 323 (1983)

Subject classification: 1.6 and 14.3.3; 22.6

Central Laboratory for Solar Energy and New Energy Sources, Bulgarian Academy of Sciences, Sofia[1])

An Electrochemical Evidence of the O_2 Photodesorption from n-TiO_2 Surface

By

G. S. Popkirov

Evidence of O_2 photodesorption from a n-TiO_2 surface is presented. The sample potential vs. reference electrode is monitored upon immersion of the sample into an electrolyte. The influence of illumination of the sample before the immersion, on the magnitude of the recorded voltage patterns, obtained upon immersion, is investigated. The results are discussed in terms of oxygen sorption processes and their influence on the n-TiO_2 surface energetics.

O_2-Photodesorption von n-TiO_2-Oberflächen wird gefunden. Dazu wird das Probenpotential bezüglich einer Bezugselektrode beim Eintauchen der Probe in einen Elektrolyten gemessen. Der Einfluß der Beleuchtung der Probe vor dem Eintauchen auf die Größe der beim Eintauchen registrierten Spannungsimpulse wird untersucht. Die Ergebnisse werden im Zusammenhang mit den Sauerstoffsorptionsprozessen und ihrem Einfluß auf die n-TiO_2-Oberflächenenergetik diskutiert.

1. Introduction

Much work has been done to investigate the surface photochemistry of TiO_2 in respect to the oxygen sorption processes and the effect of illumination [1 to 5]. Most investigations have been made on dielectric powder samples (in rutile or anatase modification). The observed oxygen photosorption processes are usually connected with the formation and action of photogenerated excitons [3, 4].

On the other hand, recently investigations on the n-type semiconducting TiO_2–electrolyte junction have achieved a growing interest, owing to the possible application for solar energy conversion [6, 7]. In this respect it is valuable to investigate the oxygen sorption processes on the surface of semiconducting TiO_2. The predominant role of the photodesorption of O_2^- from the surface, due to the separation of the photogenerated holes and electrons, is to be expected in this case [2].

In this paper an electrochemical evidence of the photodesorption of oxygen from the surface of n-TiO_2 single crystal is presented.

2. Experimental

0.5 mm thick samples were cut perpendicular to the *c*-axis from a TiO_2 single crystal. The samples were mechanically polished and reduced in H_2 atmosphere at 560 °C for 1 h. Silver paste was used to obtain an Ohmic contact and to fix a copper wire. Epoxy resin was used to cover the electrode except 2 mm² on its front side. The TiO_2 electrode potential was measured with a 20 MΩ input resistance voltmeter versus reference saturated calomel electrode (SCE). The transient voltage pulses, obtained upon immersion of the sample into the electrolyte were controlled by means of a storage oscilloscope and recorded with a *Y–t* chart recorder. The electrochemical cell used was a quartz vessel with a plastic lid, so that gases (O_2 and N_2) could be flushed through

[1]) bulv. Lenina 72, 1184 Sofia, Bulgaria.

it, and the sample could be illuminated either before, or after being immersed into the electrolyte. The electrolyte was 1 N NaOH. Prior to each measurement, i.e. immersion of the sample into the electrolyte and recording the obtained voltage change, the sample was rinsed with bidistilled water, immersed into an ultrasonic water bath, rinsed again, and finally dried with a nitrogen jet. Focused light from a 150 W xenon lamp was used to illuminate the TiO_2 electrode.

3. Results

If a n-TiO_2 electrode is immersed into the electrolyte the initial potential is measured with respect to the reference electrode. Under the given experimental conditions (voltmeter input resistance etc.) this potential decreases approaching its steady state value (in our case ≈ 0 V) in about 30 s. It was established that the magnitude of the initial voltage is strongly dependent on the pretreatment of the sample.

Fig. 1, curve a shows the voltage pulse recorded when a freshly prepared TiO_2 sample was immersed into the electrolyte (point A on Fig. 1). After steady state is established the sample was taken out (point B on Fig. 1), cleaned and dried, as explained in the experimental section. Subsequently the sample was again immersed and curve b was recorded. Curves c and d were obtained in the same way, but with a delay of 15 min and 20 h, respectively, between the cleaning procedure and the immersion.

The influence of preillumination of the TiO_2 electrode is shown on Fig. 2. After cleaning and drying the sample was illuminated for 10 s in air, without any contact to the electrolyte. It was immersed into the electrolyte after various time intervals up to 15 min after illumination. The respective voltage pulses are recorded, curves b

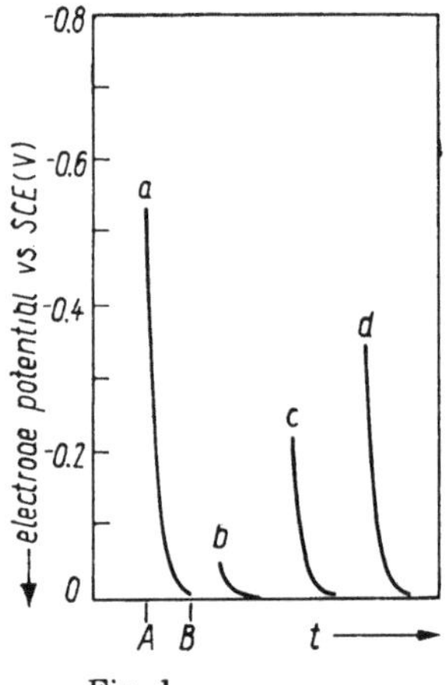

Fig. 1

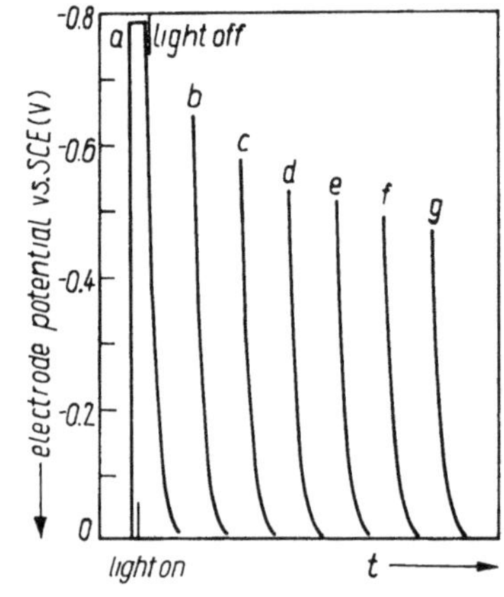

Fig. 2

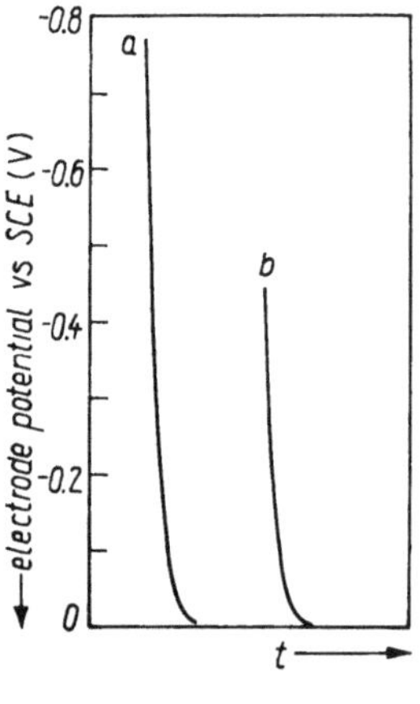

Fig. 3

Fig. 1. Electrode potential vs. SCE as a function of time, measured upon immersion of a n-TiO_2 electrode into the electrolyte: (a) freshly prepared sample; point A marks the moment of immersion, point B the moment of taking out from the electrolyte, time interval A–B = 20 s. (b, c, d) the sample is taken out of the electrolyte, cleaned and dried, and after (b) 0, (c) 15 min, and (d) 20 h immersed into the electrolyte

Fig. 2. Electrode potential vs. SCE as a function of time: (a) photovoltage of a cell: n-TiO_2–electrolyte–SCE, (b) to (g) the sample is taken out of the electrolyte, cleaned and dried, illuminated for 10 s, and after 0, 1, 3, 5, 10, and 15 min, respectively, immersed into the electrolyte. The measurements are made in air atmosphere

Fig. 3. Electrode potential vs. SCE as a function of time. Measurement conditions as in Fig. 2 curve g, but before immersion the electrode was kept in (a) N_2 atmosphere, (b) O_2 atmosphere to g. Curve a is shown for comparison and represents the voltage pulse obtained upon illumination of the sample immersed into the electrolyte, i.e. the photovoltage of the n-TiO_2–electrolyte junction.

To explain these results (Fig. 1, 2), it is suggested that oxygen sorption processes take place on the TiO_2 surface. That is why the influence of the ambient atmosphere was investigated. Fig. 3 shows the voltage pulses recorded under the same experimental conditions as used in Fig. 2, curve g, but in O_2, curve a, and in N_2 atmosphere, curve b, respectively.

4. Discussion

Upon contact of the n-TiO_2 electrode with the electrolyte a depletion layer for the electrons close to the semiconductor surface, as well as a Helmholtz double layer in the electrolyte are formed. An energy band bending at the surface of the semiconductor occurs and steady state is established [6, 7].

Let us assume the bands of the semiconductor electrode are flat before immersion. Upon immersion one should measure a voltage pulse with a magnitude representing the difference between the Fermi levels of the semiconductor before and after the immersion, i.e. the band bending V_{bb} at the surface of the semiconductor. On the other hand, V_{bb} could be determined measuring the maximum photovoltage of a cell, that contains a TiO_2–electrolyte junction. Values of about 0.8 V are usually obtained (cf. Fig. 2, curve a). Thus, it is reasonable to accept that after immersion and after the steady state is established, the band bending is $V_{bb} \approx 0.8$ V. We compare this value of V_{bb} with the voltage pulses measured in our experiments.

When a freshly prepared sample is immersed into the electrolyte, a voltage pulse of 0.53 V is recorded, Fig. 1 curve a. This leads us to the suggestion that some bending of the energy bands of the n-TiO_2 electrode exists already before the immersion. One can further suggest that some negatively charged species are adsorbed on the TiO_2 surface and cause this initial band bending. Upon immersion of the sample into the electrolyte additional band bending occurs until steady state is established and the quantity of the adsorbed, negatively charged species increases. These additional species are probably supplied from the electrolyte. If the electrode is now taken out of the electrolyte, cleaned and dried, and immediately immersed again a very low voltage pulse is recorded, Fig. 1 curve b. The recorded voltage pulse increases, however, if the sample is kept in air for a certain time after cleaning and before immersion, Fig. 1 curves c and d. We suggest here that the re-establishment of the equilibrium with the air atmosphere proceeds by means of desorption of a part of the negatively charged species from the surface. Since no free holes exist in the unilluminated TiO_2, this desorption process is very slow.

We now discuss the influence of the illumination. As is seen from Fig. 2, curve b, pre-illumination of the sample with subsequent immediate immersion into the electrolyte leads to a voltage pulse ≈ 0.7 V. This relatively great value indicates that the band bending of the semiconductor has considerably decreased during the illumination. This is possibly due to a fast neutralization or even photodesorption of the adsorbed, negatively charged species, as discussed above. Increasing the time intervals between illumination and immersion from 0 to 15 min, curves b to g, the measured voltage pulses decrease, approaching the steady state value for the TiO_2–air system. This could be explained with a slow re-adsorption of negatively charged species onto the semiconductor surface. The important role of water molecules adsorbed on the TiO_2 surface, in these processes is suggested in [8].

It is not possible from these experiments to conclude on the identity of the adsorbed species, however, a suggestion that oxygen molecules play an important role can be made, if one takes into account the results shown in Fig. 3. The voltage pulse measured after the sample has been kept in N_2 for 15 min after the illumination was switched off, is considerably greater than the result obtained under oxygen atmosphere. To explain this experiment, we suggest that oxygen molecules, adsorbed and negatively charged, can lead to an initial band bending at the TiO_2 surface. In so far, we consider the observed photoeffects, Fig. 2 curves b to g, as due to a fast photodesorption of oxygen during the illumination, subsequently followed by slow re-adsorption of oxygen from the ambient atmosphere after the illumination is switched off.

Acknowledgements

The experimental part of this work is performed in the Institute of Physical Chemistry, Kiel, FRG. The author is indebted to Prof. R. N. Schindler (University of Kiel) for his kind cooperation and critical discussions, and to Prof. Yu. V. Pleskov (Institute of Electrochemistry, Moscow, USSR) for critical reading of the manuscript. Gratefully acknowledged is the fellowship attributed to the author by DAAD.

References

[1] R. I. Bickley, in: Chemical Physics and Their Surfaces, Vol. 7, The Chemical Society, London 1978.
[2] S. Morrison, The Chemical Physics of Surfaces, Plenum Press, New York 1977.
[3] G. Manuera, Rives-Arnau, and A. Saucedo, J. Chem. Soc. Faraday Trans. I 75, 736 (1979).
[4] H. G. Voelz, G. Kaempf, H. G. Fitzki, and A. Klaeren, in: ACS Symp. Ser. No. 151, Photodegradation and Photostabilisation of Coatings, Ed. S. Papas and F. H. Winslow, Washington (D. C.) 1981.
[5] E. Serwicka, R. N. Schindler, and R. Schumacher, Ber. Bunsenges. phys. Chem. 85, 192 (1981).
E. Serwicka, M. W. Schlierkamp, and R. N. Schindler, Z. Naturf. **36a**, 226 (1981).
[6] H. Gerischer, in: Photovoltaic and Photoelectrochemical Solar Energy Conversion, Ed. F. Cardon, W. P. Gomes, and W. Dekeyser, Plenum Press, New York 1981.
[7] R. Memming, Electrochim. Acta (London) **25**, 77 (1980).
[8] R. Schumacher, J. electroanal. Chem., to be published.

(Received April 8, 1983)

phys. stat. sol. (a) 79, 327 (1983)

Subject classification: 13.4 and 14.1; 18.2; 21.1

Institute of Physics, Academia Sinica, Taipeh[1]) *(a),*
Department of Physics, Clarkson College of Technology, Potsdam[2]),
and The University of Alabama in Huntsville[3])

Deviations from Matthiessen's Rule in the Nickel–Chromium System

By

Y. D. YAO (a), S. ARAJS (b), and E. E. ANDERSON (c)

The deviations from Matthiessen's rule are studied between 4 and 300 K for Ni–Cr alloys containing Cr up to 27 at%. For Ni–Cr alloys with Cr concentration smaller than about 5 at%, the deviations from Matthiessen's rule of the resistivity can be explained by a two-current model. However, this model is not valid for Ni–Cr alloys with Cr concentration larger than about 11 at%.

Die Abweichungen von der Matthiessenschen Regel werden zwischen 4 und 300 K für Ni–Cr-Legierungen mit bis zu 27 At% Cr untersucht. Für Ni–Cr-Legierungen mit Cr-Konzentrationen von kleiner als 5 At% lassen sich die Abweichungen des Widerstands von der Matthiessenschen Regel durch ein Zwei-Strom-Modell erklären. Jedoch ist dieses Modell nicht gültig für die Ni–Cr-Legierungen mit Cr-Konzentrationen größer als etwa 11 At%.

1. Introduction

The concept of two-current conduction in ferromagnetic materials was first suggested by Mott [1]. It was used to explain the deviations from Matthiessen's rule [2] of the electrical resistivity measurements of dilute iron-based alloys by Campbell et al. [3]. Schwerer and Conroy [4] later reported that their experimental data of dilute Ni–Cr alloys with Cr concentration smaller than 5 at% can also be analyzed with this model. In this paper we shall examine the experimental data of Ni–Cr alloys reported by Yao et al. [5] to analyze the two-current conduction in the Ni–Cr alloy system.

Under the assumption that there is only one type of current carrier and that the scattering processes which cause the resistance are independent, the Matthiessen rule can be expressed as

$$\varrho_\mathrm{a}(c, T) = \varrho_\mathrm{p}(T) + \varrho_0(c)\,, \tag{1}$$

where $\varrho_0(c)$ is the electrical resistivity produced by the impurity at 0 K, $\varrho_\mathrm{p}(T)$ the resistivity of an ideal pure metal, and $\varrho_\mathrm{a}(c, T)$ the resistivity of a dilute alloy containing an impurity of concentration c.

Numerous experimental studies clearly show that there are always deviations from Matthiessen's rule. However, in many cases these deviations are very small, and Matthiessen's rule represents a quite good approximation to the experimental data. We can always write $\varrho_\mathrm{a}(c, T)$ exactly as

$$\varrho_\mathrm{a}(c, T) - \varrho_\mathrm{p}(T) + \varrho_0(c) + \Delta(c, T)\,. \tag{2}$$

Pressuming that $\varrho_\mathrm{p}(T)$, $\varrho_\mathrm{a}(c, T)$, and $\varrho_0(c)$ can be determined experimentally, (2) provides a definition for the quantity $\Delta(c, T)$. We call this experimental quantity

[1]) Taipeh, Taiwan.

[2]) Potsdam, New York 13676, USA.

[3]) Huntsville, Alabama 35899, USA.

$\Delta(c, T)$ the "deviation from Matthiessen's rule" (DMR) associated with the solute concentration c in the particular solvent. The existence of a non-zero $\Delta(c, T)$ has been known for a long time. However, with the exception of a flurry of interest in the early 1930s, quantitative investigations of the form and magnitude of $\Delta(c, T)$ proceeded rather slowly until quite recently. From the previous work devoted to this topic [6 to 12], the existence of the deviation from Matthiessen's rule in systems involving defects other than substitutional impurities is demonstrated.

According to the two-current model in which spin-up and spin-down electrons conduct in parallel, the electrical resistivity, $\varrho_a(c, T)$, can be represented as follows [3]:

$$\varrho_a(c, T) = \\ = \frac{[\varrho_\uparrow(0) + \varrho_{a\uparrow}(T)]\,[\varrho_\downarrow(0) + \varrho_{a\downarrow}(T)] + \frac{1}{2}\varrho_{\uparrow\downarrow}(T)\,[\varrho_\uparrow(0) + \varrho_\downarrow(0) + \varrho_{a\uparrow}(T) + \varrho_{a\downarrow}(T)]}{\varrho_\uparrow(0) + \varrho_\downarrow(0) + \varrho_{a\uparrow}(T) + \varrho_{a\downarrow}(T) + 2\varrho_{\uparrow\downarrow}(T)}, \tag{3}$$

where $\varrho_\uparrow(0)$ and $\varrho_\downarrow(0)$ are the impurity residual resistivities for the two spin directions, $\varrho_{\uparrow\downarrow}(T)$ is the resistivity due to the spin-mixing which is characterized by a relaxation time $\tau_{\uparrow\downarrow}$. $\varrho_{a\uparrow}(T)$ and $\varrho_{a\downarrow}(T)$ are the thermal electrical resistivities of the alloy for each spin direction. In general, $\varrho_{a\uparrow}(T) \neq \varrho_{a\downarrow}(T)$. Let $\varrho_{p\uparrow}(T)$ and $\varrho_{p\downarrow}(T)$ be the thermal electrical resistivities of the pure host metal for each spin direction, and $\varrho_h(T)$ the resistivity of the pure host metal at temperature T. At low temperatures, both $\varrho_a(T)$ and $\varrho_{\uparrow\downarrow}$ are small. For the simplest case,

$$\varrho_{a\uparrow}(T) = \varrho_{a\downarrow}(T) = \varrho_{p\uparrow}(T) = \varrho_{p\downarrow}(T) \equiv \varrho(T)\,.$$

Defining

$$\varrho(0) = \frac{\varrho_\uparrow(0)\,\varrho_\downarrow(0)}{\varrho_\uparrow(0) + \varrho_\downarrow(0)}, \tag{4}$$

we get for low temperatures

$$\varrho_a(c, T) = \varrho(0) + \varrho_h(T) + \frac{1}{2}\,\frac{[\varrho_\uparrow(0) - \varrho_\downarrow(0)]^2\,[\varrho(T) + \varrho_{\uparrow\downarrow}(T)]}{[\varrho_\uparrow(0) + \varrho_\downarrow(0)]\,[\varrho_\uparrow(0) + \varrho_\downarrow(0) + 2\varrho(T) + 2\varrho_{\uparrow\downarrow}(T)]}$$

$$\approx \varrho(0) + \varrho_h(0) + \frac{1}{2}\left[\frac{\varrho_\uparrow(0) - \varrho_\downarrow(0)}{\varrho_\uparrow(0) + \varrho_\downarrow(0)}\right]^2 [\varrho(T) + \varrho_{\uparrow\downarrow}(T)]\,. \tag{5}$$

The third term in (5) is the DMR. For the general case, the deviation does not separate out quite as simply; however, it is always positive. Physically this means that the electrons in the low-resistance spin direction, which carry most of the current at low temperature, are "braked" as the temperature rises and they are brought into contact with slower electrons. At high temperatures the DMR should saturate. Then we have

$$\varrho_a(c, T) = \tfrac{1}{2}\,[\varrho_\uparrow(0) + \varrho_\downarrow(0) + \varrho_{a\uparrow}(T) + \varrho_{a\downarrow}(T)]\,. \tag{6}$$

2. Analysis and Discussion

A detailed list of possible sources for the DMR is given in the review article by Bass [11]. In this paper, we only explore the behavior of the DMR using the data by Yao et al. [5]. Due to the experimental difficulties, the value of $\varrho_0(c)$ is replaced by the value of the electrical resistivity at 4 K. Also, we define a deviation parameter $\Delta(c, T)$ by

$$\Delta(c, T) = [\varrho_m(T) - \varrho_m(4)]_{\text{alloy}} - [\varrho_m(T) - \varrho_m(4)]_{\text{pure}} \equiv \Delta\varrho(c, T) - \varrho_c\,, \tag{7}$$

where

$$\Delta\varrho(c, T) \equiv [\varrho_{\text{alloy}}(T) - \varrho_{\text{pure}}(T)]_{\text{measured}}$$
$$\equiv [\varrho_{\text{m}}(T)]_{\text{alloy}} - [\varrho_{\text{m}}(T)]_{\text{pure}}$$
$$\varrho_{\text{c}} \equiv [\varrho_{\text{m}}(4)]_{\text{alloy}} - [\varrho_{\text{m}}(4)]_{\text{pure}}\,, \tag{8}$$

and $\varrho_{\text{m}}(T)$ is the measured resistivity at temperature T.

Fig. 1 shows the temperature-dependent behavior of $\Delta\varrho(c, T)$, which is the difference between the Ni alloy electrical resistivity of the Ni alloy and that of "pure" Ni. A plot of the measured DMR, $\Delta(c, T)$, for the Ni–Cr system as a function of T between 4 and 300 K is presented in Fig. 2. These two figures tell us that every curve, except that containing 5.5 at% Cr, exhibits a peak at $T = T_{\text{m}}$. This means that the behavior of the DMR for Ni–Cr alloys containing Cr in amounts greater than 11.3 at% does not agree with the two-current model. In other words, the two-current model is based on the spin polarization of the conduction electrons due to the ferromagnetic state of host metal. For high concentration Ni–Cr alloys (say, larger than 11.3 at% Cr), the alloys do not show strong ferromagnetic ordering in the experimental temperature range. According to Sadron [13], Marian [14], Besnus et al. [15], and Gregory and Moody [16], the ferromagnetic state in the Ni–Cr system disappears at about 13 at% Cr concentration. The decrease of $\Delta(c, T)$ with increasing temperature after the peak for the Ni–Cr alloy system could mean that above the temperature T_{m}, where the DMR reaches a maximum, it is definitely in the paramagnetic state. However, in the Ni–Cr system, T_{m} does not correspond to the Curie temperature T_{C} exactly. This is so because, besides the electron–magnon scattering effect, there are a number of other effects which contribute to the DMR, as mentioned, for example, by Kus and Taylor [17] and by Engquist and Grimvall [18]. In our Ni–Cr alloy system, T_{m} cannot be easily traced back to simple physical parameters of general interest.

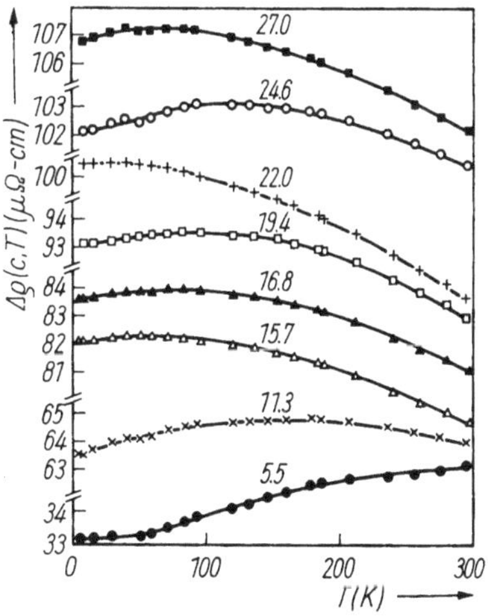

Fig. 1

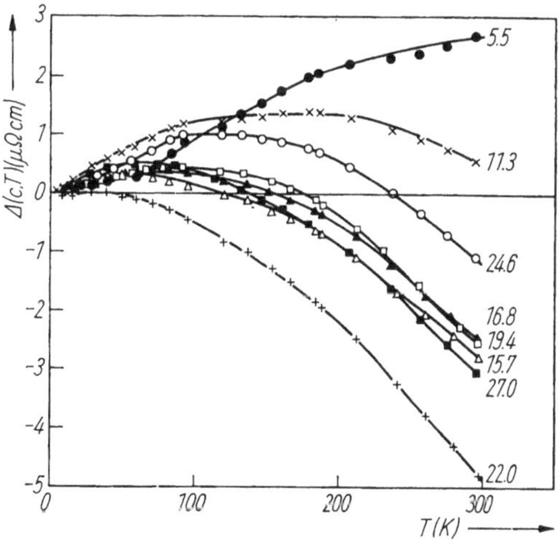

Fig. 2

Fig. 1. Temperature dependence of $\Delta\varrho(c, T)$. Parameter: Cr content (at%)

Fig. 2. Plot of the measured DMR, $\Delta(c, T)$, for the Ni–Cr system as a function of T. Parameter: Cr content (at%)

From the analysis of our experimental data of the Ni–Cr alloys containing Cr from 5.5 up to 27 at%. and those reported by Schwerer and Conroy [14] for Cr concentration smaller than 5 at%, we conclude that the DMR of Ni–Cr alloys with Cr concentration smaller than 5.5 at% can be explained by a two-current model in which spin-up and spin-down electrons conduct in parallel. However, the two-current model cannot be used for these purposes to describe the behavior of the DMR for Ni–Cr alloys with Cr concentration larger than 11.3 at%. It appears that the origin of the hump in the DMR of Ni–Cr alloys described in this paper is not well understood at the present time.

Acknowledgement

One of us (Y. D. Yao) is grateful to the National Science Council of R.O.C. for partial financial support of this work.

References

[1] N. F. MOTT, Proc. Roy. Soc. **153**, 699 (1936).
[2] A. MATTHIESSEN, Ann. Phys. (Germany) 7, 761, 892 (1864).
[3] I. A. CAMPBELL, A. FERT, and A. R. POMEROY, Phil. Mag. **15**, 977 (1967).
[4] F. C. SCHWERER and J. W. CONROY, J. Phys. F **1**, 877 (1971).
[5] Y. D. YAO, S. ARAJS, and E. E. ANDERSON, J. low-Temp. Phys. **21**, 369 (1975).
[6] G. BORELIUS, Hdb. Metall-Phys. Vol. 1, 1935 (p. 1).
[7] H. JONES, Hdb. Phys., Vol. 19, Springer-Verlag, 1956 (p. 227).
[8] D. K. C. MACDONALD, Hdb. Phys., Vol. 14, Springer-Verlag, 1956 (p. 137).
[9] A. N. GERRITSEN, see [7] (p. 137).
[10] F. R. FICKETT, Cryogenics **11**, 349 (1971).
[11] J. BASS, Adv. Phys. **21**, 431 (1972).
[12] M. R. CIMBERLE, G. BOBEL, and C. RIZZUTO, Adv. Phys. **23**, 639 (1974).
[13] C. SADRON, Ann. Phys. (France) **17**, 371 (1932).
[14] P. V. MARIAN, Ann. Phys. (France) 7, 459 (1937).
[15] M. J. BESNUS, Y. GOTTEHRER, and G. MUNSCHY, phys. stat. sol. (b) **49**, 597 (1972).
[16] I. P. GREGORY and D. E. MOODY, J. Phys. F **5**, 36 (1975).
[17] F. W. KUS and D. W. TAYLOR, J. Phys. F **10**, 1495 (1980).
[18] H. L. ENGQUIST and G. GRIMVALL, Phys. Rev. B **21**, 2072 (1980).

(Received March 3, 1983)

phys. stat. sol. (a) 79, 331 (1983)

Subject classification: 10.2; 20.3; 22.2.1

Odessa State University[1])

Strengthening of GaAs by Impurity Atom–Vacancy Complexes

By

O. K. GORODNICHENKO, E. A. GLUSHKOV, V. F. KOVALENKO, and N. D. VASILENKO

Measurements are made of flow limits and photoluminescence spectra of n- and p-GaAs deformed by uniaxial tension. It is shown that long-wave radiation bands of higher intensity correspond to higher flow limit values, the intensity rising is due to an increase in the concentration of impurity atom–vacancy complexes ($Te_{As}V_{Ga}$, $Sn_{As}V_{Ga}$, etc.). The data obtained are proved by calculations made according to the model of strengthening the crystals by complexes. It is noted that in p-GaAs no complexes are found and its flow limit is lower than that in crystals having electron-type conductivity.

Приьеден ырезультаты измерения пределов текучести и спектров фотолюминесценции кристаллов арсенида галлия n- и p-типа, деформированных одноосным растяжением. Показано, что в GaAs n-типа более высоким значениям предела текучести соответствуют длинноволновые полосы излучения большей интенсивности, возрастающей за счет увеличения концентрации комплексов типа атом примеси–вакансия ($Te_{As}V_{Ga}$, $Sn_{As}V_{Ga}$ и др.). Полученные данные подтверждаются расчетами, выполненными в соответствии с моделью упрочнения кристаллов комплексами. Отмечается, что в p-GaAs комплексы обнаружены не были, и предел текучести его ниже, чем кристаллов с электронным типов проводимости.

1. Introduction

When investigating the influence of impurities on plastic deformation of gallium arsenide in [1 to 3] it was found that in single-crystalline GaAs the presence of donor impurities led to an increase and the presence of acceptor impurities to a slight decrease in the flow limit. The observed phenomenon was associated with the impact of charged impurities both on the properties of the crystal matrix and on those of the dislocations [1, 2]. The impurity influence was also considered in the study of GaAs creep [4]. It has been determined that the maximum deformation rate in GaAs doped with Te is higher than that in undoped or Zn-doped GaAs. On analysing the velocities of α, β, and screw dislocations the authors of [4] think that the doping influence is chiefly accounted for by the influence of dislocation charge on the double kink formation energy. The experiments in [1 to 4] were carried out at temperatures $T \leqq 0.5T_{max}$.

This work presents the results of a further study of the influence of impurities on plastic deformation in gallium arsenide at a temperature approximating the temperatures of epitaxial growth of layers ($T \approx 0.6T_{max}$). It has been shown experimentally that there is a good agreement, quantitatively described by the Gilman model [5], between the process of strengthening and the formation of complexes in GaAs crystals.

2. Experimental

Crystals of both undoped and doped gallium arsenide grown in ⟨111⟩ direction by the Czochralski method were investigated. The electron concentrations in differently doped GaAs were: undoped, $n \approx (4 \text{ to } 5) \times 10^{15}\ cm^{-3}$; Te-doped, $n \approx (1 \text{ to } 3) \times$

[1]) 270000 Odessa, USSR.

$\times\ 10^{18}\ cm^{-3}$; Sn-doped, $n \approx (2$ to $3) \times 10^{18}\ cm^{-3}$, and Zn-doped, $p \approx (3$ to $4) \times \times 10^{19}\ cm^{-3}$. The density of growth dislocations was $N_d \lessapprox 5 \times 10^4\ cm^{-2}$. The samples in [110], [112], and [111] edge orientations were deformed by uniaxial tension which created a practically uniform tensile-stress state. The tensile force was applied along a ⟨111⟩ axis at a relative deformation rate of $3.7 \times 10^{-5}\ s^{-1}$. The tests were carried out at $T = 1073$ K under a layer of B_2O_3 flux in order to prevent arsenic sublimation from the crystal surface. The techniques of making and testing the samples are described in [6]. Dislocations on the (110) surface were detected by selective and "projecting" etching [7]. The spectra of stationary photoluminescence (PL) were measured at 77 K on the samples before and after deformation, and on the control samples which had undergone the same thermal treatment as the deformed ones. PL was excited by light from a helium–neon laser. The concentration of charge carriers was estimated from measurements of the Hall coefficient and PL spectra. As to the concentration of complexes, it was determined similarly to [8].

3. Results

A decrease in the intensity of edge radiation, I_e, occurred in the PL spectra of undoped, plastically deformed GaAs samples as compared with the intensity of the control samples, I_c. The value of the ratio I_c/I_e increased with the increase in the relative deformation degree, ε, and when $\varepsilon \approx (3.6$ to $4)$ %, then $I_c/I_e \approx 3$ to 8. The spectral composition of the long-wavelength region of the spectrum depended on the relative deformation rate. When $\varepsilon \approx 1\%$, a band with $h\nu_{max} \approx 1.32$ to 1.33 eV was observed in the long-wavelength region of the spectrum, its intensity being $I_l \ll I_e$. With the increase in the value ε the intensity of this band was rising, and in some samples its maximum shifted to $h\nu_{max} \approx 1.30$ eV. A band with $h\nu_{max} \approx 0.98$ to 1.02 eV also appeared in the radiation spectra when $\varepsilon \approx 4\%$ (Fig. 1). The PL spectra of plastically deformed GaAs⟨Te⟩ and GaAs⟨Sn⟩ were characterized by a decrease of I_e by 1.3 to 1.4 times, an increase of I_l with $h\nu_{max} \approx 1.23$ eV in GaAs⟨Te⟩ and with $h\nu_{max} \approx \approx 1.265$ eV in GaAs⟨Sn⟩ by about 1.5 times as compared with the control samples, and the appearance of a band with $h\nu_{max} \approx 1.0$ eV in the PL spectra of GaAs⟨Te⟩ (Fig. 1).

The electron concentration decreased in all the deformed samples of n-GaAs (see Table 1). The presence of Cu in the samples under investigation may be associated with its diffusion in the process of heating and deformation. The mass-spectroscopic analysis revealed the presence of Cu in the samples. The comparison between the values n^{def} and N_{Cu}^{def} shows that the decrease in the electron concentration in all the deformed crystals cannot be due to the compensation of Cu.

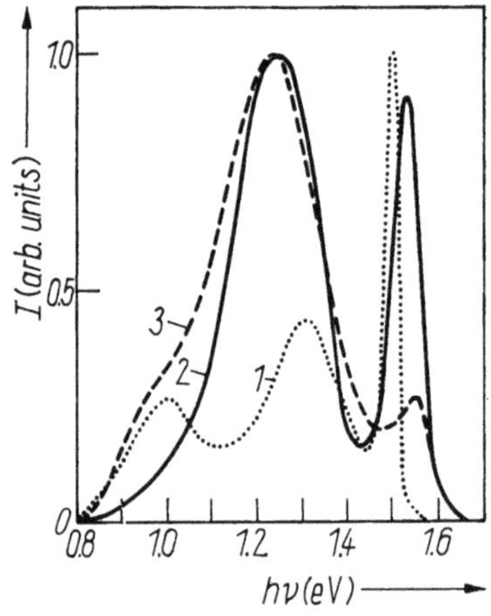

Fig. 1. Photoluminescence spectra of deformed samples of differently doped GaAs: (1) undoped (2) Sn-doped, (3) Te-doped. All spectra are reduced in intensity

Table 1

Impurity concentration in n- and p-GaAs control and plastically deformed ($\varepsilon \approx 4\%$) samples

doping impurity	n, p (cm^{-3})		N_{Cu} (cm^{-3})	
	control samples	deformed samples	control samples	deformed samples
—	4.2×10^{15}	1.1×10^{14}	1.4×10^{15}	2.6×10^{15}
Te	2.6×10^{18}	1.8×10^{18}	1.3×10^{17}	2.1×10^{17}
Sn	2.2×10^{18}	1.7×10^{18}	3.2×10^{16}	5.6×10^{16}
Zn	3.2×10^{19}	4.7×10^{19}	1.2×10^{17}	3.0×10^{17}

Note: Dislocation density in deformed samples is $N_d \lesssim 10^7$ cm^{-2}.

In both the deformed samples of GaAs⟨Zn⟩ and in the n-type samples a decrease of I_e was observed. It is noteworthy, however, that when $\varepsilon \approx 4\%$, the value $I_c/I_e \approx \approx 1.7$ to 4, which is somewhat lower than that for the undoped samples. The PL long-wavelength region of the spectrum of deformed p-GaAs contained no bands. The concentration of holes increased after deformation.

Meanwhile, the flow limit for p-GaAs ($\sigma_f \approx 0.65 \times 10^7$ Pa) estimated from tension–deformation curves was lower than that for undoped ($\sigma_f \approx 0.70 \times 10^7$ Pa) and n-GaAs ($\sigma_f \approx 1.25 \times 10^7$ Pa).

4. Discussion

The presence of long-wavelength radiation bands with $h\nu_{max} \approx 0.98$ to 1.02, 1.23, 1.265, 1.30 to 1.35 eV in the PL spectra of n-GaAs is due to the presence of complexes which include the following structure defects and impurity atoms: ($V_{As}V_{Ga}$) and ($V_{As}Cu_{Ga}$), ($Te_{As}V_{Ga}$), ($Sn_{As}V_{Ga}$), ($Si_{As}Cu_{Ga}$), etc., respectively [9, 10].

The change in the energy position of the radiation band with $h\nu_{max} \approx 1.30$ to 1.35 eV is probably caused by a change in the radiation centre structure at different values of ε [11].

It is known that vacancies of gallium and arsenic as well as intrinsic and impurity interstitial atoms may be present in GaAs crystals simultaneously [12, 13].

In n-GaAs V_{Ga} readily form complexes with donor impurity atoms, which substitute interstitials in a sublattice of arsenic, especially with Te [13, 14]. The appearance of bands with $h\nu_{max} \approx 0.98$ to 1.02, 1.0, 1.30 to 1.35 eV in the PL spectrum and the increase of the bands with $h\nu_{max} \approx 1.23$ and 1.265 eV due to an increase in the concentration of complexes suggest the possibility of generating both V_{Ga} and V_{As} in the process of deformation. It is also to be noted that a change in the internal quantum efficiency of long-wavelength radiation after plastic deformation may play a certain role in rising the efficiency of the band with $h\nu_{max} \approx 1.23$ eV [15]. A decrease in the intensity of I_c and in the concentration of electrons is presumably associated to a great extent with the transition of impurity atoms into the electrically inactive state during the formation of complexes. However, the main reason for the decrease of I_e in the deformed n-type samples is an increase in the concentration of radiationless centres. In p-GaAs it may be difficult for Zn atoms to form stable complexes even at high concentrations of the acceptor impurity ($p \approx 10^{19}$ cm^{-3}). In the deformed p-type crystals the increasing concentration of holes is possible associated with the increasing concentration of V_{Ga}, the latter being singly charged acceptors [16]. The assumption about a high positive charge on dislocations in deformed p-GaAs [4] may be accounted

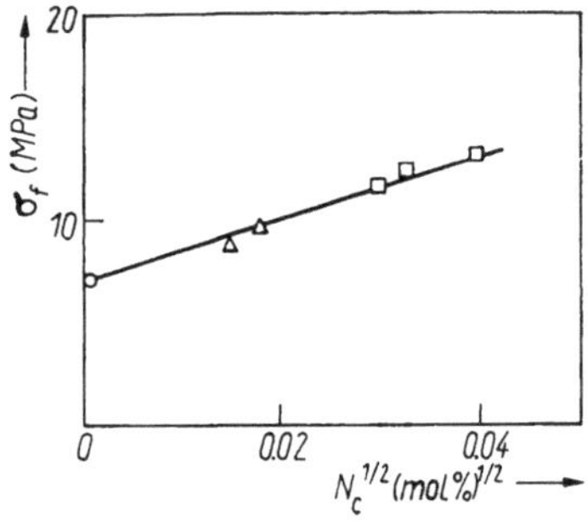

Fig. 2. Concentration dependence of the flow limit in n-GaAs: ○ undoped, △ Sn-doped, □ Te-doped

for by diffusion of V_{Ga} towards dislocations where they form regions of increased concentration of acceptors, since the diffusion coefficient of V_{Ga} is rather high, substantially exceeding that of V_{As} [17].

The observed connection between strengthening and the formation of complexes in GaAs is in agreement with the Gilman model [5] according to which strengthening is due to a decrease of the electrostatic energy during the passage of dislocations through the complex in the process of gliding. As is seen from Fig. 2, the flow limit σ_f, of n-type crystals is proportional to $N_c^{1/2}$ (N_c is the concentration of complexes in mol%). Then

$$\sigma_f = A + BN_c^{1/2}\,, \tag{1}$$

where A is the flow limit of undoped gallium arsenide and B is the doubled shear strength of the complex. The strengthening coefficient can be estimated in the following way [5]: find the difference between the electrostatic energy states before and after cutting the complexes by dislocations and equate it to the work performed during the passage of dislocations through the complex. In this case

$$B = \frac{4.7e^{*2}}{4\pi KK_0a^4}\,, \tag{2}$$

where $e^* = 0.47e$ is the effective charge of atoms in GaAs, K the relative dielectric permeability of GaAs, K_0 the absolute dielectric permeability of vacuum, and a the lattice constant of GaAs. The dependence of coefficient B on temperature is determined by the temperature dependences of the parameters K and a, which are terms of (2). Having replaced these parameters with their values we obtain: $B \approx 1.9 \times 10^8$ Pa. It was estimated experimentally that $B \approx 1.5 \times 10^8$ Pa. This agreement between theory and experiment supports the assumption that gallium arsenide is strengthened by impurity atom–vacancy complexes. Thus, from the comparison between the flow limit and the PL spectral composition it follows that in n-type crystals the presence of long-wavelength radiation bands corresponds to higher values of flow limits, which fact is associated with an increase in the concentration of impurity atom–gallium vacancy complexes. The obtained results are proved by the model of strengthening due to the formation of complexes. However, in p-GaAs, in which complexes were not detected even at a hole concentration $p \approx 10^{19}$ cm^{-3}, flow limits are lower than in the crystals with electron-type conductivity.

References

[1] V. B. OSVENSKII, O. G. STOLYAROV, and M. G. MILVIDSKII, Fiz. tverd. Tela **10**, 3208 (1968).
[2] N. N. SHERSHAKOVA, V. B. OSVENSKII, and O. G. STOLYAROV, Izv. Akad. Nauk SSSR, Ser. neorg. Mater. **6**, 457 (1970).

[3] V. SWAMINATHAN and S. M. COPLEY, J. Amer. Soc. **58**, 482 (1975).
[4] H. STEINHARDT and P. HAASEN, phys. stat. sol. (a) **49**, 93 (1978).
[5] J. J. GILMAN, J. appl. Phys. **45**, 508 (1974).
[6] N. D. VASILENKO, O. K. GORODNICHENKO, and I. E. MARONCHUK, Poluchenie i svoistva tonkikh plyonok, Izd. IPM, Akad. Nauk USSR, Kiev 1981 (p. 71).
[7] S. S. SHIFRIN, M. G. MILVIDSKII, and V. B. OSVENSKII, Kristallografiya **27**, 712 (1982).
[8] K. D. GLINCHUK and A. V. PROKHOROVICH, phys. stat. sol. (a) **44**, 777 (1977).
[9] V. I. VOVNENKO, K. D. GLINCHUK, and A. V. PROKHOROVICH, Fiz. Tekh. Poluprov. **10**, 2167 (1976); **13**, 285 (1979).
K. D. GLINCHUK, A. V. PROKHOROVICH, V. E. RODIONOV, and V. I. VOVNENKO, phys. stat. sol. (a) **41**, 659 (1977).
[10] K. D. GLINCHUK, K. LUKAT, and A. V. PROKHOROVICH, Optoelektronika i poluprovodnikovaya tekhnika, Izd. Naukova Dumka, Kiev 1982, N1 (p. 39).
[11] V. I. VOVNENKO, K. D. GLINCHUK, K. LUKAT, and A. V. PROKHOROVICH, Fiz. Tekh. Poluprov. **14**, 1003 (1980).
[12] M. G. MILVIDSKII, O. V. PELEVIN, and B. A. SAKHAROV, Fiziko-khimicheskie osnovy polucheniya razlagayushchikhsya poluprovodnikovykh soedinenii, Izd. Metallurgiya, Moskva 1974.
[13] B. T. BUBLIK, M. G. MILVIDSKII, and V. B. OSVENSKII, Izv. vuzov (SSSR), Ser. fiz. No. 1, 7 (1980).
[14] V. I. FISTUL, Raspad peresyshchennykh tvyordykh rastvorov, Izd. Metallurgiya, Moskva 1977.
[15] V. I. VOVNENKO, K. D. GLINCHUK, and A. V. PROKHOROVICH, Fiz. Tekh. Poluprov. **13**, 1929 (1979).
[16] G. M. BLOM, J. Crystal Growth **36**, 125 (1976).
[17] S. Y. CHIANG and G. L. PEARSON J. appl. Phys. **46**, 2986 (1975).

(Received May 3, 1983)

See Short Note by A. Boeuf et al.

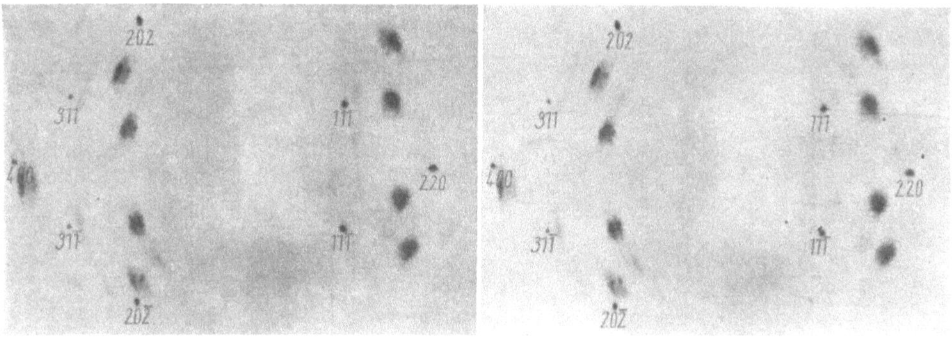

a b

Fig. 1. Neutron diffraction patterns obtained for an oscillation of $\pm 45°$ and exposure time of 9 h: a) $T = 60$ K, b) $T = 6$ K. Small Bragg angle region

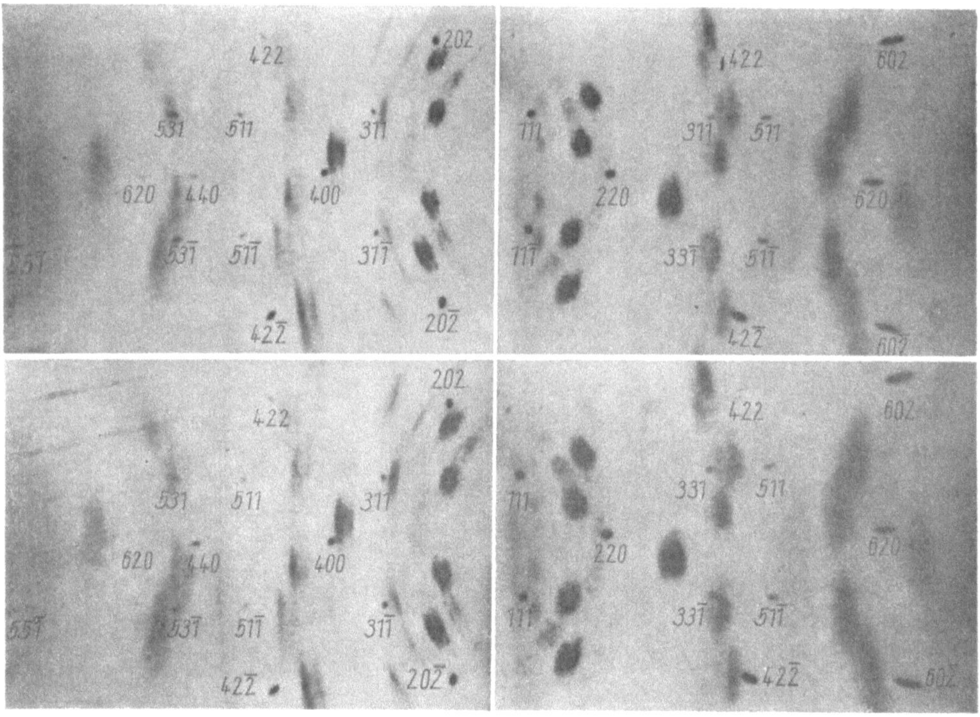

a

b

Fig. 2. Neutron diffraction patterns obtained for an oscillation of $\pm 50°$ and exposure time of 24 h: a) $T = 60$ K, b) $T = 6$ K. Large Bragg angle region

See Short Note by L. Nanu et al.

Fig. 2. RHEED of B^+ implanted Si sample: $\Phi = 5 \times 10^{18}/m^{-2}$

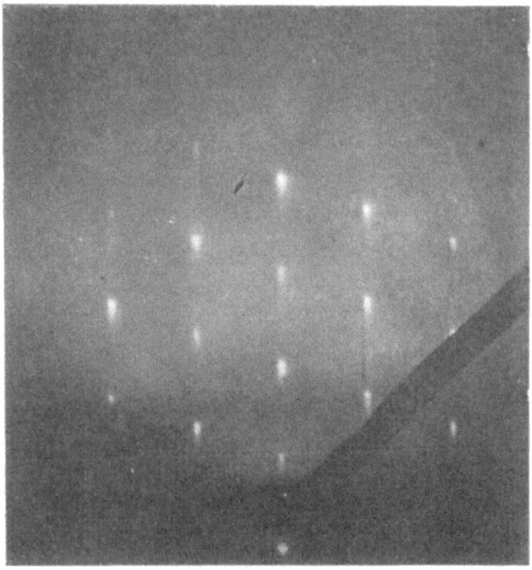

Fig. 3. RHEED of the implanted sample after exposure to a single 2J TEA-CO_2 laser pulse

See Short Note by G. K. BARANOVA

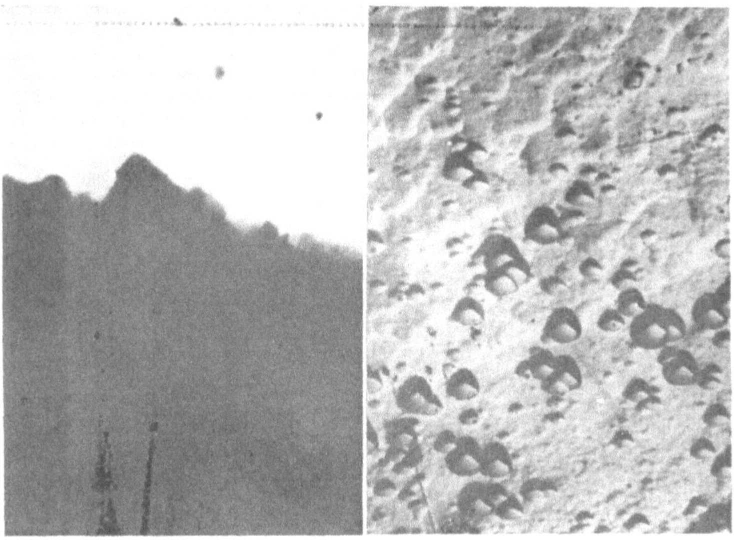

a b

Fig. 1. View of the (115) plane of Nb a) after anodizing in KBr solution, b) following dissolution in the mixture $HNO_3:HF:H_2SO_4 = 10:10:1$

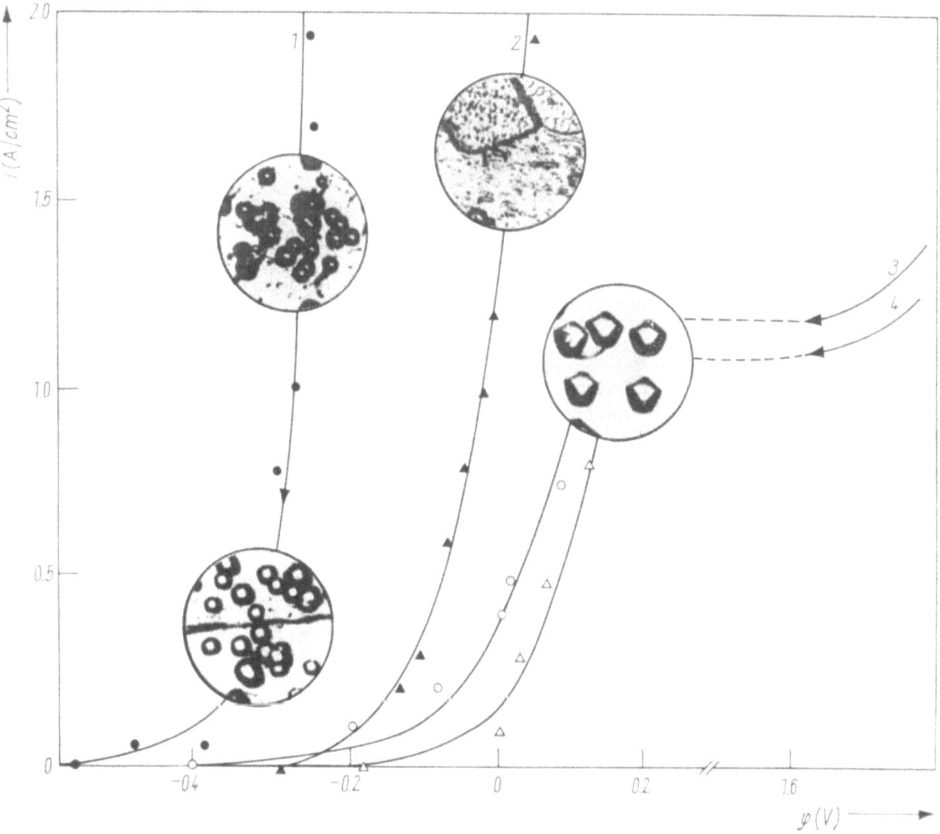

Fig. 2. Amperostatic polarization curves and view of the surface: (1) (100) plane of Fe–Si, 20% KNO solution, pH = 3; (2) (112) plane of Fe–Ni, mixture $HNO_3:HCl:H_2O = 1:3:50$; (3) (100) plane of Fe–Si, 20% KNO_3 solution, pH = 8; (4) (112) plane of Fe–Ni, mixture prepared according to [8]

See Short Note by G. K. Baranova

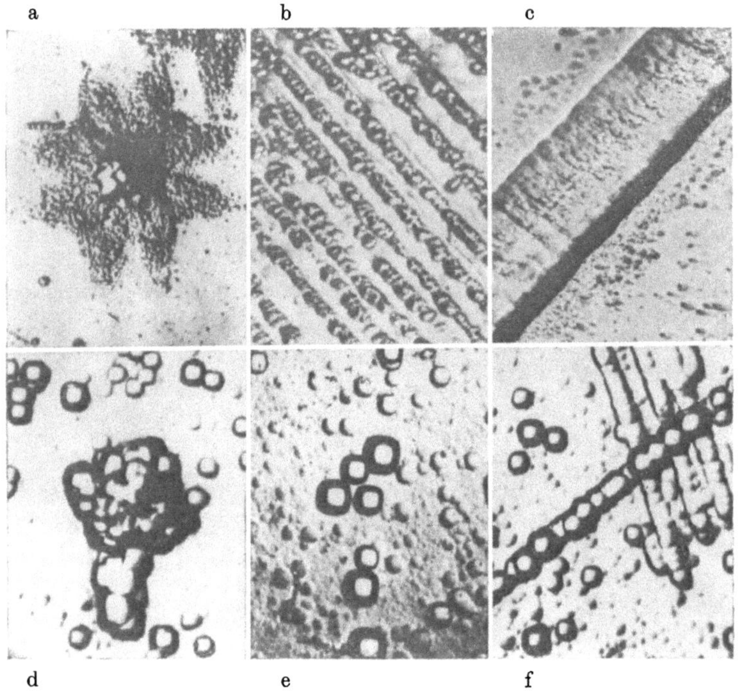

Fig. 3. Dislocation (top array) and orientation (bottom array) etching of the (100) plane of Nb after deformation by a) and d) indentation with a load of 0.6 N, b) and e) compression with 10^4 N/m^2, c) and f) scratching

See Short Note by K. BISKUPSKA et al.

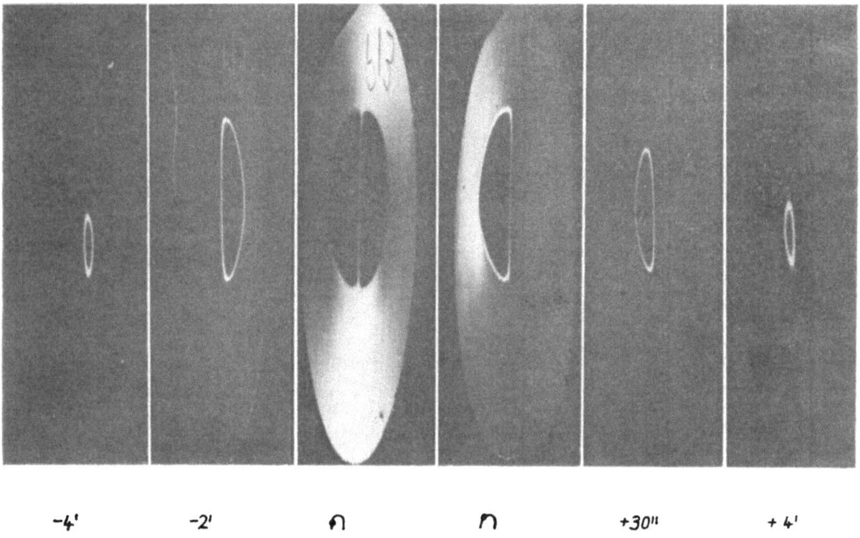

Fig. 1. Double crystal spectrometer topograms of isochronally annealed proton bombarded silicon

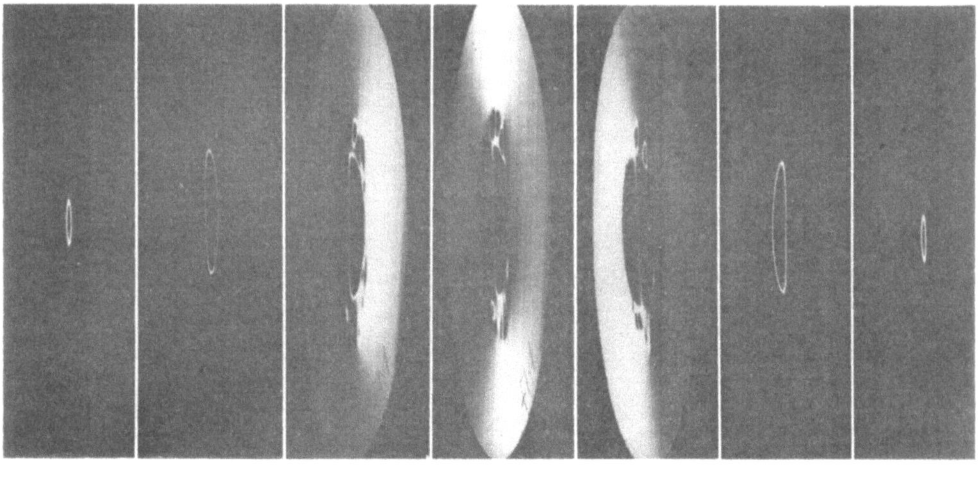

Fig. 3. Double crystal spectrometer topograms of electron irradiated proton bombarded silicon

Short Notes

phys. stat. sol. (a) 79, K1 (1983)

Subject classification: 1.2; 22.6

Joint Research Center, Ispra, and Institut Laue-Langevin, Grenoble[1] (a), Centre d'Etudes Nucléaires, Grenoble (b), Kernforschungszentrum Karlsruhe (c), Institut Laue-Langevin, Grenoble (d), and Università di Ancona (e)

Study of the 25 K NpO_2 Phase Transition by a Neutron Oscillation Camera

By

A. BOEUF (a), R. CACIUFFO (a), J.M. FOURNIER (b), L. MANES (c), J. REBIZANT (c), F. RUSTICHELLI (d), J.C. SPIRLET (e), and A. WRIGHT (e)

The existence of a phase transition in NpO_2 below 25 K was deduced from measurements of macroscopic physical quantities but no evidence based on observations of microscopic quantities has till now confirmed the presence of this phase transition. The observation of a λ peak in heat capacity measurements /1/ and subsequent magnetic susceptibility measurements /2/ showing a maximum at the transition temperature, suggested an antiferromagnetic transition. On the other hand, Mössbauer investigation, /3/ powder neutron diffraction studies /4/, and a single crystal neutron diffraction study /5/ failed to detect at low temperature either a magnetic ordering or a structural distortion. In particular the single crystal experiment showed an upper limit for the ordered moment corresponding to a type I antiferromagnetism $\mu_{ord} < 0.1\ \mu_B$, which appears compatible with the magnetic moment obtained from Mössbauer investigation. Moreover, a search for two possible kinds of distortion of the oxygen sublattice showed an upper limit of the displacement $\Delta o \approx 0.001$ nm. These upper limits were obtained by following through the transition a limited number of reflections on a four-circle neutron diffractometer. In this note we report the results of an investigation extended to a very large region of the reciprocal space performed by a neutron photographic technique.

The neutron Weissenberg camera D12 installed at the High Flux Reactor of I.L.L. Grenoble, was used in the oscillation mode. The instrument uses mono-

1) avenue des Martyrs 156X, 38042 Grenoble Cédex, France.

chromatic radiation (λ = 0.146 nm, a neutron-sensitive scintillator and X-ray film housed in a cylindrical camera /6/). The sample was a small NpO_2 single crystal (0.5 mm^3) which was grown at the J.R.C. Karlsruhe Establishment /7/ and encapsulated in an aluminium container. NpO_2 crystallizes in the face centered cubic fluorite structure and has a lattice constant a = 0.5434 nm at room temperature.

The sample was rotated around the [100] vertical axis during the exposure time. At first, oscillations of $\mp 45^{\circ}$ were performed during 9 h in order to explore the small Bragg angle region to detect eventual appearance of magnetic reflections. Fig. 1a and b (see on the photo pages before the Short Notes part) show the diffraction map obtained at 60 and 6 K, respectively. The minium observable q value, due to the presence of neutron beam catcher being q = = 0.06 nm^{-1}. No appreciable change is observed between the two diagrams above and below the 25 K transition. The films record sharp nuclear reflections as well as some streaks due to the aluminium container. No magnetic ordering is observed. Furthermore several exposures under different conditions were performed at 60 and 6 K in order to explore the large Bragg angle region to detect eventual satellites associated with Jahn-Teller distortions. Fig. 2a and b (see on the photo pages before the Short Notes part) show diffraction maps obtained for an oscillation of $\pm 50^{\circ}$, 24 h exposure time at temperature above and below the 25 K phase transition. No extra Bragg reflections appear below 25 K. Though the punctual sensitivity of the classical four-circle neutron experiment /5/ is better than the sensitivity of this photographic technique, this technique allows exploration of very large regions of the reciprocal space.

In conclusion, it appears that exploring a very large region of reciprocal space neither magnetic ordering nor lattice distortions are observed to be associated with the phase transition in NpO_2 below 25 K. Although the present results do not contradict the model proposed by Erdös et al. /8/, the NpO_2 phase transition appears to retain and even to strenghten its mysterious character.

References

/1/ E.F. WESTRUM,Jr., J.B. HATCHER, and D.W. OSBORNE, J. chem. Phys. 21, 419 (1953).

/2/ J.W. ROSS and D.J. LAM, J. appl. Phys. 38, 1451 (1967).

/3/ B.D. DUNLAP, G.M. KALVIUS, D.J. LAM, and M.B. BRODSKY, J. Phys. Chem. Solids 29, 1365 (1968).

/4/ L. HEATON, M.H. MUELLER, and J.M. WILLIAMS, J. Phys. Chem. Solids 28, 1651 (1967).

/5/ A. BOEUF, J.M. FOURNIER, G. HEGER, N. LEHNER, L. MANES, J. REBIZANT, F. RUSTICHELLI, and J.C. SPIRLET, J. Physique, Lettres 42, L-401 (1981).

/6/ D. HOHLWEIN and A.F. WRIGHT, J. appl. Cryst. 14, 82 (1981).

/7/ J.C. SPIRLET, E. BEDNARCZYK, I. RAY, and W. MÜLLER, J. Physique 40, C4-108 (1979).

/8/ P. ERDÖS, G. SOLT, Z. ZOLNIEREK, A. BLAISE, and J.M. FOURNIER, Physica (Utrecht) 102B, 164 (1980).

(Received March 1, 1983)

phys. stat. sol. (a) 79, K5 (1983)
Subject classification: 1.5 and 11; 22.1.2

Central Institute of Physics, Bucharest[1)]

Annealing of Implanted Silicon under the Action of Microsecond Pulsed TEA-CO_2 Laser Radiation

By
L. NANU, DOINA BĂRBULESCU, I.N. MIHĂILESCU, V. TEODORESCU, and LEONA C. NISTOR

A great interest arose in the last few years by the recrystallization of amorphous layers on the top surface of implanted semiconductor materials under the action of powerful laser radiation. The cw or pulsed laser irradiation actually results in either a solid-phase epitaxial regrowth or a liquid-phase process, respectively, depending on the temporal characteristic of the incident power. It was thus shown that ruby, Nd:YAG, and argon ion lasers can be successfully used in order to obtain the annealing of the damaged layers, as the laser light is strongly absorbed onto the surface of implanted semiconductors /1 to 3/.

Recent experiments were performed with the aid of cw CO_2 lasers. A good recrystallization was obtained, even though the energy of the incident photons is lower than the energy band gap of silicon; the light absorption then takes place by free carrier excitation /4, 5/.

In this note new results on annealing of implanted silicon samples, under the action of microsecond pulsed CO_2 laser radiation, are firstly reported.

At the beginning, we estimated the incident energy which is required in the laser pulse to melt the whole amorphous layer. For this purpose one has to solve the well-known heat conduction equation

$$c\varrho \frac{\partial T}{\partial t} = k \frac{\partial^2 T}{\partial x^2} \; . \qquad (1)$$

Here c, ϱ are the material's specific heat and density, respectively, while k is the thermal conductivity of amorphous silicon. We consider to have uniform distribution across the laser cross-section and, because of the relatively short duration of the interaction process we neglect the temperature dependence of

---

1) POB MG-6, R-76900 Bucharest V, Romania.

the optical and thermal properties of the material. The boundary and initial conditions are

$$-k\,\frac{\partial T}{\partial x} = I(t)\,, \qquad \text{for } x = 0\,, \qquad t > 0\,,$$

$$T(x,\,0) = T_0 = 300\text{ K}, \qquad \text{for } 0 \leqq x < \infty, \qquad t = 0\,, \tag{2}$$

where I(t) is the actual temporal evolution of the incident laser intensity. We used typical microsecond TEA-CO_2 laser pulses with a first peak (accounting for ≈ 30% from the whole laser output energy) of 150 ns FWHM followed by a tail of significantly lower power and lasting 2 μs. As shown in our paper /6/, target melting or vaporization under the action of such laser pulses is to be achieved either during the first peak or not at all. That is why we further supposed a linear evolution of the incident laser intensity

$$I(t) = (1 - R)at, \qquad \text{for } 0 \leqq t \leqq t_1\,, \tag{3}$$

where t_1 was taken ≈40 ns, i.e. the rising time of the first peak during the laser pulse. Here R is the target reflectivity corresponding to the wavelength of the incident laser radiation.

Equations (1) to (3) were solved using a Laplace transformation /7, 8/; finally, the following solution was obtained:

$$\Delta T(x,\;t) = (1 - R)(2\,ad^{1/2}/k)(4t)^{3/2}\;i^3\;\text{erfc}\left[x/2(dt)^{1/2}\right]; \tag{4}$$

here $\Delta T = T - T_0$ is the target heating under the action of powerful laser radiation, d stands for the material diffusivity while

$$i^n\;\text{erfc}\;z = \int_z^{\infty} i^{n-1}\;\text{erfc}\,\xi\,d\xi\,, \qquad n = 1,\,2,\,\dots\,,$$

$$i^0\,\text{erfc}\,z = \text{erfc}\;z = \frac{2}{\sqrt{\pi}}\int_z^{\infty} e^{-\xi^2}\,d\xi\,. \tag{5}$$

The time dependence of the sample surface temperature as derived from

$$\Delta T(0,\;t) = (1 - R)(a/3d)(d/\pi)^{1/2}(4t)^{3/2} \tag{6}$$

using the following constants for amorphous silicon: c = 836 J/kg K, ρ = = 2330 kg/m^3, k = 123.44 W/m K, R = 0.43, and a focal spot diameter of 3 mm, was plotted in Fig. 1 for different incident laser energies. It should be observed that the energy densities for which the surface melting is obtained

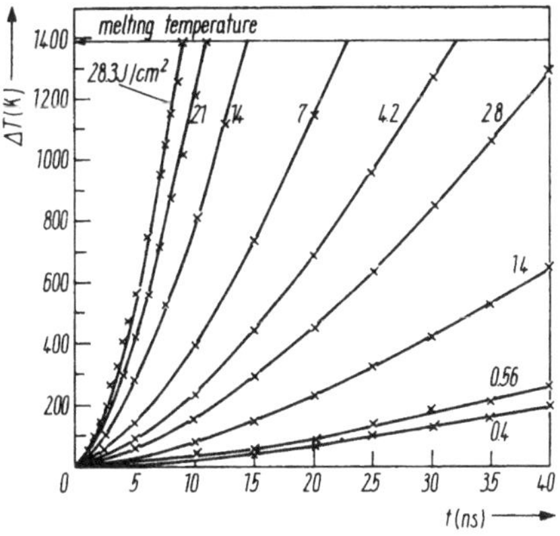

Fig. 1. The time dependence of sample surface temperature for different energy densities of the laser pulse

during the rising front of the first peak keep lower than 50 J/cm^2, i.e. the surface damage threshold of implanted silicon /9/.

Hence, annealing of amorphous implanted silicon samples can be induced as a result of a few Joules, microsecond pulsed, TEA-CO_2 laser irradiation.

With the view of verifying the afore inferred results we have focused microsecond TEA-CO_2 laser pulses of ≈ 2 J in spots of about 3 mm diameter. The samples were silicon single crystals implanted onto ⟨111⟩ surface with 50 kV boron ions at a dose of $5x10^{18}/m^2$.

Reflection high energy electron diffraction (RHEED) evidences an amorphous layer on the samples top surface (Fig. 2, see on the photo pages before the Short Notes part). Following laser annealing (even after one laser short irradiation) a complete epitaxy was evidenced (Fig. 3, see on the photo pages before the Short Notes part).

In conclusion, annealing of boron implanted silicon crystals was predicted and experimentally induced as a result of a few Joules, microsecond pulsed, TEA-CO_2 laser irradiation. However, as the RHEED technique gives information onlv about the epitaxial regrowth but cannot provide data on the dislocation density inside the implanted layer, new experimental investigations are now in progress which will allow us further checking up of the quality of the laser annealed material.

References

/1/ C.W. WHITE, Y. NARAYAN, and R.T. YOUNG, Science 204, 461 (1979) and references therein.

/2/ A. GAT, T.J. MAGEE, J. PENG, V.R. DELIVE, and C.A. EVANS, Jr., Solid State Technol. 59 (November 1979).

/3/ E. COJOCARU, N. COMANICIU, I.N. MIHAILESCU, L. NANU, L.C. NISTOR, and V. TEODORESCU, Appl. Phys. A26, 243 (1981).

/4/ M. TAKAI, P.H. TSIEN, S.C. TSOU, D. ROSCHENTHALER, M. RAMIN, H. RYSSEL, and I. RUGE, Appl. Phys. 22, 129 (1980); 23, 163 (1980).

/5/ M. MIYAO, K. OHYU, and T. TOKUYAMA, Appl. Phys. Letter 35, 227 (1979).

/6/ I. URSU, I. APOSTOL, D. BĂRBULESCU, I.N. MIHĂILESCU, A.M. PROKHOROV, P.V. AGEEV, A.A. GORBUNOV, and V.I. KONOV, Optics Commun. 39, 180 (1981).

/7/ N.S. CARSLAW and I.C. JAEGER, Conduction of Heat in Solids, Oxford University Press, Oxford 1959.

/8/ M.R. SPIEGEL, Theory and Problems of Laplace Transforms, McGraw-Hill, New York 1965.

/9/ D.C. SMITH and R.T. BROWN, J. appl. Phys. 46, 1146 (1975).

(Received June 17, 1983)

phys. stat. sol. (a) 79, K9 (1983)
Subject classification: 1.6; 10.1; 21

Solid State Physics Institute, Academy of Sciences of the USSR, Chernogolovka[1])

On the Role of Surface Oxides in the Processes of Metal Etching

By
G.K. BARANOVA

Introduction Etching techniques for revealing dislocations and determining crystallographic orientation by etch figure shapes have been widely used to study the structure of metals /1/. As is known /2, 3/, oxidation plays an important role in the processes of local dissolution. During the last years the author has developed new methods of orientation and dislocation etching for b.c.c. refractory metals /4 to 6/, silicon-iron /7/, and iron-nickel alloys /8/, in which the surface was especially oxidized before etching.

In the present note principal regularities of the processes of orientation and dislocation etching of metals suited for oxidation are considered and comparison characteristics of orientation figures and dislocation etch pits are adduced.

Experimental results and discussion As it has been shown previously /5 to 8/, orientation-dependent figures form fairly easily on metals provided their surface is oxidized prior to etching. The choice of an oxidation method is dictated by the electrochemical properties of metals. Three cases are considered below: anodizing, local precipitation of conductive oxides, and momentary passivation of the entire surface.

Anodizing is the most convenient method for oxidation of b.c.c. refractory metals /9/. It has been found /5, 6/ that sharp contrasted orientation figures on all planes of Nb, Ta, W, and V can be produced if their surface at first is completely coated with the non-conducting oxide film Me_2O_5, 0.4 to 0.5 μm in thickness and then this film and the metal are dissolved at a dissolution rate of 0.05 to 0.005 μm/s. This is clearly seen in Fig. 1 (see on the photo pages before the Short Notes part). A specimen with a (115) plane of Nb was polished, then one part of the specimen was coated with a blue oxide film by anodizing in a 3% solution of KBr (50 V, 5 s), the other one being left polished (Fig. 1a). Then the specimen completely immersed in an etching solution $HF:HNO_3:H_2SO_4$ =

---

1) Chernogolovka, Moscow District, USSR.

10:10:1 for 60 s revealed good contrasted orientation figures only on the surface part where the oxide film was found to be before etching (Fig. 1b).

Local precipitation of conducting oxides and their subsequent dissolution are used for orientation etching of Fe-Si in an acid medium /7/. In those cases dissolution conditions (composition of electrolyte; current density) are chosen so that oxides do not form a continuous layer but precipitate only on active parts of the surface in the form of separate black spots 20 to 30 μm in size. These oxides proved to transform to distinct orientation figures when being subsequently dissolved in the active region at a lower current density. Fig. 2 (see on the photo pages before the Short Notes part) (curve 1) displays a polarization curve in a 20% solution of KNO_3 and the view of the (100) plane of Fe-Si. The oxide spots formed at i = 1.3 A/cm^2 transform to distinct square orientation figures during 30 s at decreasing current density down to 0.5 A/cm^2.

The third feasible mechanism of preliminary oxidation is a short term passivation of the whole surface. It is realized at the orientation etching of Fe-Si in alkali medium /7/ and of Fe-Ni alloys in acids /8/. In this case such electrolytes for etching are selected in which the critical current density has a relatively low value, approximately 2 to 3 A/cm^2. To produce orientation figures the metal at first is kept for several seconds in a passive region at the critical current density and the potential +1.6 to 1.8 V, where oxygen deposition takes place. Then the current density is reduced by 0.2 to 0.4 A/cm^2 and the metal is dissolved in the prepassive region at an unstable value of the potential.

Fig. 2 (curves 3, 4) exhibits polarisation curves and the surface pattern after etching Fe-Si in a 20% solution of KNO_3 (curve 3), at pH = 8, and a Fe-Ni alloy in a mixture, prepared according to /8/ (curve 4 and etch figures). In both cases etch figures are produced after a 5 to 10 s passivation at i = = 1.7 A/cm^2 followed by dissolution at i = 1.2 to 1.4 A/cm^2. For comparison in Fig. 2 (curve 2) the polarization curve of the Fe-Ni alloy is given in the mixture HNO_3:HCl:H_2O = 1:3:50 /10/ where the value of critical current density exceeds 5 A/cm^2. Even at a relatively high current density (≈ 2 A/cm^2) one can observe only an indefinite relief on the surface.

Thus to obtain distinct orientation-dependent figures the following processing is required: polishing-oxidizing-etching. In this case oxidation should be deep - either up to the formation of oxide phases or to the moment the full passivation and oxygen deposition takes place. It should be noted that preliminary oxidation standardizes the surface. The techniques mentioned above are distinguished by their good reproducibility, a high quality of etch figures, and universality. It is common for etching to occur successfully on all crystallographic planes of single- and polycrystals of metals of any purity.

Dislocation etching is commonly performed by chemical or electrochemical dissolution in the active region with a low rate. Thus dislocation etch pits on the (100) plane of Fe-Si may be obtained by anode dissolution in a 20% solution of KNO_3, pH = 3, i = 0.2 to 0.4 A/cm^2 /7/. For revealing dislocations on the (100) plane of Nb chemical dissolving in a mixture involving oxidizers, activators, and complexing agents - H_2O_2:H_2SO_4:HF:NH_4F is used. Here etch pits occur only if hydrogen peroxide is introduced into a heated solution after the specimen has been immersed /4/.

Hence, dislocation etching requires no oxidation as an independent stage. The formation of etch pits at dislocations takes place in an oxidable medium. The role of the oxidizer is likely to amount to a short-term adsorption at sites of the dislocation yield, as the so-called "surface decoration", which is responsible for the occurrence of dissolution centres at dislocations.

Differences in etching mechanisms involve discrepancies in size, form and, mainly, in the arrangement of dislocation pits and orientation figures.

Fig. 3 (see on the photo pages before the Short Notes part) illustrates a view of the (100) plane of Nb after dislocation (top array) and orientation (bottom array) etching. After polishing the crystals were deformed by indentation, compression, and scratching and then etched to reveal dislocations /4/ and produce orientation figures /5/. In all cases no coincidence has been observed in the arrangement of the orientation figures and etch pits at dislocations.

After dislocation etching clearly one can see wings of the indentation rosette (a), slip lines (b), a wide band of dislocations on each side of the scratch (c). Dislocation etch pits are shallow, shapeless. They form, as a rule, selectively on a certain plane. Their arrangement complies with the geometry of that of dislocations in a deformed crystal. The etching pattern is found to be reproducible after repolishing.

At orientation etching on the same crystals wings of the indentation rosette, slip lines, and block boundaries are absent (d, e), large individual figures form along the scratch (f). Orientation figures form on all planes at a time. They locate, as shown above, on more rough concentrators of stresses such as scratches, phase inclusions, or other inhomogeneities of the composition. Their arrangement is random and after repolishing the etch picture changes.

Thus, etch orientation figures reveal no dislocation structure of the metals in question and may be used to define their orientation only. The fact that etch pits belong to dislocations one may safely ascertain only if the etch pattern corresponds to the deformation geometry.

References

/1/ YU. P. PSHENICHNOV, Vyyavlenie tonkoi strukturii kristallov, Metallurgizdat, Moscow 1974 (p. 101).

/2/ H. RINGPFEIL and L. ILLGEN, Prakt. Metallogr. 2, 107 (1965).

/3/ R. FUGUEN-MOLANO, N. P. SUH, and A. P. L. TURNER, Scripta metall. 8, 1033 (1974).

/4/ G.K. BARANOVA, Scripta metall. 11, 827 (1977).

/5/ G.K. BARANOVA and L.K. FIONOVA, Scripta metall. 14, 167 (1980).

/6/ G.K. BARANOVA, Zavodsk. Labor. 46, 1128 (1980).

/7/ G.K. BARANOVA and E.M. NADGORNYI, Scripta metall. 9, 845 (1975).

/8/ G.K. BARANOVA, and R.F. KHATALLAKH, Zavodsk. Labor. 48, 65 (1982).

/9/ A. KELLY and J. ARORA, J. Electrochem. Soc. 115, 243 (1968).

/10/ H. RINGPFEIL and T. FRITSCHE, Abh. DAW Berlin, Kl. Math. und Tech., Beiträge zur Metallphyik, Akademie-Verlag, Berlin 1962 (p. 31).

(Received July 19, 1983)

phys. stat. sol. (a) 79, K13 (1983)
Subject classification: 11; 10.2; 20.1; 22.1.2

Institute of Nuclear Research, Świerk[1)]

Comparison of Fast Electron and Thermal Annealing of Proton Bombarded Silicon - X-Ray and IR Spectroscopy Study[2)]

By
K. BISKUPSKA, J. TATARKIEWICZ, and K. WIETESKA

This note is the continuation of our previous X-ray and IR spectroscopy study /1/ of thermal annealing of some silane band intensities and changes in contrast shape of proton bombarded silicon. We have noticed that a 600 °C annealing step leads to the macroscopic deformation of the sample. Here we present more details of our results as compared with fast electron annealing - see /2/ where electron dose effects are discussed.

Dislocation-free, 5 Ωcm, P-doped silicon single crystals were used. 9 MeV proton implantations were carried out at the INR linear accelerator ANDRZEJ. The details of implantations were described elsewhere /3/. The total proton dose in the present study was about 2.5×10^{17} cm^{-2}. Isochronal thermal annealing (20 min steps with 100 °C temperature increments) were done in nitrogen atmosphere. 1.5 MeV electron irradiation was carried out with a van de Graaff accelerator to a dose of 2.5×10^{18} cm^{-2}. For relative absorbance measurements a Perkin-Elmer 580B FIR spectrometer was used (RT measurements only). The details of this technique were given elsewhere /3/. To obtain the crystal lattice strain image, X-ray double crystal spectrometer topography and Berg-Barrett topography were applied.

The topograms were obtained using CuKα radiation. The spectrometer was used in (1, -1) setting with (111) sample and monochromator orientations.

After annealing at 600 °C we made several X-ray topograms for different angular positions on the rocking curve (see Fig. 1, on the photo pages before the Short Notes part). The observed double circular shape, which changes with angular setting, is due to the curvature of the sample surface corresponding to the implanted area. This deformation has azimuthal and axis symmetries. The

---

1) 05-400 Świerk/Otwock, Poland.

2) Partly supported by the Institute of Physics, Polish Academy of Sciences, Warszawa, under contract MR.I - 4

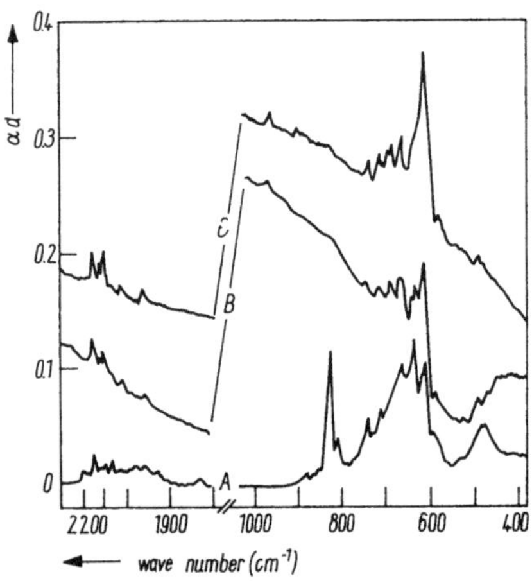

Fig. 2. IR spectra of proton bombarded ($2.5x10^{17}$ cm^{-2}) silicon (A) as implanted, (B) isochronal annealing till 600 °C, (C) 1.5 MeV electron irradiation

axis symmetry can be seen on the topograms obtained for different points of the rocking curve. The azimuthal symmetry has been confirmed by a series of topograms (not show here) obtained for different crystal positions revolving around the diffraction vector.

IR spectra of thermally and electron annealed samples are compared in Fig. 2. We observe near-edge slanting absorption background which is typical for defect agglomeration /1/. The observed spectra are quite similar even in the details of weak absorption lines at about 700 cm^{-1}. This similarity can also be seen on X-ray topograms made in the same setting for thermally and electron annealed samples (see Fig. 3 on the photo pages before the Short Notes part). For the electron irradiated sample one can see additional contrast on the side of the double circular shape. It is not clear if this effect is due to additional strain introduced by radiation defects near the sample surface during electron irradiation or to the inhomogeneity of the proton beam.

We made Berg-Barrett topograms of the electron irradiated sample and found the same contrast as for the thermally annealed sample /1/.

We were able to measure the macroscopic change of samples thickness due to surface bending which is about 5 μm. This roughly corresponds to the curvature of about 10 m as determined by the X-ray measurements. Taking into account the thickness of the proton damaged silicon layer, which is about 50 μm /3/, one obtains good agreement with 10% smaller density of amorphous silicon /4/.

The results obtained show additional evidence that despite the process of annealing of proton bombarded silicon the lattice is highly deformed by the growth of defect clusters filled with hydrogen as was suggested in the previous studies /1/ and /2/. The correlation of X-ray topograms with IR spectra intensities suggests that the deformation of the sample plays an active role in the creation of clusters of defects with hydrogen having smaller density than the virgin silicon crystal.

More details of the complex hydrogen-deformation interaction are needed for fully understanding the observed effects.

References

/1/ J. TATARKIEWICZ and K. WIETESKA, phys. stat. sol. (a) 66, K101 (1981).

/2/ J. TATARKIEWICZ, to be published.

/3/ J. TATARKIEWICZ, phys. stat. sol. (a) 63, 423 (1981).

/4/ A. SEEGER and K. P. CHIK, phys. stat. sol. 29, 455 (1968).

(Received July 11, 1983)

phys. stat. sol. (a) 79, K17 (1983)
Subject classification: 10.2 and 12.2; 22.5.2

Physico-Technical Institute, Academy of Sciences of the Ukrainian SSR, Donetsk[1])

Dislocation Mobility in Hydrostatically Compressed KBr Crystals

By
V.I. BARBASHOV and YU.B. TKACHENKO

The effect of high hydrostatic pressure on the dislocation mobility was studied in different alkali-halide crystals /1 to 3/. It was stated that with pressure rise up to 1.0 GPa the mean rate of thermoactivated dislocation motion decreases by about an order of magnitude for pure NaCl and impure LiF crystals and by over than three orders of magnitude for pure KCl crystals. The detailed studies carried out for KCl /4/ show that the observed effect is increased as both the concentration and the aggregation degree of divalent impurities increase. When analyzing the results the question arises, whether there are some other alkali-halide crystals which, like KCl, have the same high sensitivity of dislocation velocity to hydroatic pressure and what parameters determine the value of this effect.

The object of the present study is to investigate weakly doped KBr crystals with yield point of about (2.2 ± 0.1)MPa. This choice is due to the fact that KBr crystals are close to KCl concerning some physical characteristics (energy of ion bond, polarizability, energy of stacking fault, behaviour of elastic constant C_{44} with pressure).

Our study was carried out on isolated screw dislocation loops obtained from scratches by subsequent application of the methods of shock loading and layer-by-layer polishing of samples /5/. The means of sample loading and measuring the dislocation velocity as well as the equipment for creating hydrostatic pressure did not differ from those described previously /1 to 3/.

The dependences of dislocation velocity v on applied stress τ, given in Fig. 1, show that with pressure increase they shift towards higher values of τ, while the magnitude of the effect is close to that previously obtained for impure KCl crystals /3, 4/.

---

1) 340048 Donetsk, USSR.

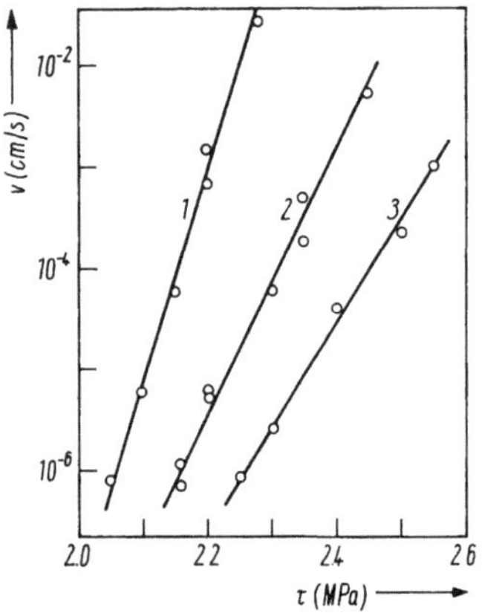

Fig. 1. Dependence of screw dislocation velocity on the applied stress. (1) $P = 10^{-4}$, (2) 0.3, (3) 0.6 GPa

Taking into account the linear character of the curves lg v versus τ, we obtain the dependences of activation parameters V_τ and V_P, respectively, on the applied pressure and stress within the frame of thermoactivation analysis. The activation volumes V_τ and V_P are determined by the expressions

$$V_\tau = kT \frac{\partial \ln v}{\partial \tau} ; \qquad V_P = -kT \frac{\partial \ln v}{\partial P} .$$

Fig. 2 illustrates the dependences $V_\tau(P)$ for the investigated KBr crystals, which are compared with the same curves obtained for the impure KCl crystals. It is seen that the character of the curves is unchanged, while the magnitude of the effect is somewhat greater for KBr crystals. V_P, characterizing the sensitivity of dislocation velocity to pressure change, equals 90 and 75 cm^3/mol for KBr and KCl, respectively, at $P \rightarrow 0$ and $v(0) = 10^{-2}$ cm/s.

The analysis of the obtained curves as well as their comparison with the results of studies for some other alkali-halide crystals allows us to state some regularities. The polarizability of KBr and KCl is higher than that of NaCl and especially LiF. The calculated stacking fault energy of a split dislocation core (a/2) [110] $(1\bar{1}0)$ is lowest for KBr and greatest for LiF /6/. Therefore, dislocations in KBr have the greatest splitting. At last, among the crystals studied, only KBr and KCl reveal a decrease of the elastic constant $C_{44} \sim \mu$ with pressure, where μ is the shear modulus /7/. An account of the above factors will apparently lead to a more complete understanding of the character of the hydrostatic pressure effect on the dislocation mobility in alkali-halide crystals.

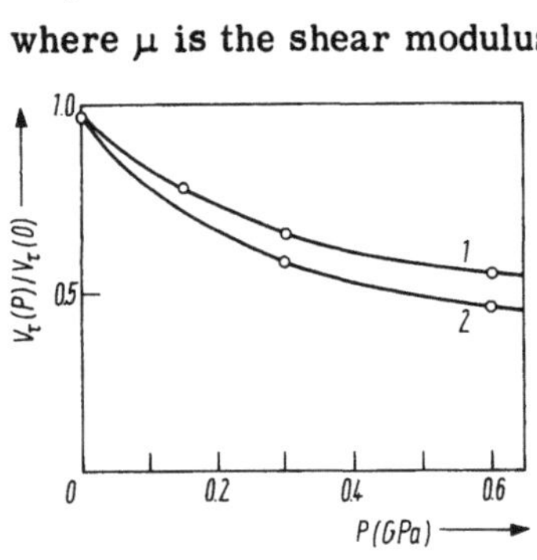

Fig. 2. Change of the activation volume V_τ with pressure. (1) KCl, $V_\tau(0) = 10^5$ cm^3/mol, (2) KBr, 1.15×10^5 cm^3/mol

The authors are grateful to Prof. V.I. Zaitsev for initiation of the present work.

References

/1/ V.A. STRELTSOV, V.I. BARBASHOV, and V.I. ZAITSEV, phys. stat. sol. (a) 58, 517 (1980).

/2/ V.I. BARBASHOV and YU.B. TKACHENKO, phys. stat. sol. (a) 62, K103 (1980).

/3/ V.A. STRELTSOV, V.I. BARBASHOV, and YU.B. TKACHENKO, phys. stat. sol. (a) 69, 497 (1982).

/4/ V.I. BARBASHOV, V.A. STRELTSOV, and YU.B. TKACHENKO, Fiz. tverd. Tela 24, 1187 (1982).

/5/ O.I. MARTYSHENKO, N.N. NOVIKOV, and A.I. LANDAU, Fiz. tverd. Tela 23, 2298 (1981).

/6/ G. FONTAINE, J. Phys. Chem. Solids 29, 209 (1968).

/7/ G.B. BARSCH and Z.R. CHANG, phys. stat. sol. 19, 139 (1967).

(Received June 21, 1983)

phys. stat. sol. (a) 79, K21 (1983)
Subject classification: 14.3.2 and 14.3.3; 22.1.2

Department of Physics, Federal University of Technology, Bauchi[1)]

Electrical Properties of Junctions between Sputter Deposited Silicon Films and Monocrystalline Silicon

By
J. KOSHY

Junctions between amorphous semiconductor films and crystalline semiconductors have been studied by several authors. Early work of Grigorovici et al. /1 to 5/ described the more or less rectifying properties of contacts between evaporated a-Ge (amorphous germanium) and a-Si (amorphous silicon) films and the monocrystals. The electrical properties of junctions between Ge films and monocrystalline Si were reported by Tove et al. /6/. Yoshida et al. /7/ determined the density of states at the Fermi level of evaporated amorphous Ge films from the electrical characteristics of heterojunctions composed of amorphous and crystalline Ge. An attempt towards a theoretical description of the amorphous-crystalline heterojunction was made by Brodsky et al. /8/. They developed a model for such a junction where they concluded that the forward characteristics could not deviate markedly from that of metal-semiconductor junctions, but in the reverse direction band bending on the amorphous side could cause the reverse current to increase exponentially with bias. The increase becomes larger for a low ratio of density of localized states in the film to the density of states in the bulk.

In this paper, we will discuss the properties of junctions formed by sputter deposited a-Si films onto chemically etched n and p-Si crystals. I-U and C-U measurements on the junctions were used to calculate the barrier height and the density of states at the Fermi level of sputter deposited a-Si films for the first time.

Highly polished n- and p-type silicon wafers were cut in appropriate size and were cleaned in trichloroethylene, acetone, and alcohol in an ultrasonic bath. The wafers were then etched in a 2:1:1 parts by volume etching solution, $2HNO_3$ (80%), 1HF (48%), 1 acetic acid (98%) for 3 min followed by rinsing in deionised water. In order to eliminate as far as possible the residual layer

1) PMB 0248, Bauchi, Nigeria.

of silicon oxide that might have formed during etching in a solution containing HNO_3, the samples were given a rapid etch in buffered HF prior to insertion in the vacuum system. The amorphous silicon films were deposited by means of rf sputtering in an MRC sputtering unit type 8622 rf. When the pressure was 10^{-5} Pa, argon gas was let in and the silicon target material was cleaned by rf sputtering at 400 W for 15 min. During this process the substrate wafers were protected by a shield. When the chamber is cooled the substrates were placed under the target and sputter deposition of Si was done at an argon pressure of 0.667 Pa to a thickness of about 100 nm. The voltage of the applied rf potential was about 3.2 kV. During deposition no bias was applied to the substrate and the substrates were kept at room temperature. On the top of the film, Ag contacts were evaporated to get ohmic contacts. The ohmicity was checked by I-U measurements on Si films contacted on both sides by evaporated silver. Indium-mercury contacts to the back of the silicon wafers were made by the usual method /9/. The diodes were annealed in argon atmosphere for 30 min at 300 ^oC. The diodes were circular in shape of area approximately 0.07 cm^2.

The a-Si - p-Si heterojunctions were investigated by I-U and C-U measurements. The capacitance was measured at 1 MHz with a Boonton 75C bridge. Fig. 1 shows the reverse measurements made at room temperature for samples deposited on p-Si of resistivity of 10 Ωcm and 7 kΩcm. The curve for low resistivity shows much smaller tendency towards saturation and this behaviour is similar to that reported by Brodsky et al. /8/.

The $1/C^2$ versus U plot in Fig. 2 is linear which verifies that abrupt junction behaviour governed by a depletion layer in the Si wafer is observed. The doping concentration obtained from the slope is found to be $N_A = 1.350 \times 10^{15}$ cm^{-3} which corresponds to a p-type resistivity of 10 Ωcm. This value agrees very closely with the manufacturer's data. The intercept on the horizontal axis of the plot, 0.66 eV, however, does not give the diffusion voltage but a rather nearly equal quantity $(\Psi_o + (N_A/2N_F))$ where N_F is the density of states in the amorphous film at the Fermi level, N_A is the acceptor concentration on the crystalline material, and Ψ_o the difference between Fermi levels in both sides /7/.

For determining N_F a plot of /7/ ln I versus $(\Psi_o + (N_A/2N_F) - qU)^{1/2}$ $(\Psi_o + (N_A/2N_F))^{1/2}$ is made and the slope of this straight line is given by

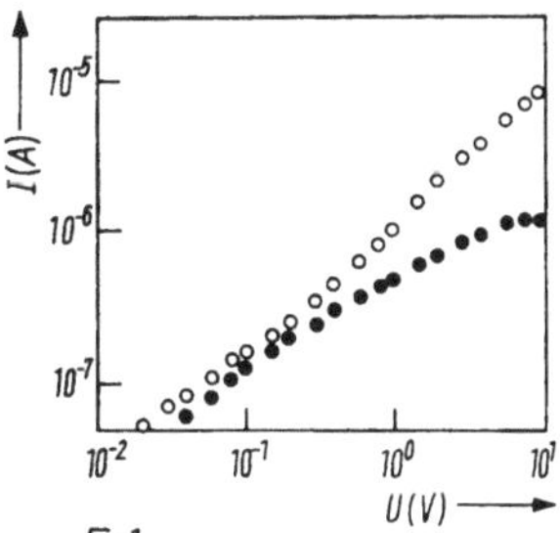

Fig.1

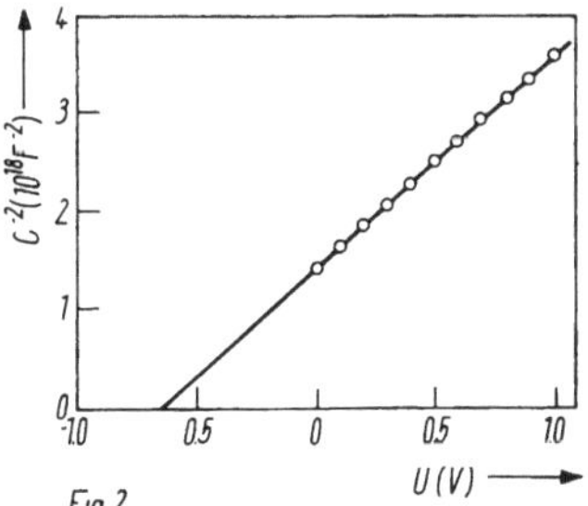

Fig.2

Fig. 1. Reverse I-U characteristics for a-Si - p-Si structures for different silicon resistivity; ● 7 kΩcm, ○ 10 Ωcm

Fig. 2. $1/C^2$ versus U_{rev} plot for a-Si - p-Si structure

$(1/kT)(2\,N_A/N_F)^{1/2}$. Such a plot for different temperatures is shown in Fig. 3 where the densities of states for different temperatures are indicated. The extrapolation of the straight lines determines the values of saturation current I_o. From the temperature dependence of the intercepts I_o, the activation curve of $\ln(I_o/T^2)$ versus $1/T$ was plotted as shown in Fig. 4. The barrier height $V^b(0)$ was found to be 0.77 eV.

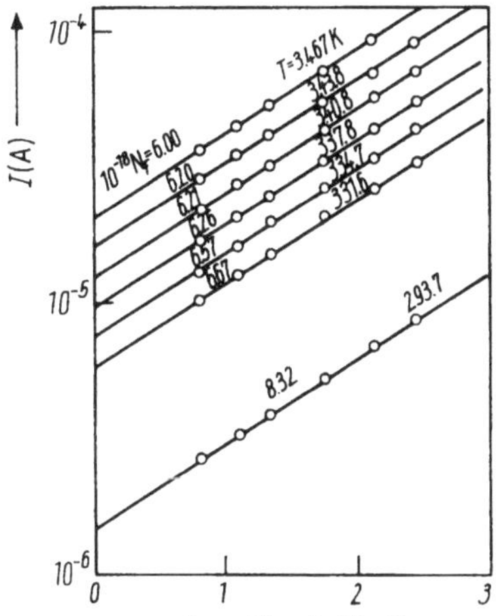

Fig.3

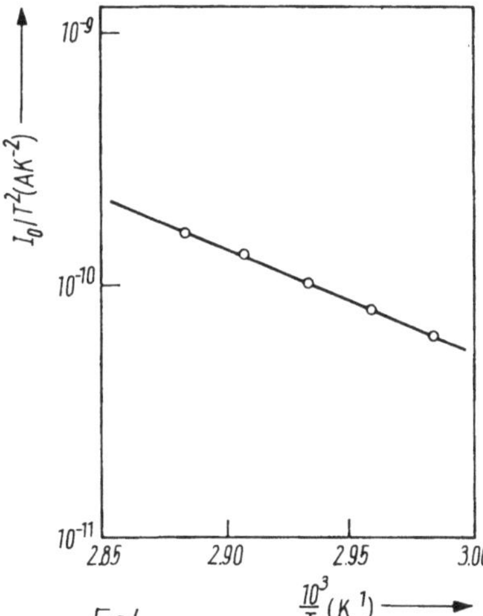

Fig.4

Fig. 3. Curves of ln I versus $(\Psi_o + (N_A/2\,N_F) - qU)^{1/2} - (\Psi_o + (N_A/2\,N_F))^{1/2}$ for different temperatures

Fig. 4. Activation energy plot for saturation current data obtained from Fig. 3

The barrier height measurements were also made by forward measurements derived from Norde plots /10/ and the barrier height was calculated as 0.76 eV. The electron barriers also were determined by depositing a-Si films on n-type silicon substrate which gave a barrier height of 0.34 eV showing ohmic behaviour of the junction.

The measurements obey closely the model proposed by Brodsky et al. /8/ and give information about the density of states in the amorphous film at the position of the Fermi level which is about 0.77 eV above the valence band of the silicon. The density of states for sputter deposited a-Si film is found to be $8x10^{18}$ cm^{-3} eV^{-1} which is smaller than that reported for evaporated a-Si films. In the case of sputter deposited contacts the barrier on p-Si was always larger than n-Si. This offers technological application by giving blocking contact to p-Si and ohmic to n-Si.

The experimental part of the work has been done at the Institute of Technology, University of Uppsala, Sweden, and the author is grateful to Prof. P.A. Tove for providing laboratory facilities.

References

/1/ R. GRIGOROVICI, N. CROITORU, E. TELEMAN, and M. MARINA, Rev. Roumaine Phys. 10, 641 (1965).

/2/ R. GRIGOROVICI, N. CROITORU, M. MARINA, and L. NATASI, Rev. Roumaine Phys. 13, 317 (1968).

/3/ R. GRIGOROVICI, N. CROITORU, L. VESCAN, and M. MARINA, phys. stat. sol. 24, K17 (1967).

/4/ R. GRIGOROVICI, N. CROITORU, A. DEVENYI, and E. TELEMAN, Proc. Conf. Phys. Semicond., Dunod, Paris 1968 (p. 423).

/5/ R. GRIGOROVICI, N. CROITORU, A. DEVENYI, and L. VESCAN, Rev. Roumaine Phys. 10, 649 (1965).

/6/ P.A. TOVE, M.P. ALI, M.M. IBRAHIM, and H. NORDE, phys. stat. sol. (a) 51, 491 (1979).

/7/ A. YOSHIDA, T. IDO, and T. ARIZUMI, Conf. Amorphous and Liquid Semicond., Leningrad 1975.

/8/ M.H. BRODSKY, G.H. DÖHLER, and P.J. STEINHARDT, phys. stat. sol. (b) 72, 761 (1975).

/9/ R. HILL, D. RICHARDSON, and S. WILSON, J. Phys. D 5, 185 (1972).

/10/ A. NORDE, J. appl. Phys. 50, 5052 (1979).

(Received July 4, 1983)

phys. stat. sol. (a) 79, K25 (1983)
Subject classification: 14.3.3 and 14.3.4; 22.1.2
Institute of Physics, Hanoi[1)]

Stress Effects on the Switching Voltage of Metal-SnO_2-Si(n)-Si(p^+) Devices

By

H.K. PHAN, L.H. PHU, and C.V. CHIEM

The phenomenon of switching observed in metal-thin insulator Si(n)-Si(p^+) structures (MISS) has been the subject of several recent experimental and theoretical investigations. The experiments reported in /1 to 3/ have shown that this phenomenon can be governed by heating, illuminating, or using a third extra-polarized terminal. These properties of MISS structures have been found to be a convenient tool for special electrical and optical switching applications. Little information is available, however, on the behaviour of these devices under stress. In the present work we report a first experimental study of the stress effects on the current-voltage characteristics of metal-SnO_2-Si(n)-Si(p^+) devices. Among the observed effects the most noticeable one is the decrease of the switching voltage with increasing stress. This effect is reversible.

A typical structure investigated with the arrangement of an applied stress head is shown in Fig. 1. The structures were fabricated by electron-beam evaporation of tin oxide on the surface of n-type epitaxial layers grown on p^+ silicon substrates. The thickness and resistivity of the epitaxial layers were 31 μm and 4.9 Ωcm and those of the p^+ substrates were 200 μm and 0.02 Ωcm, respectively. The SnO_2 powder was placed in a graphite liner and the powder was heated to the sublimation temperature by an YBH-75P-I electron beam source. The electron energy used for the evaporation was 7 keV with a beam current of 30 mA. The pressure during the evaporation was 2×10^{-5} Torr. The distance between the SnO_2 source and the substrates was 25 cm. During the deposition the samples were heated to about 80 °C by a quartz lamp Kr 220-2000-3. These yielded a deposition rate of about 18 Ås^{-1}. The thickness of the tin oxide layers varied from 1000 to 2000 Å. After deposition of the tin oxide some of the coated samples were heated in air at 500 °C for 30 min.

---

1) Hanoi, Nghia Do, Tu Liem, Vietnam.

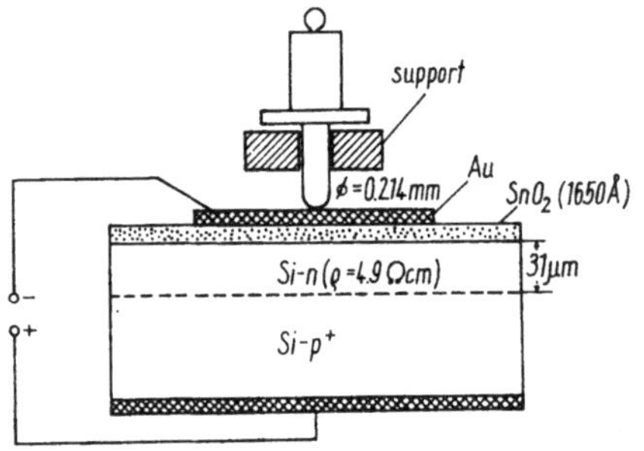

Fig.1

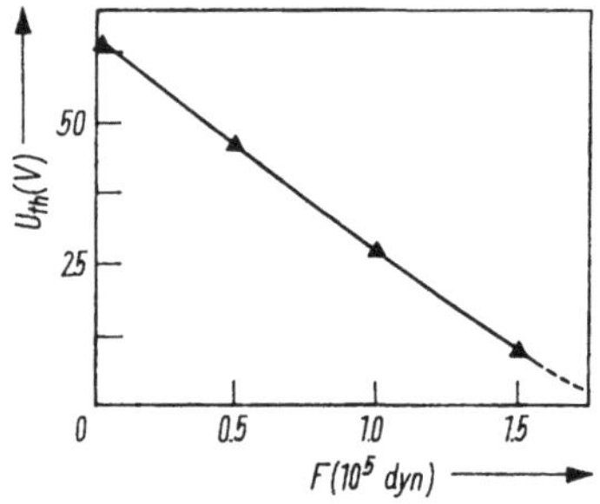

Fig.2

Fig. 1. Experimental configuration for studying the stress effect on the switching voltage of MISS devices

Fig. 2. The switching voltage of a typical MISS device as a function of stress

Thereafter, aluminium or gold contacts were deposited through a metal mask on the surface of tin oxide and on the underside (p^+) surface of the substrates. The area of the contacts was 2.27×10^{-2} cm^2. It is very interesting to note that variation in the deposition rates, heat treatment condition, thickness of SnO_2 films, and other fabrication parameters modify only quantitatively the current-voltage characteristics, but do not damp the effect of switching. In addition, heat treatment resulted in a substantial decrease of the resistivity of the tin oxide films. In this case the SnO_2 film could be used as an electrical contact.

The current-voltage characteristics of the fabricated devices are similar to those reported by Kroger and Wegner /4/ for MISS structures with 20 to 200 Å silicon nitride layers. The threshold voltage U_{th} for switching out of the high impedance state was found to be 65 V. This voltage is less than that required to completely deplete the n-region, the punch-through voltage U_p. The other parameters of the devices were measured: the threshold current $I_{th} = 100\ \mu A$ and the holding voltage $U_H = 2$ V.

Applying stress to the metal contacts or directly to the surface of the SnO_2 layers, if the latter was annealed at high enough temperatures (above 500 °C), we have observed a very interesting phenomenon, a substantial reduction of the threshold voltage U_{th}. A higher applied stress resulted in a lower threshold voltage. This relationship between U_{th} and F is presented in

Fig. 2. In addition, as reported in a recent work /5/, the high photosensitive effect on the switching voltage was also observed in our experiments. This sensitive suppression of the threshold voltage U_{th} by external excitation strongly supports the model of Majkusiak et al. of the switching mechanism in MIS-n-p^+ devices /6/. According to this model the threshold condition for switching is defined by the equation

$$i_c = \frac{i_{pd} + i_{pb}}{1 - (\gamma_{nt} + \beta_o \gamma_{pj})} , \tag{1}$$

where i_c is the total current flowing through the device and is limited by the load resistance; i_{pd} and i_{pb} are the currents of holes generated in the depletion region and in the base of the epitaxial layer, γ_{nt} is the coefficient of tunnel injection of majority carriers from metal to semiconductor, β_o the coefficient of transport of holes through the base, and γ_{pj} the coefficient of hole injection from the p^+ substrate to the n-epitaxial region. It is clear from the above equation that the current i_c will increase abruptly when the sum $\gamma_{nt} + \beta_o \gamma_{pj}$ reaches unity. This determines the threshold condition for switching.

Since γ_{nt}, β_o, and γ_{pj} depend on several factors such as the doping concentration, the thickness of the epitaxial layers, the optical and other external excitations, etc. it is possible to use these factors to govern the switching voltage for a special application. The fact that the uniaxial stress changes obviously the I-U characteristics of p-n junctions and MIS diodes /7, 8/ suggests the validity of this explanation.

The authors wish to acknowledge N.T. Ha for her help in preparing MISS devices.

References

/1/ T. YAMAMOTO, K. KAWAMURA, and H. SHIMIZU, Solid State Electronics 19, 701 (1976).

/2/ J. BUXO, A.R. OWEN, G. SARRABAYROUSE, and J.P. SEBBAA, Rev. Phys. appl. 13, 767 (1978).

/3/ K.C. CHIKT and J.G. SIMMONS, Solid State Electronics 22, 589 (1979).

/4/ H. KROGER and H.A.R. WEGNER, Appl. Phys. Letters 23, 397 (1973).

/5/ A. ADAN and K. DOBOS, Solid State Electronics 23, 17 (1980).

/6/ B. MAJKUSIAK, A. JAKUBOWSKI, and J. RUŻYŁŁO, phys. stat. sol. (a) 60, K171 (1980).

/7/ S.M. SZE, Physics of Semiconductor Devices, John Wiley and Sons, New York/London/Sidney/Toronto 1969.

/8/ B.A. MALAKHOV, S. A. TEROSHIN, and V.I. POKALIAKIN, Mikroelektronika 11, 369 (1982).

(Received July 15, 1983)

phys. stat. sol. (a) 79, K29 (1983)
Subject classification: 14.3; 22.8

First Laboratory of Physics, Aristoteles University of Thessaloniki[1)]

Cation Mixing and the AC Conductivity of $CdInGaS_4$

By
A.N. ANAGNOSTOPOULOS and A. KAROUTIS

Introduction In a previous paper we reported the observation of a mutual exchange of 1/3 Ga and 1/3 Cd cations between the layers of $CdInGaS_4$ /1/. They were considered responsible for the observed compensation /2/ and for the high photosensitivity of this compound /3/.

In the present note we will show that the exchange of 1/3 Ga and 1/3 Cd cations also explains the frequency-dependent ac conductivity of $CdInGaS_4$, measured parallel to the c-axis of this compound.

Ac conductivity measurements $CdInGaS_4$ single crystals were grown following the method described in /1/.

Thin plates used for the electric measurements parallel to the c-axis of this compound were cleaved from larger ones. The cleavage plane was perpendicular to the c-axis of the samples. Then two indium electrodes A and B were evaporated on the upper and lower surface of the plates. The dimensions of the samples were approximately 2x2x0.1 mm^3.

The prepared specimens were placed in a cryostat, enabling measurements in a vacuum of $5x10^{-6}$ Torr and within a temperature range between 10 and 360 K. The applied ac electric field had a peak-to-peak value of 40 V/mm. The frequency range was betwen 10^{-2} and $2x10^5$ Hz. The variations of the peak-to-peak current $j_{pp\|}$ with frequency were measured at fixed temperatures. A Tektronix 5A. 18N oscilloscope was used with a Data Pulse 410 ac voltage source. Care was taken to avoid the influence of the oscilloscope input impedance and the coaxial cable impedances.

Prior to the ac measurements and during them the $CdInGaS_4$ samples were kept in darkness since they appear to be very photosensitive /4, 5/.

Fig. 1 shows the measured ac current densit $j_{pp\|}$ as a function of the frequency ω of the applied ac electric field for 170 and 300 K (full and open circles, respectively). At very low and very high frequencies, $j_{pp\|}$ is nearly frequency

1) Thessaloniki, Greece.

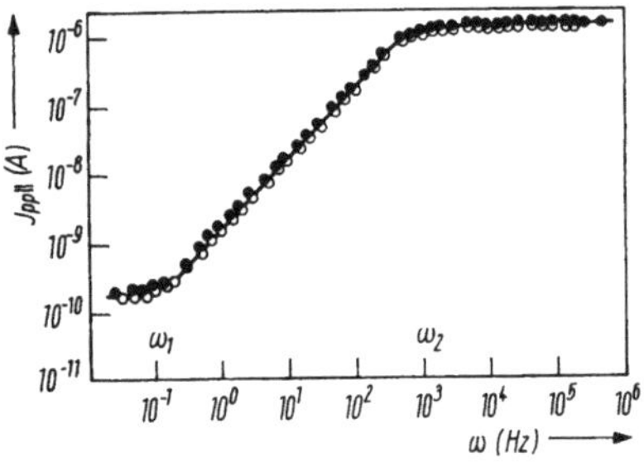

Fig. 1. Measured currents densities $j_{pp\parallel}$ as a function of the frequency ω of the applied ac field at 300 K (open circles) and at 170 K (full circles). The full line is calculated for the model in Fig. 2 with $r_s = 2x10^7 \Omega$, $NR = 2x10^{11} \Omega$, $C/N = 50$ pF

independent. In the region between 0.1 and 10^3 Hz, $j_{pp\parallel}$ is an increasing linear function of ω, with slope ≈ 1. There is scarcely any temperature dependence of the ac conductivity. This is true both for the dc /2/ and the ac conductivity in the range between 300 and 90 K.

Interpretation of the measurements In a previous paper /6/ we attributed the ac conductivity of $ZnIn_2S_4$(III) to the composition faults observed with the electron microscope in this compound /7/. Such composition faults, consisting of an extra S-Zn layer between the layers of $ZnIn_2S_4$(III) give rise to potential barriers. Electrons can move freely in all directions between them, but they have to tunnel through the barriers in order to contribute to the conductivity parallel to the c-axis. So, the ac conductivity of $ZnIn_2S_4$(III) was explained assuming that the composition faults act as capacitors connected in parallel with a large resistance representing the low transparency of the barriers.

In the present work following a similar idea we consider the $Cd_{2/3}Ga_{1/3}$ and $Cd_{1/3}Ga_{2/3}$ layers of $CdInGaS_4$ as potential barriers restraining the motion of electrons with energies lower than the barrier height along the c-axis of this compound.

In contrary to composition faults in $ZnIn_2S_4$(III) crystals, the mutual exchange of cations between the Cd and Ga layers is rather an intrinsic property of the ordered $CdInGaS_4$ structure.

In Fig. 2a are sketched schematically the $Cd_{2/3}Ga_{1/3}$ and $Cd_{1/3}Ga_{2/3}$ planes in the $CdInGaS_4$ structure. In Fig. 2b the equivalent electric circuit is sketched; for the dc conductivity parallel to the c-axis the capacitors C act as an off-switch and the dc conductivity is determined by the large resistance R. For the ac conductivity parallel to the c-axis, the capacitors C act as an off-switch at very low frequencies and as on-switch at very high frequencies. Thus the ac conductivity is a function of the frequency ω. The low resistances r_i represent the ohmic behaviour of layers between successive potential barriers.

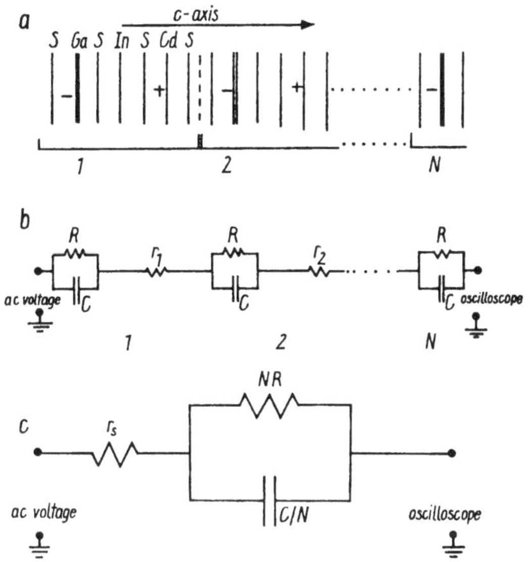

Fig. 2. a) Schematic representation of a $CdInGaS_4$ layered crystal with N barriers acting as capacitors C connected in parallel with a resistance R; b) resistance r_i represents the ohmic behaviour of $CdInGaS_4$ between successive barriers. The equivalent circuit to that of b) is shown in c)

Assuming that the sample contains N potential barriers the circuit of Fig. 2b is equivalent to that of Fig. 2c. This circuit consists of a parallel combination of an ohmic resistance NR and a capacitor C/N connected in series with an ohmic resistance $r_s = \sum_{i=1}^{N} r_i$. This circuit is really a high pass filter and the curves of Fig. 1 represent the corresponding Bode plot /8/.

The low- and high-corner frequencies ω_1 and ω_2 are given by the following expressions /9/:

$$\omega_1 = \frac{1}{NR}\,\frac{C}{N} \tag{1}$$

and

$$\omega_2 = \frac{1}{r_s}\,\frac{C}{N}\ . \tag{2}$$

With a value of $\omega_1 \approx 0.1$ Hz from Fig. 1 and for $NR = 2\mathrm{x}10^{11}\ \Omega$ as indicated from dc measurements, (1) gives

$$C/N \approx 50\ \text{pF}\ . \tag{3}$$

Each unit cell contains three barriers. Hence, with c = 37.02 Å /1, 2/ there are $\approx 10^5$ barriers in the specimen. Thus

$$C \approx 5\ \mu F\ . \tag{4}$$

With a value of $\omega_2 = 1$ kHz in Fig. 1 and with $C/N \approx 50\ \mu F$ we obtain for r_s

$$r_s \approx 2 \times 10^7\ \Omega . \tag{5}$$

Discussion The value of 5 μF for the capacity C derived in the previous section can be estimated in the following way:

Each barrier is considered as a plane capacitor with an electrode area A and a plane distance d. Thus

$$C = \varepsilon \varepsilon_0 \frac{A}{d}\ , \tag{6}$$

where ε is the relative dielectric constant and ε_0 the absolute dielectric constant; with $A \approx 2\ mm^2$, d = 2 Å /2/, and a value of ≈ 10 for ε, we can estimate C as

$$C \approx 1\ \mu F\ . \tag{7}$$

This value is not very different from the value of C derived from (4).

The full line in Fig. 2 is calculated for this model with the above parameters. The agreement with the measurements is perfect.

In Fig. 2 the measured resistance R does not depend on the temperature. This is only possible if the electrons do not pass over the barriers, but they rather have to tunnel through them /10/. In order to fulfil this condition the barriers must have a height greater than 10 meV /11/. A rough estimation of the barrier height can be made in the following way:

Each barrier considered as a plane capacitor has a voltage U between its plates, given by

$$U = \frac{Q}{C}\ , \tag{8}$$

C as estimated from (4) has a value of $\approx 5\ \mu F$. Assuming that each exchanged ion in the layers carries one elementary charge and for specimens with an area of 2 mm^2, Q is $\approx 10^{-6}$ C. Then U has a value of 0.2 V and the barriers must have a height of the order of some 0.1 eV.

We thank Prof. Dr. F. Stöckmann for critical reading of the manuscript.

References

/1/ A.N. ANAGNOSTOPOULOS, C. MANOLICAS, and PAPADOPOULOS, phys. stat. sol. (a) 77, 595 (1983).

/2/ A.N. ANAGNOSTOPOULOS and C. MANOLICAS, Proc. 4th Greek-Bulgarian Symp. on Semicond. Phys., Thessaloniki (Greece) 1983.

/3/ A.N. ANAGNOSTOPOULOS and CHR. LIOUTAS, Proc. 4th Greek-Bulgarian Symp. on Semicond. Phys., Thessaloniki (Greece) 1983).

/4/ G.B. ABDULLAEV, N.D. DZURAEV, V.B. ANTONOV, R.KH. NANI, and E.YU. SALAEV, Soviet Phys. - Semicond. 7, 1428 (1974).

/5/ A. KAROUTIS, Thesis, Thessaloniki 1983.

/6/ A.N. ANAGNOSTOPOULOS, phys. stat. sol. (a) 75, 595 (1983).

/7/ A.N. ANAGNOSTOPOULOS, C. MANOLICAS, D. PAPADOPOULOS, and J. SPYRIDELIS, phys. stat. sol. (a) 72, 731 (1982).

/8/ P. CUTLER and H. HOOVER, Electronic Circuit Analysis, Vol. I, McGraw-Hill Publ. Co., New York 1960 (p. 88).

/9/ M. HOROWITZ, How to Built Solid State Audio Circuits, Foulsham-Tab Ltd., Slough Bucks 1973 (p. 194).

/10/ A.N. ANAGNOSTOPOULOS and B. PLOSS, phys. stat. sol. (a) 73, 91 (1982).

/11/ CHR. LIOUTAS and A.N. ANAGNOSTOPOULOS, to be published.

(Received July 12, 1983)

phys. stat. sol. (a) 79, K35 (1983)
Subject classification: 1.1 and 14.3; 22.9

Department of Chemistry, Sheffield City Polytechnic[1)]

Electrical and Structural Characterisation of $(DBzBP)(TCNQ)_4$

By
G.J. ASHWELL and K. OSBORNE

Introduction The properties of $(DBzBP)(TCNQ)_2$ and $(DBzBP)(TCNQ)_4$ have been investigated as part of a series of studies on viologen-TCNQ salts /1 to 3/, our interest in these materials being renewed by the recent discovery of the highly conductive simple salt, $(4,4\text{-bipyridinium})^{2+}(TCNQ)_2^{2-}$ /3/ and its conductive congeners /4/. $(4,4'\text{-bipyridinium})^{2+}(TCNQ)_2^{2-}$ is one of only a few examples of conductive TCNQ simple salts known /4, 5/, the majority being Mott insulators. In these the conductivity is limited by a high on-site electron-electron Coulomb repulsion associated with placing two electrons on the same TCNQ site during the conduction process. In $(4,4\text{-bipyridinium})^{2+}(TCNQ)_2^{2-}$ the anomalous electrical properties are attributed to mixed-valence behaviour arising from a minimal random occupation of the dication sites by neutral 4,4'-bipyridyl /3/.

In this note we report the crystal structure and conductivity of $(DBzBP)(TCNQ)_4$ and also the electrical properties of the simple salt $(DBzBP)(TCNQ)_2$. The crystal structure was first determined by Sundaresan and Wallwork /6/. In their study poor intensity data prevented an accurate determination of the molecular dimensions. The structure has been reinvestigated in order to determine the degree of back charge transfer, if any, and the degree of electron localisation on the crystallographically independent TCNQ sites.

Synthesis of the TCNQ salts Purple needles of the simple salt $(DBzBP)(TCNQ)_2$ were prepared by the metathesis of 1,1'-bis(benzyl)-4,4'-bipyridinium dichloride and LiTCNQ (1:2 mole ratio) in aqueous acetonitrile. The crystals were washed with ethanol and ether. Analysis: calc.: C 77.21, H 4.02, N 18.77%, $\varepsilon(395)/\varepsilon(842) = 0.59$; found: C 77.01, H 4.22, N 18.69%, $\varepsilon(395)/\varepsilon(842) = 0.59$.

---

1) Pond Street, Sheffield S1 1WB, Great Britain.

Black parallelpiped crystals of the mixed-valent salt $(DBzBP)(TCNQ)_4$ were obtained when a hot aqueous acetonitrile solution of 1,1'-bis(benzyl)-4,4'-bipyridinium dichloride, LiTCNQ, and TCNQ (1:2:2 mole ratio) was allowed to cool slowly to ambient temperature. The crystals were washed with toluene, until the washings were colourless, to remove unreacted neutral TCNQ. Analysis: calc.: C 74.87, H 3.29, N 21.84%, ε(395)/ε(842) = 2.2; found: C 74.68, H 3.50, N 21.60%, ε(395) = 1.68×10^4 mol^{-1} m^2, ε(842) = 7.7×10^3 $mol^{-1} m^2$, ε(395)/ε(842) = 2.1.

Structure of $(DBzBP)(TCNQ)_4$ Crystal data: $(DBzBP)(TCNQ)_4$, $(C_{24}H_{22}N_2)^{2+}(C_{12}H_4N_4)_4^{2-}$, M_r = 1155.2, triclinic, a = 7.11(1), p = 16.83(1), c = 12.16(1) Å, α = 99.40(5)°, β = 83.57(4)°, γ = 90.47(4)°; U = 1427 Å^3, D_m = = 1.32 mg m^{-3}, Z = 1, D_c = 1.34 mg m^{-3}. Space group $P\bar{1}$.

The unit cell parameters of $(DBzBP)(TCNQ)_4$ were obtained from oscillation and Weissenberg photographs and subsequently refined on an Enraf-Nonius Cad 4 four-circle diffractometer using MoKα radiation. Intensity data were collected from a crystal of approximate dimensions 0.5x0.3x0.1 mm^3 using MoKα radiation. 2977 reflections had significant counts $[I > 3\sigma(I)]$. The intensities were corrected for Lorentz and polarisation factors but not for absorption. Using the atomic non-hydrogen parameters of Sundaresan and Wallwork /6/ and the hydrogen parameters, calculated from the molecular geometries, and using isotropic and anisotropic temperature factors, respectively, the structure refined in $P\bar{1}$ to a final R = 0.083.

In $(DBzBP)(TCNQ)_4$ the TCNQ's stack along b, in stoichiometric groups of four, with an exocyclic bond to ring overlap of molecules within tetrads and poor overlap between. Within each tetrad the mean interplanar spacings are 3.16 Å between TCNQ(A)-TCNQ(B), 3.23 Å between TCNQ(A)-TCNQ(A'), and 3.62 Å between TCNQ(B)-TCNQ(B'). The closest inter-tetrad contact is N(5)-N(5') = 3.174 Å, between tetrads in neighbouring stacks, the stacks forming sheets in the ab plane. The sheets are interleaved along c by the bipyridinium cations.

The dimensions of the chemically similar bonds in the crystallographically independent types of TCNQ are listed in Table 1. The dimensions of TCNQ(A) closely resemble those of the radical anion as found, for example, in Na^+TCNQ^- /7/, K^+TCNQ^- /8/, and Rb^+TCNQ^- /9/, whereas the dimensions of TCNQ(B) compare more favourably with those of neutral TCNQ /10/. The exocyclic bond

Table 1

Comparison of the mean bond lengths (in Å) in $TCNQ^-$ /7 to 9/, $TCNQ^0$ /10/, and $(DBzBP)(TCNQ)_4$, all uncorrected for libration

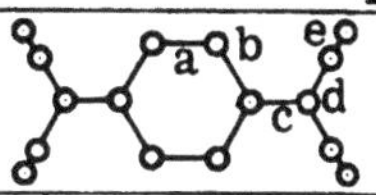

	a	b	c	d	e
$TCNQ^-$	1.363	1.422	1.420	1.417	1.145
$TCNQ^0$	1.344	1.442	1.373	1.435	1.139
$(T^- - T^0)$	0.019	-0.020	0.047	-0.018	0.006
TCNQ(A)	1.359	1.425	1.404	1.421	1.147
TCNQ(B)	1.350	1.433	1.384	1.432	1.139
(A - B)	0.009	-0.008	0.020	-0.011	0.008

of TCNQ(A) is significantly shorter and the molecule shows quinonoid character. By applying the method of Flandrois and Chasseau /11/ to the molecular dimensions, the mean charge on TCNQ(A) is estimated as 0.7e and on TCNQ(B) as 0.3e. The cation charge, obtained by summing the charges on the TCNQ's within the stoichiometric unit, is 2+. There is no evidence of back charge transfer from $TCNQ^-$ to dication.

There is indication of the negative charge being localised on specific sites within the TCNQ lattice but, in view of the sensitivity of the calculation to relatively minor changes in the bond lengths, the data should be treated with caution. However, it is of interest to note that TCNQ(A) makes the closest contact with the cation and that this contact is to the α position of the pyridinium ring. The closest non-hydrogen contacts are N(2)--C(28) = 3.276 Å for TCNQ(A) and N(8)--C(27) = 3.338 Å for TCNQ(B) (see Fig. 1). Similar results have also been obtained for the congeneric 1,1'-bis(alkyl)-4,4'-bipyridinium and 1,1'-bis (p-cyanophenyl)-4,4'-bipyridinium TCNQ salts /2/.

Electrical properties Electrical conductivities of the 1:2 and 1:4 TCNQ salts were determined by a four-probe technique with Acheson 915 Electrodag as electrodes. The conductivity of $(DBzBP)(TCNQ)_4$, along b, varies as $\sigma = \sigma_0 \exp(-E_a/kT)$ where $\sigma(RT) = 1.3 \times 10^{-3}$ S cm^{-1} and E_a = 0.20 eV. The

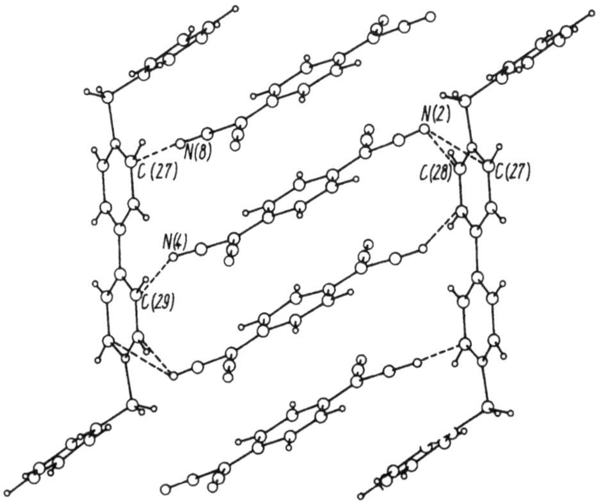

Fig. 1. Crystal structure of $(DBzBP)(TCNQ)_4$ showing the closest non-hydrogen TCNQ to cation contacts: TCNQ(A), N(2)--C(27) = 3.309 Å, N(2)--C(28) = = 3.276 Å, N(4)--C(29) = 3.326 Å; TCNQ(B), N(8)--C(27) = 3.338 Å

intermediate conductivity is typical of periodically distorted TCNQ salts, whereas the conductivity of the simple salt, $(DBzBP)(TCNQ)_2$, is significantly lower, at all temperatures, with $\sigma(RT) = 5\times10^{-10}$ S cm^{-1} and $E_a = 0.40$ eV. The low conductivity arises from the need to place two electrons on the TCNQ lattice sites during the conduction process, whereas in the 1:4 complex salt the conduction electrons can correlate their movement along the stacks so as to avoid double occupancy. Differences in the on-site (U') and adjacent-site (U") electron-electron Coulomb repulsions, typically U'- U" = 1 eV for TCNQ salts /12/, contribute in part to the factor of 10^7 difference in the room temperature conductivities of the 1:2 and 1:4 salts. It should be noted that the conductivity of $(DBzBP)(TCNQ)_2$ is low compared with, for example σ (RT) = 1×10^{-6} S cm^{-1} of [N, N'-bis(methyl)-4, 4'-bipyridinium]$(TCNQ)_2$ and 6×10^{-6} S cm^{-1} of [N, N'-bis(phenyl)-4, 4'-bipyridinium] $(TCNQ)_2$ /3/. In these salts, steric effects play an important role in determining the electrical properties and we assume that the bulky benzyl groups affect the TCNQ packing.

We acknowledge the SERC and Royal Society of Chemistry Research Fund for equipment grants in support of this work and Dr. M.C. Jones for data collection facilities.

References

/1/ G.J. ASHWELL, phys. stat. sol. (b) 86, 705 (1978).

/2/ G.J. ASHWELL, G.H. CROSS, D.A. KENNEDY, I.W. NOWELL, and J.G. ALLEN, J.C.S. Perkin II, in the press (1983).

/3/ G.J. ASHWELL and J.G. ALLEN, J. Physique, (1983) in the press.

/4/ G.J. ASHWELL, J.G. ALLEN, E.P. GOODINGS, and I.W. NOWELL, phys. stat. sol. (a) 79, No. 2 (1983).

/5/ L.B. COLEMAN, J.A. COHEN, A.F. GARITO, and A.J. HEEGER, Phys. Rev. B 7, 2122 (1973).

/6/ T. SUNDARESAN and S.C. WALLWORK, Acta cryst. B28, 2474 (1972).

/7/ M. KONNO and Y. SAITO, Acta cryst. B30, 1294 (1974).

/8/ M. KONNO, T. ISHII, and Y. SAITO, Acta cryst. B33, 763 (1977).

/9/ A. HOEKSTRA, T. SPOEDLER, and A. VOS, Acta cryst. B28, 14 (1972).

/10/ R.E. LONG, R.A. SPARKS, and K.N. TRUEBLOOD, Acta cryst. 18, 932 (1965).

/11/ S. FLANDROIS and D. CHASSEAU, Acta cryst. B33, 2744 (1977).

/12/ J.B. TORRANCE, Ann. New York Acad. Sci. 313, 210 (1978).

(Received June 10, 1983)

phys. stat. sol. (a) 79, K41 (1983)

Subject classification: 11 and 14.3; 22.2.1

Department of Microelectronics, Slovak Technical University, Bratislava[1)]

Pulse Radiation Annealing of Capless GaAs

By

J. GRŇO, Š. BEDERKA, and M. VESELÝ

Introduction The purpose of semiconductor sample annealing after implantation is to repair the disturbed crystallographic structure and to activate electrically the dopants in the implanted region. Contrary to elemental semiconductor annealing, compound semiconductor annealing is more complicated. This is caused mainly by the fact that the evaporation rate of the individual components is usually not the same and this leads to distortion of the surface layer stoichiometry. Also during annealing of semi-insulating GaAs samples a compensating impurity out-diffusion near the surface is often observed (this depends on the quality of the material used). Thus thermal conversion of this originally high resistance layer to n-type conduction takes place. These phenomena could be suppressed by GaAs encapsulation with a passivation layer or by GaAs annealing in an As atmosphere /1, 2/. Arai et al. /3/ have shown that thermal conversion of semi-insulating GaAs could be minimized also when radiation annealing by halogen lamps for several seconds is used. The aim of this note was to find out the effect of radiation pulse annealing by halogen lamps on the dissociation of GaAs surface layers.

Annealing apparatus The experiments were carried out with chromium-doped semi-insulating (100) GaAs samples of 10^8 Ωcm resistivity. Eight 1 kW halogen lamps were used in our radiation annealing apparatus. The optimal geometrical configuration of these lamps as far as homogeneous temperature distribution in the oven is concerned was based on a calculation. The sample to be annealed is placed in a quartz tube. This tube can be either pumped down to a low pressure or filled up with a desired gas. The required temperature can be set up and stabilized by an electronic control system. The measured time dependences of the sample temperature are given in Fig. 1 with radiation time as a parameter. The temperature distribution homogeneity on a sample of 5 cm in diameter was better than 2%.

---

1) Vazovova 5, 81219 Bratislava, Czechoslovakia.

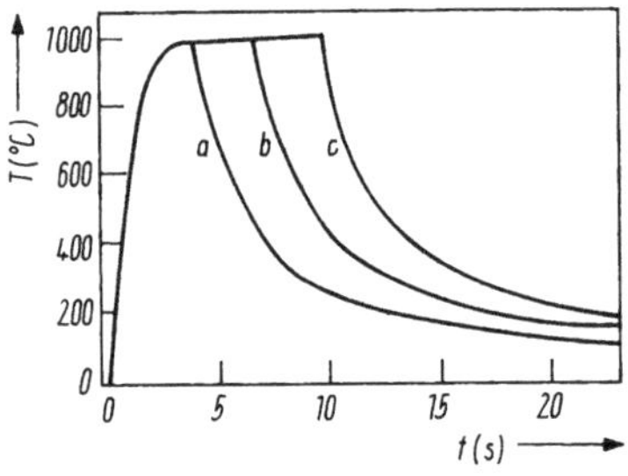

Fig. 1. The measured time dependences of the sample temperature with radiation time τ as a parameter: (a)τ = 3, (b) 6, (c) 9 s

Dissociation of annealed capless GaAs All investigated samples were capless, the working pressure was 1 Pa, the temperature and the radiation time varied. The procedure was as follows: at a given temperature the radiation time was changed and after each annealing cycle the sample surface was checked visually and compared with the non-annealed sample surface. Up to a certain critical radiation time no changes in the annealed sample surface were observed. Above the critical radiation time the GaAs surface showed remarkable changes. The sample surface was dull and small drops of Ga could be detected under the microscope. These were apparently the consequence of faster evaporation of As from the GaAs surface. The samples under investigation were annealed for 2 to 8 s at different temperatures in the range from 850 to 1200 °C. Each of the sample annealing regimes is given as a point in Fig. 2 with annealing temperature T and radiation time τ as coordinates. If no surface changes were observed, the annealing regimes are marked with circles; if dissociation occurred, crosses label these annealing regimes. In Fig. 2, two groups of points can be clearly distinguished and a boundary can be drawn between them (dashed line) that corresponds to the critical radiation time at a given temperature. Thus, there will be no dissociation if such annealing regime is chosen that corresponds to a point below the above described boundary in Fig. 2.

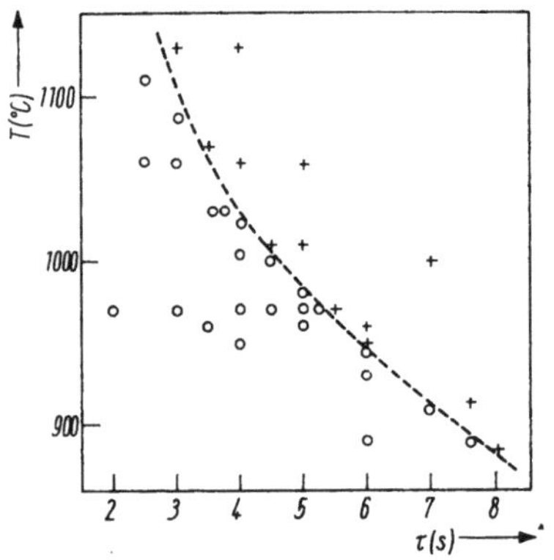

Fig. 2. Sample surface dissociation after annealing at temperature T for a radiation time τ; ---dissociation boundary, o non-dissociated, + dissociated sample

AES analysis Visually observed changes on the sample surface were compared with the stoichiometry measurements by Auger electron spectrometry (AES).

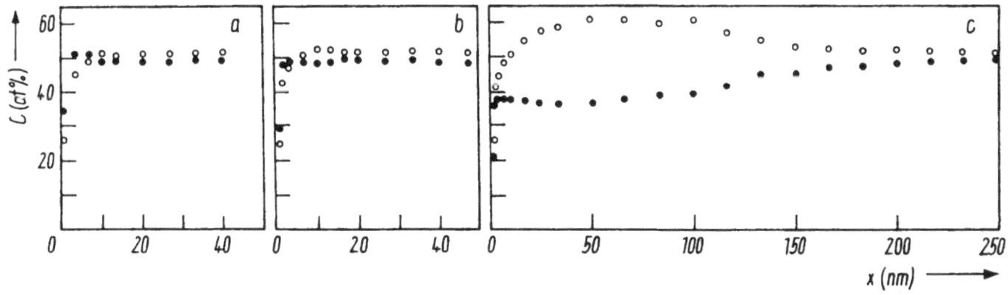

Fig. 3. Chemical depth profiles of Ga (o) and As (•) in GaAs surface layers as determined by AES: a) non-annealed samples, b) annealed sample at T = 970 °C and τ = 5 s (non-dissociated sample), c) annealed sample at T = 1010 °C and τ = 4.5 s (dissociated sample)

Chemical depth profiles were obtained from the samples by combining the AES with ion beam sputtering. Chemical depth profiles of dissociated and non-dissociated annealed samples were compared with those of non-annealed ones. Regimes close to the critical radiation time boundary were chosen for these analyses. Some typical depth profiles obtained are given in Fig. 3. It can be seen from Fig. 3a and b that annealed samples without any visually observed dissociation show the same depth profiles of As and Ga as non-annealed samples. Dissociated samples with annealing regimes above the critical radiation time boundary (Fig. 3c) showed no stoichiometry in the surface layers of GaAs, the degree of which was dependent on the degree of dissociation.

Electrical measurements In order to find out if the annealing regimes below "the dissociation boundary" can produce a sufficient electrical activity of the implanted impurities, also the dependences of electrical properties of implantation doped GaAs on thermal annealing regimes were studied. These measurements were also done on semi-insulating (100) GaAs with resistivity of (0.9 to 3)x10^8 Ω cm. In order to find out the effect of conversion of a semi-insulating substrate, test measurements were performed just prior to annealing. Non-implanted GaAs substrates were annealed under the same conditions as samples with implanted dopants. Test samples did not show an active layer formation. GaAs samples were implanted with 200 keV $^{28}Si^+$ ions with a total dose of 6x10^{12} cm^{-2} at room temperature. Such doped capless samples were annealed in the described apparatus at 970 °C with different radiation times. Sheet resistance and Hall voltage were measured on annealed samples. Sheet

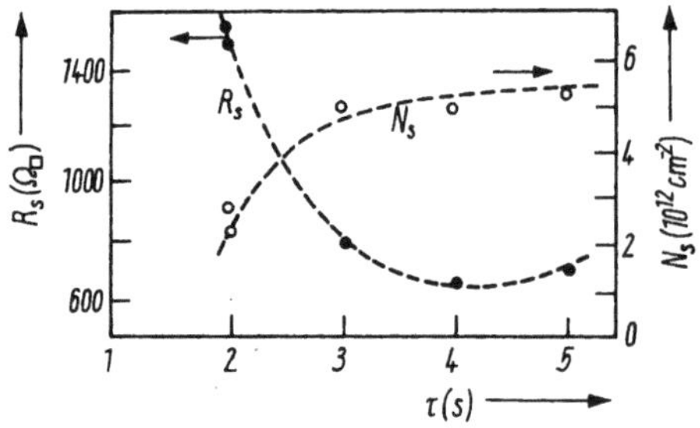

Fig. 4. Sheet resistance R_s and sheet carrier concentration N_s of semi-insulating GaAs implanted by Si^+ ions (200 keV, $6x10^{12}$ cm^{-2}) in dependence on radiation time τ

carrier concentration and effective mobility were determined from experimental data. Also the dependences of these properties on time were studied. The experimental results are given in Fig. 4. As can be seen in this figure, a remarkable decrease of sheet resistance occurs already after 2 s of radiation. The effective carrier mobility calculated from the experimental data reached a maximum value of 2720 cm^2/Vs. The calculated dopant electrical activity was 85%. The real implanted dopant electrical activity was probably higher than the calculated value because the $^{28}Si^+$ ion beam was partially contaminated by molecular nitrogen ions.

Based on the above given results the pulse radiation annealing of capless GaAs in vacuum can be used for thermal heat treatment without observable dissociation of the sample surface layer. The relationship of annealing temperature and radiation time is critical for this dissociation.

References

/1/ J.S. HARRIS, F.H. EISEN, B. WELCH, J.D. HASKEL, R.D. PASHLEY, and J.W. MAYER, Appl. Phys. Letters 21, 601 (1972).

/2/ J. KASAHARA, M. ARAI, and N. WATANABE, J. appl. Phys. 50, 541 (1979).

/3/ M. ARAI, K. NISHIYAMA, and N. WATANABE, Japan. J. appl. Phys. 20, L124 (1981).

(Received April 19, 1983)

phys. stat. sol. (a) 79, K45 (1983)
Subject classification: 10.2 and 14.3; 11; 22.1.1

Belorussian State University, Minsk[1)]

Influence of Growth Dislocations on Low Temperature Conductivity in Irradiated p-Germanium

By
V.P. DOBREGO and A.K. FEDOTOV

It has been shown earlier /1/ that low temperature, T = 77 K, irradiation of gold doped germanium by relatively small fluences of γ-quanta by 4 MeV electrons and in a reactor brings about notable conductivity over the range of 80 > T > 4.2 K. The sample becomes inhomogeneous in conductivity at intermediate stages of irradiation and annealing. It has been proposed that the conductivity is realized through a net of conduction channels, the physical basis of these could be dislocations with impurity atmospheres, which change substantially while interacting with radiation-induced defects. In this note the correlation between the low temperature conductivity of irradiated samples and the concentration and distribution of growth dislocations is examined.

We studied germanium single crystals, gold doped, $N_{Au} = (0.7$ to $1.0) \times 10^{15}\ cm^{-3}$ and weakly compensated by antimony. The growth dislocation density was $D = 10^3$ to $10^4\ cm^{-2}$. The samples studied were irradiated by γ-quanta of ^{60}Co with a fluence of $8 \times 10^{15}\ cm^{-2}$. The sample dimensions were close to $2 \times 1 \times 0.8\ mm^3$. The largest face was on the (111) plane. The density of dislocations D was determined from etch pits in the (111) plane according to the technique given in /2/. Larger samples were taken to consider the effect of dislocation density distribution on the electrical properties. The etch pits were calculated in squares of size $370 \times 370\ \mu m^2$ into which the sample face ((111) plane) has been divided. The peculiarities of the low temperature experiment were described in /1/.

Fig. 1 shows the temperature dependence of the resistivity of irradiated samples differing by the mean value of the growth dislocation density, the samples being cut from the same ingot section. All the samples are characterized by a termination of the sharp resistivity increase corresponding to the activation energy of the gold impurity E = 0.15 eV at temperature lowering.

---

1) prospekt Lenina 4, 220080 Minsk, USSR.

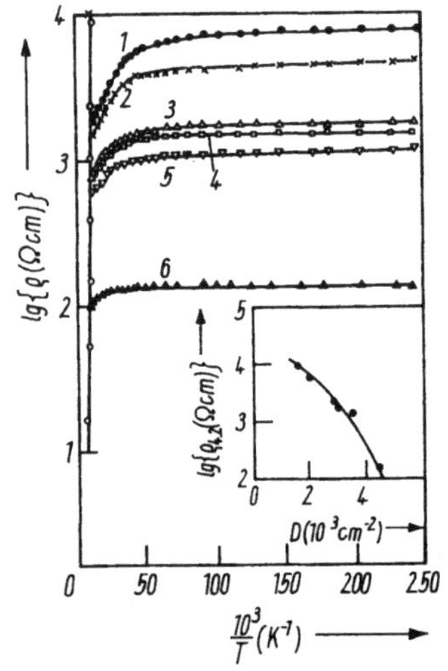

Fig. 1. Resistivity of the irradiated samples with various growth dislocation densities versus temperature. (1) D = 1.6, (2) 2.0, (3) 2.9, (4) 3.0, (5) 3.6, (6) 4.5×10^3 cm^{-2}. Insert: Sample resistivity at T = 4.2 K versus growth dislocation density

Below 80 to 100 K a weak temperature dependence of resistivity is evident, which allows to extend the measurements to helium temperatures.[2)] In Fig. 1 a monotonous dependence of the resistivity on the mean dislocation density shows up in the low temperature region.

For the larger samples a correlation between the resistivity of irradiated samples, measured at 77 K, and the dislocation density distributions in them is observed. Fig. 2 shows a histogram for the dislocation density distribution on two typical samples having about the same mean dislocation densities - (5 to 6)$\times 10^3$ cm^{-2}. These samples show equal changes in resistivity after the irradiation. Samples having squares without etch pits vary their resistivity weaker as compared to those having a more uniform dislocation distribution. These results support the assumption /1/ that the low temperature conductivity of irradiated p-Ge is closely connected with the presence of dislocations in the samples.

In earlier works, see reviews /3, 4/, the electrical activity of dislocations is associated with the presence of dangling bonds in a dislocation core as well as with the surrounding field of microstresses. Later on /5 to 7/ it has been found necessary to take into account the presence and state of point defects in the dislocation microstress field for explaining the electrical properties of dislocations. The important role of dislocation atmospheres for the interaction between point defects of radiative origin and dislocations in Ge has been earlier indicated in an interesting paper /8/. It is natural to suggest that in our experiments the interaction of primary radiation-induced defects with dislocations is to a considerable extent connected with the large atmospheres of growth

---

2) The upper temperature limit of the measurements is 100 to 120 K, beyond which a distinct annealing process begins.

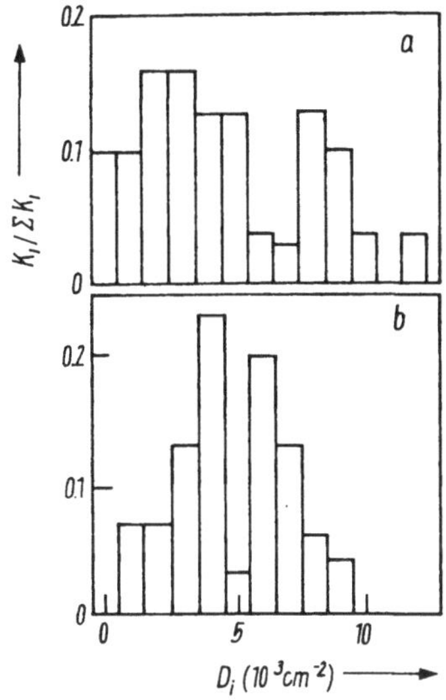

Fig. 2. Histograms of the dislocation density distribution; a) sample with a more homogeneous, b) less homogeneous dislocation distribution. On the abscissa the mean dislocation densities D in the calculation square are plotted, on the ordinate the number of squares K_i with mean dislocation density D_i which is normalized to the total number of squares

dislocations if taking into consideration the rather high mobility of Frenkel pair components even at low temperature irradiation of p-germanium /9/. Supplementary conductivity in samples with high dislocation density is described in /10/, dislocations being introduced by plastic deformation of pure Ge. With a compression deformation of 20 to 30% conductivity of hopping type is evident, which is close by absolute value to that observed in our experiments. The difference between our samples and those in /10/ is the manyfold lower dislocation density and probably the presence of larger impurity atmospheres around the growth dislocations.

In order to explain the mechanism of low temperature conductivity in our study it could be suggested from the above that interaction between primary radiation-induced defects and dislocation impurity atmospheres is important in the occurrence and shorting of possible conduction paths. The importance of dislocation atmospheres in these processes is also evidenced by our experiments on heat treatment of samples before irradiation. Heat treatment at 400 °C for 10 min, not affecting the growth dislocation density but capable of changing dislocation atmospheres, causes a substantial increase of the effect of low temperature irradiation on the lowering of the sample resistivity in the low temperature region. Further experiments are needed to make a more definite conclusion as to the effect of dislocation atmospheres on the low temperature conductivity of irradiated samples.

References

/1/ V.P. DOBREGO, A.V. KOLYADKO, and YU.E. MAKUSHOK, Fiz. Tekh. Poluprov. 15, 301 (1981).

/2/ V.N. VASILEVSKAYA and E.G. MISELYUK, Fiz. tverd. Tela 3, 429 (1961).

/3/ Y. BARDSLEY, Uspekhi fiz. Nauk 73, 121 (1961).

/4/ H.F. MATARÉ, Defect Electronics in Semiconductors, Wiley-Interscience, New York 1974.

/5/ V.G. EREMENKO, V.I. NIKITENKO, and E.B. YAKIMOV, Zh. eksper. teor. Fiz. 73, 1129 (1977).

/6/ V.G. EREMENKO, V.I. NIKITENKO, E.B. YAKIMOV, and N.A. YARKIN, Fiz. Tekh. Poluprov. 12, 273 (1978).

/7/ W. SCHRÖTER, E. SCHEIBE, and H. SCHOEN, J. Microscopy 118, 23 (1980).

/8/ V.V. KHAINOVSKAYA and L.S. SMIRNOV, Fiz. tverd. Tela 10, 1549 (1968).

/9/ T.M. FLANAGAN and E.E. KLONTZ, Phys. Rev. 167, 789 (1968).

/10/ I.V. KLYATSKINA, M.L. KOZHUKH, S.R. RIVKIN, V.A. TRUNOV, and I.S. SHLIMAK, Fiz. Tekh. Poluprov. 13, 1089 (1979).

(Received May 6, 1983)

phys. stat. sol. (a) 79, K49 (1983)

Subject classification: 2 and 8; 22

Wissenschaftsbereich Tieftemperaturphysik, Sektion Physik der Technischen Universität Dresden[1)]

Low-Temperature Specific Heat of Vapour-Deposited Ge_xSe_{1-x} Glass (x = 0.3)

By

M. MERTIG, G. POMPE, and E. HEGENBARTH

The binary system Ge_xSe_{1-x} is an amorphous semiconductor with a wide glass-forming region. Bulk samples can be prepared in the concentration range x = 0 to about x = 0.42 by quenching from the melt /1/. Vapour-deposited films exist in the whole composition range /2/. From this point of view the Ge_xSe_{1-x} system is very attractive to study the dependence of the thermal properties on the parameter x and furthermore for comparison of the properties of bulk and film samples.

The influence of structure on the low-temperature specific heat c of bulk amorphous samples ($x \leqq 0.4$) was investigated by Thomas /3/. At helium temperature Ge_xSe_{1-x} glass shows a peak in c/T^3, which becomes larger and moves to lower temperatures as the Ge content decreases. An analogous behaviour was observed in the Ge_xS_{1-x} glass system /4/.

In this note, we present the first measurement of the specific heat of an amorphous $Ge_{0.3}Se_{0.7}$ film system. The results will be compared with results of an equivalent bulk sample /3/ in order to investigate the influence of the method of preparation. Up to now analogous investigations were only made for amorphous As /9, 13, 14/, which show considerable differences in the data for film and bulk samples.

The compacted film ($d \approx 10\ \mu m$) was prepared by flash-evaporation of crushed bulk $Ge_{0.3}Se_{0.7}$ glass onto a sapphire substrate (d = 0.3 mm, $A \approx 1\ cm^2$) with a deposition rate of approximately 10 nm/s. The base pressure was 5×10^{-5} Torr. The temperature of the sapphire substrate was kept below 100 °C. The starting bulk material was prepared by quenching the melt into cold water /1/. The evaporated film exhibits no X-ray diffraction lines.

---

1) Mommsenstr. 13, DDR-8027 Dresden, GDR.

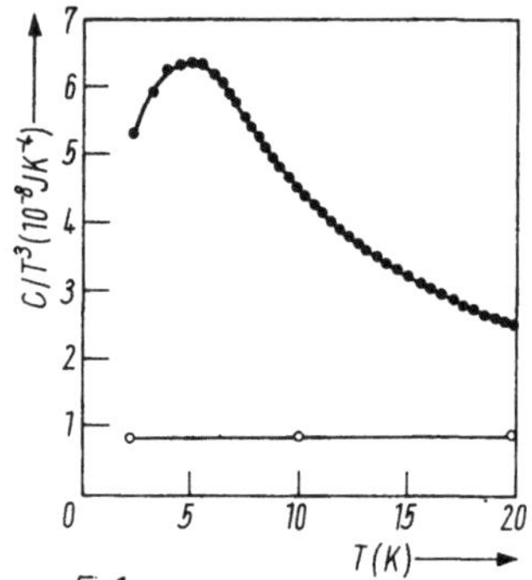

Fig.1

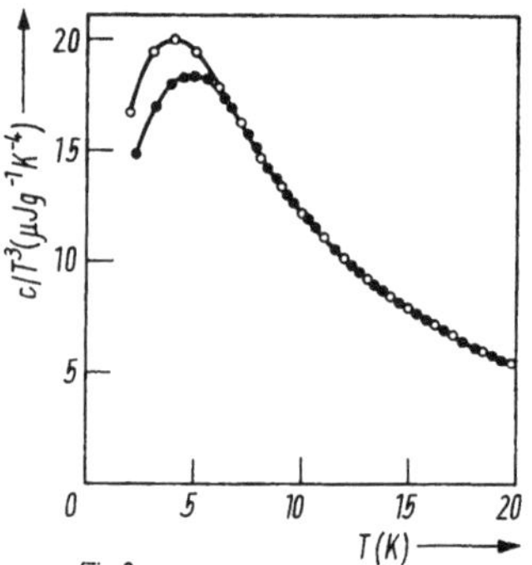

Fig.2

Fig. 1. Heat capacity divided by T^3 against T of the "sandwich" system (● total heat capacity, o heat capacity of the sapphire substrate) in the temperature range from 2 to 20 K

Fig. 2. Specific heat divided by T^3 against T of $Ge_{0.3}Se_{0.7}$ (● specific heat of the amorphous film, o specific heat of bulk amorphous $Ge_{0.33}Se_{0.67}$ /3/) in the temperature range from 2 to 20 K

The "sandwich" sample consisting of the sapphire substrate (100.0 mg) with the amorphous $Ge_{0.3}Se_{0.7}$ film (3.0 mg) was glued with a small amount of Apiezon-N-grease to a sapphire sample holder with an evaporated Bi heater and an Au(+0.03 at% Fe)-Cu thermocouple. The total heat capacity in the temperature range between 2 and 20 K was determined by an ac calorimetry technique /5/. The sinusoidal heat input has a frequency of $\nu_g = 3$ Hz. The inner thermal relaxation time τ_{inn} of the sample system was < 2 ms. The heat capacity of the sample holder including the Apiezon-N-grease was previously determined in a separate run and the heat capacity of the "sandwich" system was obtained by subtraction.

The heat capacity of the "sandwich" is shown in Fig. 1 as C/T^3 versus T. The accuracy of the measurement is about 2%. The heat capacity of the sapphire substrate was calculated from results by Fugate and Swenson /6/. The ratio of the heat capacity of the $Ge_{0.3}Se_{0.7}$ film to that of the sapphire substrate varied from 7.4 at 5 K, to 2.8 at 20 K. Finally we obtain the specific heat of the amorphous film by means of subtraction.

The results, in the form of c/T^3 versus T, are shown in Fig. 2 together with results for bulk $Ge_{0.33}Se_{0.67}$ glass /3/. For the film system we observe a peak in c/T^3 at $T \approx 5$ K. This peak is a general feature of the amorphous state. In amorphous Ge /7/ and in vitreous silica /8/ the peak has been attributed to a peak in $g(\omega)$ analogous to the transverse-acoustic (TA) peaks observed in crystalline Ge and cristobalite, respectively. It seems probable that the modes responsible for the peak in c/T^3 in the glass involve the same sort of bond-bending motion as the crystalline TA modes. This picture is supported by the theoretical work of Weaire and Alben /10/. There it was shown that the TA modes in both crystalline and amorphous Ge have a zero frequency when the bond-bending force is zero. From this general point of view we suppose that the observed peak in c/T^3 of $Ge_{0.3}Se_{0.7}$ indicating a marked peak in the phonon spectrum $g(\omega)$ at $\omega \approx 17$ cm^{-1} is caused by very flat TA modes. But a detailed discussion for this system is not possible. On the one hand, the measured values of the sound velocity differ considerably, e.g. the values of the longitudinal velocity vary between 2.2 and 2.8 km/s /11, 12/, so that a quantitative analysis of the heat capacity data is not possible. On the other hand, it is very difficult to discuss the results in comparison with Raman measurements because the frequency dependence of the phonon-photon coupling constant $K(\omega)$ at very low frequencies is not clear. Raman measurements of $Ge_{0.33}Se_{0.67}$ by Nemanich /11/ show a marked peak in $I^R(\omega)/\omega^2 = K(\omega)g(\omega)/\omega^2$, where $I^R(\omega)$ is the reduced Raman spectrum and ω is the frequency shift. Nemanich /11/ supposes that this peak at frequencies of 15 to 20 cm^{-1} is due to a maximum in the coupling constant $K(\omega)$ and not to a maximum in the density of states $g(\omega)$.

A comparison of the film and bulk sample shows, that the heat capacity data for both samples are similar. For $T > 6$ K the data are the same. Below 6 K the heat capacity of the film is about 10 to 15% lower than the heat capacity of the bulk sample. The maximum in c/T^3 of the amorphous film occurs at a little bit higher temperature. These results are in contrast to heat capacity measurements of film and bulk amorphous As /9, 13, 14/. These measurements show differences in the data of both specimens by a factor 2, indicating that the properties of thin films are very sensitive to the method of preparation. Jones et al. /13/ suggest that these differences may be a property of the unusual bonding of the As atoms (the effective coordination number of crystalline As is six and for the amorphous modification three), which is connected with the different mass densities of the film and bulk sample. Such effects will not be expected for $GeSe_2$ because the coordination number of the Ge atoms and Se atoms is assumed to be equal in the amorphous and crystalline modification /15/.

We wish to thank Prof. P. Süptitz for helpful discussion, Dr. N. Mattern for X-ray diffraction measurements, and N. Sebastian for sample preparation.

References

/1/ R. AZOULAY, H. THIBIERGE, and A. BRENAC, J. non-crystall. Solids 18, 33 (1975).

/2/ B.J. MOLNAR and D.B. DOVE, J. non-crystall. Solids 16, 149 (1974).

/3/ N. THOMAS, Ph.D. Thesis, University of Cambridge, 1978 (unpublished).

/4/ N. THOMAS and W.A. PHILLIPS, J. Physique 39, C6-980 (1978).

/5/ P.F. SULLIVAN and G. SEIDEL, Phys. Rev. 173, 679 (1968).

/6/ R.Q. FUGATE and C.A. SWENSON, J. appl. Phys. 40, 3034 (1969).

/7/ C.N. KING, W.A. PHILLIPS, and J.P. DE NEUFVILLE, Phys. Rev. Letters 32, 538 (1974).

/8/ N. BILIR and W.A. PHILLIPS, Phil. Mag. 32, 113 (1975).

/9/ J.S. LANNIN, H.F. ENO, and H.L. LUO, Solid State Commun. 25, 81 (1978).

/10/ D. WEAIRE and R. ALBEN, Phys. Rev. Letters 29, 1505 (1972).

/11/ R.J. NEMANICH, Phys. Rev. B 16, 1655 (1977).

/12/ R. OTA, T. YAMATE, N. SOGA, and M. KUNUGI, J. non-crystall. Solids 29, 67 (1978).

/13/ D.P. JONES, N. THOMAS, and W.A. PHILLIPS, Phil. Mag. B 38, 271 (1978).

/14/ C.T. WU and H.L. LUO, J. non-crystall. Solids 13, 437 (1974).

/15/ O. UEMURA, Y. SAGARA, and T. SATOW, phys. stat. sol. (a) 26, 99 (1974).

(Received June 9, 1983)

phys. stat. sol. (a) 79, K53 (1983)

Subject classification: 14.3.4; 22.8

Institute of Physics, Academy of Sciences of the Azerbaidzhan SSR, Baku[1])

Effect of Illumination on Current-Voltage Characteristics of $GdIn_3S_6$ Single Crystals

By

G.M. GUSEINOV, G.M. NIFTIEV, and M.G. SHAKHTAKHTINSKII

In the previous report /1/ it was shown that the presence of heterogeneities in $GdIn_3S_6$ crystals exerted a substantial effect on the spectrum of thermostimulated conductivity (TSC). Of interest is a study of the current-voltage characteristics (CVC) of heterogeneous crystals usually exhibiting both N- and S-like CVC's /2/, low-frequency oscillations of current /3/, etc.

The present note deals with an investigation of the CVC's of $GdIn_3S_6$ compounds in three states (dark state, residual-conductivity (RC) state, i.e. dark state after preliminary bias lighting, and illuminated state).

The procedure used to obtain $GdIn_3S_6$ single crystals is described elsewhere /4/. The CVC's were measured on the same crystals as those used in the TSC study /1/.

The specimens possessed n-type conductivity with a resistivity of 7×10^{10} Ωcm at room temperature. The contacts for the specimens were produced by fusing-in indium with a subsequent annealing in vacuum at a temperature of 500 K. The distance between the contacts amounted to 3.2×10^{-2} cm.

Fig. 1 gives the CVC's of In-$GdIn_3S_6$-In structures in the dark (curve 1) and under exposure to 200 lx white light (curve 2) at a temperature of 131 K. When the specimen passes from a high-resistance to a low-resistance state the resistance decreases by a factor 10^4. Such a state of the specimen is maintained for a long time, i.e. switching with an electric memory takes place.

Fig. 2 shows dark CVC's after a preliminary bias illumination at various exciting electric fields applied to the specimen during light exposure. The CVC's were taken in the dark after a stable state of RC had been established.

It should be noted that under the effect of an electric field of strength 4.4×10^3 V/cm the specimen passes from a more photoconductive to a less

1) prospekt Narimanova 33, 370143 Baku, USSR.

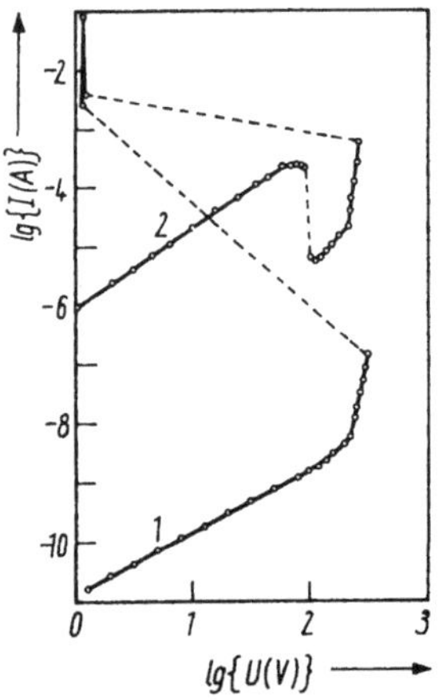

Fig. 1. CVC's of In-$GdIn_3S_6$-In structures in the dark (curve 1) and under exposure to white light (curve 2) at a temperature of 131 K

photoconductive state, i.e. an N-like CVC is observed, and as this takes place, one can see a growth of the currents corresponding to the points of transition from the high-conductivity to the low-conductivity state.

The CVC's have been studied at various wavelengths and temperatures, too. In the wavelength region of 580 to 630 nm and at temperatures of 77 to 160 K the specimens possess a substantial photosensitivity, and the CVC's of In-$GdIn_3S_6$-In structures have N-like appearance. As the temperature grows (T $\geqq$ 160 K) the N-like section of the CVC disappears both in the crystals pre-exposed to light and in those exposed to light, and a quadratic section, an $I \sim U^n$ relationship (where n = 2 to 2.5), a linear section, and a quasi-saturation region are observed. The existence of an RC and N-like CVC can be explained on the basis of a model suggested in /5/ and based on the assumption that the specimens contain high-resistance inclusions in a low-resistance matrix.

In order to determine the energetic positions of the levels responsible for the RC and the N-like CVC, TSC has been studied following exposure to various quenching electric fields. Two groups of traps are revealed in the TSC spectrum, namely shallow traps manifesting themselves in the temperature range from 77 to 160 K, and deep traps showing up at temperatures of 160 to 260 K.

The investigation of TSC after switching the specimen from the high-conductivity to the low-conductivity state has shown that the peaks observed earlier in the temperature range from 77 to 160 K in the RC state, fail to manifest themselves. Hence, the levels giving the TSC peaks in the 77 to 160 K range are responsible for the RC and the N-like CVC. The energetic depths of these levels have been determined from thermal-glow curves on the basis of the initial growth of the TSC current and are in the range from 0.01 to 0.16 eV, the charge-carrier trapping cross-sections being equal to 8×10^{-30} to 5×10^{-24} cm^2, as determined with formula (22) of an earlier report

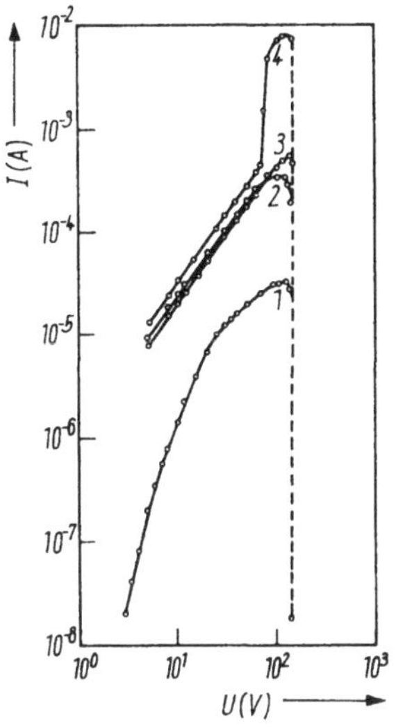

Fig. 2. Dark CVC's after preliminary bias illumination at different exciting electric fields applied to the specimen during light exposure. (1) 6.2x10^2, (2) 1.2x10^3, (3) 1.8x10^3, (4) 2.5x10^3 V/cm

/6/. The existence of a quadratic section and a charge current growth region in the CVC, the fulfilment of the proportionalities $j \sim L^{-3}$, $U_{tfl} \sim L^2$ and $\sqrt{U_{thr}} \sim T$, where j is the current density, L is the specimen thickness, U_{tfl} is the voltage corresponding to the switching of the specimen from the low-conductivity to the high-conductivity state (S-section)), the U_{thr}-independence of the contact material, as well as the metallic nature of conductivity in the low-resistance state, point to the fact that the S-type switching with electric memory is associated with the filling of trap levels by monopolar injection by a strong electric field, and a subsequent separation of the metallic component from the switching cell between the electrodes.

References

/1/ G.M. GUSEINOV, G.M. NIFTIEV, and M.G. SHAKHTAKHTINSKII, Fiz. Tekh. Poluprov. 16, 1332 (1982).

/2/ M.K. SHEINKMAN and A.YA. SHIK, Fiz. Tekh. Poluprov. 10, 209 (1976).

/3/ S.G. KALASHNIKOV, G.S. PADO, V.I. PASTOVOIT, and E.F. TOKAREV, Fiz. Tekh. Poluprov. 3, 1028 (1969).

/4/ S.G. DEMETRASHVILI, P.G. RUSTAMOV, O.M. ALIEV, and E.M. NANOBASHVILI, Rep. Acad. Sci. Georg. SSR 89, No. 1 (1978).

/5/ S.M. RYVKIN and I.S. SHLIMAK, phys. stat. sol. (a) 16, 515 (1973).

/6/ G.A. BORDOVSKII, V.G. BOITSOV, and B.A. DEMIDOV, Fiz. Tekh. Poluprov. 8, 1918 (1974).

(Received July 26, 1983)

phys. stat. sol. (a) 79, K57 (1983)

Subject classification: 11; 14.2; 21; 21.1

Institute of Metal Physics, Academy of Sciences of the USSR, Ural Scientific Centre, Sverdlovsk[1])

Simulation of Collision Cascades in Intermetallic V_2Hf Compounds

By

V.G. CHUDINOV, V.P. GOGOLIN, B.N. GOSHCHITSKII, and V.I. PROTASOV

Introduction Together with good superconducting parameters the V_2Hf compound with C-15 structure is characterised by a high radiation stability. The superconducting transition temperature T_c is shown /1/ to change by 5% under irradiation with fast neutrons up to 1.6×10^{19} cm^{-2}. As for the compounds with A-15 structure which have found wide-spread use in practice, T_c decreases by 25% at the same doses.

Calculation technique and discussion In the present note the molecular dynamics technique is used to simulate a collision cascade in V_2Hf. The calculations were carried out on a crystallite of almost spherical shape which consisted of 3800 atoms, 2000 of which being located on the immobile wall. This wall allowed one to keep atoms of the mobile part during the calculation at the sites of the C-15 structure. The energy was removed from ≈ 400 atoms of the intermediate layer. The kinetic energy of these atoms could be decreased by a definite value at very time step. The rate of removal was taken equal to 0.9 taking into account the absence of highly energetic atoms from the hard wall.

Pair interaction potentials in Buckingham form were calculated for V and Hf using the technique described in /2/:

$$\varphi(r) = b\exp(-a\,r) - \frac{c}{r^6}\ ,$$

where r is the distance between the atoms; a, b, c are constants the values of which are given in Table 1. Mixed potentials were calculated by the following formula:

$$\varphi_{12}(r) - \sqrt{b_1 b_2}\exp\left(-\frac{a_1 + a_2}{2}\,r\right) - \frac{\sqrt{c_1 c_2}}{r^6}\ .$$

1) ul. S. Kovalevskoi 18, 620219 Sverdlovsk GSP-170, USSR.

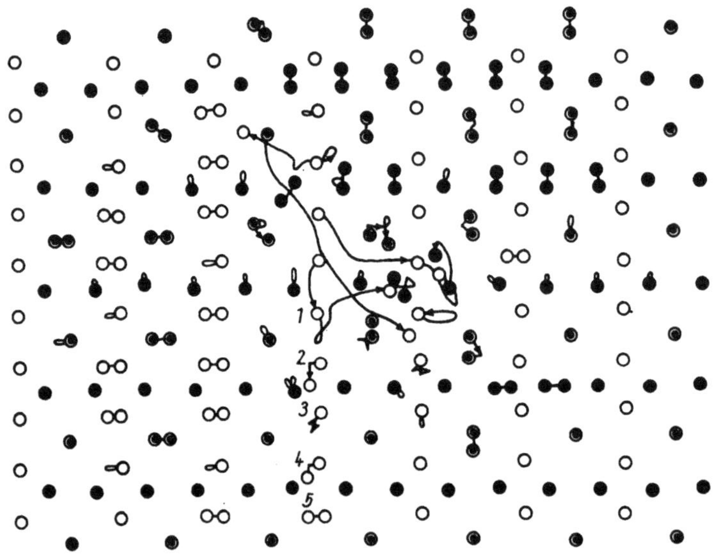

Fig. 1. A collision cascade in V_2Hf with the initial energy of 450 eV at the instant of 2×10^{-13} s ((100) plane); ⊙ Hf atoms; •; o V atoms which lie on 1/8 of the lattice constant above the plane of the figure; ⟶ trajectories of atom movements

With V-V, Hf-Hf, and V-Hf interatomic distances of 0.25, 0.175, and 0.21 nm, respectively, the potentials were laced with repulsive branches in Born-Meyer form, assuming that the first and the second derivatives are equal:

$$\varphi(r) = 2 Z_1 Z_2 E_R \frac{a}{r} \exp\left(-\frac{r}{a^*}\right) + A \exp\left(-\frac{r}{b^*}\right) ;$$

$$A(r) = 2.58\times10^{-5} (Z_1 Z_2)^{11/4} ; \quad b^* = \frac{1.5\, a_o}{(Z_1 Z_2)^{1/6}} ,$$

where Z_1, Z_2 are the atomic charges of atoms, E_R is the Rydberg constant, a_o is Bohr's radius.

Table 1

	a(nm^{-1})	b(eV)	c(eV nm^{-1})	Z	a_o(nm)	E_R(eV)
V	63.8	0.60042×10^{7}	0.22679×10^{4}	23	0.053	13.6
Hf	97.71	0.53147×10^{13}	0.71277×10^{4}	72	0.053	13.6

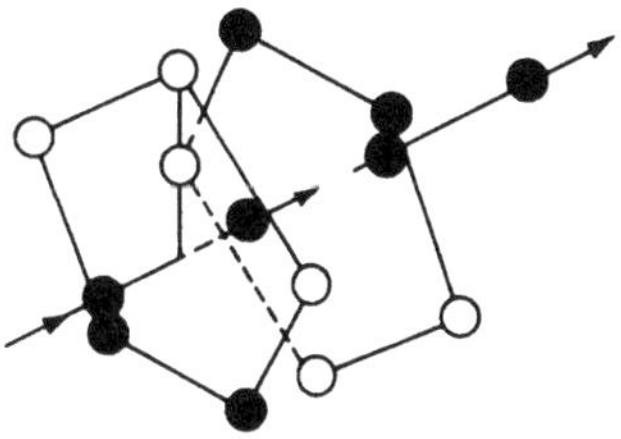

Fig.2

Fig.3

Fig. 2. The defocusing lens in the direction of closest packing of V atoms (o Hf atoms, • V atoms)

Fig. 3. The defocusing lens in the direction of closest packing of Hf atoms (the same signatures as in Fig. 2)

Fig. 1 shows the calculated results for the collision cascade, the energy of the primarily knocked-out atom (PKA) being equal to 450 eV per one instant of 2×10^{-13} s. The vector of the initial velocity of PKA was in the (100) plane, forming with the ⟨110⟩ direction an angle of 15°. The most important feature of the cascade development is as follows: the final stage is characterised by the absence of chains of focusing collisions. In media with simple structures it is the main mechanism of energy removal from a cascade region (CR). Even at random coincidence of the velocity vector with the direction of closest packing defocusing over a space of 3 to 4 interatomic distances was observed. For example, such a defocusing chain is seen along atoms 1, 2, 3, 4, 5 in Fig. 1.

For a closer examination of the problem collision cascades with low PKA energy have been calculated, the velocity vectors coinciding with the directions of closest packing of atoms of V (Fig. 2) and Hf (Fig. 3). In both cases defocusing has been observed resulting from the fact that atoms pass through an asymmetrical lens which consists of different types of atoms. The threshold energies needed for Frenkel pairs in these directions turned out to be ≈25 and ≈ 55 eV, respectively.

At the instant of 2×10^{-13} s four vacancies, four interstitials, and one antisite defect (Hf substitutes V) were formed in the CR. Simultaneously, due to the strong localized disturbance a compressive shock wave was formed. It removes energy from the CR, when all the atoms reach an energy lower than the threshold one. This wave propagates mostly along the ⟨100⟩ direction. In /3/ it has been pointed out that under radiation these waves can possibly form in a solid. This mechanism can lead to a disturbance of the surface /4/.

It should be noted that the mechanism of energy dissipation in a shock wave differs from that in the chains of focusing collisions. The release of energy takes place only at some structural defects, inclusions of another phase, admixtures, etc. or due to the presence of viscosity in a system. Therefore, in real situation a considerable portion of the PKA energy would be removed far from the CR and hence, would not affect further evolution of the CR during which a great number of defects can also be formed as in Nb_3Sn /5/. It is also the reason why the sizes of CR in compounds with C-15 structure should be comparatively smaller. It is natural to suppose that because of this reason all the dose dependences of the material parameters (which are determined by the volume of the new phase) would be weaker.

In conclusion it should be noted that in our previous paper on simulation of a collision cascade in Nb_3Sn /2/ nucleation of shock waves also takes place, but much later. Here the waves are generated by the chains of Nb atoms along which focusing collisions propagate.

References

/1/ B.S. BROWN, M.J.W. HATSTROM, and T.E.J. KLIPPERT, Appl. Phys. 48, 1759 (1977).

/2/ V.G. CHUDINOV, V.P. GOGOLIN, B.N. GOSHCHITSKII, and V.I. PROTASOV, phys. stat. sol. (a) 67, 61 (1981).

/3/ M. CRUINAN, J. nuclear Mater. 53, 171 (1974).

/4/ V.G. CHUDINOV, V.P. GOGOLIN, B.N. GOSHCHITSKII, and V.I. PROTASOV, to be published.

/5/ Y. HAYASHINCHI, Y. KITAZOE, T. SEKIYA, and Y. YAMAMUR, J. nuclear Mater. 71, 181 (1971).

(Received March 23, 1983)

phys. stat. sol. (a) 79, K61 (1983)

Subject classification: 13.4 and 20.3; 11; 22.2.4

Wissenschaftsbereich Halbleiterphysik,
Sektion Physik der Karl-Marx-Universität Leipzig[1)]

P_{Ga} Antisite Defect in Zn-Implanted $GaAs_{0.1}P_{0.9}$ LED

By

R. PICKENHAIN[2)], B. RHEINLÄNDER, and S. BREHME

Recently, Killoran et al. /1/ and O'Donnell et al. /2/ indicated the ODMR observation of a 1.1 eV emission band in GaP:Zn related to the two-electron state of the P_{Ga}^{3+} antisite defect. This IR-radiative recombination competes with the pair processes giving the visible luminescence. The 1.1 eV emission is produced by electron transition from the P_{Ga} level to the Zn acceptor level.

In this note we report on injection luminescence (EL), DLTS, and optic DLTS (ODLTS) /3/ studies on yellow emitting, Zn-implanted VPE-$GaAs_{0.1}P_{0.9}$:N LED. The net donor doping level in the non-doped n-region is determined by C-V analysis to be $N_D - N_A \approx 2\times10^{16}$ cm^{-3}. The diodes were stressed during 1000 h using a current density $j_s = 40$ A cm^{-2}. The degradation behavior of these diodes is described in detail by Rheinländer et al. /4/. After stressing a small increase of the net donor doping level is observed.

The DLTS measurement exhibited the emergence of a deep level having a density of about 10^{13} cm^{-3} and a thermal activation energy $E_c - E_T = 930$ meV below the conduction band. In contrast, Ferenczi /5/ has found only a donor level $E_c - E_T = 490$ meV appearing due to an similar stress treatment.

The high-temperature part of the DLTS curve of the 930 meV level is deeper than the expected one. This indicates the contribution of hole emission to the DLTS emptying of the deep donor level.

We used the possibility of influencing this emptying process by additional monochromatic-light illumination at two temperatures Θ_1 and Θ_2. In Fig. 1 the experimental results are given: The illumination forces electron transitions into the conduction band. The threshold of the optical cross-section is found to be $E_c - E_T^O \approx 760$ meV. It should be pointed out that the ODLTS signal at $T = \Theta_1$

1) Linnéstr. 5, DDR-7010 Leipzig, GDR.

2) Applikationsgruppe "Halbleiterdiagnostik".

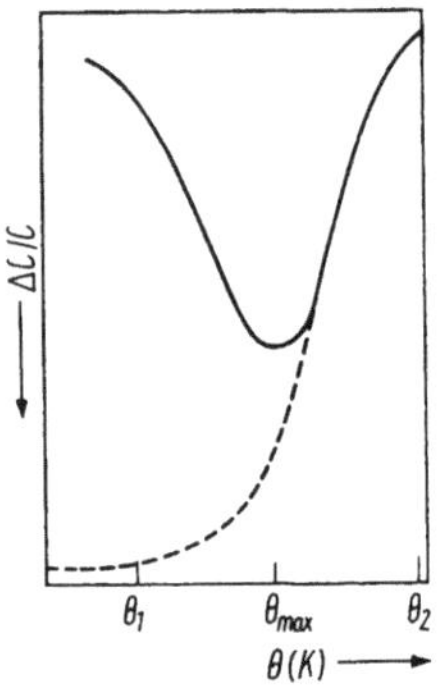

Fig. 1. Spectral parts of the 930 meV DLTS signal (full line) and the respective ODLTS signal (dashed line) of a stressed sample

is higher than the maximum non-optical DLTS signal. Using a period of duration $2t_p$ the ODLTS signal is given as

$$S = \frac{1}{2t_p}\left(1 - \frac{e_p^t + e_p^o}{e_n^t + e_n^o + e_p^t + e_p^o}\right)\frac{N_{TT}}{e_n^t + e_n^o + e_p^t + e_p^o} \times$$

$$\times \left(1 - \exp(-(e_n^t + e_n^o + e_p^t + e_p^o)t_p\right)^2 , \qquad (1)$$

where N_{TT} is the level density and $e_n^{o,t}$ and $e_p^{o,t}$ are the optical and the thermal emission rates of electrons and holes, respectively.

The observed feature $S_{ODLTS}(\Theta < \Theta_{max}) > S_{DLTS}(\Theta_{max})$ can be explained only by $e_p \gg e_n$. In this case the DLTS signal does not reflect the true value N_{TT}. Furthermore, regarding (1), in this case we can recognize the failure of the Arrhenius-plot analysis in order to determine the exact thermal activation energy because the deduced value $E_c - E_T$ depends on the hole emission rate.

The EL spectra ($\lambda = 0.5$ to 1.8 μm) yielded the emergence of a 1.1 eV emission band during stress accompanied by the diminishing of the bands at $h\nu =$ = 1.82 and 2.17 eV (e.g. Fig. 2). It should be remarked that purely diffused diodes obtained from the same epitaxy wafer did not show the 1.1 eV band as well as the 930 meV deep level. Therefore, the DLTS and EL behavior after the stressing should be induced by the implantation process.

Taking into consideration (i) the energy position of the deep level found by DLTS measurements after the stressing, (ii) the high hole emission rate, and (iii) the correlation to the 1.1 eV EL band and taking into account the ODMR results /1, 2/ we suggest this deep level to be created by the P_{Ga} antisite defect. Because this defect type exists without doubt after the Zn implantation, the degradation of the n-region should be effected partly by the diffusion of antisite defects into the junction region. Moreover, the degradation of the device can be

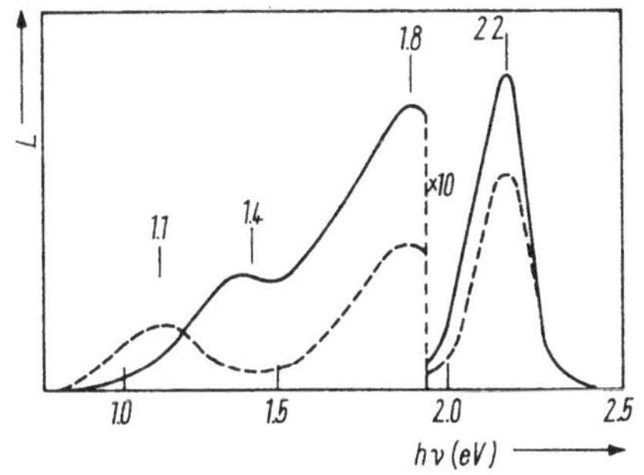

Fig. 2. Room temperature injection luminescence spectra of non-stressed diode (full line) and a diode stressed at $j = 40\ A\ cm^{-2}$ during 1000 h (dashed line)

influenced by the change of the Zn profile as well as by the carrier injection ratio /4/.

The authors are greatly indebted to Dr. sc. G. Oelgart, Humboldt-Universität Berlin, for stimulating discussions.

References

/1/ N. KILLORAN, B.C. CAVENETT, M. GODLEWSKI, T.A. KENNEDY, and N.D. WILSEY, J. Phys. C 15, L723 (1982).

/2/ K.P.O'DONNELL, K.H. LEE, and C.D. WATKINS, Solid State Commun. 44, 1015 (1982).

/3/ A. CHANTRE, G. VINCENT, and D. BOIS, Phys. Rev. B 23, 5335 (1981).

/4/ B. RHEINLÄNDER, G. OELGART, R. REULKE, H. HAEFNER, and R. STEGMANN, phys. stat. sol. (a) 77, 527 (1983).

/5/ G. FERENCZI, (unpublished).

(Received April 28, 1983)

phys. stat. sol. (a) 79, K65 (1983)
Subject classification: 14.3; 22.8

Departamento de Física, Facultad de Ciencias, Universidad de Los Andes, Mérida[1)]

On the Hole Mobility of $CuGaSe_2$

By
S.M. WASIM and G. SÁNCHEZ PORRAS

The growth, electrical and optical properties of $CuGaSe_2$ which belongs to the I-III-VI_2 family of semiconducting compounds and whose binary analog is ZnSe, has been reported /1 to 6/. This compound, which has invariably been produced as p-type with electrical resistivity ranging from 10^{-2} to $10^7\,\Omega$cm, appears to be a possible candidate for photovoltaic devices /7/. Three different acceptor states with ionization energies of about 30 to 50 meV, 150 meV, and 550 meV obtained from different experimental techniques have been reported /8/. A theoretical estimate of the density of states effective mass of the highest valence band in $CuGaSe_2$ has been made which gives $m_p^* = 1.2\ m_o$ /4/. The mobility data of bulk /4/ and thin film samples on GaAs substrates /5/ have been reported. Although their magnitude and temperature dependent behaviour is very similar to what has been reported for other ternary compounds /9/, no detailed analysis has been made. In the present work we analyze the mobility data, already published, to account for the different scattering mechanisms of the charge carriers in $CuGaSe_2$ which also gives an estimate of certain band parameters.

The hole mobility as a function of temperature of bulk and thin film sample No. 83 of $CuGaSe_2$ taken from /4/ and /5/, respectively, are given in Fig. 1. From the temperature dependent behaviour it can be observed that in the bulk sample at lower temperatures the charge carriers are predominantly scattered by ionized impurities, whereas both in the thin film and at higher temperatures in the bulk samples, the main scattering mechanisms are due to acoustic and optical modes. Although in the analysis of the mobility data of p-type semiconductors simultaneous contributions from both the polar and nonpolar optical mode scattering have been considered /10, 11/, it has been

1) Mérida 5101, Venezuela.

suggested /12/ that in this type of material polar optical mode scattering plays a very minor role. Hence, we have only considered the scattering of the holes by ionized impurities, acoustic and nonpolar optical modes.

For the ionized impurity scattering we have used the expression derived by Brooks-Herring for non-degenerate semiconductors /13/:

$$\mu_I = \frac{2}{300} \frac{2^{7/2} \varepsilon_o^2 (kT)^{3/2}}{\pi^{3/2} e^3 (m_n^*)^{1/2} N_I f(x)} (cm^2/Vs) , \tag{1}$$

where the factor two in the numerator is from the overlap integrals for the p-like valence band, N_I is the ionized impurity density, ε_o the low frequency dielectric constant, and m_p^* the density of states effective mass expressed in the units of free-electron mass.

The function f(x) is given by the relation $\ln(1 + x) - x/(1 + x)$ where

$$x = \frac{6 \varepsilon_o m_p^* (kT)^2}{\pi e^2 \hbar^2 p} . \tag{2}$$

By combining the acoustical-mode contribution with the nonpolar optical mode, Wiley and Di Domenico /12/ have obtained an expression given by

$$\mu_{acnpo} = \mu_{ac} S(\Theta, \eta, T) , \tag{3}$$

where Θ is the optical-phonon characteristic temperature and $\eta \approx (E_{npo}/E_{ac})^2$, E_{npo} and E_{ac} being the nonpolar optical and acoustic mode deformation potentials. An analytical approximation to $S(\Theta, \eta, T)$ is found to be

$$S(\Theta, \eta, T) \approx \left\{1 + \left(\frac{\Theta}{T}\right) \eta H / \left[\exp\left(\frac{\Theta}{T}\right) - D\right]\right\}^{-1} , \tag{4}$$

where the constants H and D have been tabulated for each value of η.

For the acoustic-mode scattering mobility, the expression used is /11/

$$\mu_{ac} = \frac{2}{300} \frac{(8\pi)^{1/2} e \hbar^4 \varrho u^2}{3 E_{ac}^2 (m_p^*)^{5/2} (kT)^{3/2}} , \tag{5}$$

where ϱ is the density and u the longitudinal velocity of sound.

The total mobility has been calculated by using Matthiessen's approximation $\mu^{-1} = \mu_I^{-1} + \mu_{acnpo}^{-1}$. For the calculation of μ_I and μ_{acnpo} we have taken for $CuGaSe_2$ $\varrho = 5.45$ g/cm^3, $m_p^* = 1.2\ m_o$, $\varepsilon_o = 13.6$, and $u = 2.47 \times 10^5$ cm/s,

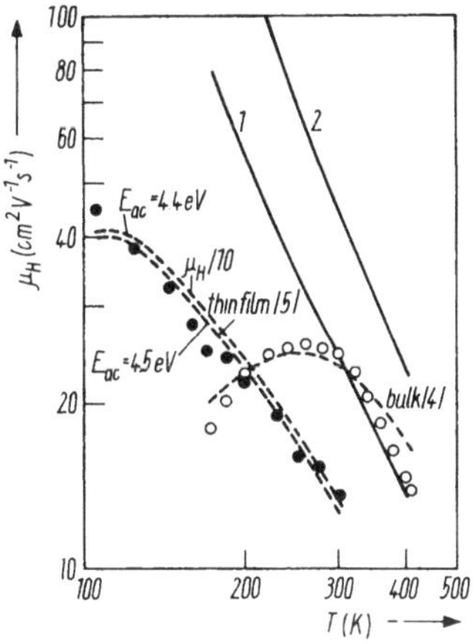

Fig. 1. The hole mobility of $CuGaSe_2$ as a function of temperature with symbols o and • taken from /4/ and /5/, respectively. The curves 1 and 2 represent the calculated values of μ_{acnpo} with E_{ac} = = 5.5 eV, Θ = 360 K, η = 4.0, and E_{ac} = 4.0 eV, Θ = 600 K, η = 6.5, respectively. The dashed lines represent a theoretical fit for the mobility data of /4/ and /5/ with the parameters given in the text

obtained by using the formula $u = (k\Theta_D/\hbar)(V/6\pi^2)^{1/3}$ with Θ_D = 262 K /14/. In the absence of published data for E_{ac} and Θ, the values for its binary analog /15/ were used. Hence, using a rough rule of thumb, the valence band deformation potential has been taken as E_{ac} (valence band) $\approx E_{ac}$(conduction band) - - 6 eV = 5.5 eV. Thus, taking E_{ac} = 5.5 eV, Θ = 360 K, η = 4, and the corresponding values of H and D as 1.34 and 0.914, respectively /12/, the combined mobility μ_{acnpo} was calculated. This is shown in Fig. 1 by the dashed line. It can be observed that in the high temperature region, the calculated value is very nearly identical to what has been reported for bulk $CuGaSe_2$. However, these values of E_{ac} and Θ fail to give any reasonable fit when the ionized impurity scattering is included to explain the low temperature data. For this reason, to get an acceptable fit in the entire temperature range, we have used E_{ac}, Θ, η, and N_I as adjustable parameters. The best result obtained with E_{ac} = 4.0 eV, Θ = 600 K, η = 6.5, and N_I = 1.55×10^{19} cm^{-3} is also shown in the same figure. It should be mentioned that before varying E_{ac} and Θ, the scattering of holes by neutral impurities and space-change effects were also included in the analysis of the data, but the results were not satisfactory. Using the relation $\eta \approx (E_{npo}/E_{ac})^2$ we find $E_{npo} \approx 10.1$ eV, which agrees quite well with 12 eV reported for $CuInSe_2$ /10/. On the other hand, the mobility data of sample No. 83 can be explained without taking into account the nonpolar optical mode scattering with E_{ac} = 4.5 eV and $N_I = 10^{17}$ cm^{-3}. These results suggest that the deformation potential is smaller and the optical-phonon characteristic temperature is higher in $CuGaSe_2$ as compared to its binary analog ZnSe.

The work reported in this paper was supported by the Consejo Nacional de Investigaciones Científicas y Tecnológicas (CONICIT) under project number SI-0253, and by the Consejo de Desarrollo Científico y Humanístico (C.D.C.H.) of the Universidad de Los Andes.

References

/1/ L.S. LERNER, J. Phys. Chem. Solids 27, 1 (1966).

/2/ L. MANDEL, R.D. TOMLINSON, and M.J. HAMPSHIRE, J. Crystal Growth 36, 152 (1976).

/3/ J. STANKIEWICZ, W. GIRIAT, J. RAMOS, and M. VECCHI, Solar Energy Mater. 1, 369 (1979).

/4/ L. MANDEL, R.D. TOMLINSON, M.J. HAMPSHIRE, and H. NEUMANN, Solid State Commun. 32, 201 (1979).

/5/ B. SCHUMANN, A. TEMPEL, G. KÜHN, H. NEUMANN, NGUYEN VAN NAM, and T. HÄNSEL, Kristall und Technik 13, 1285 (1978).

/6/ B. TELL, J.L. SHAY, and M.H. KASPER, J. appl. Phys. 43, 2469 (1972).

/7/ N. ROMEO, G. SHERVEGLIERI, L. TARRICONE, and C. PAORICE, Appl. Phys. Letters 30, 108 (1977).

/8/ M.P. VECCHI, J. RAMOS, and W. GIRIAT, Solid State Electronics 21, 1609 (1978).

/9/ S.M. WASIM, G.S. PORRAS, and R.D. TOMLINSON, phys. stat. sol. (a) 71, 523 (1982).

/10/ T. IRIE, S. ENDO, and S. KIMURA, Japan. J. appl. Phys. 18, 1303 (1979).

/11/ D.C. LOOK and J.C. MANTHURUTHIL, J. Phys. Chem. Solids 37, 173 (1976).

/12/ J.D. WILEY and M. DI DOMENICO, Jr., Phys. Rev. B 2, 427 (1970).

/13/ S.M. WASIM, G. MARCANO, and G.S. PORRAS, Japan. J. appl. Phys. 19, Suppl. 19-3, 77 (1980).

/14/ K. BOHMHAMMEL, P. DEUS, G. KÜHN, and W. MÖLLER, phys. stat. sol. (a) 71, 505 (1982).

/15/ D.L. RODE, Phys. Rev. B 2, 1012 (1970).

(Received June 17, 1983)

phys. stat. sol. (a) 79, K69 (1983)

Subject classification: 14.3.3; 22.2

Institute of Radio Engineering and Electronics, Czechoslovak Academy of Sciences, Prague[1])

Determination of p-n Junctions in InP/GaInAsP Laser Heterostructures Using a Profilometer

By

J. WALACHOVÁ and D. NOHAVICA

The knowledge of the correct position of the p-n junction in InP/GaInAsP laser heterostructures is very important for the fabrication of good lasers emitting near 1.3 μm /1 to 3/. The position of the p-n junction in this heterostructure has been studied in /2/ by a scanning electron microscope. The method presented here is less exact, but it is simpler. It utilizes measurements of the doping and/or the composition profile of the heterostructure /1, 4/. When profiling a lightly-loaded metal probe is tracked slowly down the bevelled heterostructure. A fixed bias is applied between the probe and the bevelled heterostructure and the current through the probe is measured. This current is displayed on a chart recorder. The profiles in /1, 4/ have been measured only for one polarity (negative) of the probe. We find that the profiling for both polarities of the probe is very useful and by comparing these two profiles one can find the position of the p-n junction in the heterostructure.

Our samples of the laser heterostructures are prepared by liquid-phase epitaxy on (100) n-type InP substrates ($N_D = 2\times10^{18}$ cm^{-3}). On the substrate the following layers are grown from the substrate side: n^+-InP (tellurium-doped), n-GaInAsP (tin-doped or no specially doped active layer), and p-InP (zinc-doped). This structure has two heterojunctions. We denote as first and second heterojunction that nearer and farther from the substrate, respectively. It is expected that the p-n junction in the grown structure coincides with the second heterojunction, but the p-n junction can be displaced to the direction of the substrate because of zinc contamination of the n-type layers during the growth /1, 2/.

We use in our experiments a 3×10^{-3} N loaded tungsten carbide probe (diameter ≈10 μm), a fixed bias of 0.8 V, and the samples are bevelled at 2°. The

1) Lumumbova 1, 182 51 Praha 8, Czechoslovakia.

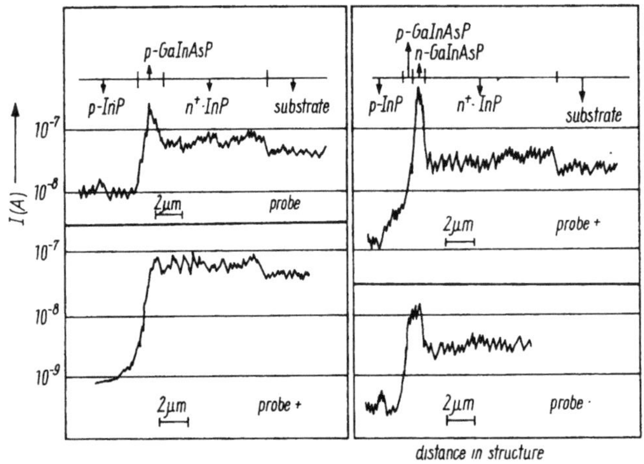

Fig. 1. The profiles of the current through the probe tracked down the bevelled InP/GaInAsP heterostructure. The different behaviour of the profiles of the n-type and p-type parts of the heterostructure for the two probe polarities is apparent. a) The p-n junction coincides with the first heterojunction from the substrate side, b) the p-n junction is in the GaInAsP layer

bevel is prepared by polishing the sample on a glass plate covered by a tissue using 3 µm diamond paste. The layers of the heterostructure are electrically connected with the substrate by a silver paste. The probe leaves a shallow track approximately 3 µm wide in InP.

We find that the probe forms a diode with the semiconductor which has forward and reverse characteristics like a Schottky diode. This explains the behaviour of the measured profiles when the polarity of the probe is changed from minus to plus. The profile is going down for the p-type part of the heterostructure and it is going up for the n-type part. Hence we can distinguish the n-type and the p-type parts of the heterostructure and the position of the p-n junction can be approximately determined. Two examples of such measurements are in Fig. 1a and b. There the profiles of two laser heterostructures are shown. As we examine the profiles in Fig. 1a from the substrate side we can see that the change of the profile with the change of the polarity of the probe for the substrate and for the n^+-InP layer is of the same order as the noise. The profile of the following layers - GaInAsP and p-InP is very sensitive to the change of the polarity of the probe and the profile is going down when the polarity is changed from minus to plus. Therefore we can conclude that the GaInAsP layer is also p-type. The p-n junction in this case is displaced from the expected position and coincides with the first heterojunction. We explain it with contamination of the growing solution of GaInAsP with zinc vapor from the growing solution of p-InP /1, 2/. This contamination is heavy because the GaInAsP layer is grown as not specially doped and the p-type layer is grown as zinc-doped to a concentration as high as $N_A \approx 2x10^{18}$ cm^{-3}. The profiles of the second laser heterostructure are shown in Fig. 1b. There the profiles of the n^+-InP layer and p-InP layer behave like the profiles in Fig. 1a discussed above, but the profiles of the GaInAsP layer behave differently. The part of the GaInAsP layer profile nearer to the substrate is going up and the remaining part is going down when the polarity of the probe is changed from minus to plus. This means that the p-n junction is in the GaInAsP layer. The displacement of the p-n junction from the expected position is not so large as in Fig. 1a because the concentration of zinc in p-InP is smaller and the GaInAsP layer is tin-doped. In this case we explain the displacement of the p-n junction with diffusion of zinc from the p-InP layer during its growth. The composition of the active layer in Fig. 1a and b is $Ga_{0.23}In_{0.77}As_{0.54}P_{0.46}$ and $Ga_{0.27}In_{0.73}As_{0.62}P_{0.38}$, respectively. The thickness resolution for abrupt junctions and used geometry (3 μm wide track and 2° bevel) was estimated as better than 0.2 μm but the penetration of the probe into the semiconductor and the depletion layer under the contact of the probe with the semiconductor can make it worse. The best observed thickness resolution was 0.23 μm. The compensated region of the p-n junction can be well resolved in some cases.

For determination of the p-n junction in heterostructures with thinner layers a 1° or $30'$ bevel is needed.

The method described above can be used in a modified version also for the determination of p-n junctions in GaAlAs/GaAs laser heterostructures.

We would like to thank J. Těminová for technical assistance.

References

/1/ S.D. HERSEE, A.C. CARTER, R.C. GOODFELLOW, G. HAWKINS, and I. GRIFFITH, Solid State Electron. Devices 3, 179 (1979).

/2/ J.J. COLEMAN and F.R. NASH, Electronics Letters 14, 558 (1978).

/3/ Y. ITAYA, Y. SUEMATSU, S. KATAYAMA, K. KISHINO, and S. ARAI, Japan. J. appl. Phys. 18, 1795 (1979).

/4/ R.C. GOODFELLOW, A.C. CARTER, R. DAVIS, and C. HILL, Electronics Letters 14, 328 (1978).

(Received July 11, 1983)

phys. stat. sol. (a) 79, K73 (1983)
Subject classification: 12.2 and 14.4.1; 22.8.1

Institute of Semiconductors, Academy of Sciences of the Ukrainian SSR, Kiev[1]

Determination of Nonlinear Elastic Coefficients by Resonant Methods

By

A. A. DEMIDENKO, A. P. ZDEBSKII, V. N. PISKOVOI, and A. U. SAVCHUK

Investigations of elastic nonlinearities in ferroelectric crystals are interesting due to the large amount of information about lattice dynamics, on the one hand, and the wide applications of nonlinear properties of crystals in signal processing schemes, on the other hand. There are different methods for studying elastic nonlinearities of solids /1/. From the point of view of experimental sensitivity resonant methods are more preferable and, in particular, measurements of the fundamental frequencies of specimens versus alternating current amplitude in the so-called resonance-antiresonance methods /2/.

In this note the method of calculation is proposed and experimental investigations of third-order elastic coefficients are carried out for polycrystalline $BaTiO_3$ with additions of CeO_2 (temperature of ferroelectric phase transition $T_c = 33\ ^oC$) in the paraelectric region where this material is non-piezoelectric. After applying a constant electric field E_o to the specimen it may be regarded as a transverse-isotropic one with axis of symmetry (index 3) parallel to E_o.

The ac field $\vec{E}_3 \parallel \vec{E}_o$ excites two modes: a) longitudinal oscillations of a bar in the transverse electric field, and b) longitudinal thickness oscillations of thin layers (plates).

For a theoretical description of this experimental situation we start from the equations of state

$$u_i = S^E_{ij}(1 + i/Q)\sigma_j + S^E_{ijk}\sigma_j\sigma_k + d_{3j}(E_o)E_3 \ , \tag{1}$$

where u_i and σ_i (i = 1 to 6) are components of the strain and stress tensors, respectively, S^E_{ij} and S^E_{ijk} are components of the tensor of elastic coefficients of second and third order, respectively, $d_{3j}(E_o)$ are components of the piezo-

---

1) prospekt Nauki 115, 252650 Kiev 28, USSR.

module tensor induced by a constant electric field, they are approximately proportional to E_o /3/, Q is the mechanical quality factor. Here, only the physical nonlinearity determined by third-order elastic coefficients in (1) is taken into account. The geometric nonlinearity is neglected (in particular u_i - the usual strain tensor in the linear theory). Such an approximation is justified because in the vicinity of the temperature of the ferroelectric phase transition (see /2/ and our experimental results below) $|S_{111}| \gg S_{11}^2$ whereas in the other cases they are usually of the same order. From the equations of motion of the resonator and the equations of state (1) it is possible to get relations which allow the determination of the nonlinear elastic coefficients as a function of the resonance (antiresonance) frequency shift $\Delta\omega_r$ ($\Delta\omega_a$) in an ac field of amplitude E_3^o. The geometrical resonance conditions in the resonance-antiresonance method allow to find solutions of the nonlinear wave equation assuming the spatial dependence of alternating strains and fields of the linear theory to be preserved /4/.

The component of the third-order elastic coefficient is determined from longitudinal oscillations of the bar in a transverse field:

$$S_{111}^E = \frac{3\pi^2 \sqrt{3}\,(S_{11}^E)^2}{32\,Q\,E_3^o\,d_{31}\,(E_o)} \sqrt{\frac{\Delta\omega_r}{\omega_r}}\,. \tag{2}$$

Similar problems have been solved for piezoelectric crystals with constant value of the piezomodulus, and for a non-piezoelectric medium with electrostrictive electromechanical coupling in the absence of a constant electric field /2/. Our formula (2) differs from an analogous one obtained in /2/ by the sign of the resonance frequency shift $\Delta\omega_r$ apart from an unessential numerical multiplier. The positive sign of $\Delta\omega_r$ in formula (2) is confirmed by our experimental results.

From the longitudinal thickness oscillations of thin plates one finds

$$S_{tt} = \frac{3\pi\sqrt{3}\,S_t^2\,A\,\omega_a\,\varepsilon_t}{128\,Q\,I\,d_t(E_o)} \sqrt{\frac{\Delta\omega_a}{\omega_a}}\,, \tag{3}$$

where A is the area of the plate, I the amplitude of the alternating current of the resonator at the antiresonance frequency. The index t in (3) marks the considered mode of plate thickness oscillations. Their coupling with the tensor

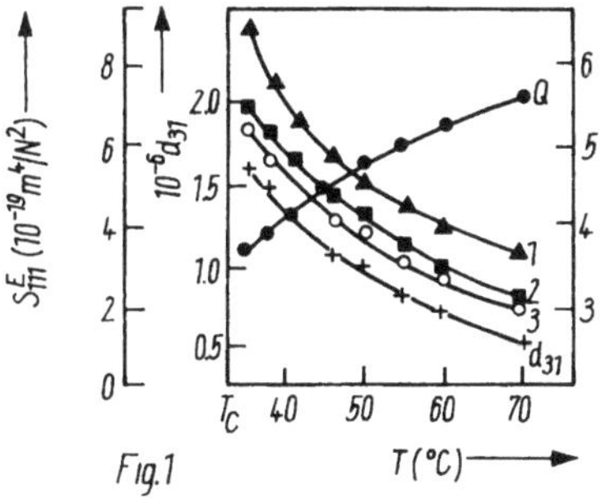

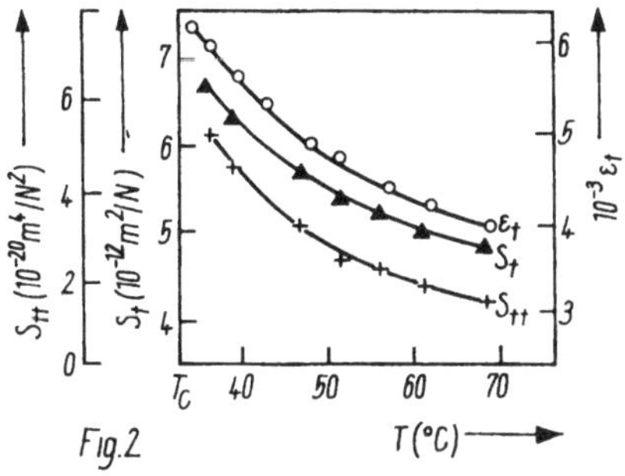

Fig. 1. Temperature dependence of S^E_{111} for different values of electrical fields E_o: (1) 2.5×10^5, (2) 5×10^5, (3) 7.5×10^5 V/m; Q quality factor; piezomodulus d_{31} for an electric field $E_o = 5\times10^5$ V/m

Fig. 2. Temperature dependence of second-order S_t and third-order elastic compleances S_{tt} and dielectric permittivity ($E_o = 5\times10^5$ V/m)

components in (1) for the case of a transverse isotropic medium is

$$S_t = S_{33} - 2\,\frac{S_{13}^2}{S_{11} + S_{12}} \; ;$$

$$S_{tt} = S_{333} - 4S_{331}\,\frac{S_{13}}{S_{11} + S_{12}} + 2\,\frac{S_{13}^2(S_{311} + S_{312})}{(S_{11} + S_{12})^2} \; ,$$

$$d_t = d_{33} - \frac{2S_{13}d_{31}}{S_{11} + S_{12}} \; ; \qquad \varepsilon_t = \varepsilon_3 - \frac{8\pi d_{31}^2}{S_{11} + S_{12}} \; . \tag{4}$$

ε_3 is the zz component of the dielectric permittivity. The third-order coefficients in (2) and (3) were determined from measured values of the resonance ω_r and antiresonance ω_a frequencies and values of the impedance at these frequencies. The corresponding formulas for these quantities were received in /3/ and /5/.

The temperature dependence of S^E_{111} is plotted in Fig. 1 for different electric fields together with the piezomodulus d_{31} and the mechanical quality factor Q.

On Fig. 2 the temperature dependence of S_t and S_{tt} and the dielectric constant ε_t at a field $E_o = 5x10^5$ V m^{-1} are shown.

From these data one can see that the third-order elastic constants strongly depend on temperature near the ferroelectric phase transition as the linear characteristics of materials do.

References

/1/ L.K. ZAREMBO and V.A. KRASILNIKOV, Uspekhi fiz. Nauk 102, 549 (1970).

/2/ H. BAIGE and G. SCHMIDT, Izv. Akad. Nauk SSSR, Ser. fiz. 39, 970 (1975).

/3/ A.A. DEMIDENKO, A.P. ZDEBSKII, V.N. PISKOVOI, and N.S. CHERNAYA, Radiotekhnika i Elektronika 21, 285 (1976).

/4/ YU.A. MITROPOLSKII and B.I. MOSEENKOV, Asimptoticheskie resheniya uravnenii v chastnikh proisvodnik, Vishaya shkola, 1976.

/5/ A.A. DEMIDENKO, V.N. PISKOVOI, and NGUYEN TCHI CHI, Ukr. fiz. Zh. 20, 1028 (1975).

(Received May 25, 1983)

phys. stat. sol. (a) 79, K77 (1983)
Subject classification: 13.4 and 20.3; 22.5.2

Health Physics Division, Bhabha Atomic Research Centre, Bombay[1)]

Fractional Order of Kinetics and Its Variation with Trap Occupancy in LiF TLD-100

By

S. P. KATHURIA and C.M. SUNTA

Moharil /1/ has used our data /2/ and recomputed the TL (thermoluminescence) parameters. Based on his computations he agreed with the fractional order proposed by us but contradicted our conclusion that the numerical value of the order of kinetics changes with trap occupancy, until the intensity of the glow peak has fallen to less than 20% of its saturated value at 165 $^{\circ}$C. In this note we point out the errors in Moharil's treatment /1/ which led him to the paradoxial result like negative value of activation energy and thereby contradict our conclusion.

The TL intensity (I) equation for a glow peak following general order kinetics as proposed by May and Partridge /3/ is

$$I = C\, s'\, n^{b} \exp(-E/kT) \ , \tag{1}$$

where C is a constant, E the activation energy, s' the pre-exponential constant, b the order of kinetics, n the number of filled traps at the temperature T, and k the Boltzmann constant. On differentiating (1) with respect to time, the following relation is obtained /1/:

$$\frac{bI^2}{dI/dt} = C\, n \ . \tag{2}$$

Using the initial (0 to 3 min) isothermal decay data from our results /2/ for the saturation exposure of 10^5 R (when b is varying), Moharil calculated n/n_s values at various intervals of time with the help of (2). n_s is the number of filled traps with saturation exposure. He also obtained the following empirical relation between b and n/n_s:

$$b = b_o \left[1 - \frac{b_o - 1}{b_o} \exp\left\{ \frac{1 - n_s/n}{5} \right\}^{6} \right] , \tag{3}$$

where $b_o = 1.6$ is the steady value of the order of kinetics /2/.

1) Bombay 400 085, India.

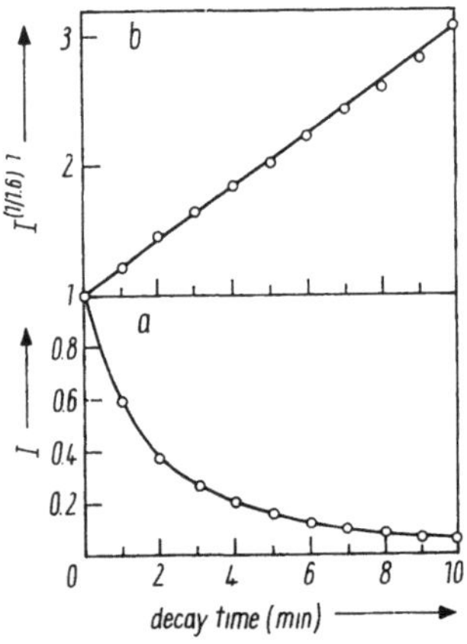

Fig. 1. a) Isothermal decay of glow peak V at 165 ^{0}C for γ-exposure of 10^4 R and b) plot of data of A in the form $I^{(1/b)-1}$ versus time for b = 1.6

It must be pointed out here that (2) is obtained by assuming b to be constant. Moharil /1/ has, however, used this equation for treating the isothermal decay data in which b is varying. The n/n_s values thus calculated by him and hence the empirical relation between b and n/n_s (equation (3) above) are both erroneous.

Equation (2) can, however, be applied to the data obtained with non-saturation exposures which are less than 20% of saturation exposure. Such data were given by us in our earlier paper /2/. Fig. 1 shows the isothermal decay of peak V at 165 ^{0}C for such an exposure (10^4 R) after correcting for the contribution from peak VI /4/. Also shown in the figure is the straight line plot between $I^{(1/b)-1}$ and time for b = 1.6 indicating that the peak V in TLD-100 has the constant b value 1.6 for the exposure of 10^4 R. Using data of Fig. 1 and equation (2), n/n_0 values are calculated at every 1 min interval and given in Table 1. n_0 is the number of filled traps at t = 0. It is seen from the table that I/I_0 and n/n_0 do not vary proportional to each other with time as is the case in the first-order kinetics. I_0 is the TL intensity at t = 0. The non-proportionality between I/I_0 and n/n_0 may be attributed to progressive increase in retrapping as the trap occupancy (n/n_0) decreases. It may be noted from the table that after 4 min when the TL intensity has fallen to almost 20%, 62% of the initially trapped charges are released.

Variation of the order of kinetics during the TL run has been reported by some other workers, too /3, 5/. De Muer /5/ suggested the change in kinetics to be due to variation in the occupancy of the traps while the sample is being heated for measuring the TL. According to him the order of kinetics is 1 in the beginning of the peak when $n/n_s \leqq 1$ and it changes slowly to 2 when $n/n_s \ll 1$ at the end of the peak. However, it has not been easy to detect experimentally this variation in the order of kinetics. We could observe the variation in the order of kinetics in TLD-100 only for the saturation exposure of 10^5 R by the isothermal decay method /2/. This variation could not be observed from the

Table 1

Relative TL intensity of peak V and fractions of trapped charges remaining after different intervals of time at 165 °C

time (min)	TL intensity I/I_o	n/n_o	$\frac{I/I_o}{n/n_o}$
0	1	1	1
1	0.590	0.542	1.09
2	0.370	0.475	0.78
3	0.265	0.426	0.62
4	0.197	0.381	0.52
5	0.153	0.348	0.44

glow peak shape method which gives only an average or effective order of kinetics during the TL run. The glow peak shape has another disadvantage that the peak shape can be distorted by the effect of thermal quenching /6/. The TL intensity with non-first-order kinetics is represented by the empirical relation given in equation (1). In this relation b represents the effective order of kinetics /7/ and is assumed to be constant during the TL run. This relation cannot be applied to the data representing varying order of kinetics as stated clearly by the original authors May and Partridge /3/. Therefore, getting a paradoxical negative E value for peak V /1/ by substituting varying b values in equation (1) is not surprising.

The suggestion of Moharil /1/ that the initial concentration of filled traps may not be constant due to formation/dissociation of impurity aggregates during isothermal decay study resulting in an apparent variation in the order of kinetics is also ruled out as clearly shown by us /4/ that there is no change in TL response upto 20 min of thermal treatment at 165 °C. Thus the arguments presented above make it clear that all the objections raised by Moharil /1/ against the variation in the order of kinetics are not valid.

References

/1/ S.V. MOHARIL, phys. stat. sol. (a) 68, 413 (1981).
/2/ S.P. KATHURIA and C.M. SUNTA, J. Phys D 12, 1573 (1979).
/3/ C.E. MAY and J. PARTRIDGE, J. chem. Phys. 40, 1401 (1964).

/4/ S.P. KATHURIA and C.M. SUNTA, J. Phys. D 15, 497 (1982).

/5/ D. DE MUER, Physica 48, 1 (1970).

/6/ C.M. SUNTA and V.N. BAPAT, Proc. 2nd Specialist Seminar on TL Dating, Oxford, September 1980, published in PACT Journal, Council of Europe, Strasbourg 1982.

/7/ R. CHEN, M. KRISTIANPOLLER, and Z. DAVIDSON, J. Lum. 23, 293 (1981).

(Received June 30, 1982)

phys. stat. sol. (a) 79, K81 (1983)
Subject classification: 13.4 and 20.3; 22.5.1

Department of Physics, Nagpur University[1)]

Reply to "Fractional Order of Kinetics and Its Variation with Trap Occupancy in LiF TLD-100" by S. P. KATHURIA and C.M. SUNTA /3/

By
S.V. MOHARIL

Kathuria and Sunta /1/ (hereinafter abbreviated as KS) concluded from the isothermal decay of peak V of LiF TLD-100, that when traps are filled to saturation concentration, the 'order of kinetics' (b) varied from 1 to 1.6 for decay times upto 3 min. A steady value (1.6) is obtained thereafter. This variation was not apparent in the glow curve, when Chen's method was used for calculating the trapping parameters. The reason given was that a sufficient number of traps were emptied and a constant b value was attained before any of the temperatures T_1, T_m, or T_2 could be reached.

In /2/ I have examined these results in greater details and shown that the variation of b should have been observed at least upto temperature T_1 (b at T_1 was calculated to be 1.187). The purpose of /2/ was to show that the results on isothermal decay imply a similar variation at least upto temperature T_1' in the glow curve, and to show that these changes were not observed; the constant value of b (1.55 ± 0.05) was observed down to temperature T_1' (quarter intensity point on low temperature side of T_m), and hence the anomaly.

KS have raised some objections against the procedure in /2/. This will be taken up later. First it will be shown that arguments of /2/ can be substantiated by following an alternative procedure. Instead of calculating the exact variation of b with temperature, the lowest temperature at which a substantial variation from b = 1.6 should be observed will be computed.

We start with the fact that b = 1.3 (significantly less than the constant value 1.6) was observed till 1.75 min /1/. Thus, the variation in b should be observed at least upto such n/n_s values as prevailing at 1.75 min are reached. Now, after about 16 min of decay, the intensity goes down to 1/50-th of the initial value. Had the system followed first-order kinetics, this would have meant that 98% of the initially present traps are emptied by this time. In the

1) Nagpur 440010, India.

second-order kinetics, this corresponds to emptying 86% of the originally filled traps. As the order of kinetics was always between 1 and 2, the number of traps emptied is between 86 and 98% of the initial number (n_s). Thus, the area A_o under the decay curve (plotted between I and t on linear scale) upto 16 min is such that

$$c' A_o/0.98 \leqq n_s \leqq c' A_o/0.86 \quad , \tag{1}$$

where c' is an arbitrary constant.

The area A_s which would have been proportional to n_s ($n_s = c' A_s$) is such that

$$A_o/0.98 \leqq A_s \leqq A_o/0.86 \ . \tag{2}$$

From curve of KS /1/, the decay was re-drawn to linear scale. It gave $A_o =$ $= 204\ cm^2$. Hence from (2)

$$208.16 \leqq A_s \leqq 237.21 \ . \tag{3}$$

Again the area under the curve between 0 to 1.75 min was 104.25 cm^2, hence the remaining area after 1.75 min (A_s area upto 1.75 min) is such that

$$103.91 \leqq A_{1.75} \leqq 132.96 \ . \tag{4}$$

From (3) and (4)

$$0.499 \leqq A_{1.75}/A_s \leqq 0.560 \ . \tag{5}$$

As the number of traps filled at instant t is proportional to the remaining area after t, it is clear that $n/n_s = A_t/A_s$, and hence

$$0.499 \leqq (n/n_s)_{1.75} \leqq 0.560. \tag{6}$$

Thus, a deviation from b = 1.6 should be observed at least till $n/n_s \geqq 0.499$. From the area under glow curve measurements /2/ we find that (6) is satisfied for T = 460.5 to 463 K. Thus, b = 1.3 should have been observed at least upto 460.5 K. Hence, the statement of /2/ that a substantial variation from constant value b = 1.6 should have been observed at least upto T_1 (450.5) is valid. In /2/ it has already been proved that no variation is observed down to temperature T_1' .

The point raised in /2/ is, thus, made here, using an alternative method. KS have objected /3/ against the procedure of /2/. Thus, in /2/, the starting equation was

$$I = - c\,\frac{dn}{dt} = c s' n^{b} \exp(-E/kT) \qquad (7)$$

after differentiating and re-arranging it was shown

$$b\,I^{2}\,dI/dt = c\,n \ . \qquad (8)$$

It was implied that b represents an average value over some period and thus is constant (or is very slowly varying function of time compared to n). Ratios n/n_s for four discrete t values were obtained by substituting the b values for these times as given in /1/. KS object /3/ that (8) is inadequate to handle the variation in b because b was treated constant during differentiating (7). Equation (8) they argue, is erroneous, and so are the conclusions drawn from it.

It is very interesting to note that KS also have used a parallel procedure in analysis of their data /1/, and the objection they have raised applied to their analysis as well. Thus, after integrating (7) between fixed limits, obtaining the value of n in terms of n_o and then substituting back in (7) one can obtain an equation of the form

$$I^{1/b\,-1} = At + B \ . \qquad (9)$$

KS have calculated b values from the isothermal decay of TL by making use of (9) and not the original equation of May and Partridge (7). Starting from (9) they show that b varies. It is to be noted that while deriving (9) from (7), b was treated constant during the integration. If b is variable, as KS have later claimed then one does not obtain (9) at all. Thus, in that case the quantity KS have measured, and which they have shown to vary does not correspond to the order of kinetics of (7).

Thus, it is shown that the procedures used in /1/ and /2/ are equally valid or equally erroneous.

In my opinion, the controversy over the 'order of kinetics' reflects on the shortcomings of (7) which is only an empirical relation which often fits to the experimental results. It has been shown elsewhere /4/, that it is more useful to make the equation of Antonov-Romanovski a starting point. The retrapping ratio obtained from this has much more direct connection with the physical quantities than the parameter b of (7).

References

/1/ S.P. KATHURIA and C.M. SUNTA, J. Phys. D 12, 1573 (1979).
/2/ S.V. MOHARIL, phys. stat. sol. (a) 68, 413 (1981).
/3/ S.P. KATHURIA and C.M. SUNTA, phys. stat. sol. (a) 79, K77 (1983).
/4/ S.V. MOHARIL, phys. stat. sol. (a) 73, 509 (1982); J. Phys. D, in the press.

(Received November 18, 1982)

phys. stat. sol. (a) 79, K85 (1983)
Subject classification: 14.3; 20.1; 22.1.2

Sektion Physik der Friedrich-Schiller-Universität Jena[1)]

Dependence of the Ambipolar Diffusion in Silicon on the Carrier Density

By
H. BERGNER and V. BRÜCKNER

The interband absorption of ultrashort light pulses can lead to an extremely high conduction band population by free non-equilibrium charge carriers which results in changes of the optical properties of the semiconductors. The dynamics of the carriers has been the subject of experimental and theoretical investigations /1/. If the carrier density is initially produced in a shallow layer it becomes important in many cases to take into account the spatial and temporal evolution of the carrier distribution through diffusion. Using the excite-and-probe technique it is possible to perform time-resolved reflectivity measurements on a picosecond time scale where recombination in undoped monocrystalline silicon is negligible /2/. We have utilized this property to study experimentally for the first time the diffusion process of the carriers into the bulk in silicon at extremely high carrier densities.

The silicon samples used in our experiments are disks of about 200 μm thickness. The acceptor concentration was less than 10^{13} cm^{-3}, i.e. the silicon was of high resistivity (about 2000 Ωcm). The 25 ps Nd-YAG laser system as well as the experimental set-up were described elsewhere /2/.

The temporal behaviour of the reflectivity $R(\Delta t)$ at the excitation energy density of about 2.9 mJ cm^{-2} is shown in Fig. 1. In agreement with the Drude model /3/ the minimum reflectivity depends on the excitation energy density. In order to compare the time behaviour of the sample reflectivity at different excitation levels we have calculated the normalized reflectivity

$$\bar{R}(\Delta t) = \frac{1 - R(\Delta t)}{1 - R_{min}} , \qquad (1)$$

where R_{min} is the minimum reflectivity of the sample. The dependence of the normalized reflectivity $\bar{R}$ on the delay time between excite and probe pulses at three different excitation levels is shown in Fig. 2.

1) Max-Wien-Pl. 1, DDR-6900 Jena, GDR.

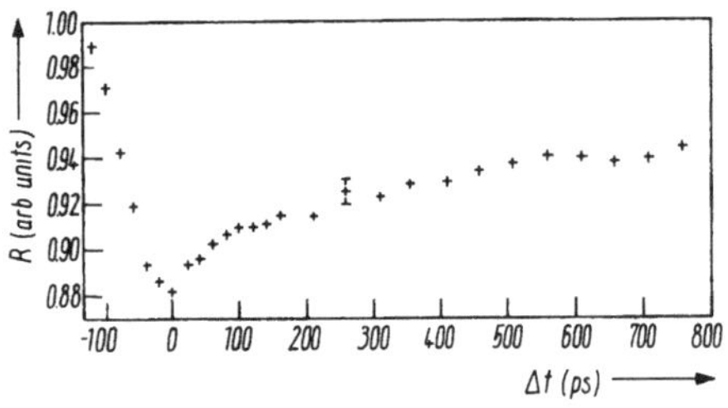

Fig. 1. Time-resolved reflectivity of monocrystalline silicon (excitation energy density 2.9 mJ/cm^2)

In order to explain these results numerical calculations were performed. The temporal and spatial behaviour of the non-equilibrium charge carriers can be described by the continuity equation considering generation and diffusion /2/. Taking into account the pulse duration of the exciting pulse of the second harmonic (20 ps) and the typical absorption coefficient $\alpha = 8\times10^3$ cm^{-1}, numerical calculations were performed for different diffusion constants D. Comparing these results with the experimentally obtained normalized reflectivity the best fit was found at diffusion constants D of 18, 30, and 60 cm^2 s^{-1} (solid lines in Fig. 2) at excitation energy densities of 1.4, 2.9, and 6.6 mJ/cm^2, respectively. At an excitation energy density of 0.3 mJ/cm^2 no difference in the temporal behaviour in comparison with 1.4 mJ/cm^2 excitation was found, therefore D = 18 cm^{2}s^{-1} is obtained.

It should be mentioned that diffusion occurs in the presence of density gradients, and a given experiment will not be able to obtain the diffusion coefficient for a particular density. Therefore, our calculated ambipolar diffusion coefficient D is an effective value. For an exact solution we have to solve the inhomogeneous continuity equation /2/ with a carrier concentration dependence of the current density vector.

Taking into account Fresnel reflection and using the value of the absorption constant mentioned above the generated carrier density can be calculated from the excitation energy density. It was shown in a previous paper /2/ that there is no influence of Auger recombination on the subnanosecond time scale up to carrier densities in silicon of about 5×10^{19} cm^{-3}. The influence of Auger recombination at the highest excitation energy density used (corresponding to $N = 8\times10^{19}$ cm^{-3}) results in an effective diffusion constant less than our calculated one. In order to compare our experimental results with the theoretical ones we will use the dependence of the ambipolar diffusion coefficient on the carrier density as presented in /1, 4, 5/. At high excitation ($N \approx 10^{20}$ cm^{-3}) the semiconductor is degenerated /3/, therefore, the ambipolar diffusion co-

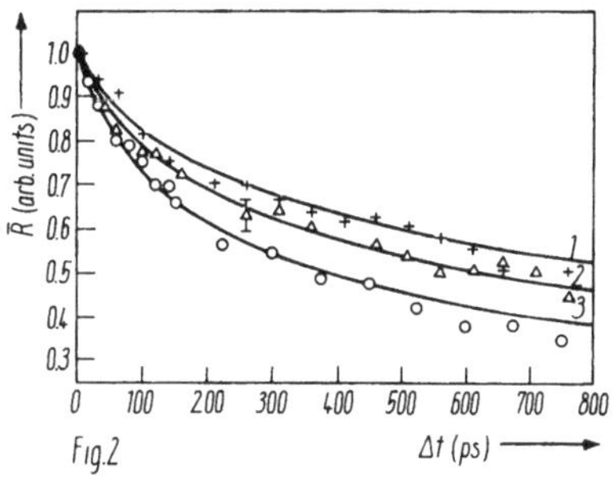

Fig. 2

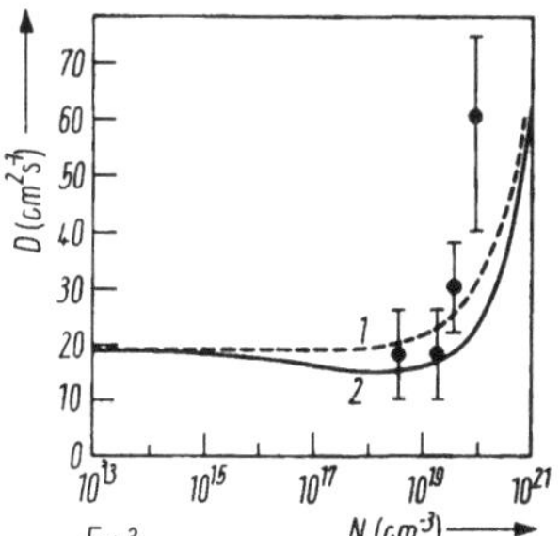

Fig. 3

Fig. 2. Normalized reflectivity dependence on excitation energy density; + $E = 1.4\ mJ/cm^2$; Δ $E = 2.9\ mJ/cm^2$; o $E = 6.6\ mJ/cm^2$. Solid lines represent the numerical calculation (1) $D = 18.1\ cm^2s^{-1}$; (2) $30\ cm^2\ s^{-1}$; (3) $60\ cm^2\ s^{-1}$

Fig. 3. Density dependence of the ambipolar diffusivity in Si at T = 300 K. The lines represent the calculations of Young and van Driel /5/

efficient in the degenerated case D_{deg} differs from the non-degenerated one (D_{nond}). If the carriers can be considered to move independently, the Boltzmann transport theory in the relaxation-time approximation /3/ predicts /1/

$$D_{deg}/D_{nond} \propto N^{1/3}$$

for $N \gg 10^{18}\ cm^{-3}$ at room temperature (curve 1 in Fig. 3).

At high carrier densities the independent-carrier description will break down as carrier exchange and correlation effects become important /4, 5/. Taking into account many-body effects the dependence of the ambipolar diffusion coefficient on the carrier density (up to $10^{21}\ cm^{-3}$) was calculated for Si at room temperature /5/ (curve 2 in Fig. 3). Experimental difficulties to obtain many-body carrier effects were pointed out in /5/. The dependence of the ambipolar diffusion coefficient fitted from our experimental results to the carrier density at the surface is shown in Fig. 3 in comparison with the theoretically predicted ones. As it is seen from Fig. 3, the experimental results are in a reasonable agreement with the theoretical ones taking into account all difficulties of the experimental and theoretical procedure.

References

/1/ D.H. AUSTON and C.V. SHANK, Phys. Rev. Letters 32, 1120 (1974).

/2/ H. BERGNER and V. BRÜCKNER, Opt. Quantum Electronics 15, in the press.

/3/ V.L. BONCH-BRUEVICH and S.G. KALASHNIKOV, Halbleiterphysik, Verlag der Wissenschaften, Berlin 1982.

/4/ M. WAUTELET, J. Phys. C 14, 881 (1981).

/5/ J.F. YOUNG and H.M. VAN DRIEL, Phys. Rev. B 26, 2147 (1982).

(Received June 7, 1983)

phys. stat. sol. (a) 79, K89 (1983)
Subject classification: 10.2 and 20.3; 2; 22.3

A.F. Ioffe Physico-Technical Institute, Academy of Sciences of the USSR, Leningrad[1] (a) and
N.S. Kurnakov Institute for General and Inorganic Chemistry, Academy of Sciences of the USSR, Moscow (b)

The Effect of Low Oxygen Concentrations on the Photoluminescence of Vitreous Selenium

By
B.T. KOLOMIETS (a), T.N. MAMONTOVA (a), A.V. CHERNYSHOV (a), G.Z. VINOGRADOVA (b), and N.V. DEMOKRITOVA (b)

Investigation of chalcogenide vitreous semiconductors (ChVS) started with studies of binary, ternary, and more complex compounds, with As_2S_3 and As_2Se_3 considered as model ChVS. There are ample grounds to class glassy selenium (g-Se) among the model ChVS, its study being of particular interest since in contrast to other ChVS it is atomic in structure and, besides, it makes part of many vitreous semiconductors. It is possibly this consideration that has initiated investigation of the photoluminescence (PL) phenomena in this material. While photoluminescence was shown to exist in g-Se in the very first experiments /1, 2/, these results have not obtained unambiguous interpretation of the probable mechanism involved because of discrepancies in the experimental data. Considering that the disagreement between both the data and their interpretation could be due to differences in the concentration of impurities present, in particular, of oxygen, which could be implied from the afore mentioned papers /1, 2/, we have undertaken a study of PL in g-Se samples containing an exactly known amount of oxygen. The dynamic version of vacuum distillation was used for efficient purification of g-Se from oxygen. Preset amounts of oxygen were introduced into g-Se during the vacuum preparation of samples with different O_2 content by alloying purified industrial grade selenium containing 5 to 8 ppm O_2 and SeO_2 at 300 °C for 48 h under stirring with a subsequent quenching in air /3/.

The oxygen content in selenium was determined from the variation of the SeO_2 absorption band intensity at 930 cm^{-1} /4, 5/.

The PL was studied on bulk g-Se samples with oxygen content ranging from 2×10^{-4} to 2×10^{-2} wt%.

---

1) ul. Politekhnicheskaya 2, 194021 Leningrad, USSR.

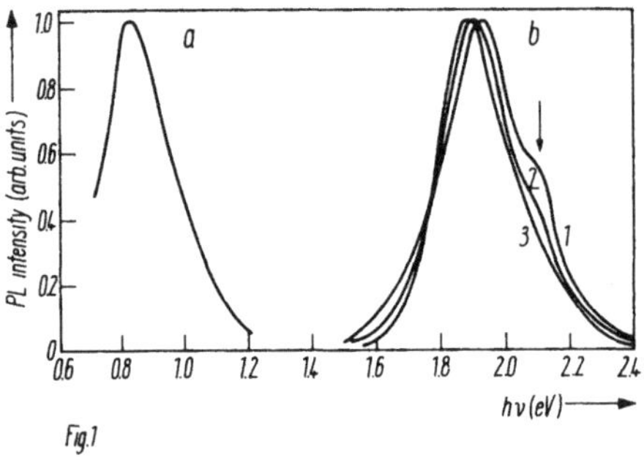

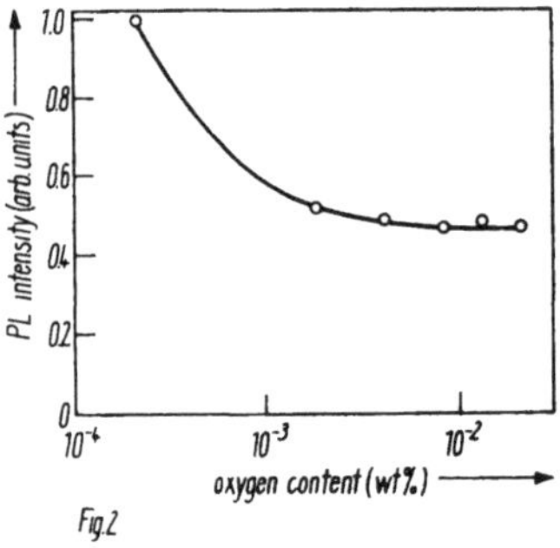

Fig. 1. PL emission and excitation spectra of oxygen doped glassy selenium. (1) 2×10^{-4} wt%, (2) 2.5×10^{-3} wt%, (3) 8.7×10^{-3} wt%

Fig. 2. PL intensity of glassy selenium versus oxygen content

Depending on the actual oxygen content, we studied PL excitation spectra $E_{exc}(h\nu)$, PL emission spectra $E_{pl}(h\nu)$, the temperature dependence of PL intensity under quasi-steady-state emission, and PL fatigue.

The very first experiments showed that introducing oxygen at low concentrations into purified selenium reduces dramatically the PL intensity (Fig. 2) and brings about noticeable changes in the excitation spectrum.

The excitation spectra of the PL are presented in Fig. 1b. One readily sees that as the oxygen concentration increases, the low energy part of the spectrum becomes less steep, the peak of the PL excitation band shifts toward higher energies and, which is particularly interesting, an additional band appears as a wing at $h\nu\approx2.1$ eV.

The emission spectra of the samples studied always revealed only one band peaking at 0.8 eV, its shape and position being unaffected by the oxygen content (Fig. 1a). We have not been able to detect in purified selenium the additional 0.57 eV emission band observed earlier /2/. This band is apparently associated with the specific thermal treatment of the material (abrupt quenching from 700 to 0 °C) leading to the formation of another phase, rather than with purification of g-Se from oxygen.

A study of the temperature behavior of PL displayed in Fig. 3 revealed a difference between the PL temperature quenching rate in the initial purified selenium and the oxygen doped material. Our data evidence an increase of the quenching rate with the introduction of oxygen into selenium. At the same time,

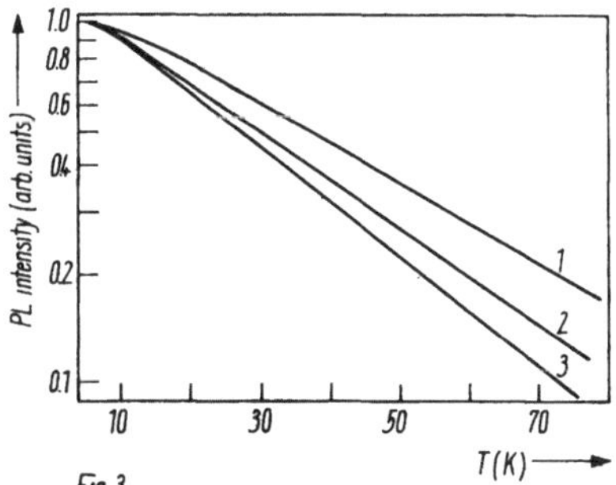

Fig.3

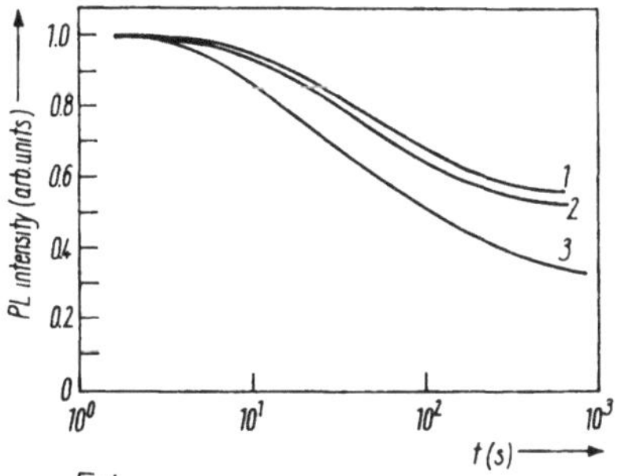

Fig.4

Fig. 3. PL intensity of oxygen doped glassy selenium versus temperature. (1) 2×10^{-4} wt%; (2) 2.5×10^{-3} wt%, (3) 8.7×10^{-3} wt%

Fig. 4. PL fatigue of oxygen doped glassy selenium. (1) 1.7×10^{-3} wt%, (2) 2.5×10^{-3} wt%, (3) 8.7×10^{-3} wt%

the effect of PL fatigue becomes more pronounced in oxygen doped selenium which is seen in Fig. 4. Note also a quantitative correlation between the change in fatigue and the oxygen content in a sample.

An analysis of these experimental data leads us to the conclusion that doping glassy selenium with oxygen even at low concentrations (down to 10^{-3}%) affects drastically the PL characteristics.

The observed shift of the PL excitation maximum toward higher energies originates from a shift of the optical absorption edge. The smoother low energy edge of the excitation band is connected with the absorption edge being diffuse. The first maximum in the excitation spectrum lying at energies corresponding to an absorption coefficient $\alpha\approx10\ \text{cm}^{-1}$ can be attributed, in our opinion, to direct excitation of the defect center. The second PL excitation maximum lies in the high energy wing at energies corresponding to $\alpha\approx10^3\ \text{cm}^{-1}$. Absorption of a photon of such energy creates an electron-hole pair which diffuses to the center and excites it. It is the increase in the defect excitation probability under the formation of electrons and holes (relative to the probability of direct defect excitation) which is associated with a noticeable increase in the electric conductivity of the materials under study /6/ that can account for the presence of the high energy wing in the excitation spectrum.

The existence of two channels of excitation of the same centers in oxygen doped selenium is indicated by the only emission band observed.

The decrease of the intensity of this emission band in oxygen doped selenium is connected with the enhanced probability of nonradiative recombination, which can be accounted for by a higher probability of the diffusion of electrons and holes and their recombination in a way other than via the radiative center. The more pronounced fatigue of the PL of doped g-Se could be explained by the increase of the probability of diffusion or tunneling of non-equilibrium carriers away from the radiative center excluding the latest from a radiative recombination process. This is evidenced also by the behavior of the PL temperature quenching: indeed, as the oxygen concentration increases, the probability of non-equilibrium carrier tunneling from a radiative center increases resulting in a larger slope of the $I = I_0 \exp(-T/T_0)$ curve.

References

/1/ R.A. STREET, T.M. SEARLE, and I.G. AUSTIN, Proc. 12th Internat. Conf. Semicond., 1974 (p. 1037).

/2/ S.G. BISHOP and P.C. TAYLOR, Proc. 8th Internat. Conf. Amorphous Semicond., Cambridge (USA) 1979 (p. 909).

/3/ G.Z. VINOGRADOVA, N.V. DEMOKRITOVA, L.M. ILIZAROV, S.A. DEMBOVSKII, and N.K. BELSKII, Proc. Conf. Amorphous Semicond., Bucharest 1982 (p. 308).

/4/ A. VASKO, phys. stat. sol. 8, K41 (1965).

/5/ B. VOIGT and G. DRESLER, Analit. Chim. Acta 127, 87 (1981).

/6/ V.A. TWADELL, W.C. LA COURSE, and J.P. MACKENZIE, J. non-crystall. Solids 8/10, 831 (1972).

(Received August 9, 1983)

phys. stat. sol. (a) 79, K93 (1983)
Subject classification: 14.1 and 18.4, 21.5

Department of Nuclear Engineering, Nagoya University[1)]

Critical Composition of Antiferromagnetic Ordering in UN_xC_{1-x}

By

M. TAMAKI, H. MATSUI, G. MATSUMOTO, and T. KIRIHARA

Introduction Dilution of antiferromagnetic UN (uranium mononitride) by paramagnetic UC (uranium monocarbide) diminishes the magnetic ordering. Three competing results have been reported for the critical composition of the antiferromagnetism of UN_xC_{1-x}. De Novion and Costa /1/ suggested the critical composition at x = 0.90 from measurements of the magnetic susceptibility, the specific heat, and the electrical resistivity, whereas Ohmichi et al. /2 to 4/ proposed it at x = 0.40 from the results of magnetic and electric properties of U(N, C). Another critical composition, x = 0.86, was reported by Lander et al. /5/, using a neutron diffraction technique. However, all samples in the above studies were powdered or sintered which had high porosity, impurities, and inhomogeneity, thus being away from a precise determination of T_N. In the present investigation, we prepared fused and comparatively pure uranium carbonitrides. Precise measurement of the low-temperature electrical resistivity was carried out with a particular emphasis of the critical composition of the antiferromagnetic ordering in U(N, C).

Experimental Solid solutions of uranium carbonitrides with various compositions of non-metal atoms were prepared by melting of UN and UC in a nitrogen plasma jet furnace under suitable gas pressures between 0.05 and 0.5 MPa. The original UN and UC were obtained also by plasma melting /6/. Sliced specimens, 1x1x15 mm^3 in dimension, were provided for the resistivity measurement. They were annealed at 1300 °C for 6 h in vacuo before starting the measurement. Material characterizations were done by chemical analysis for metal and non-metals, including oxygen (≈ 800 ppm). In addition, an evaluation of the chemical composition was tried using the relation between lattice parameter and (N+C)/N ratio reported by Cordfunke /7/.

The electrical resistivity was measured in a LHe cryostat in a temperature range between 4 and 100 K, applying the conventional dc four-probe method.

1) Furo-cho, Chikusa-ku, Nagoya 464, Japan.

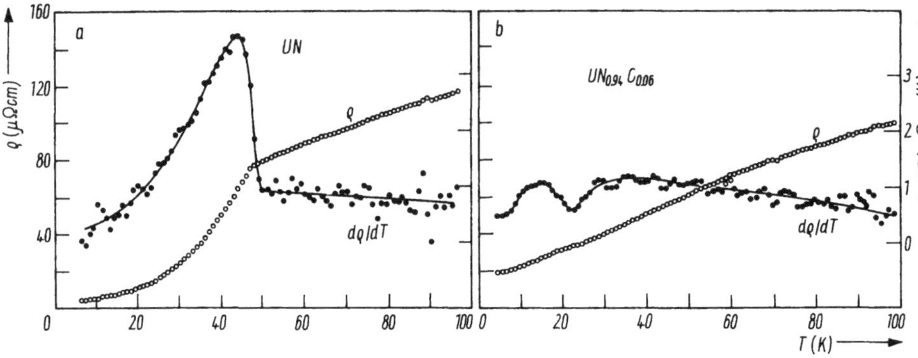

Fig. 1. Electrical resistivity and the derivatives (dϱ/dT) of a) UN and b) $UN_{0.94}C_{0.06}$ as a function of temperature

The measurement was performed at every 0.1 K interval and the data were analyzed by a computer (FACOM 230-60) at the Nagoya University. A Ge thermometer was used for temperature control. Errors of the measured values were within 0.2 μΩcm and 0.1 K for the resistivity and the temperature, respectively.

Results and discussion The electrical resistivity of pure UN is illustrated in Fig. 1a, together with the first derivatives of the resistivity with respect to temperature (dϱ/dT). In the dϱ/dT versus T plot, a maximum was observed at 44 K. In a previous paper /8/, we found also a maximum in dϱ/dT at 47.5 K in UN single crystals, but those were not real magnetic transition points. According to a procedure proposed by van Doorn and du Plessis /9/, the Néel temperatures were determined to be 50.1 K in the present case and 50.5 K in UN single crystals /8/. In the magnetic susceptibility measurement /6/, on the other hand, a clear transition point was revealed at 51.6 K for the same sample studied here.

The paramagnetic UC showed a lower electrical resistivity with a metallic temperature dependence (dϱ/dT = 0.1 μΩcm/K) between 4 and 100 K (see Fig. 2), being in good agreement with literature data /10/.

Fig. 1b shows the electrical resistivity and the derivatives of $UN_{0.94}C_{0.06}$ in the same manner as in Fig. 1a. A small maximum was observed in the dϱ/dT versus T plot, and the Néel temperature was determined to be 18.3 K in this sample. The value of the maximum dϱ/dT became small by the substitution of nitrogen with carbon in UN_xC_{1-x}. However, in $UN_{0.91}C_{0.09}$, there was no

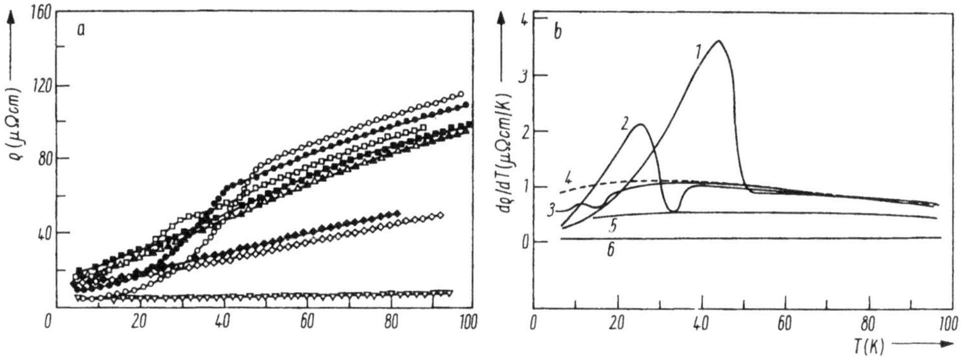

Fig. 2. a) Low-temperature electrical resistivities of various uranium carbonitrides, UN_xC_{1-x} (○ UN, ● $UN_{0.99}C_{0.01}$, □ $UN_{0.96}C_{0.04}$, ■ $UN_{0.94}C_{0.6}$, ▲ $UN_{0.93}C_{0.07}$, △ $UN_{0.91}C_{0.09}$, ◆ $UN_{0.82}C_{0.18}$, ◇ $UN_{0.70}C_{0.30}$, ▽ UC);

b) The first derivatives of the resistivity with respect to temperature (dρ/dT) of some UN_xC_{1-x} illustrated in a): (1) UN, (2) $UN_{0.96}C_{0.04}$, (3) $UN_{0.93}C_{0.07}$, (4) $UN_{0.91}C_{0.09}$; (5) $UN_{0.82}C_{0.18}$, (6) UC

more indication of such a maximum down to 4 K. Two samples, $UN_{0.82}C_{0.18}$ and $UN_{0.70}C_{0.30}$, showed a nearly temperature independent electrical resistivity, indicating no magnetic transition. All results of the low-temperature electrical resistivity and the derivatives (d ρ/dT) of UN_xC_{1-x} investigated here are shown in Fig. 2a and b, respectively. Several outstanding features in the resistivity measurements were as follows:

1) The most pronounced behavior was a shift of the magnetic transition temperature (T_N) by dilution of UN with UC.

2) If the sample was an antiferromagnet, the electrical resistivity above the transition point showed a parallel shift to lower level with increasing carbon content (for instance, compare the resistivity at room temperature, ρ_{RT}, in Table 1), while being opposite below the Néel temperature.

3) The evaluated resistivity due to the magnetic ordering at T_N ($\rho_{MO}(T_N)$) decreased by dilution and disappeared finally for $UN_{0.91}C_{0.09}$.

Various parameters derived from the electrical resistivity measurements are listed in Table 1 for a variety of x in UN_xC_{1-x}. In this table, we also present the change of the lattice parameter (a_o) and the electrical resistivity extrapolated to 0 K (ρ_o).

Table 1

Magnetic-correlated parameters obtained from the resistivity of U(N, C)

carbonitride	a_0 (nm)	T_N (K)	ρ_{RT} ($\mu\Omega$cm)	$\rho_{MO}(T_N)$ ($\mu\Omega$cm)	ρ_0 ($\mu\Omega$cm)
UN	0.48885	50.1	162	74	5
$UN_{0.99}C_{0.01}$	0.48891	41.1	157	59	9
$UN_{0.96}C_{0.04}$	0.48922	29.2	158	32	16
$UN_{0.94}C_{0.06}$	0.48937	18.3	154	16	18
$UN_{0.93}C_{0.07}$	0.48945	12.5	154	10	16
$UN_{0.91}C_{0.09}$	0.48954	4.2	152	?	17
$UN_{0.82}C_{0.18}$	0.49041	-	132	0	11
$UN_{0.70}C_{0.30}$	0.49149	-	123	0	11
UC	0.49590	-	33	0	4
deviation (±)	0.00005	0.5	5	5	2

In $UN_{0.91}C_{0.09}$, we could not detect the magnetic transition, but as seen in Table 1, ρ_0 and ρ_{RT} were quite the same as in other antiferromagnetic samples. In addition, $d\rho/dT$ was also coincident to that found in the magnetically ordered samples, and those parameters changed clearly their values after that composition. Therefore, the magnetic transition might occur probably below 4 K in that sample, $UN_{0.91}C_{0.09}$.

Changes of the Néel temperature (T_N) and the magnetic-ordered resistivity ($\rho_{MO}(T_N)$) are illustrated in Fig. 3 as a function of the chemical composition. Extrapolations of both parameters to zero yielded x = 0.90 ± ± 0.05, which corresponded to the critical composition of antiferromagnetism for UN_xC_{1-x}. As shown in Fig. 3, de Novion and Costa /1/ reported that the transition point appeared at 30 K in $UN_{0.91}C_{0.09}$, but not observed it in $UN_{0.88}C_{0.12}$. On the other hand, $UN_{0.86}C_{0.14}$ did not show any transition after $UN_{0.88}C_{0.12}$ (T_N = 30 K) in the neutron diffraction study /5/. The large drops in the magnetic transition point revealed in the above two investigations might be due to the inhomogeneity of the samples (sintered U(N, C)). In the present study, the Néel temperature shifted systematically by

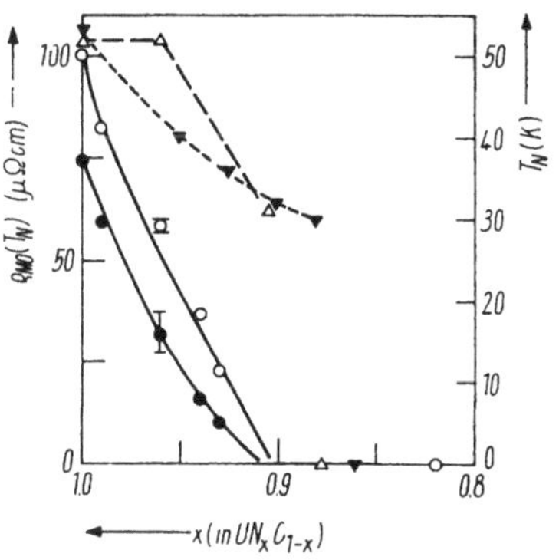

Fig. 3. Changes of Néel temperature (T_N)(o) and magnetic-ordered resistivity ($\varrho_{MO}(T_N)$)(●) as a function of x in UN_xC_{1-x}. Former studies (shift of T_N) by de Novion and Costa /1/ (Δ) and by Lander et al. (▼) /5/ are also represented for comparison

dilution of antiferromagnetic UN with paramagnetic UC. The wide range of x ($0.4 < x < 1.0$) for the antiferromagnetic state of UN_xC_{1-x} reported by Ohmichi et al. /2 to 4/ was in open discussions /11, 12/. Again, the homogeneity of the samples is a large problem in carbonitrides.

In conclusion, the critical composition of the antiferromagnetic structure of UN_xC_{1-x} was determined to be 0.90 ± 0.05, by precise and careful resistivity measurements, using fused and well characterized U(N, C). This was derived by an extrapolation of diminishing magnetic-correlated parameters (T_N and the magnetic-ordered resistivity) obtained in resistivity measurements.

References

/1/ C.H. DE NOVION and P. COSTA, J. Physique 33, 257 (1972).

/2/ T. OHMICHI and S. NASU, J. nuclear Sci. Technol. 7, 268 (1970).

/3/ T. OHMICHI, S. NASU, and T. KIKUCHI, J. nuclear Sci. Technol. 9, 11 (1972).

/4/ T. OHMICHI, T. KIKUCHI, and S. NASU, J. nuclear Sci. Technol. 9, 77 (1972).

/5/ G.H. LANDER, D.J. LAM, J.F. REDDY, and M.H. MUELLER, CONF-731104-5 (1973).

/6/ Y. IKEDA, M. TAMAKI, and G. MATSUMOTO, J. nuclear Mater. 59, 103 (1978).

/7/ E.H.P. CORDFUNKE, J. nuclear Mater. 56, 319 (1975).

/8/ S. NASU, T. KURASAWA, H. MATSUI, M. TAMAKI, and M. OKUDA, Pu-1975 and Other Actinides, North-Holland, 1976 (p. 515).

/9/ C.F. VAN DOORN and P. DE V. DU PLESSIS, J. low-Temp. Phys. 28, 391 (1977).

/10/ H. MATSUI, M. TAMAKI, S. NASU, and T. KURASAWA, J. Phys. Chem. Solids 41, 351 (1980).

/11/ M. KUZNIETZ, J. nuclear Sci. Technol. 8, 51 (1971).

/12/ T. OHMICHI and S. NASU, J. nuclear Sci. Technol. 8, 54 (1971).

(Received April 27, 1983)

phys. stat. sol. (a) 79, K99 (1983)
Subject classification: 14.4.1; 22.8.1

Institute of Experimental Physics, University of Wrocław[1]

The Hampered, Sideways Motion of a Single Domain Wall in Rochelle Salt Crystals

By
R. STYRKOWIEC

When electric field or mechanical stress applied to a sample quasi-statically increases, a hampered motion of the domain walls is observed. Moreover, when two walls are close to each other a significant repulsion occurs between them. This repulsion leads to a reduction of the velocity of approaching walls and makes coalescence of them difficult /1, 2/.

There are two possible suggestions: the interaction between moving walls is mechanical or electrostatical in nature. The first follows from that domain walls move together with some dislocations existing near the domain boundary. These dislocations produce a stress field around the walls which leads to interaction between them /3 to 7/.

The second suggestion may be connected with existence of a net surface charge density of a domain wall and hence approaching walls could repel each other /8, 9/. Recently Tikhomirova et al. have shown that the interactions between moving walls in TGS become significant at distances between domains comparable with their dimension and probably were electrostatic in nature /10/.

As it was shown earlier, the hampered motion of domain walls results from the change of the depolarization energy of two adjacent domains caused by the moving wall /11 to 13/. Such a situation takes place when the domain walls are far apart from each other (more than about 20 μm in the case of RS crystals). But when two domain walls approach each other a change in the interaction mechanism is observed.

In this experiment domain wall displacement was caused in two ways: first by applying an external electric field (along the polarization direction) and second by applying unidirectional stress (in (001) direction). The electric field and external stress were gradually increasing. After each rise of

1) ul. Cybulskiego 36, 50-205 Wrocław, Poland.

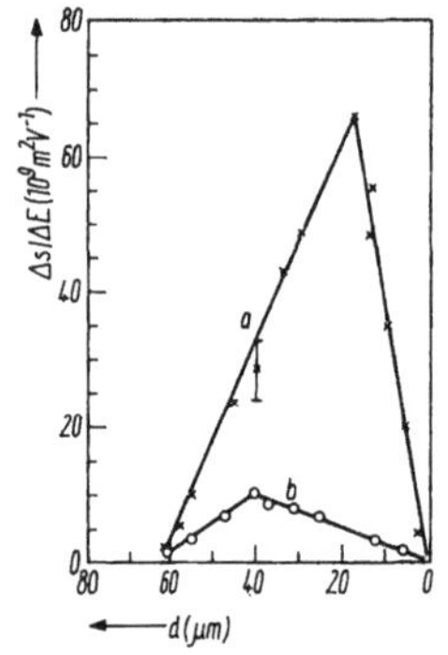

Fig.1

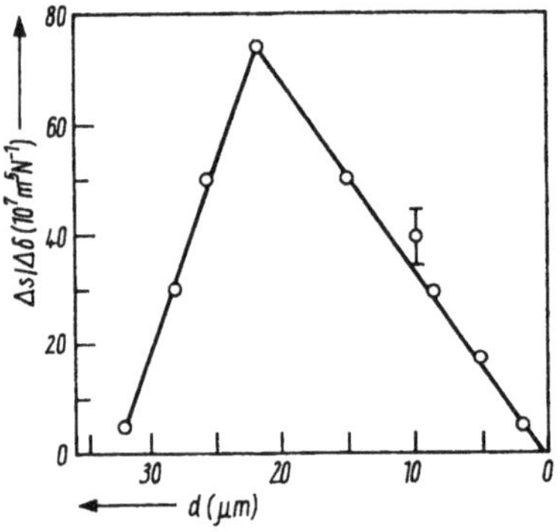

Fig.2

Fig. 1. The displacement of a single domain wall per unit intensity of electric field E as a function distance d between approaching domain walls (a) before irradiation and (b) the same wall after irradiation of the sample

Fig. 2. The displacement of a single domain wall per unit applied stress δ as a function of distance d between approaching domain walls

electric field (or stress) the domain walls shifted by some distance and then stopped. In both cases slow changes of domain structures were observed. The displacement of walls was measured by means of a calibrated eyepiece.

To define the interaction between approaching domain walls the displacement of the wall per unit intensity of the electric field (and unidirectional stress) was studied as a function of the distance between them. It was found that these dependences at first increased and then decreased. At a certain distance between the walls a maximum was observed (Fig. 1 and 2).

Then the same sample was irradiated using a 20 kV X-ray lamp with copper anode. The radiation dose was about 0.3 Mrd. It is known that the X-ray irradiation changes the fundamental properties of ferroelectric crystals in the whole and the dynamics of domain walls in particular /14 to 23/. In our case the irradiation has changed the dynamic properties of the domain structure in the following ways:

a) the threshold field for the moving walls increased,

b) the maximum value of $\Delta s/\Delta E$ for irradiated samples was many times smaller than for pure ones (Fig. 1b),

c) the interaction distance between approaching walls was longer than for the non-irradiated sample.

The soft X-ray irradiation of RS crystals leads to the creation of ionisation defects in the sample /24 to 27/. The experimental data mentioned above may indicate the electrostatical mechanism of a long-range interaction between domain walls.

As it can be seen from Fig. 1 and 2, the dependences of $\Delta s/\Delta E$ and $\Delta s/\Delta \delta$ on the distance between walls may be described in the form $y = k x$. Moreover, at a certain distance between the approaching walls a change of value and reversal of sign of the coefficient k is observed. This may indicate the competitive effect of the electrostatical and mechanical mechanism.

It seems that the simultaneous effect of electric field and mechanical stress applied to the sample permits to decide which factor is controlling the interaction between domain walls.

References

/1/ H.H. WIEDER and D.A. COLLINS, Phys. Rev. 120, 725 (1960).
/2/ R.C. MILLER, Phys. Rev. 111, 736 (1958).
/3/ J. BORNAREL and J. LAJZEROWICZ, J. appl. Phys. 39, 4339 (1968).
/4/ J. BORNAREL and J. LAJZEROWICZ, J. Phys. Soc. Japan, Suppl. 28, 360 (1970).
/5/ J. BORNAREL, J. appl. Phys. 43, 845 (1972).
/6/ J. BORNAREL and J. LAJZEROWICZ, Ferroelectrics 4, 177 (1972).
/7/ J. FOUSEK and B. BREZINA, Czech. J. Phys. B10, 511 (1960).
/8/ R.C. MILLER, Phys. Rev. 111, 736 (1958).
/9/ M.E. DROUGARD and R. LANDAUER, J. appl. Phys. 30, 1663 (1959).
/10/ N.A. TIKHOMIROVA et al., Ferroelectrics 29, 145 (1980).
/11/ A. JAŚKIEWICZ, S. DACKO, and VO DUY DANH, Acta Phys. Polon. A45, 209 (1974).
/12/ A. JAŚKIEWICZ and R. STYRKOWIEC, Acta Univ. Wratislaviensis 231, 3 (1974).
/13/ A. JAŚKIEWICZ, Acta Univ. Wratislaviensis 341, 91 (1977).
/14/ W. A. YURIN, Izv. Akad. Nauk SSSR, Ser. fiz. 21, 329 (1957).
/15/ N.A. ROMANIUK and I.S. ZHELUDEV, Kristallografiya 9, 38 (1964).
/16/ T. REWAJ, phys. stat. sol. 2, 1151 (1962).
/17/ J. EISNER, Izv. Akad. Nauk SSSR, Ser. fiz. 29, 1848 (1965).
/18/ B. HILCZER, Izv. Akad. Nauk SSSR, Ser. fiz. 33, 1105 (1969).

/19/ B. HILCZER, Internat. Meeting Ferroelectrics, Prague 1966.
/20/ E.W. PESHIKOV, Kristallografiya 14, 1074 (1969).
/21/ N.A. ROMANIUK, Kristallografiya 15, 647 (1971).
/22/ B. HILCZER, Ferroelectrics 7, 319 (1974).
/23/ B. HILCZER, Proc. II Summer Meeting Phase Transitions and Critical Phenomena, Polish Acad. Sci., Wrocław 1979.
/24/ D.W. OVENAL and K.A. MULLER, Helv. Phys. Acta 34, 766 (1961).
/25/ J. SUZUKI and R. ABE, J. Phys. Soc. Japan 23, 1049 (1967).
/26/ T. KATO and R. ABE, J. Phys. Soc. Japan 26, 948 (1969).
/27/ K. OKADA, Japan. J. appl. Phys. 2, 613 (1963).

(Received August 1, 1983)

phys. stat. sol. (a) 79, K103 (1983)
Subject classification: 2 and 14.3; 22.4.4

Department of Physics, Sri Venkateswara University, Tirupati[1)]

Electrical Properties of Vapour Phase Grown $(ZnSe)_x(CdTe)_{1-x}$ Single Crystals

By

P. CHANDRASEKHARAM[2)], D. RAJA REDDY[3)], and B.K. REDDY

Introduction In recent years much attention was drawn towards the alloy semiconductors mainly to tailor the optoelectronic properties. Although some work has been reported on band-gap studies of the $(ZnSe)_x(CdTe)_{1-x}$ pseudobinary II-VI alloy system /1/ the electrical properties are reported for the first time in this note.

Experimental procedure Single crystals of a few cm length and 5 mm diameter of the $(ZnSe)_x(CdTe)_{1-x}$ system have been grown using a vapour phase growth technique in the entire range of composition x ($0 \leqq x \leqq 1$) /2/. The samples of 5x5x0.5 mm^3 were cut from the boules, ground with fine carborundum powder, polished over chamois leather with chromium oxide, and etched using suitable etches. Indium contacts were used and their ohmic behaviour was confirmed by I-U characteristics. A cold finger type cryostat was used for temperature studies in the range 80 to 303 K. The as-grown crystals were resistive and were made conducting by annealing in Zn/Cd at 900 $^{\circ}C$ for 48 h. The conductivity and Hall effect studies were carried out simultaneously on the same samples using the dc van der Pauw technique /3/. However, for as-grown crystals the two-probe method was used. The type of conductivity was determined by the sign of Hall voltage/hot probe test /4/.

Results and discussion The temperature dependence of the electrical conductivity of as-grown $(ZnSe)_x(CdTe)_{1-x}$ single crystals is shown in Fig. 1. The crystals with $x \leqq 0.60$ were p-type and the rest were n-type. The conductivity activation energies calculated from these data are given in Table 1. These energies are found to decreases as the ZnSe content is increased in the

---

1) Tirupati 517502, India.

2) Present address: Department of Physics, P. Venkata Krishnaiah Naidu Government College, Chittoor 517002 (A.P.), India.

3) Present address: Department of Physics, Indian Institute of Technology, Madras 600036, India.

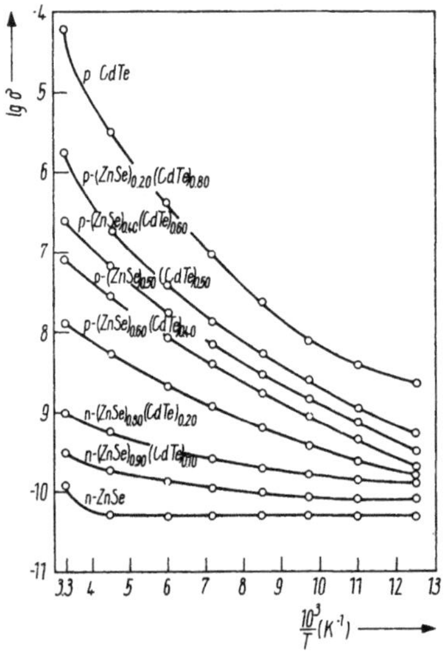

Fig. 1. Temperature variation of electrical conductivities of as-grown $(ZnSe)_x(CdTe)_{1-x}$ single crystals

mixed crystals. In general the activation energy of a particular donor/acceptor increases with the band gap and effective mass of electrons/holes. However, it decreases with the dielectric constant. This behaviour was observed in crystals of cadmium chalcogenides /4/ and their mixed crystals /5/. If the foreign donors or acceptors are responsible for the observed conductivity behaviour, the activation energy should increase as x is increased because the energy gap, effective mass of either electrons or holes of ZnSe are greater than those for CdTe and also the dielectric constant of the former is less than that of the latter. Therefore the activity of foreign donors or acceptors does not seem to be active in controlling the electrical properties of these as-grown single crystals. In the case of p-type crystals the native defects like Zn/Cd vacancies should be responsible for the observed conductivity behaviour. When the first and second ionisation energies of Zn (0.48, 0.14 eV) and Cd (0.05, 0.84 eV) vacancies in their tellurides are compared /5/,

Table 1

Activation energies of $(ZnSe)_x(CdTe)_{1-x}$ single crystals

x	activation energy (eV)
0.00	0.024
0.20	0.019
0.40	0.016
0.50	0.013
0.60	0.011
0.80	0.005
0.90	0.004
1.00	0.002

Table 2

Electrical conductivities σ (in $(\Omega cm)^{-1}$) of $(ZnSe)_x(CdTe)_{1-x}$ single crystals

x	as-grown crystals	crystals annealed in molten Zn at 900 °C for 48 h	crystals annealed in molten Cd at 900 °C for 48 h	μ_H (cm^2/Vs) 303 K	80 K
0.00	$6.8x10^{-5}$	$2.0x10^{-4}$	0.96	600	6000
0.20	$2.5x10^{-6}$	$3.0x10^{-5}$	0.64	620	4000
0.40	$3.3x10^{-7}$	$5.0x10^{-6}$	0.38	400	3500
0.50	$8.3x10^{-8}$	$6.0x10^{-5}$	0.25	360	3500
0.60	$2.0x10^{-8}$	0.22	$2.0x10^{-6}$	350	3000
0.80	$1.0x10^{-9}$	0.17	$3.0x10^{-7}$	340	2000
0.90	$6.3x10^{-10}$	0.11	$5.0x10^{-8}$	320	1500
1.00	$1.8x10^{-10}$	0.05	$3.0x10^{-9}$	300	1000

it is clear that the activation energies involved in Cd vacancies are greater than those of Zn. Therefore the activation energy decrease with increase of ZnSe content in the mixed crystals observed in the present investigation is justified because it is highly likely that crystals grown at high temperatures by the vapour phase method will have native defects. In the case of n-type crystals the donors are very shallow and showed a similar behaviour.

Table 2 shows the room temperature conductivities of as-grown and Zn/Cd annealed crystals. In the annealing process self-purification /7/ takes place and the conductivities increase. In this process the electrically active vacancies are filled and may be interstitials are also added. The conductivity type observed after this treatment for all the crystals is n-type. This is due to the decrease in acceptor density and an increase in donor (interstitial) density. Fig. 2 show typical temperature variations of electrical conductivity, Hall mobility, and carrier concentration for A, B, and C crystals (x = 0.00, 0.40, 0.80). The 303 and 80 K Hall mobilities for all the compositions are included in Table 2. It is clear that the low temperature mobility is always greater than that of the room temperature one for all the crystals. The temperature dependence of mobility is very similar to those of the end compositions and hence

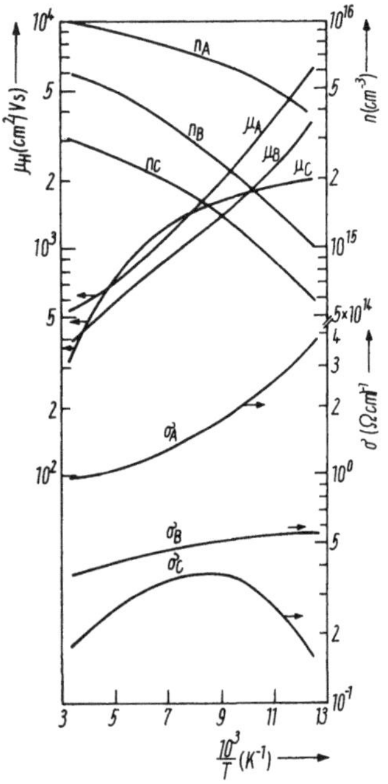

Fig. 2. Temperature variation of electrical conductivity (σ), carrier concentration (n), and Hall mobility μ_H of A (x = 0.00), B (x = 0.40), and C (x = 0.80) crystals

the scattering mechanisms active in the end compositions, namely optical mode and impurity scatterings, seem to be active in the mixed crystals, too /6/. There is a regular increase in mobility, conductivity, and carrier concentration as the composition is varied from ZnSe to CdTe. The donor activation energies calculated from the carrier concentration data are shown in Table 3. There is no regular variation in these energies with composition. This may be due to the strong dependence of the activation energy on the degree of compensation /8/. The conductivity peaks observed in some of the compositions (x = 0.50, 0.60, 0.80, 0.90, and 1.00) at intermediate temperatures is a result of mobility increase and carrier concentration decrease with lowering temperature.

Table 3

Activation energies of Cd/Zn annealed $(ZnSe)_x(CdTe)_{1-x}$ single crystals

x	activation energy (eV)
0.00	0.006
0.20	0.013
0.40	0.015
0.50	0.014
0.60	0.013
0.80	0.014
0.90	0.012
1.00	0.006

One of us (P.C.S) is grateful to the authorities of Sri Venkateswara University and University Grants Commission for the award of a Teacher Fellowship under the Faculty Improvement Programme.

References

/1/ P. CHANDRASEKHARAM, D. RAJA REDDY, and B.K. REDDY, phys. stat. sol. (b) 118, K5 (1983).

/2/ P. CHANDRASEKHARAM, Ph.D. Thesis, S.V. University (India) 1982, to be published in J. Crystal Growth.

/3/ L.J. VAN DER PAUW, Philips tech. Rev. 20, 220 (1958/59).

/4/ W.R. RUNYAN, Semiconductor Measurements and Instrumentation, McGraw-Hill, Kogakusha Ltd., Tokyo 1975 (p. 145).

/5/ S.S. DEVLIN, in: Physics and Chemistry of II-VI Compounds, Ed. M. AVEN and J.S. PRENER, North-Holland, Amsterdam 1967 (p. 604).

/6/ D. RAJA REDDY, B.K. REDDY, and P.J. REDDY, Bull. Mater. Sci., in the press.

/7/ M. AVEN and B. SEGALL, Phys. Rev. 130, 81 (1963).

/8/ H.H. WOODBURY and M. AVEN, Phys. Rev. B 9, 5195 (1974).

(Received June 23, 1983)

phys. stat. sol. (a) 79, K109 (1983)

Subject classification: 14.3.3 and 16; 22.9

Instrumentation Engineering, Faculty of Engineering, Keio University, Yokohama[1)]

Photovoltaic Behaviour of Rhodamine 6G Schottky-Type Cells

By

J. MUTO and Y. TSUNEKAWA[2)]

Rhodamine 6G (Rh6G) is an organic dyestuff, belonging to the xanthene family and is widely used as a laser dye material. We have fabricated Rh6G thin film sandwitch cells and examined their photovoltaic properties. Rh6G was evaporated in vacuum from a Ta crucible on a transparent conducting coating (SnO_2) glass. A film of Al was subsequently evaporated on top of the organic solid. Photovoltaic behaviours of the cells were examined in vacuum of 10^{-2} Torr and the irradiation of light was done through the glass electrode of the cell unless otherwise mentioned.

An absorption maximum of evaporated Rh6G film exists around 550 nm. The evaporated film was readily dissolved in ethanol, yielding the usual Rh6G/ethanol absorption spectrum with its peak at about 530 nm (see Fig. 1). A 550 nm irradiation source, obtained by a tungsten lamp and an interference filter, was utilized in the following experiment.

When the incident radiation was directed through the Al electrode, the photocurrent action spectrum of the cell closely matched the film absorption spectrum. When the irradiation was directed through the glass electrode, however, the action spectrum was similar to the film transmittance spectrum. Whereas, under light illumination through the glass electrode, the action

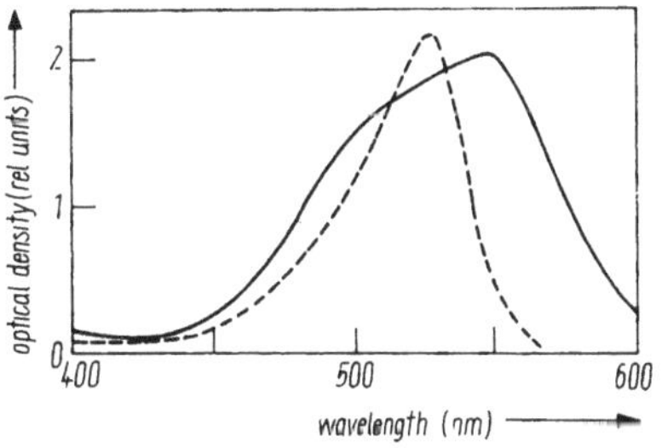

Fig. 1. Absorption spectra of Rh6G. Solid line: evaporated Rh6G film, broken line: Rh6G once evaporated and dissolved in ethanol

1) 3-14-1, Hiyoshi, Kohoku, Yokohama 223, Japan.

2) Present address: Suwa Seikosha, 3-3-5, Yamato, Suwa 392, Japan.

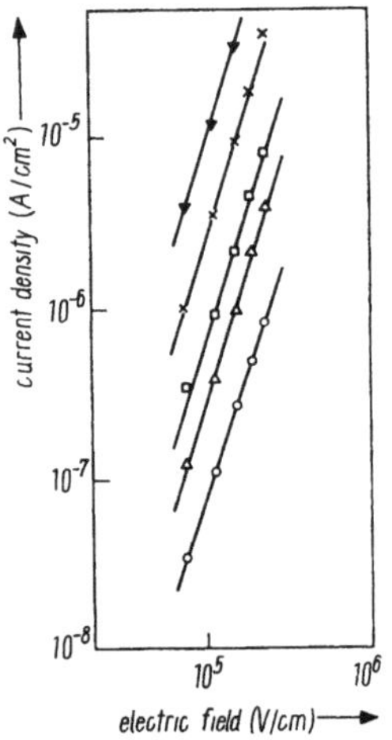

Fig. 2. Current density versus applied electric field of Rh6G cell at different light intensities. Cell thickness: 550 Å. Light intensity: Circles: 9.6×10^{-5} mW/cm^2; triangles: 3.5×10^{-4} mW/cm^2; squares: 7.5×10^{-4} mW/cm^2, crosses: 2.7×10^{-3} mW/cm^2, and black triangles: 4.9×10^{-3} mW/cm^2

spectrum of thinner cells (≦ 550 Å thick) was similar to the absorption spectrum. The above results, in addition to the capacitance-voltage relationship of the cell described below, suggest that the photovoltaic effect of the cell has its origin in the Schottky barrier formed at the junction between the Al and Rh6G.

At dark, the cell showed the usual J (current density)-U (applied voltage) relationship of p-n junctions or Schottky barriers, i.e., ln J varies linearly with U, while at light irradiation, $J \propto U^3$ was observed for the forward biased direction (Al electrode positive), as shown in Fig. 2. The super quadratic voltage dependence of the space charge limited current of the form $J \propto U^{m+1}$ ($m > 1$) has usually been interpreted assuming current carriers to be trapped by exponentially distributed local states /1/. Now that $m = T_c/T$, where T_c and T are characteristic and cell temperatures, respectively, we have $T_c =$ = 600 K for the examined sample.

For the transmitted light of 5.1 mW/cm^2, the cell generated an open circuit voltage of 0.56 V, while the short circuit current was 1.6×10^{-7} A/cm^2. The fill factor, which is defined as the maximum power output of the cell divided by the product of the open voltage and short circuit current, had a value of 0.36. The power conversion efficiency of the cell was calculated as 6×10^{-4}%. A low power conversion efficiency of 10^{-4}% was also obtained for the tetracene Schottky barrier cell /2/.

The Schottky barrier capacitance C is described as

$$1/C^2 = 2(U_d - U)/e\ \varepsilon_0\ \varepsilon_s N\ , \qquad (1)$$

where U denotes applied bias voltage as already mentioned, U_d Schottky barrier voltage between Al and Rh6G, e electronic charge, ε_0 permittivity of free space, ε_s the dielectric constant, and N is the carrier concentration /3/. The width of the depletion region, W, will change under bias aso that

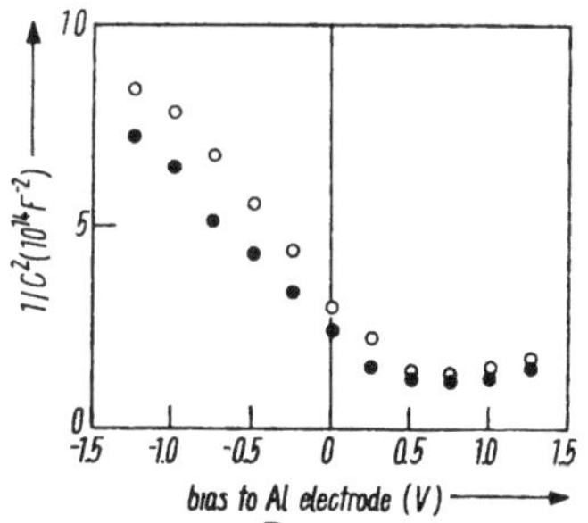

Fig. 3. $1/C^2$ versus U plots of Rh6G cell. Cell thickness; 550 Å. White circles; at dark: black circles: under illumination of $4.9x10^{-3}$ mW/cm^2

$$W = \left[2 \varepsilon_o \varepsilon_s (U_d - U)/eN\right]^{1/2} . \tag{2}$$

A plot of $1/C^2$ versus U should result in a straight line if the capacitance is attributable to a Schottky depletion region. From the intercepts with the two axes we can calculate U_d (when $1/C^2 = 0$) and W at zero bias, W_o (when U = 0) and furthermore, from the slope of the line a value of N is determined, given the dielectric constant of the depletion region.

The capacitance measurements of the cells were performed by the low frequency oscillographic method /4/. At higher frequencies (> 1 Hz), C is U independent and becomes to be a geometric capacitance. At lower frequencies (< 1 Hz), $1/C^2$ of the thinner cells (≦ 550 Å) was observed to vary linearly with U (see Fig. 3). At highly reverse biased voltage, the depletion region stretches to the counter conducting glass electrode and the voltage independent geometric capacitor is realized, from which we can determine ε_s of the depletion region. The obtained ε_s of the depleted Rh6G film is 3.4.

Typical calculated values of N, U_d, and W_o of the cell at different frequencies without illumination are summarized in Table 1. A scarce change in the $1/C^2$ versus U characteristic was observed in case when the cell was examined under higher evacuation of 10^{-4} Torr. W_o of several hundred Å, as determined for our cell, was also obtained for organic Schottky diodes of X-metal free phthalocyanine /5/ and Mg-phtalocyanine /6/.

Table 1

frequencies (Hz)	N (10^{23} m^{-1})	U_d (V)	W_o (Å)
0.1	2.0	0.58	330
0.05	2.1	0.60	330
0.02	2.2	0.60	320
0.01	2.2	0.61	310

The relationship $1/C^2 \propto U$ still holds when the Rh6G cell is under illumination. The cell showed approximately 30% increase of N, 5% increase of U_d, and 10% decrease of W_o, at 5.1 mW/cm^2 irradiation and for frequencies below 0.1 Hz.

References

/1/ P. MARK and W. HELFNISCH, J. appl. Phys. 33, 205 (1962).
/2/ L.E. LYONS and O.M.G. NEWMAN, Austral. J. Chem. 24, 13 (1971).
/3/ A. VAN DER ZIEL, Solid State Physical Electronics, Prentice Hall, New Jersey 1976 (p. 284).
/4/ H.K. HENISCH, Rectifying Semiconductor Contacts, Clarendon, Oxford 1957 (p. 152).
A.J. TWORSKI and A.C. ALBRECHT, J. chem. Phys. 70, 2255 (1979).
/5/ R.O. LOUFTY, J.H. SHARP, C.K. HSIAO, and R. HO, J. appl. Phys. 52, 5218 (1981).
/6/ A.K. GHOSH, D.L. MOREL, F. FENG, R.F. SHAW, and C.A. ROWE, Jr., J. appl. Phys. 45, 230 (1974).

(Received June 8, 1983)

Pre-Printed Titles

Papers to be published
in "physica status solidi (a) - applied research"
Vol. 79, No. 2

Original Papers

W. BECK and H. KRONMÜLLER, Stuttgart
The Influence of Surface Roughness on the Coercive Field of the Nearly Non-Magnetostrictive Amorphous Ferromagnetic Alloy $Co_{58}Ni_{10}Fe_{5}Si_{11}B_{16}$

A.P. TYUTNEV, G.S. MINGALEEV, V.S. SAENKO, and A.V. VANNIKOV, Moscow
Radiation Induced Conductivity of Polyethyleneterephtalate and Polystyrene

H. PFEIFFER, Jena
Analytical Theory of the Growth of Crystal Faces

A.D. REDDY, S.G. SATHYANARAYAN, and G.S. SASTRY, Hyderabad
Thermally Stimulated Depolarisation Current Studies on Some Hydrogen-Bonded Crystals

V.G. CHUDINOV, V.P. GOGOLIN, B.N. GOSHCHITSKII, and V.I. PROTASOV, Sverdlovsk
Evolution of the Cascade Region in Nb_3Sn during the Thermal Spike under Fast-Neutron Irradiation

B. RAUSCHENBACH and K. HOHMUTH, Dresden
Analysis of Heavy-Ion Sputtering Based on the Shock Wave Model

I. GREGORA, V. VORLÍČEK, and R. UECKER, Prague and Berlin
Determination of the Fresnel Indicatrix of the Monoclinic $NaVO_3$ Crystal in the Visible and Near Infrared Spectral Range

A. CHARAI and C. BOULESTEIX, La Garde and Marseille
Visualization and Study of the Shape and Size of Very Small Coherent Precipitates by a Phase Contrast Method

S. A. ABO-NAMOUS, Y. ZAKA, and R. W. FANE, Birmingham
Dependence of the Electronic and Optical Properties of Unhydrogenated a-Si on Preparation Conditions

M. KATO, Yokohama
Activation Energy for Dislocation Motion in a One-Dimensional Periodic Field

S. S. SIMEONOV and M. D. IVANOVITCH, Sofia
Determination of Quasi-Fermi Level at the Metal-Semiconductor Interface in Forward-Biased Schottky Barrier Diode

J. ŠTERNBERK, E. KRATOCHVÍLOVÁ, J. HŘEBÍK, and A. GEMPERLE, Prague
Coercivity and Microstructure of Low-Alloy Cr-Mo Steel

P. DEUS and H. A. SCHNEIDER, Freiberg
Thermal Expansion of the Diamond-Like Semiconducting Compound $ZnSiP_2$ within the Temperature Range of 20 to 300 K

L. KOMITOV and A. G. PETROV, Sofia
Quasihomotropic Reversely Pretilted Nematic Liquid Crystal Layers

I. TOMÁŠ and YU. A. KUZIN, Donetsk and Prague
Internal Forces in Magnetic Bubble Lattices

E. I. GOLDMAN, YU. V. MARKIN, P. S. SULZHENKO, and A. G. ZHDAN, Moscow
Grain Boundary Tunneling Conductivity of Semiconductor Poly- and Bicrystals

M. GASGNIER, L. NEVOT, P. BAILLIF, and J. BARDOLLE, Orsay and Orléans
Characterization and Crystalline Structures of Tungsten Thin Films

B. M. WILAMOWSKI, Tucson, Arizona
The Punch-Through Transistor with MOS Controlled Gate

G. A. SCHOLZ and R. F. FRINDT, Burnaby, British Columbia
Electrical Resistivity of Silver Intercalated $2H\text{-}TaS_2$

H. HAHN, G. FROHBERG, and H. WEVER, Berlin (West)
Self-Diffusion in the Intermetallic B2 Electron Compound PdIn

L.N. VASILEV, V.M. GRABOV, A.V. GOLUBKOV, A.G. GOROBETS, V.S. OSKOTSKII, I.A. SMIRNOV, and V.V. TIKHONOV, Leningrad

Physical Properties and Phase Transitions of the Rare Earth Monosulphides in the Homogeneity Range

S.E. DONNELLY, A.A. LUCAS, PH. LAMBIN, and J.P. VIGNERON, Clayton, Victoria, and Namur

A Possible Mechanism for Electron-Bombardement-Induced Loop Punching in Helium-Implanted Materials

J.C. BERNÈDE, Nantes

Influence du métal de l'anode sur l'existence d'une résistance différentielle négative dans des structures MIM

V. ATZROTH and H. LANGE, Berlin

Principal Component Analysis as a Method for Silicide Investigation with Auger Electron Spectroscopy

H. BAHADUR and R. PARSHAD, New Delhi

Some New Results on Irradiation Characteristics of Synthetic Quartz Crystals and Their Application to Radiation Hardening

K.H. GEORGY and R. KAMEL, Dokki and Cairo

The Transient Creep of Cobalt near the Temperature of Transformation

A.N. GEORGOBIANI, A.V. MIKULYONOK, I.G. STOYANOVA, and I.M. TIGINYANO, Moscow

Non-Equilibrium Carrier Radiative Recombination in Indium Phosphide Single Crystals

G.L. WHITTLE and A.M. STEWART, Canberra

Mössbauer Effect Study of $Fe_{85-x}Ni_xB_{15}$ Metallic Glasses

F. FENSKE, Berlin

Influence of Lateral Charge Inhomogeneities in MOS Structures on the Determination of the Interface State Density Distribution from HF CV Curves

I.D. HENNING and H. THOMAS, Cardiff

Process-Induced Defect States in GaAsP Light Emitting Diodes

Papers to be published
in "physica status solidi (b) - basic research"
Vol. 119, No. 2

Original Papers

G. MÜTZENICH, R. FAYMONVILLE, and P. GROSSE, Aachen
One- and Multiphonon Absorption in EuSe and $Eu_{1-x}Sr_xS$ $(0<x<1)$

B. PÖDÖR, Budapest
Zone Edge Phonons in Gallium Phosphide

N.S. ORLOVA, Minsk
Variation of Phonon Dispersion Curves with Temperature in Indium Arsenide Measured by X-Ray Thermal Diffuse Scattering

T. SOMA and H.-M. KAGAYA, Akita
Local Heine-Abarenko Model Potential for III-V and II-VI Covalent Compounds

I. MUŠEVIČ, B. ŽEKŠ, R. BLINC, TH. RASING, and P. WYDER, Ljubljana and Nijmegen
Dielectric Study of the Modulated Smectic C*-Uniform Smectic C Transition in a Magnetic Field

M.C. OLIVEROS and D.R. TILLEY, Natal and Colchester, Essex
Quantization Procedure for Phonons in an Isotropic Elastic Half-Space

B. LORENZ, Leipzig
Thermodynamic Properties of Uniform and Dimerized Hubbard Chains

B.A. VAID, K.C. SHARMA, and V.K. SYAL, Simla
A Disorder Model of Melting

A.Y. BELOV, V.A. CHAMROV, V.L. INDENBOM, and J. LOTHE, Moscow and Oslo
Elastic Fields of Dislocations Piercing the Interface of an Anisotropic Bicrystal

P. PETELENZ, Cracow
Optical Detrapping in Molecular Crystals (I)

L. BOSI, P. GAGLIARDELLI, and M. NIMIS, Milano
More on the F-Centre Decay in Additively Colored NaF:LiF

J. IMHOF, Budapest
Texture Analysis by Iteration (I)

M. WEGER and O. ENTIN-WOHLMAN, Jerusalem
The Ground State Energy of Small Polaron Gas

N.M. KHAMIDULLINA, Moscow
Carrier Mobility in Doped Nondegenerate Semiconductors

G. POMPE, M. GAAFAR, P. BÜTTNER, and TH. FRANCKE, Dresden
Low-Temperature Thermal Conductivity of Amorphous (Fe, Ni, Co)(P, B, Si) Alloys and Their Change by Heat Treatment

M. ORTUNO, Barcelona
Impurity Potentials in Diamond

P. BHATTACHARYA and A.R. CHETAL, Dhanbad
Chemical Shift and EXAFS Studies of Some Copper Complexes

V. NAGIRNYI, T. SOOVIK, S. ZAZUBOVICH, and N. JANSON, Tartu
Polarized A_T Emission of In^+ Centres in KI:In

D.J. GONZÁLEZ, M.L. ZORITA, J.A. ALONSO, and M.P. INIGUEZ, Valladolid
Density Functional Pseudopotential Calculation of the Cohesive Properties of Disordered Solid Alloys of Nontransition Metals

V.K. JAIN, V.P. SETH, R.K. MALHOTRA, and S.K. YADAV, Rohtak
Electron Paramagnetic Resonance of Mn^{2+} and VO^{2+} in $Ba_2Zn(HCOO)_6 \cdot 4\,H_2O$

G. IRMER, V.V. TOPOROV, B.H. BAIRAMOV, and J. MONECKE, Leningrad
Determination of the Charge Carrier Concentration and Mobility in n-GaP by Raman Spectroscopy

A. MILCHEV and I. AVRAMOV, Sofia
On the Influence of Amorphization on Atomic Diffusion in Condensed Systems

K. KÖHLER, H.P. GESERICH, TH. WOLF, and G. MÜLLER-VOGT, Karlsruhe
Optical Properties of TiN Single Crystals

P. A. ALEKSEEV, E. A. GOREMYCHKIN, B. LIPPOLD, E. MÜHLE, and I. P. SADIKOV, Moscow, Dubna, and Leipzig
Crystal Field Splitting in $NdAl_3$ Studied by Inelastic Neutron Scattering

A. F. SADREEV, I. S. SANDALOV, and A. N. PODMARKOV, Krasnoyarsk
The Spin-Wave Theory of Electron and S-F Exchange Magnetostriction in the Rare Earth Metals

R. POŽÉRA, A. SUTKUS, and R. TOLUTIS, Vilnius
The Influence of Hydrostatic Pressure on the Effective Concentration and Cyclotron Mobility of Electrons in Semiconducting $Bi_{1-x}Sb_x$ Alloys

M. SCHIENLE, A. KASTEN, and P. H. MÜLLER, Karlsruhe
Magnetic Relaxation in Dysprosium Phosphate ($DyPO_4$)

J. KOWALSKA and J. ZIELIŃSKI, Katowice
Spinless Anderson Model with a Local Phonon Mode

H. S. KIM and S. S. SHEININ, Edmonton, Alberta
Analysis of Critical Voltage Data Using the Dynamical Theory of Electron Diffraction

V. HIZHNYAKOV, G. LIIDYA, V. NAGIRNYI, T. SOOVIK, and S. ZAZUBOVICH, Tartu
Effect of Magnetic Field on the Decay Kinetics and Polarization of the A_T Emission of KCl:Ga at 0.38 K

19. Magnetic Resonance

20. Optical Properties
    20.1 Spectra. Optical Constants (X-Ray Spectra See 20)
    20.2 Lasers
    20.3 Luminescence (X-Ray Spectra See 20)

21. Metals. Alloys
    21.1 Transition Metals (Group Sc to Zn) and Their Alloys
        21.1.1 Fe and Fe Alloys
    21.2 Alkali Metals
    21.3 Alkali-Earth Metals
    21.4 Rare-Earth Metals
    21.5 Actinides
    21.6 Noble Metals
    21.7 Semi-Metals (Semiconducting Alloys 22.7)

22. Semiconductors and Ionic Crystals
    22.1 Elements
        22.1.1 Germanium
        22.1.2 Silicon
        22.1.3 Group VI Elements
    22.2 III–V Compounds
        22.2.1 Arsenides
        22.2.2 Phosphides
        22.2.3 Antimonides
        22.2.4 Mixed Crystals
    22.3 IV–IV, II–V, and III–VI Compounds
        22.3.1 Mixed Crystals
    22.4 II–VI Compounds
        22.4.1 Sulphides
        22.4.2 Selenides
        22.4.3 Tellurides
        22.4.4 Mixed Crystals
    22.5 Halides
        22.5.1 Silver Halides
        22.5.2 Alkali Halides
        22.5.3 Alkali-Earth Halides
        22.5.4 Mixed Crystals
    22.6 Simple Oxides (CdO, ZnO See 22.4)
        22.6.1 Mixed Crystals
    22.7 Semiconducting Intermetallic Compounds
        22.7.1 Mixed Crystals
    22.8 Tertiary and Higher Compounds
        22.8.1 Salts of Oxy-Acids (Carbonates, Phosphates, Silicates Including Mica etc.)
        22.8.2 Spinels, Garnets, Ferrites
    22.9 Organic Semiconductors

23. Solidified Gases

Attention

The preceding classification scheme is closely akin to the scheme used by Physics Abstracts to classify the various fields of solid state physics. A paper will either be classified under one or more of the subsections or, if none of these are relevant, it will just appear under the general section heading (i. e. sections and subsections are equivalent in this sense).

The *subject classification* given at the head of each paper begins with the main subject matter of the paper, and this is followed, if necessary, by figures referring to subsidiary subject matter and the type of solids investigated.

The papers published in this issue are listed in the order of their subject classification at the end of the table of contents.

82914

Manuscripts and letters for physica status solidi (b) — basic research and physica status solidi (a) — applied research

should be sent to

DDR-1020 Berlin, Neue Schönhauser Str. 20

or

Internationaler Buchversand GmbH, 1000 Berlin 19, Eschenallee 21.

The object of "physica status solidi (b)" and "physica status solidi (a)" is to publish rapidly important and topical results concerning fundamental research in the field of solid state physics and aspects of application and experimental techniques of fundamental solid state research, respectively. The issues of "physica status solidi (b)" and "physica status solidi (a)" will be published on the 1st and the 16th day of each month, respectively. Either series will contain the following categories:

1. Review articles written by arrangement with the editors. They will be remunerated.
2. Original papers selected from those which have not been submitted for publication elsewhere, after being reviewed. (The latest date for acceptance of contributions is the 8th and the 23rd day of each month for "physica status solidi (b)" and "physica status solidi (a)", respectively, 50 days before the delivery of the journal.)
3. Short notes published in the off-set part of the journal. They are, if necessary after being reviewed, selected by the editorial staff. (The latest date for acceptance of manuscripts for this part is the 12th and the 27th day of each month for "physica status solidi (b)" and "physica status solidi (a)", respectively, 16 days before the delivery of the journal.)
4. Pre-printed titles of papers to be published in the next issue of both series.

Editorial Information

Manuscripts should be submitted preferentially in English. Besides, papers in German, French, and Russian are admitted. The final decision on the incorporation of the papers in the series (b) or (a) is taken by the editorial staff after consultation of the referee. Authors are asked to observe the following rules:

1. Manuscripts have to be written as concisely as possible. As most, 20 and 4 pages (28 lines per page; abstracts, tables, references, figures, and captions of figures included, where 3 figures are counted as about one page) are admitted for original papers and short notes, respectively. All manuscripts should be type-written on one side only. Two copies are required. Badly arranged manuscripts are returned.
2. The author's name, initials, and institution should be given.
3. One abstract of about ten lines in English and one in German, French, or Russian are required for original papers.
4. Because of lack of space conclusions (if any) must not have the character of a summary (i. e. substantially a repetition of the contents of the abstract).
5. Formulae should not be written too small and not with pencil. Separate lines for formulae are desirable. Indistinct writing will lead to a return of the manuscript.
6. Drawings should be enclosed on separate sheets of paper, not exceeding the maximum and minimum sizes of 20×30 and 10×15 cm^2, respectively. Photographs should be submitted, on white paper, only for review articles and original papers. Duplicates are required.
7. Captions for the figures should be submitted on separate sheets of paper, and special instructions with regard to the size of illustrations should be given on these.
8. References should be numbered in order of appearance and should be given, on a separate sheet, in the form used in physica status solidi.
9. SI-units should be used.
10. Each paper should contain a proposal for the subject classification according to the classification scheme appearing on the second and third cover of each issue.

The authors will be sent galley proofs of review articles and original papers. Proof reading of short notes and pre-printed titles will be done by the editorial staff.

For each original paper and short note a total of 75 reprints is available free of charge. The number of reprints free of charge per review article is 25. Requests for additional reprints (against payment) should be given together with the submitted manuscripts. **The deadline for orders of additional reprints is the 15th and the 30th of each month** for "physica status solidi (b)" and "physica status solidi (a)" respectively, two weeks before the delivery of the journal.

Bei Fragen zur Produktsicherheit wenden Sie sich bitte an:
If you have any questions regarding product safety,
please contact:

Walter de Gruyter GmbH
Genthiner Straße 13
10785 Berlin
productsafety@degruyterbrill.com